Energy-Efficient Chemistry

Energy-Efficient Chemistry

Collection Editors

Gabriella Fiorentino
Maria Laura Parisi
Riccardo Basosi
Sergio Ulgiati

MDPI

Basel • Beijing • Wuhan • Barcelona • Belgrade • Novi Sad • Cluj • Manchester

Collection Editors
Gabriella Fiorentino
Department for Sustainability
ENEA
Rome
Italy

Maria Laura Parisi
Department of Biotechnology,
Chemistry and Pharmacy
University of Siena
Siena
Italy

Riccardo Basosi
Department of Biotechnology,
Chemistry and Pharmacy
University of Siena
Siena
Italy

Sergio Ulgiati
Department of Science
and Technology
Parthenope University
of Napoli
Napoli
Italy

Editorial Office
MDPI AG
Grosspeteranlage 5
4052 Basel, Switzerland

This is a reprint of the Topical Collection, published open access by the journal *Energies* (ISSN 1996-1073), freely accessible at: https://www.mdpi.com/journal/energies/topical_collections/Energy_Efficient_Chemistry.

For citation purposes, cite each article independently as indicated on the article page online and as indicated below:

Lastname, A.A.; Lastname, B.B. Article Title. *Journal Name* **Year**, *Volume Number*, Page Range.

ISBN 978-3-7258-5787-6 (Hbk)
ISBN 978-3-7258-5788-3 (PDF)
https://doi.org/10.3390/books978-3-7258-5788-3

Contents

About the Editors

Gabriella Fiorentino

Gabriella Fiorentino, Department for Sustainability – ENEA (Italian National Agency for New Technologies, Energy and Sustainable Economic Development). Her main research interests include the valorisation and sustainable use of resources from both perspectives of laboratory-scale tests and holistic evaluations, based on the Life Cycle Thinking methodology. Recently, her research activity has mainly focused on the evaluation of environmental sustainability of second-generation biorefineries and bio-based production pathways, in the framework of an efficient exploitation of residual substrates. Her current research topics include Life Cycle Assessment and Life Cycle Costing evaluations, circular economy models, pathways towards resource efficiency and closing the loops approach. She has wide experience in the management of national and international projects. She is a supervisor of undergraduates and PhD students and she has published 37 Scopus-indexed articles and over 50 book chapters, technical reports, conference proceedings and reviewed national articles.

Maria Laura Parisi

Maria Laura Parisi, Department of Biotechnology, Chemistry and Pharmacy – University of Siena. Her research activity spans from the environmental profile characterization of advanced materials and technologies for energy and resource efficiency implementation to the integrated life cycle thinking methodological approach for studying renewable energy conversion and storage systems. She is a co-founder of the R^2ES Lab—Research on Renewable Energy and Sustainability Laboratory that boasts collaborations with national and international research groups and companies that are active in the energy system, chemical production, and many other industrial sectors. Through LifeCARES srl, a consulting company of which she is vice-president, Maria Laura's strong interest and engagement in sustainability, eco-innovation, and reliable environmental communication lay the foundation for her commitment to supporting enterprises in eco-designing their value chains.

Riccardo Basosi

Riccardo Basosi, Department of Biotechnology, Chemistry and Pharmacy – University of Siena. His research interests were expressed in more than 330 papers published in International Journals of Physical Chemistry and/or Energy and several books. In the last 20 years, his research activity has mostly been devoted to the LCA of Energy Processes, particularly involving renewable sources and Complex Systems. He used the LCA approach in thermodynamic analysis of transport systems, in the environmental impact of Geothermal power plants also comparing their environmental performances with other renewable energy systems, and in the environmental analysis of a biodiesel production line from sunflower. In the last several years, he was particularly interested in the LCA of novel photovoltaics (Organic Dyes, DSSC, Micromorf, etc.) and in defining the environmental profile of Perovskite photovoltaics. He lead a strong group at the University of Siena, developing a variety of International Projects. He acted as Deputy Rector for Energy at the University of Siena from 1995 to 2017, a Scientific Advisor of the Italian Presidency of the Minister Council (2007–08), an Advisor for Energy of the MIUR Ministry at G7 (Berlin 2015, Kyoto 2016 and Torino 2017), a Member of the Technical Table of the Presidency of the Minister Council on "De-carbonization of Italian Economy" 2015–2017, a Member and Vice President of ENEA National Techno-Scientific Board (CTS) from 2017 to 2024, an Italian Permanent Representative in the Energy Committee of EU HORIZON 2020 from

2013 to 2021, a National MUR Delegate in the EU Strategic Energy Technology Plan, a National Expert for IEA/EGRD and a Chairman of the VIII SET Plan Conference in the frame of the Italian EU Presidency, in Dec 2014. He was the recipient of the 2014 SCI Gold Medal for excellent achievements in the field of energy and the environment and a co-recipient of the World Energy Globe Award 2014.

Sergio Ulgiati

Sergio Ulgiati, Department of Science and Technology – Parthenope University of Naples. His research interests include Energy–Exergy–Emergy Assessment and Environmental Impact methods; integrated urban and agro-industrial energy systems, with a special focus on the valorization of local-scale energy and material resources and the recycling–reuse–conversion of residues and waste materials, towards "zero emission" and the "circular economy"; and the evaluation of efficiency and environmental sustainability of renewable and non-renewable energy sources. He runs collaboration programs with Universities and Research institutions in Italy (including ENEA, Ca' Foscari University of a Venice, University of Bologna) and abroad (Europe, South America, United States, China). He was Visiting Professor in the U.S. at the University of Florida (1992 and 1997, with Howard Odum and Mark Brown) and at Cornell University (New York, 1994, with David Pimentel); Universidad Paulista (UNIP) – Laboratorio de Produção e Mejo Ambiente, San Paolo, Brasile (2013, with professor Biagio Giannetti); Beijing Normal University within the "National One Thousand Foreign Experts Plan" program of the Chinese Government (2013–2019, with professors Zhifeng Yang and Gengyuan Liu). He has authored over 400 publications in Italian and international peer-reviewed scientific volumes and journals and Proceedings of International Scientific Conferences (20,668 citations, 290 Scopus publications, h-index 68). He has been awarded the "Research for a Better World" 2017 award by ELSEVIER, UNESCO, UNEP, and WWF. He is member of the Editorial Board of international scientific journals, including *Energy*, the *Journal of Cleaner Production*, *Ecological Modelling*, *Ecological Questions*, *Ecological Indicators*, and he is the Editor-in-Chief of the international journal *Environments* (MDPI). From 2017 to today, he is included in the 2% World Top Scientists list of Stanford University (USA).

energies

Editorial

Energy-Efficient Chemistry

Gabriella Fiorentino

ENEA, Division Circular Economy, Department for Sustainability, Research Centre of Portici, P.le E. Fermi 1, 80055 Portici, Italy; gabriella.fiorentino@enea.it

1. Introduction

The growing environmental concerns related to climate change and resource depletion urgently call for a global and profound innovation of industrial production models and energy systems to foster the adoption of environmentally efficient technologies, circular economy principles, and long-term sustainable value chains [1].

According to the latest IEA data, global energy-related CO_2 emissions increased by 900 million tonnes (Mt) between 2019 and 2023 [2]. In 2023 alone, emissions rose by an additional 410 Mt, setting a new record of 37.4 gigatonnes (Gt). Industry remains the largest energy-consuming sector, which includes chemicals, petrochemicals, iron, and steel, accounting for more than one-third (34%) of industrial energy demand worldwide [3]. The chemical industry alone is responsible for around 5% of global GHG emissions annually, including both direct and indirect emissions [4]. With energy consumption currently representing one of the most pressing challenges, the chemical sector must urgently transform to reduce energy requirements and minimize environmental impacts. Importantly, energy demand in this sector not only includes the manufacturing processes that convert feedstocks into products but also the production of petrochemical feedstocks themselves, which accounts for roughly 60% of the total energy input. Consequently, rethinking conventional chemical production models and promoting alternatives based on renewable feedstocks, energy-saving technologies, and sustainable product design have become imperative [5]. For instance, the use of biomass for chemical production offers a potential breakthrough opportunity to reduce dependence on fossil resources, promote sustainability, and foster the development of new green industrial value chains [6].

Within the circular economy, energy-efficient chemistry can be achieved only when material and energy issues are addressed in an integrated manner. This requires not only the rational use of resources but also the achievement of energy savings through technological advances that reduce electricity and fuel consumption or through the substitution of fossil fuels with renewable energy sources.

This transition towards a more energy-efficient chemistry appears slower than expected despite significant global efforts to ensure technical and economic feasibility. While technological progress continues and remaining challenges are gradually addressed, technology itself does not seem to constitute the main barrier to large-scale implementation. Rather, the environmental benefits of such innovations, particularly regarding their impacts across the entire production chain, require careful and comprehensive assessment [7].

This Reprint, titled "Energy-Efficient Chemistry", gathers select research contributions that explore innovative approaches aimed at achieving these goals across theoretical, experimental, and applied perspectives. The goal is to offer a multidisciplinary overview of recent advances in energy-efficient chemical processes, emphasizing both environmental benefits and technological feasibility.

2. An Overview of Published Articles

This Reprint covers a broad spectrum of research topics related to resource and energy efficiency in chemical processes, reflecting the urgent need to decarbonize energy systems and industrial production while advancing circular economy principles. The presented contributions illustrate the diversity of technological, methodological, and conceptual approaches being explored to reconcile economic growth and industrial development with environmental protection and climate change mitigation.

A central theme in the collected papers is the role of green hydrogen as a key enabler of the low-carbon transition. Hydrogen produced from renewable energy sources has increasingly been recognized as a cornerstone in the shift away from fossil fuel-based energy systems. Ghisellini et al. (contribution 13) provide a comprehensive assessment of the environmental and social implications of deploying green hydrogen across multiple sectors. Their study highlights not only the promise of green hydrogen in decarbonizing industry, transport, and power generation but also the necessity of considering the full life cycle impacts associated with its production and use. Such a holistic view is essential to avoid burden-shifting, whereby reductions in greenhouse gas emissions are offset by negative impacts in other environmental or social categories.

Complementing this broad sustainability analysis, Scheepers and Lehnert (contribution 9) investigate the technical dimension of hydrogen production, specifically the modeling of polymer electrolyte membrane (PEM) electrolyzers. By introducing alternative statistical methods to analyze polarization curves, they enhance the accuracy of kinetic descriptions and provide tools for improving electrolyzer efficiency. These advances are critical because PEM electrolyzers represent one of the most mature technologies for large-scale hydrogen production, and their optimization is a prerequisite for cost-competitive and sustainable green hydrogen deployment.

The exploration of hydrogen as a clean energy vector is expanded by Żukowski et al. (contribution 8), who investigate polyolefin pyrolysis in multilayer fluidized beds. Their study identifies gaseous mixtures rich in hydrogen precursors, offering potential feedstocks for hydrogen generation. By linking plastic waste valorization with hydrogen production, this research exemplifies how circular resource flows and decarbonization strategies can be synergistically combined. Together, these studies on green hydrogen illustrate a multi-layered research landscape in which systemic sustainability assessments, advanced electrochemical modeling, and innovative feedstock pathways converge to support the development of efficient, low-carbon hydrogen technologies.

Beyond hydrogen, many contributions in this Reprint focus on the enhanced use of biomass as a renewable resource for producing valuable chemicals, energy carriers, and fertilizers. Biomass is increasingly regarded as a game changer in the transition toward sustainable manufacturing paradigms given its potential to replace fossil resources while contributing to carbon sequestration and soil health. Corinto Cavalloni et al. (contribution 12) present oxidative pyrolysis of wood pellets as a carbon-negative heating system capable of simultaneously producing useful heat and storing carbon in the form of biochar. Their work highlights oxidative pyrolysis as a simplification of conventional reactor design, demonstrating how negative emissions technologies can be implemented in practice.

Adamski et al. (contribution 11) examine the valorization of biomass for energy and fertilizer applications by investigating the influence of thermal conditions on compost transformation processes. Their study underscores the importance of optimizing composting parameters not only to ensure biological stabilization of biomass but also to support waste-to-energy strategies. In this way, biological processes become integral to circular bioeconomy models that aim to close material loops and reduce dependence on synthetic fertilizers.

In parallel, Fernández et al. (contribution 10) analyze microwave-assisted pyrolysis (MAP) of forest biomass, identifying this technology as an efficient alternative to conventional pyrolysis. MAP allows for rapid and uniform heating, improved yields of valuable bio-products, and reduced energy consumption. Such advances are particularly relevant as global energy demand rises and the need to exploit forestry residues sustainably becomes ever more pressing. Similarly, Bielecki et al. (contribution 1) focus on the densification of agricultural residues and its effects on pyrolysis behavior. Their results show how pre-treatment methods influence the composition and structure of pyrolytic products, opening pathways for tailoring biomass-derived products to specific applications. Finally, Şen et al. (contribution 5) critically review bark-based biorefinery systems, emphasizing their potential to generate both biochars and high-value extractives such as antioxidants. By mapping existing knowledge gaps and future directions, their work points to the emerging relevance of bark valorization in integrated biorefinery concepts.

Together, these contributions underscore the centrality of biomass in sustainable transitions. Whether through pyrolysis, composting, or integrated biorefineries, biomass-based strategies provide multiple benefits: renewable energy production, carbon sequestration, waste reduction, and the creation of value-added products. This multifunctionality makes biomass a cornerstone of energy-efficient and circular chemical systems.

A further thematic block in this Reprint involves the development of alternative fuels, which play an increasingly important role in reducing dependence on fossil energy sources. Sirviö et al. (contribution 6) examine the compatibility of B20 biodiesel blends with steel components, addressing a critical issue for the large-scale adoption of biodiesel in transportation. Their findings indicate that while some corrosion concerns remain, biodiesel blends made from used cooking oil methyl ester represent a viable renewable fuel option. This aligns with the broader strategy of valorizing waste oils within circular resource frameworks.

Glushkov et al. (contribution 3) provide an extensive review of gel fuels, a relatively less explored class of alternative fuels with promising safety and environmental benefits. Compared to conventional liquid and solid fuels, gel fuels offer improved combustion control, reduced emissions, and enhanced handling safety. By systematically reviewing rheological properties, atomization, ignition, and combustion behavior, their work paves the way for more widespread exploration of gel fuels in both civilian and defense applications. Both studies stress the importance of innovation in fuel formulation and performance optimization, demonstrating that the diversification of fuel options is an essential component of sustainable energy systems.

The Reprint also features contributions that address advances in materials, modeling, and systemic approaches, which are equally critical to achieving sustainable innovation. Qiu and Liu (contribution 2) review stannous tungstate semiconductors, exploring their potential applications in photocatalysis and photoelectrochemical water splitting. Their analysis highlights both the opportunities and challenges associated with this material, such as excessive electron–hole recombination and limited stability. Overcoming these issues could significantly enhance solar energy utilization, thereby contributing to clean hydrogen production and pollutant degradation.

Kačur et al. (contribution 7) shift the focus toward underground coal gasification (UCG), presenting a review of laboratory-scale investigations. While coal is a controversial resource in the context of decarbonization, UCG technologies have the potential to convert coal into syngas more efficiently and with lower environmental impacts compared to conventional mining. The reviewed studies explore process optimization, the role of gasification agents, and environmental risks, providing insights that may inform future deployment of UCG under stricter sustainability conditions.

From a systemic perspective, Ncube et al. (contribution 4) discuss the integration of circular economy and green chemistry, highlighting their combined potential to drive radical innovation in product design. By advocating renewable feedstock, waste minimization, and the extension of product lifespans, they demonstrate how chemistry can be a central driver of circular production models. Importantly, their work also emphasizes that achieving this integration requires supportive policies, adequate investment, and widespread environmental education. In this way, systemic approaches connect technical innovation with broader societal and institutional changes, ensuring that the transition to sustainability is not only technologically feasible but also socially inclusive and economically viable.

Collectively, the contributions included in this Reprint reflect the multi-dimensional nature of sustainability challenges. They cover a wide spectrum of approaches, ranging from technological innovations in hydrogen, biomass conversion, and alternative fuels, with a focus on optimizing processes and materials, to systemic strategies and policy dimensions. Complementary pathways emerge; only through such an integrated and multidisciplinary approach can energy-efficient and environmentally sound chemical processes be developed and implemented on a larger scale.

3. Conclusions

The goal of this Reprint is to offer a comprehensive overview of recent advances in resource and energy-efficient chemical processes, emphasizing not only their technical feasibility but also their environmental and social implications. By showcasing both conceptual frameworks and practical case studies, it highlights the synergies between green chemistry, circular economy, and clean energy transitions. Simultaneously, it acknowledges the complexities and trade-offs that must be managed to ensure that solutions intended to mitigate climate change do not create unintended environmental burdens elsewhere.

The global community continues to strive toward net-zero targets and sustainable development goals, and the contributions gathered here provide valuable insights into how chemistry, materials science, and engineering can converge to support this transformation. By aligning innovation with sustainability imperatives, they chart pathways toward an energy-efficient and circular future, where chemical processes contribute not only to economic development but also to environmental regeneration and societal well-being.

This Reprint is addressed to researchers, engineers, industry professionals, and policymakers interested in green chemistry, sustainable engineering, and clean energy technologies. It serves as a resource for those seeking to understand and apply innovative solutions that reduce the energy and carbon intensity of chemical manufacturing while maintaining high performance and economic viability.

We thank all the authors and reviewers whose contributions made this Reprint possible, and we hope that this collection will stimulate further research and collaboration toward a more sustainable chemical future.

Conflicts of Interest: The author declares no conflicts of interest.

List of Contributions:

1. Bielecki, M.; Zubkova, V.; Strojwas, A. Influence of Densification on the Pyrolytic Behavior of Agricultural Biomass Waste and the Characteristics of Pyrolysis Products. *Energies* **2022**, *15*, 4257. https://doi.org/10.3390/en15124257.

2. Qiu, W.; Liu, Y. Stannous Tungstate Semiconductor for Photocatalytic Degradation and Photoelectrochemical Water Splitting: A Review. *Energies* **2022**, *15*, 9194. https://doi.org/10.3390/en15239194.
3. Glushkov, D.; Paushkina, K.; Pleshko, A. Gel Fuels: Preparing, Rheology, Atomization, Combustion. *Energies* **2023**, *16*, 298. https://doi.org/10.3390/en16010298.
4. Ncube, A.; Mtetwa, S.; Bukhari, M.; Fiorentino, G.; Passaro, R. Circular Economy and Green Chemistry: The Need for Radical Innovative Approaches in the Design for New Products. *Energies* **2023**, *16*, 1752. https://doi.org/10.3390/en16041752.
5. Şen, U.; Esteves, B.; Pereira, H. Pyrolysis and Extraction of Bark in a Biorefineries Context: A Critical Review. *Energies* **2023**, *16*, 4848. https://doi.org/10.3390/en16134848.
6. Sirviö, K.; Kaivosoja, J.; Nuortila, C.; Wang-Alho, H.; Niemi, S.; Ovaska, T. B20 Fuel Compatibility with Steels in Case of Fuel Contamination. *Energies* **2023**, *16*, 5933. https://doi.org/10.3390/en16165933.
7. Kačur, J.; Laciak, M.; Durdán, M.; Flegner, P. Investigation of Underground Coal Gasification in Laboratory Conditions: A Review of Recent Research. *Energies* **2023**, *16*, 6250. https://doi.org/10.3390/en16176250.
8. Żukowski, W.; Leski, K.; Berkowicz-Płatek, G.; Wrona, J. Polyolefin Pyrolysis in Multilayer Fluidized Beds: An Innovative Approach to Obtain Valuable Alternative Fuels. *Energies* **2024**, *17*, 1034. https://doi.org/10.3390/en17051034.
9. Scheepers, F.; Lehnert, W. Investigating the Applicability of the Tafel Equation in Polymer Electrolyte Membrane Electrolyzers Via Statistical Analysis. *Energies* **2024**, *17*, 3298. https://doi.org/10.3390/en17133298.
10. Fernández, I.; Pérez, S.F.; Fernández-Ferreras, J.; Llano, T. Microwave-Assisted Pyrolysis of Forest Biomass. *Energies* **2024**, *17*, 4852. https://doi.org/10.3390/en17194852.
11. Adamski, M.; Herkowiak, M.; Marek, P.; Klimek, K.E.; Kapłan, M. Analysis of the Effects of Biomass Stabilization Under Varying Thermal Conditions with Respect to the Quality Characteristics of Compost Transformation Products. *Energies* **2024**, *17*, 6410. https://doi.org/10.3390/en17246410.
12. Corinto Cavalloni, F.; Strassburg, J.; Lustenberger, D.; Griffin, T. Oxidative Pyrolysis for Variable Heating Output with Wood Pellets. *Energies* **2025**, *18*, 1702. https://doi.org/10.3390/en18071702.
13. Ghisellini, P.; Passaro, R.; Ulgiati, S. Is Green Hydrogen an Environmentally and Socially Sound Solution for Decarbonizing Energy Systems Within a Circular Economy Transition? *Energies* **2025**, *18*, 2769. https://doi.org/10.3390/en18112769.

References

1. Ulgiati, S.; Fiorentino, G.; Raugei, M.; Schnitzer, H.; Lega, M. Cleaner production for human and environmental well-being. *J. Clean. Prod.* **2019**, *237*, 117779. [CrossRef]
2. IEA—International Energy Agency. *CO_2 Emissions in 2023. A New Record High, but Is There Light at the End of the Tunnel?* IEA Publications: Paris, France, 2024.
3. REN21. Renewables 2024 Global Status Report Collection, Renewables in Energy Demand. 2024. Available online: https://www.ren21.net/gsr-2024/modules/energy_demand/01_renewables_in_industry/#top (accessed on 1 September 2025).
4. Gabrielli, P.; Rosa, L.; Gazzani, M.; Meys, R.; Bardow, A.; Mazzotti, M.; Sansavini, G. Net-zero emissions chemical industry in a world of limited resources. *One Earth* **2023**, *6*, 682–704. [CrossRef]
5. Fiorentino, G.; Zucaro, A.; Ulgiati, S. Towards an energy efficient chemistry. Switching from fossil to bio-based products in a life cycle perspective. *Energy* **2019**, *170*, 720–729. [CrossRef]
6. IEA—International Energy Agency. *Technology Roadmap: Energy and GHG Reductions in the Chemical Industry via Catalytic Processes*; IEA Publications: Paris, France, 2013.
7. Gonella, F.; Almeida, C.M.V.B.; Fiorentino, G.; Handayani, K.; Spanò, F.; Testoni, R.; Zucaro, A. Is technology optimism justified? A discussion towards a comprehensive narrative. *J. Clean. Prod.* **2019**, *223*, 456–465. [CrossRef]

energies

MDPI

Article

Influence of Densification on the Pyrolytic Behavior of Agricultural Biomass Waste and the Characteristics of Pyrolysis Products

Marcin Bielecki, Valentina Zubkova * and Andrzej Strojwas

The Institute of Chemistry, Jan Kochanowski University in Kielce, Uniwersytecka Str. 7, 25-406 Kielce, Poland; marcin-bielecki-1994@wp.pl (M.B.); andrzej.strojwas@ujk.edu.pl (A.S.)
* Correspondence: zubkova@ujk.edu.pl or walentyna.zubkowa@ujk.edu.pl

Abstract: TG/FT-IR techniques, UV-spectroscopy, microwave extraction, XRD and SEM were used to study how densification of the three types of agricultural biomass wastes (wheat straw, soft wood, and sunflower husk) changes the composition and structure of their pyrolysis products. It was determined that densification changes the composition of volatile products of pyrolysis at the temperature of 420 °C: sunflower husk emits 4.9 times less saturated and unsaturated hydrocarbons and 1.9 times less compounds with carbonyl group; soft wood emits 1.8 times more saturated and unsaturated hydrocarbons and compounds with carbonyl groups and 1.3 times more alcohols and phenols; and wheat straw emits 2 times more compounds with carbonyl groups. These changes are probably caused by the differences in interaction of formed volatiles with the surface of chars. These differences can be caused by distinct places of cumulation of inorganic components in the densified samples. In the densified char, the inorganics cumulate on the surface of sunflower husk whereas for wheat straw they cumulate inside the sample. In the case of soft wood, the inorganics cumulate both inside and on the surface. The decreased contribution of hydrocarbons in volatiles can be connected with the morphology of nano-particles formed in inorganics.

Keywords: environmental chemistry; biomass; pyrolysis; densification; volatile-char interaction

1. Introduction

Limitations on the use of fossil fuels to obtain heat and electrical energy resulted in intensification of research on ways of their substitution by renewable sources, including various types of biomass. In order to increase the efficiency of its use, biomass was subjected to densification—pelleting, briquetting, granulation etc. [1]—before combustion. The densification of biomass made its transportation and storage easier and facilitated the work of equipment applied to its thermal conversion [2,3]. The densification is conducted both without binders [4,5] and with application of binders of various types: coal tar, coal tar sludge, paraffin [6], microalgae [7], or carboxymethyl cellulose [8,9], for appropriate biomass species. There is a wide range of works connected with the search for new ways of biomass pre-treatment before briquetting [5,10–12] and obtaining of new biomass species [13,14]. The influence of moisture content and pressure on the formation of pellets [15] and biomass/water ratio in hydrothermal carbonization [16] was analyzed. Mainly, the influence of additives of different types on physical-mechanical properties of briquettes [7,17–21] was evaluated along with the optimalization of pelleting conditions by mixing biomass of various types [22] or reheating of obtained pellets [23]. The combustion process of briquetted biomass was studied from the viewpoint of emission of particulate matter [8,24].

Research conducted by a number of authors shows that the obtained volatile products can interact with char [25–28] and tar [29] formed during biomass pyrolysis and change their composition and yield. According to Hosoya et al. [30], during biomass pyrolysis there can occur some interactions between solid and liquid phases that accelerate the formation of char from cellulose and lignin. Chen et al. suggested [31] that secondary reactions can take place between volatile products of even unmixed biomass components. These findings give good reason to suggest that the biomass densification during briquetting will hinder the removal of volatile products from the inside of briquette and in fact can enhance the aforementioned interactions. However, this suggestion was not studied completely.

During the process of pyrolysis, the basic components of biomass (hemicellulose, cellulose, and lignin) undergo the process of thermal decomposition, when solid (char), liquid (condensate, bio-oil), and gaseous products are formed. Based on mass spectrometric analysis, Penzik et al. [32] stated that, in the composition of pyrolytic gases, H_2O and CH_4, have a greater contribution and CH_4, CH3COOH, furans, phenols, and aromatic compounds are present to a lesser extent. According to Osman [33], nitrous oxide, hydrogen, methyl alcohol, acetaldehyde, acetic acid, and ethane can evaporate from lignocellulose biomass during pyrolysis. Ojha et al. [34] confirmed the results obtained by the previous authors and completed the list with a further group of compounds that are formed during the pyrolysis of bagasse: condensed organics, furans, cyclo-oxygenated compounds, N-containing organics, and aromatics. On the basis of research on thermal behavior of basic biomass components, Chen et al. [35] stated that anhydrosugars and furans were the main components of bio-oil from cellulose whereas the bio-oils from hemicellulose and lignin contained acids, phenols, aldehydes, ketones, and alcohols. Based on kinetic studies on agro-waste, Osman et al. [36] proposed a 10-step pyrolysis mechanism. In their opinion, the pyrolyses of hemicellulose and cellulose can occur in 3 steps whereas that of lignin can occur in 4 steps. The mechanism was proposed on the basis of studies on the decomposition of single components without any interaction between them.

It was established that the presence of inorganic components [37–39] affects the course of biomass pyrolysis. The inorganic components can catalyze the conversion of particular biomass components [29,40–43] and thus change the composition of formed pyrolysis products. According to Fan et al. [44], potassium salts can facilitate an increase in the yield of furans, aldehydes, ketones, and CO_2. Zhu et al. [45] suggest that potassium favors the transformation of bio-oils into gas (CO_2) and the reaction of demethylation of guaiacyl at low temperature. According to Lu et al. [46], Ca can promote the activity of Fe by the formation of $Ca_2Fe_2O_5$ wherein Fe/SiO_2, Ca/SiO_2, and Fe-Ca/SiO_2 catalysts influence selective deoxygenation. Shen et al. [47] prove that alkaline-earth-metal additives can decrease the content of acids and increase the content of hydrocarbons in bio-oil whereas additives of Mg compounds can be more favorable for production of hydrocarbons compared with the additives of Ca compounds since during biomass pyrolysis oils are formed and the presence of Mg and Ca compounds in the densified biomass can also intensify the course of such reactions.

The research described in the literature was mainly carried out on the briquettes (pellets) obtained with the use of both various binders [6–9] or moisture [15]; it thus determined the changes taking place under the influence of the two factors—presence of an additive and densification. However, the influence of one factor (biomass densification without binders) on the composition and structural chemical parameters of formed pyrolysis products has not been well described in the literature. Evaluating the influence of densification in research on switchgrass, Yang et al. [5] determined that densification influences the contents of anhydrous sugars, phenols, and guaiacols in the products of pyrolysis. Investigating switchgrass, Sarkar et al. [48] established that, under the influence of densification, the yield of H_2 increases, the yield of CH_4 decreases, and both carbon conversion and cold gas efficiencies rise. However, there is still no detailed information and comparison of pyrolytic behavior for different types of loose and densified biomasses.

This research aims to investigate the way that biomass densification without binders influences: (i) the thermal behavior and composition of volatile products of pyrolysis of wheat straw, soft wood, and sunflower husk, (ii) the thermal stability of their basic components, (iii) the structural chemical characteristics of liquid products of thermal conversion of biomass, (iv) the structure and texture of the formed char along with inorganics cumulating in it.

2. Materials and Methods

2.1. Materials

The samples of wheat straw (WS), softwood (SW), and sunflower husk (SH) were dried at room temperature, then ground and sieved through a 0.2 mm sieve, and dried at the temperature of 105 °C. The elemental composition of the samples determined by an Elementar Vario Micro Cube CHNS analyzer is presented in Table 1.

Table 1. Main characteristics of studied samples.

Main Characteristics	WS	SW	SH
C [%]	39.90 ± 0.15	45.34 ± 0.13	45.82 ± 0.08
H [%]	5.75 ± 0.02	5.86 ± 0.04	6.32 ± 0.02
N [%]	0.65 ± 0.08	0.58 ± 0.11	2.61 ± 0.05
S [%]	0.13 ± 0.05	0.17 ± 0.07	0.14 ± 0.02
O [a] [%]	41.97 ± 0.07	42.45 ± 0.04	38.31 ± 0.08
A [d] [%]	11.59 ± 0.76	5.60 ± 0.38	6.81 ± 0.51
HHV [b] [$MJ \cdot kg^{-1}$]	16.12 ± 0.19	18.23 ± 0.13	19.31 ± 0.13

[a] calculated by difference, O [%] = 100 − C − H − N − S − Ash; [b] calculated by HHV [$MJ \cdot kg^{-1}$] = 0.3491 × C + 1.1783 × H + 0.1005 × S − 0.0151 × N − 0.1034 × O − 0.0211 × Ash. [d] dry basis.

It follows from the data in the table that SH sample has the greatest content of C, H, and N elements and a greater value of the HHV parameter. WS sample has the greatest content of inorganic components—the ash content in this sample is almost twice as highas that in other samples.

Table 2 shows the data on the contents of selected inorganic elements that are present in the studied samples.

Table 2. The content of selected inorganic elements in studied biomass [%].

Elements	WS	SW	SH
Al	<0.01	0.20 ± 0.05	0.08 ± 0.05
P	0.04 ± 0.02	0.07 ± 0.01	0.36 ± 0.02
S	0.19 ± 0.01	0.08 ± 0.01	0.33 ± 0.02
Cl	0.46 ± 0.01	0.03 ± 0.01	0.10 ± 0.01
K	1.30 ± 0.04	0.36 ± 0.02	1.30 ± 0.04
Ca	0.54 ± 0.04	1.32 ± 0.06	0.76 ± 0.05
Fe	0.10 ± 0.01	0.37 ± 0.01	0.14 ± 0.01
Si	2.58 ± 0.05	0.71 ± 0.03	0.70 ± 0.05
Zn	<0.01	<0.01	<0.01

The contents of these elements were determined by the ED-XRF technique. A Thermo Scientific Niton Goldd+ analyzer was used for this purpose. The data in Table 2 show that K element is present in the studied WS and SH samples in the amount of 1.3%, and that the greatest amounts of Ca were detected in the SW sample. According to the authors [42,44,49,50] the compounds containing K, Ca, and Mg elements can be catalyzers in the process of biomass pyrolysis. In the studied samples, the following elements were detected in different amounts: Si (WS sample has most of all), P and S (SH sample has most of all), Cl (WS sample has the greatest content), and Fe (SW sample contains the greatest amount of this element).

The bulk densities of the WS, SW, and SH samples were 0.29 ± 0.03, 0.34 ± 0.02, and 0.19 ± 0.03 $g \cdot cm^{-3}$, respectively. The tablets were formed from loose biomass without binder additives. Pistons of 1.2 cm and 0.7 cm diameter were used to compress the biomass. The tablets' densities were as follows: 1.02 ± 0.03 g·cm^{-3} for the tablets from WS sample, 1.15 ± 0.01 g·cm^{-3} for the tablets from SW sample, and 1.04 ± 0.02 g·cm^{-3} for the tablets from SH sample. The obtained tablets of about 0.7 mm in thickness were studied by TG-FT-IR.

2.2. A Pyrolytic Test in a TG/FT-IR Analyzer

The ground biomass in loose and densified forms was heated in a Q50 thermobalance under nitrogen atmosphere up to a temperature of 750 °C. The samples of 25 ± 0.05 mg in weight were heated with a heating rate of 4 °C·min^{-1}. The obtained DTG curves were deconvoluted using OMNIC9 software. The deconvolution was conducted by Gaussian/Lorentzian functions, taking into account the constant base line according to the methodology presented in the work [51].

The rate of nitrogen flow through the thermobalance was 10 mL·min^{-1}. The formed products of biomass decomposition were moved by a nitrogen stream of 90 mL·min^{-1} through a transporting line and interface to a Nicolet iS10 spectrometer in order to register the FT-IR spectra. The obtained spectra were elaborated by OMNIC9 software. The FT-IR spectra of volatile products were normalized with respect to the CO_2 band.

2.3. A Test Conducted in a PyrolyticOven

Loose and densified biomass was additionally heated in quartz cuvettes that were placed in a pyrolytic oven PRC 70 × 708/110 M manufactured by Czylok company. The heating to the temperatures of 450 and 750 °C was conducted under the stream of nitrogen of high purity with the flow rate of 330 mL·min^{-1}. The average heating rate was 4 °C·min^{-1}. During the pyrolysis up to 450 °C, the volatile products were passed through a layer of methanol that was cooled by water with ice. The decomposition products condensed in methanol were regained by distillation of solvents under vacuum.

2.4. Microwave-Assisted Extraction

The biomass samples in loose and densified forms with a mass of 2.0 g were extracted by a mixture of 25 mL of methanol with 25 mL of chloroform in an Anton Paar Multiwave 3000 microwave extractor at the temperature of 150 °C under the pressure of 18 bar. After filtration, the solvents were distilled away from the extracts under vacuum.

2.5. Spectroscopic Investigations of Pyrolysis Products

Liquid and solid pyrolysis products were obtained in the pyrolytic oven during heating to the temperature of 450 °C and the extracts were investigated by the Attenuated Total Reflectance (ATR) techniques. The ATR investigations of these samples were performed using a Smart MIRacle module placed in the Nicolet iS10 spectrophotometer. The samples were placed on a ZnSe monocrystal and pressed with a stamp with great force. The ATR spectra in the range of 4000–600 cm^{-1} were registered by 32 scans. The spectra were elaborated by an automatic correction of the base line (OMNIC9 software) taking into account local optical minima near 2400, 2000, and 900 cm^{-1}. The obtained spectra of chars, condensates, and extracts were normalized with respect to the C=C band near the wavenumber of 1600 cm^{-1}.

2.6. XRD and SEM Investigations

After the addition of 10% of the internal standard (NaF) to the samples of chars pyrolyzed to the temperatures of 450 and 750 °C, they were studied using the technique of X-ray quantitative phase analysis. The diffractograms were obtained with a variable step using Cu_k radiation; the X-ray tube voltage was 40 kV, whereas the current—20 mA, the pulse counting time was 20 s. The obtained diffractograms were normalized with regard to the reflex (002) from NaF. The C_{ord} parameter, which characterizes the number of carbon atoms that participate in origination of the reflex of coherent scattering in the range of

2θ angles from 14 to 31 degrees, was determined according to the formula presented in work [51].

The chars obtained at the temperature of 750 °C were studied by a scanning electron microscope Quanta 3D FEG manufactured by FEI Company. An Everhart-Thornley Detector (ETD) was used during the SEM tests. The samples of densified biomass were investigated on the surface and inside of the tablets. The EDX analysis was made for the samples of chars.

3. Results and Discussion

3.1. An Analysis of the TG and DTG Curves of Loose and Densified Samples

Figure 1 presents the curves of mass loss and mass loss rate of the studied WS, SW, and SH samples pyrolyzed in loose (WSL, SWL, SHL, respectively) and densified forms (WSD, SWD, SHD, respectively). Taking into account only the shape of the TG and DTG curves in this figure, it could be suggested that densification does not substantially change their shape. The densification of WS and SH samples lowers the peak value of the mass loss rate and changes the yield of char at the temperature of 750 °C. During the pyrolysis of SWD sample, the mass loss rate does not change but the amount of char shows a tendency to increase. More differences in thermal behavior of loose and densified samples can be observed in the Gram–Schmidt plots (Figure 1a).

The DTG curve characterizes the changes in mass loss rate depending on the temperature of pyrolysis. These changes were caused by the decomposition of pyrolyzed biomass components and show their thermal stability at different temperatures. On the basis of published data on thermal behavior of basic biomass components [52,53], we can estimate in which temperature ranges their decomposition in the studied samples took place.

Taking into account the aforementioned temperature ranges, it is possible to calculate the contribution of distinct biomass components in formation of the DTG curve using the deconvolution method. Hence, it is possible to evaluate the changes in their thermal stability that was caused by densification. The deconvolution carried out for DTG curves shows that the general contribution of decomposition of cellulose pseudocomponents in formation of the DTG curve during pyrolysis of WSD samples tends to increase. This contribution increases from the value of 49.3% for WSL samples to the value of 52.1% in WSD samples. This is connected with a lower thermal stability of the pseudocomponent of cellulose I—the contribution of cellulose I in formation of the DTG curve increases by nearly 5%. Gong et al. [54] also pointed out that granulation can disrupt the structure of biomass cell walls, and Cao et al. [55] suggested that there is a possibility of decomposition of crystalline cellulose in pellets into amorphous carbon.

The shape of the Gram–Schmidt plots in Figure 1b suggests that starting from 80 min there are some deviations in the course of pyrolysis of SWD sample compared to SWL sample. The results of deconvolution of the DTG curves suggest that the densification of SW samples does not facilitate the decomposition of hemicellulose and cellulose pseudocomponents but only tends to increase the contribution of the pseudocomponents of moisture and light volatile components (LVS) in formation of the DTG curve. Hence, it should be presumed that the densification of SW samples does not affect the results of deconvolution of the DTG curve—the contribution of basic pseudocomponents does not substantially change. The lack of interaction between softwood components during pyrolysis was reported by Kawamoto et al. [56].

The Gram–Schmidt curves of SHL and SHD sample presented in Figure 1c differ from each other in shape in the whole time range of pyrolysis. The deconvolution results of the DTG curve presented in Figure 1c clearly show substantial changes in thermal stability of particular pseudocomponents that were caused by the densification of SH samples. Densification does not practically affect the thermal stability of the pseudocomponents of moisture, LVS, and extractives but it reduces the capacity of thermal decomposition of hemicellulose—the contribution of hemicellulose in formation of the DTG curve decreases from 45.5 to 29.6%. Moreover, under the influence of densification, the thermal stability of the pseudocomponents of cellulose II and lignin diminishes. This can be caused by mutual interaction between cellulose and lignin [57–59] that was intensified by densification. The deconvolution results of the DTG curve suggest that a new pseudocomponent decomposes during pyrolysis of SHD sample in the temperature range of 450–500 °C. Such a pseudocomponent could have been formed as a result of interaction between the components of hemicellulose material or the course of secondary reactions between volatile products of decomposition of hemicellulose, cellulose, and lignin. Possible courses of such secondary reactions were reported in a number of works [31,39,60].

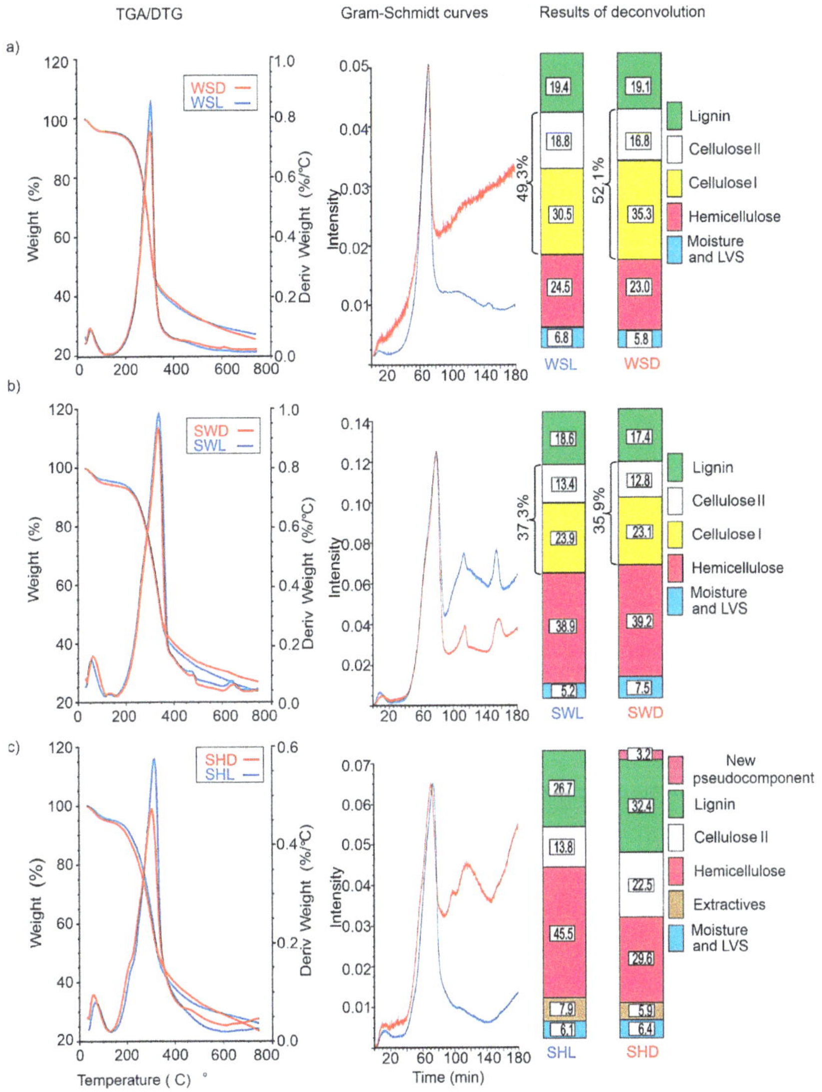

Figure 1. The TG, DTG, and Gram–Schmidt curves along with the deconvoluted DTG curves of biomass samples. (**a**) WSD and WSL samples; (**b**) SWD and SWL samples; (**c**) SHD and SHL samples.

Therefore, depending on biomass type, the densification influences the interaction between biomass components and changes the thermal stability of hemicellulose, cellulose, and lignin in different ways.

3.2. The Influence of Densification on the Composition of Volatile Products of Pyrolysis

Figure 2 presents the FT-IR spectra of volatile products of pyrolysis of WSL and WSD samples at the temperatures of 260, 300, and 420 °C. It follows from comparison of the FT-IR spectra of volatile products of WSL and WSD samples that the densification of wheat straw does not substantially change the contribution ratio of saturated and unsaturated hydrocarbons, alcohols, and phenols in the composition of volatile products of pyrolysis but it increases the contribution of compounds containing carbonyl groups at the temperatures of 260 and 420 °C.

The ratios of the surface area of these bands (range 1850–1700 cm^{-1}) normalized in relation to the CO_2 band from the densified samples were calculated with respect to appropriate bands from loose samples. The calculations prove that at the temperatures of 260, 300, and 420 °C these ratios are 1.7, 1.3, and 2.2 respectively. On the basis of research conducted by other authors [3,61–63], it can be suggested that during pyrolysis of the samples of densified wheat straw more compounds with carbonyl groups are released with volatiles, e.g., more aldehydes, ketones, acids, sugars, and aromatics.

Figure 3 presents the FT-IR spectra of volatile products emitted from SWL and SWD samples at the pyrolysis temperatures of 240, 380, and 420 °C.

The analysis of the shape of the normalized spectra and the calculations of the surface areas of bands imply that the densified sample tends to a lower emission of hydrocarbons, alcohols, and phenols only at T = 240 °C. Starting from the temperature of 380 °C, during pyrolysis process SWD sample emit 1.5 times more saturated and unsaturated hydrocarbons, 1.7 times more compounds with carbonyl groups, and 1.3 times more alcohols and phenols than SWL sample. This implies that the densification of SWD sample substantially influenced the interaction between volatiles and char.

SHD sample behave in a different way during pyrolysis (Figure 4). A contrastive analysis of the spectra of volatile products of SHL and SHD samples at the temperatures of 300, 340, 420, and 480 °C shows that much lower contribution ratios of hydrocarbons along with compounds containing carbonyl groups, alcohols, phenols, and moisture are emitted with volatiles during the entire pyrolysis process from the densified sample. At the temperature of 300 °C, SHL sample emit 1.3 times more compounds with carbonyl groups (ketones, aldehydes, acids, and ethers [3,61–64]) and 1.6 times more alcohols and phenols than SHD sample; at the temperature of 340 °C they emit 2 times more saturated and unsaturated hydrocarbons, 1.2 times more compounds with carbonyl groups, and 1.9 times more alcohols and phenols; at the temperature of 420 °C they emit 4.9 times more saturated and unsaturated hydrocarbons, 1.9 times more compounds with carbonyl groups; at the temperature of 480 °C they emit 3.5 times more saturated and unsaturated hydrocarbons, 1.7 times more compounds with carbonyl groups, and 2.7 times more alcohols and phenols [3,61–64]. These results suggest that the densification of SHD sample affected the volatile-char interactions to a greater extent.

The initial WS and SH samples have the same content of K element and similar contents of Ca element (Table 2). Taking into account the catalytic properties of these elements described in works [39,44,57], such changes in the course of the pyrolysis process in WS and SH samples was expected as well as their influence on changes in the composition of volatiles. However, the data presented in Figures 2–4 imply that the changes in composition of volatile products of pyrolysis after the densification of WS and SH samples take place in different ways. In WSD and SHD samples, the interaction between volatiles and their chars takes place in a different way.

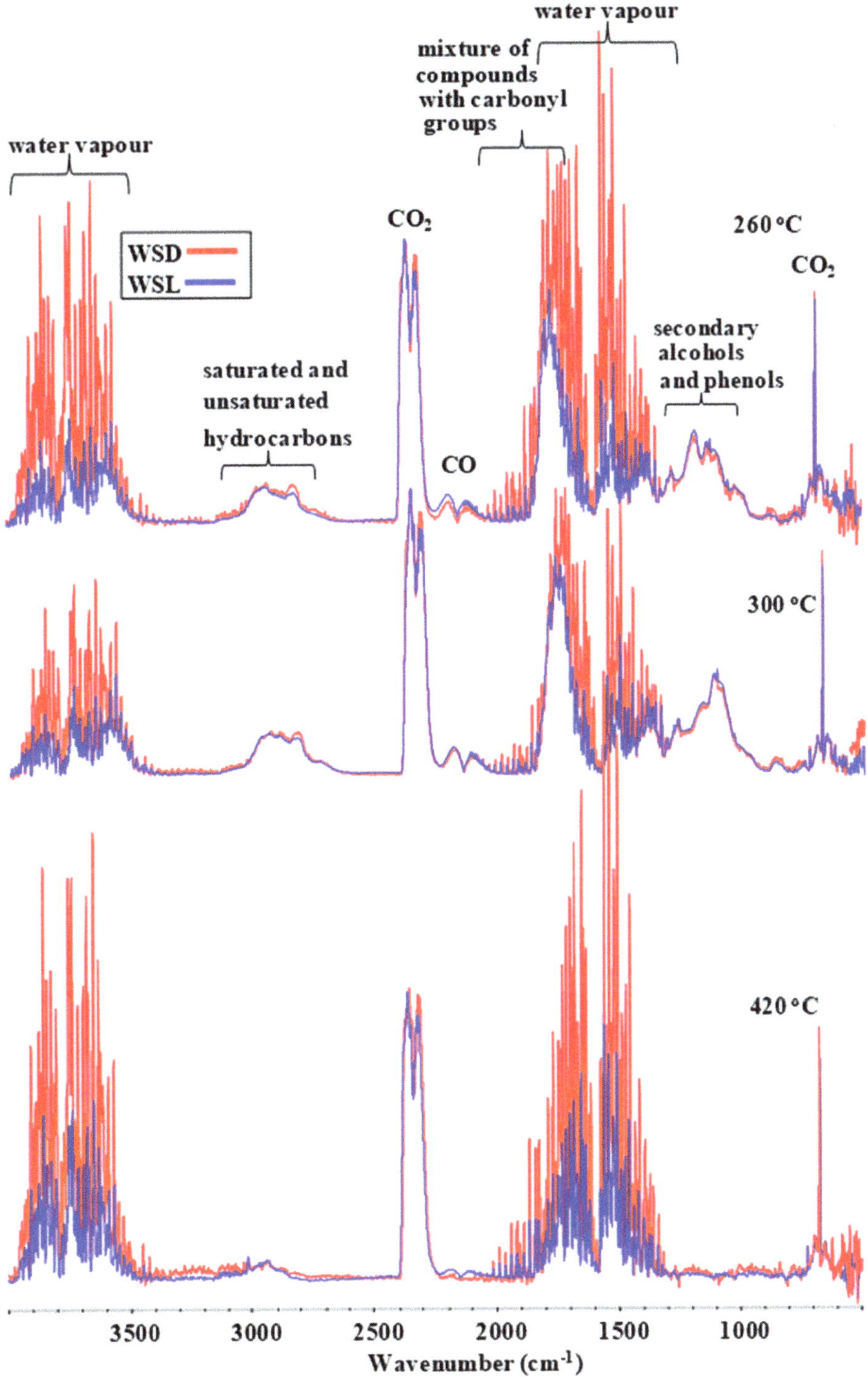

Figure 2. The FT-IR spectra of volatile products of pyrolysis of WSL and WSD samples.

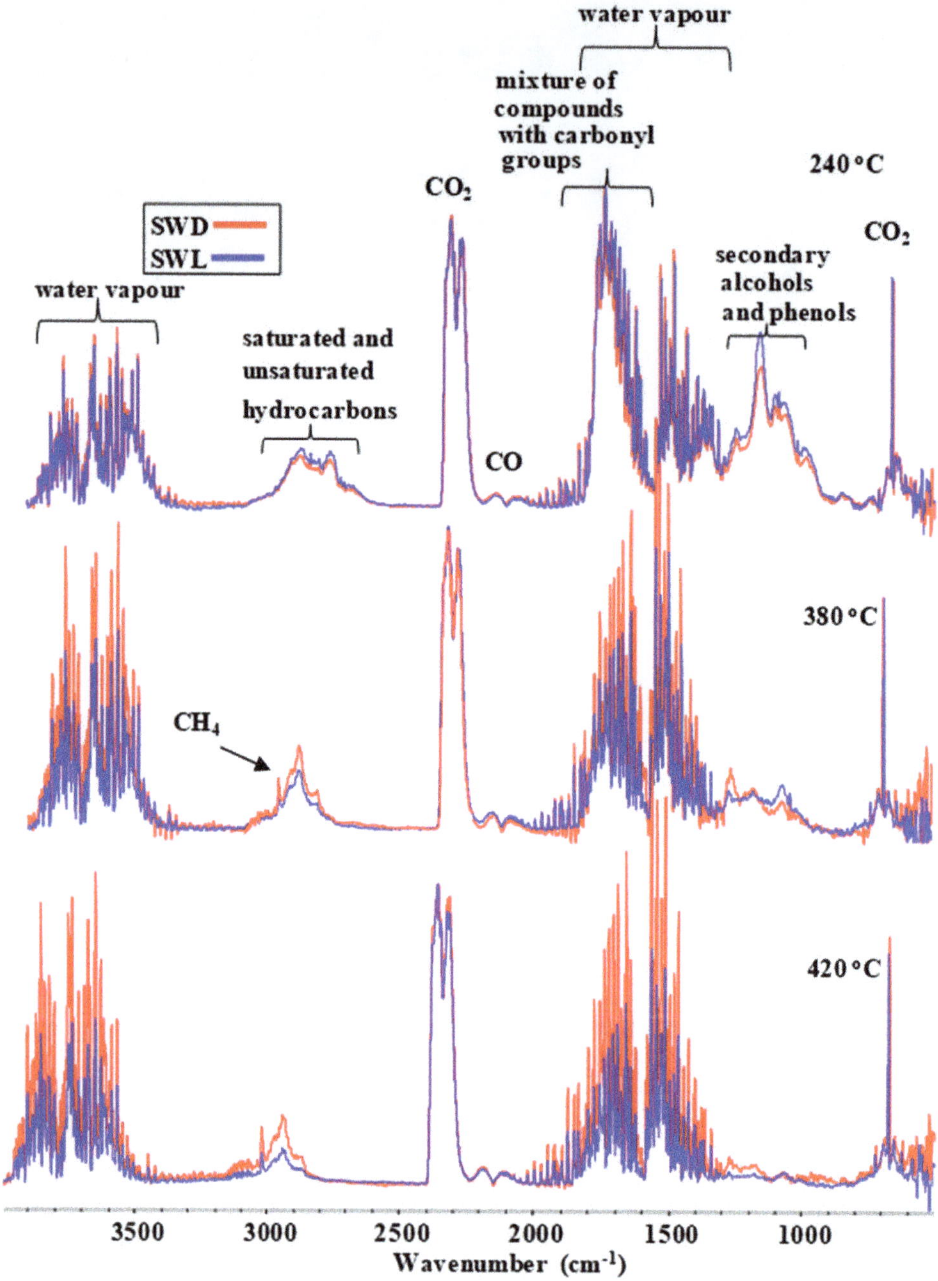

Figure 3. The FT-IR spectra of volatile products of pyrolysis of SWL and SWD samples.

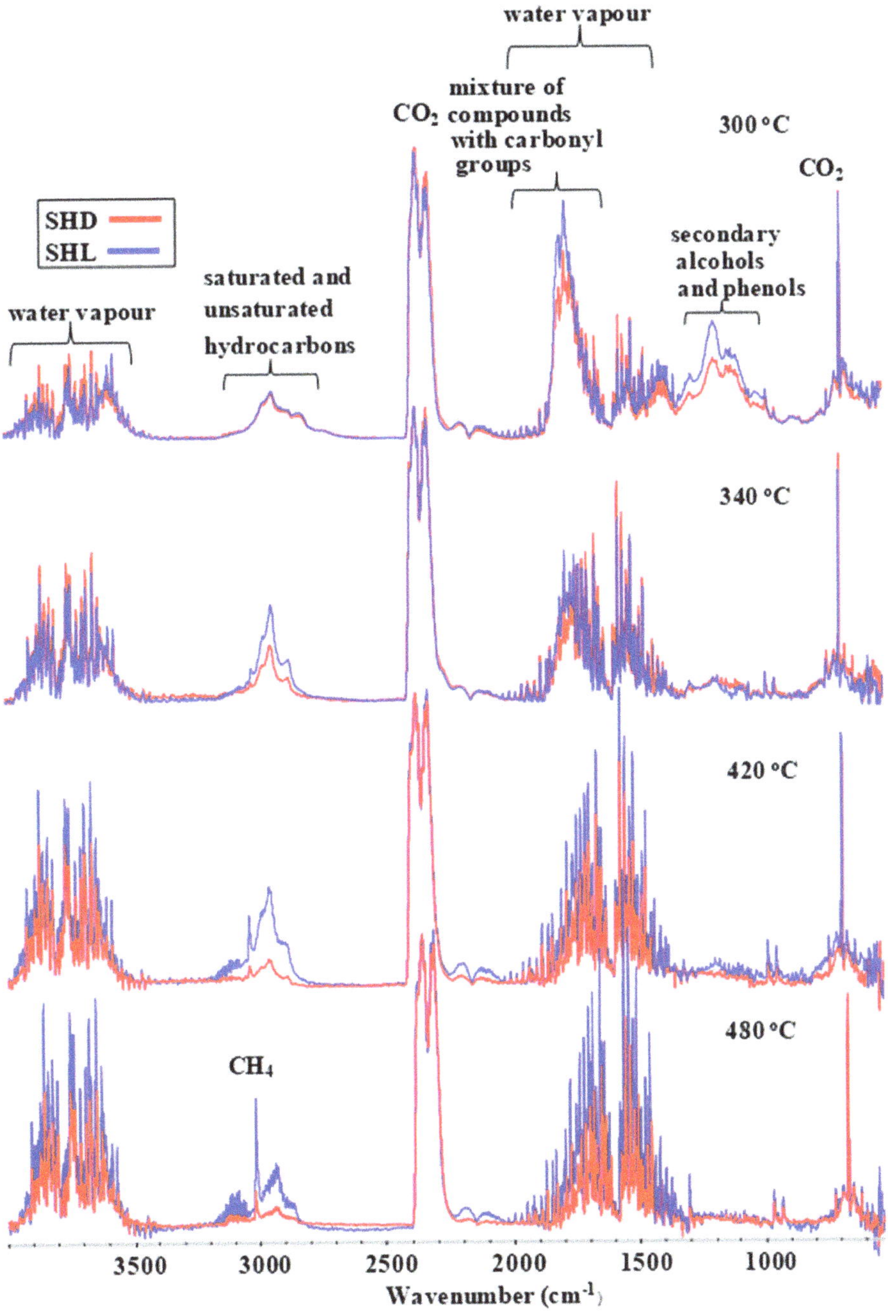

Figure 4. The FT-IR spectra of volatile products of pyrolysis of SHL and SHD samples.

3.3. The Influence of Densification on Structural-Chemical Parameters of Liquid Products of Pyrolysis

Figure 5 sets out the ATR spectra of the material extracted from the studied samples and of condensates obtained during the process of pyrolysis to the temperature of 450 °C.

In the ATR spectra of extracts and condensates, there are some spectra that correspond to the same functional groups and groups of atoms. This implies that, when performing extraction of the samples of studied biomass, some precursors, which cause the occurrence of saturated and unsaturated hydrocarbons, aldehydes, ketones, alcohols, and phenols in volatile products [65], can be partly removed from samples before pyrolysis.

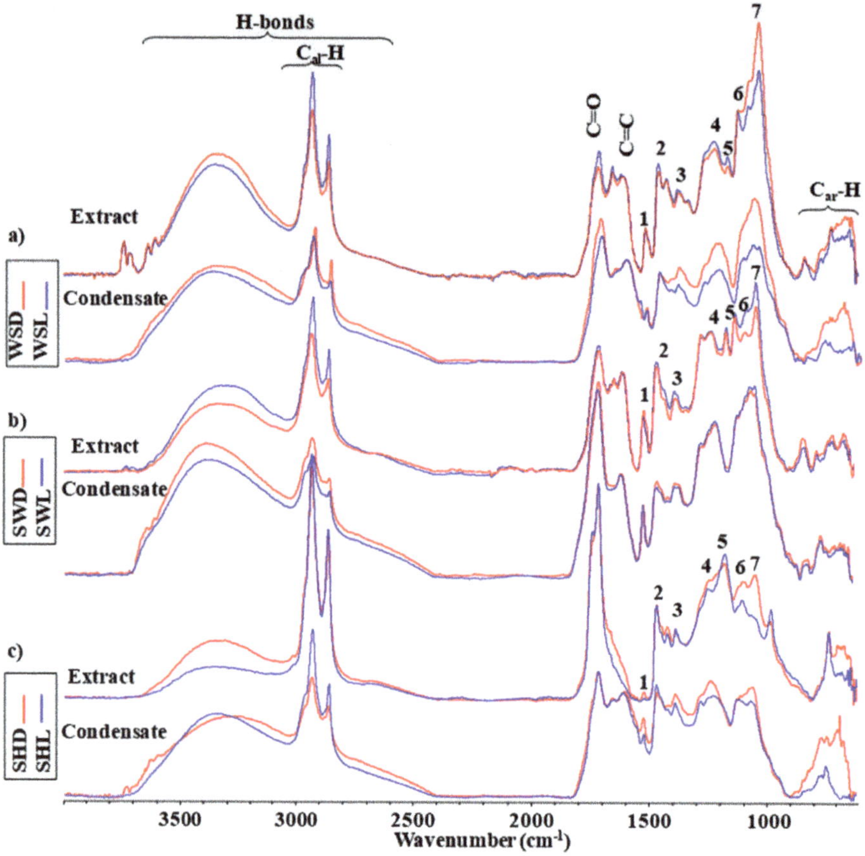

Figure 5. The ATR spectra of the obtained liquid products of pyrolysis of studied samples. (see explanation in the text). (**a**) WSD and WSL samples; (**b**) SWD and SWL samples; (**c**) SHD and SHL samples.

It follows from the comparison of the normalized spectra in Figure 5 that the densification increases the contribution of polar compounds that are able to form hydrogen bonds in the extracts from WS (Figure 5a) and SH (Figure 5c) samples and condensates from WS (Figure 5a) and SW (Figure 5b) samples but it decreases the contribution of polar compounds in the extract from SW samples (Figure 5b). The densification increases the contribution ratio of compounds that cause the appearance of band near 1030 cm^{-1} in the extracts and condensates from WS (Figure 5a) and SH (Figure 5c) samples but decreases it in the extract from SW samples (Figure 5b). This band corresponds to C-O stretching in fragments of cellulose [51,63,64]. The densification does not exert influence on the contribution ratio of groups in fingerprint range (1500–1100 cm^{-1}) in the extracts from WS (Figure 5a) and SW (Figure 5b) samples and in the condensate from SW (Figure 5b) samples. This indicates an increase in the relative contribution of cellulose and lignin decomposition fragments in the discussed samples. This is confirmed by the deconvolution results of the DTG curve presented in Figure 1. The densification increases the intensity of deformation vibrations of the C_{ar}-H type (in the range of 900–600 cm^{-1}) in the ATR spectra of the condensates from WS (Figure 5a) and SH (Figure 5c) samples.

3.4. The Influence of Densification on Changes in Structure Texture of Chars

Figure 6 presents the diffractograms of studied samples in loose and densified forms that were pyrolyzed to the temperatures of 450 and 750 °C. A wide reflex in the range of angles of 2θ 14–30 degrees is present on all diffractograms. Taking into account this range of angles, it can be suggested that γ-fraction formed C atoms of ordered aliphatic chains has a significant contribution in formation of this reflex. These chains form the so-called local molecular arrangements, which were caused by the interference of the radiation scattered by the sample [66,67]. The appearance of γ reflex on diffractograms of chars was reported in works by other researchers [68,69]. The contribution of C atoms in graphite-like sheets (the peak of which should be near 2θ = 26 degrees) is insignificant in shaping of this reflex.

It follows from the shapes of the diffractograms that WS and SH samples, both loose and densified, after heating to the temperature of 450 °C, have similar surface areas under diffractograms, and hence they do not substantially differ by integral intensity. The values of the C_{ord} parameter that determines the general degree of ordering of carbon atoms in aliphatic chains (γ-fraction) and in graphite-like sheets in their chars are 33.4 and 34.1% for WSD and WSL samples appropriately, and for SHD and SHL they are 35.9 and 32.4%, respectively (Figure 6).

However, SWD and SWL samples pyrolyzed to the temperature of 450 °C substantially differ from each other by C_{ord} parameter that amounts to 54.9 and 42.1%, respectively. The degree of ordering increases in WSD and WSL samples when the pyrolysis temperature rises to 750 °C as is evidenced by an increase in value of the C_{ord} parameter. WSD samples have a greater degree of ordering (parameter C_{ord} = 57.5%) in comparison to WSL samples (parameter C_{ord} = 46.8%). The ordering of structure takes place in a different way in SWD sample that at the temperature of 750 °C have the value of C_{ord} parameter (C_{ord} = 50.5%), less than it has at the temperature of 450 °C. The C_{ord} parameter for SWL sample at the temperature of 750 °C was equal to 42.1%, and this parameter has the same value for SWL sample at the temperature of 450 °C. The processes of transformation of structure in the temperature range 450–750 °C occur for SHD sample in a completely different way. At the temperature of 750 °C, SHD sample have a somewhat lower value of the C_{ord} parameter (C_{ord} = 34.1%) compared to the value at the temperature of 450 °C (C_{ord} = 35.9%). However, at the temperature of 750 °C there is an increase in the C_{ord} parameter up to 44.9% of that observed for SHL sample.

The diffractograms and given values of the C_{ord} parameter presented in Figure 6 imply that the densification influences transformations of structure of WS, SW, and SH samples during pyrolysis at various temperatures in different ways. The phenomenon of transformation of diffractogram profiles in the studied temperature ranges was observed by other researchers [70–72] who attributed a decrease in total intensity of reflexes in the studied range of 2θ angles to degradation of cellulose macromolecules.

Besides the reflex caused by the presence of ordered carbon atoms in ordered aliphatic chains and graphite-like sheets, there are some reflexes from inorganic components on the diffractograms of chars. The intensity of these reflexes rises with an increase in temperature from 450 °C to 750 °C. On the diffractograms of densified samples, the reflexes from inorganic components have somewhat greater intensity. This implies that volatile inorganic components can be emitted from densified samples to lesser extent. Hwang et al. [73] established that K and Ca agglomerate in higher concentration in bio-chars. Inorganic components [57] including alkali metals and alkaline earth metals can have a catalytic effect on the course of pyrolysis processes [39]. Compounds containing K can catalyze secondary reactions of cracking of volatile products of pyrolysis [43]. This would suggest that the reorganization of structure of chars in WS and SH samples would have taken place in similar way. However, the comparison of transformation of the profiles of reflexes on the diffractograms of WS and SH samples having similar content of K and Ca elements (Table 2) and the comparison of changes in their ordering degree at the temperatures of 450 and 750 °C do not confirm this.

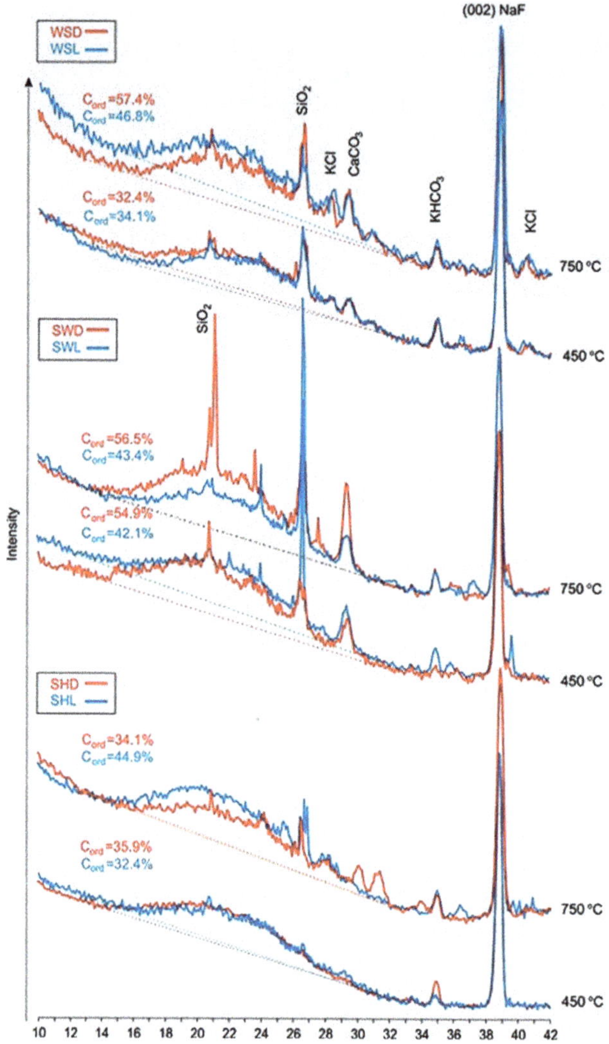

Figure 6. The diffractograms of obtained chars.

The comparison of shapes and locations of the reflexes on the diffractograms imply that inorganic components can form alloys of different composition during pyrolysis of densified samples. In biomass samples, inorganic components can influence the yield of volatile products of biomass decomposition [41] and thus the processes of structural ordering of chars with which these volatiles will interact. Moreover, the presence of alloys of miscellaneous composition in the chars of samples will affect the processes of ordering of their structure in a different way.

Figure 7 presents SEM images of chars at the temperature of 750 °C at magnification of M1 k and M25 k.

It follows from Figure 7 that the pyrolyzed WSL sample has a heterogeneous structure. The magnification of M25 k shows the differences in shape of inorganic particles and in their location in different areas of the sample. In some areas inorganic particles have an irregular shape, and in others there are some nano-particles shaped like needles of about 40 nm in diameter (magnification of M100 k) that are visible near irregularly shaped particles. There were no such nano-objects found in the SEM images of densified samples. On the surface of WSD samples, the organic material does not show any visible differences but the edges of inorganic particles have some signs of melting. Inside the pyrolyzed WSD sample there are some areas of molten inorganic particles in the form of drops visible at a magnification of M25 k. The possibility of formation of "low melting alkali silicate component" was reported by Zevenhoven–Onderwater et al. [40].

Such a situation could have taken place in biomass rich in silicon. The occurrence of such areas in the pyrolyzed WSD sample suggests that they may result from a hindered removal of volatile products from inside of the sample. The volatile organic compounds were able to interact with inorganic components inside of the tablet longer but also the inorganic components could have formed various alloys and undergone coalescence.

In contrast to WS samples, SW and SH samples manifest the signs of melting of organic particles (Figure 7). At magnifications of M1 k, it is impossible to notice any differences in relief of the pyrolyzed material in SWL and SWD samples on the surface and inside of the tablet. However, at magnification of M25 k the material of the pyrolyzed SWL sample looks like swollen molten fibres, and the pyrolyzed SWD sample manifest traces of stickiness in contact points of particles of pyrolyzed material on the tablet surface. The material inside the tablet of the pyrolyzed SWD sample shows the traces of melting too. The possibility of melting of hemicellulose and lignin, in which cellulose chains were immersed, was reported by a number of authors [29,37,74]. The location of cellulose chains in molten lignin could have facilitated the formation of areas of parallel arrangement of its chains, what caused the lowering of the background line on the diffractogram of SWD sample on the side of lower angles of scattering at the temperature of 450 °C (Figure 6). The results of the conducted microanalysis of pyrolyzed SW samples (Figures S1–S3) show a tendency toward an increase in content of Fe and Ca elements in SWD sample. A possible influence of Fe and Ca elements on increased content of hydrocarbons in the composition of volatile products of pyrolysis was reported earlier by the authors [46,61]. The char of densified SWD sample shows a tendency toward an increase in contribution of Mg and K atoms. Moreover, the presence of Na was determined in this char. The presence of these elements in a char can cause secondary reactions between the char and volatile products [75]. The increase in contribution ratio of saturated and unsaturated hydrocarbons, compounds with carbonyl groups along with aldehydes and ketones in the volatile products of densified sample observed in Figure 3 can be probably connected with the cumulation of compounds containing the aforementioned elements in the char of SW samples.

The magnification of M25 k exposes clear signs of melting of particles of the SHL and SHD samples (Figure 7). It could be expected that, similar to WSD samples, SHD sample would show a greater structural ordering at the temperature of 750 °C. This was proposed on the basis of the data from Table 2; in SW and SH samples there are similar amounts of K, Ca, and Fe elements. However, in contrast to WSD samples, at the temperature of 750 °C SHD sample had a worse ordering than SHL sample (Figure 6). WS and SH samples belong to lignocellulose biomass but there is no pseudocomponent of cellulose I in the composition of SH samples (Figure 1c). This could have been the probable reason for the differences in thermal behavior of SHD sample and the differences in course of ordering of these samples under the influence of temperature.

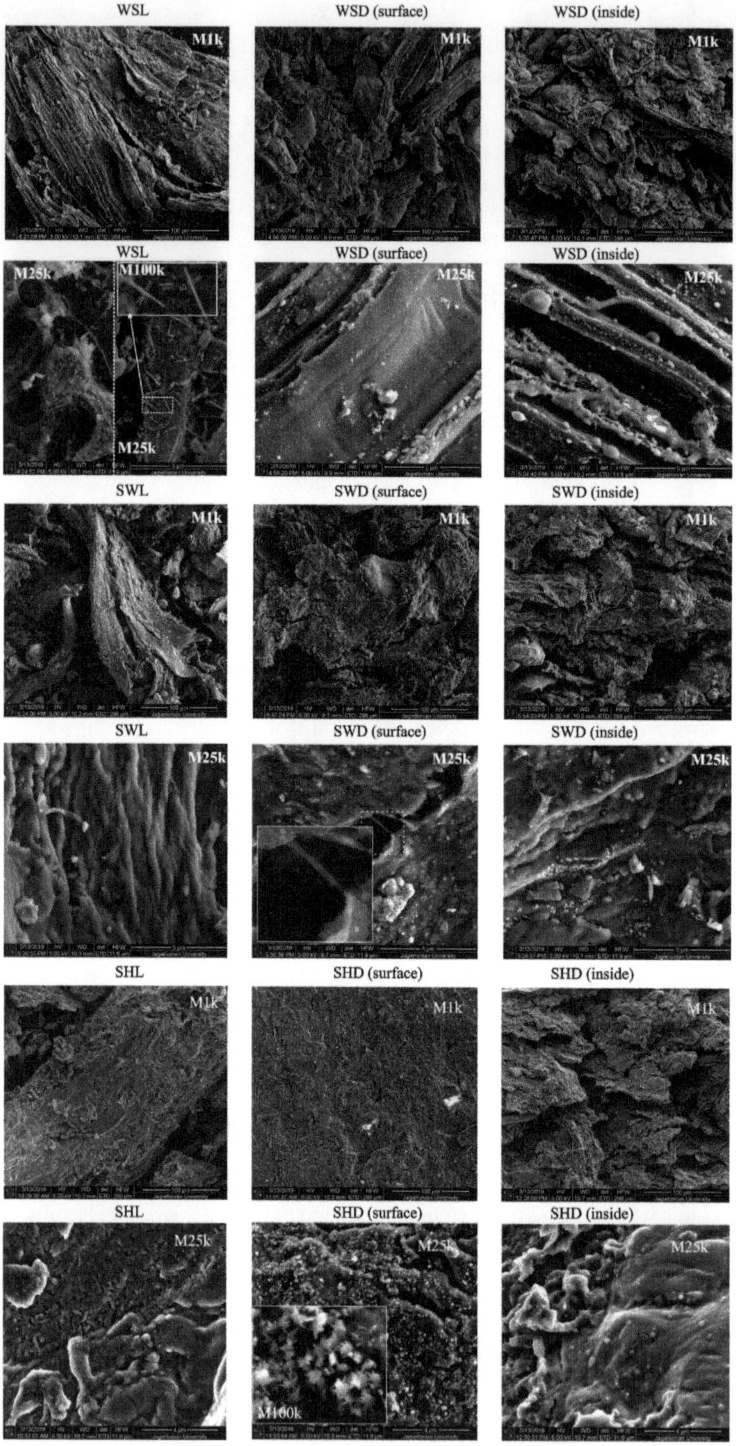

Figure 7. The SEM images of chars obtained at the temperature 750 °C.

The comparison of the results of microanalysis of SHL (Figure S4) and SHD (Figures S5 and S6) points to an increased content of such elements as Mg, P, S, and Fe. The presence of Ti was detected on the surface of the SHD sample. In the char of SHD sample, there are more K atoms cumulating than in the char of WSD. Inside of the char of WSD, there are more Ca atoms detected than in SHD sample. Moreover, the results of microanalysis of the chars of SH samples show the presence of N element in them. This element was not detected in the chars of WS and SW samples. It cannot be excluded that the presence of this element in SHD sample caused a decrease in the contribution of hydrocarbons in the composition of the volatile products of pyrolysis.

In contrast to WS and SW samples, there are greater amounts of inorganic nano-particles shaped like multi-pointed stars (M100 k) visible on the surface of SHD sample. Such particles were not found in SHL sample or inside the tablets of the pyrolyzed SHD sample. Taking into account a well-known fact that inorganic components can influence the changes in composition of volatile products of pyrolysis [43] and that the formed char can interact with volatile components [40], the differences in thermal behavior of the studied samples can be connected not only with the differences in composition of inorganic elements in the composition of the densified lignocellulose biomass but also with the location of agglomerates of inorganic components in the tablet or the type of morphology of particles in the formed agglomerates.

The agglomerates of inorganic particles were located inside the tablet, and the morphology of particles in WSD sample differed from the morphology of inorganic particles in WSL sample. The contribution ratio of compounds with carbonyl groups was greater in the composition of the volatile products emitted from WSD sample (Figure 2). On the micrographs of SWD sample (inside and on the surface of the tablet) there are greater amounts of inorganic particles than for SWL sample (Figure 7), which corresponds to a greater intensity of the reflexes on the diffractograms of SWD sample (Figure 6); i.e., there were more components that could have caused not only greater degradation of basic biomass components but also the interaction of volatile products both inside and on the surface of the tablet. This could have caused an increase in the contribution of saturated and unsaturated hydrocarbons, compounds with carbonyl groups, alcohols, and phenols in volatile products of pyrolysis of SWD sample compared to SWL sample (Figure 3). Some nano-particles in the shape of stars are visible on the surface of the tablet of SHD sample. This could have caused a lower emission of hydrocarbons in the composition of volatile products during pyrolysis of SHD sample. These results show that the location of inorganics in chars influences the volatile-char interactions.

4. Conclusions

The influence of the densification without additives of the three types of lignocellulose biomass (wheat straw, soft wood, and sunflower husk) on the composition of their solid, liquid, and gaseous pyrolysis products was investigated. It was found that the densification intensifies volatile-char interactions and causes the following changes in composition of volatile products of pyrolysis: at the temperature of 420 °C, the contribution of compounds with carbonyl groups emitted from wheat straw doubles; in case of soft wood, the contribution of saturated and unsaturated hydrocarbons along with compounds with carbonyl groups increases 1.8 times, and the contribution of emitted alcohols and phenols increases 1.3 times; however, in the case of sunflower husk, the contribution of saturated and unsaturated hydrocarbons decreases 4.9 times and the contribution of emitted compounds with carbonyl groups decreases 1.9 times.

At the temperature of 750 °C, the densification increases the degree of ordering of the tablets from pyrolyzed wheat straw and soft wood samples but lowers it in the tablets from pyrolyzed sunflower husk samples.

The densification changes the location of inorganic components in the samples of pyrolyzed tablets and, hence volatile-char interactions. In the char of wheat straw tablets, the agglomerates of inorganic particles occur inside the tablet; in the char of sunflower husk tablets, these agglomerates are observed only on the surface; and in the char of soft wood tablets, the inorganic components do not show any particular location disposition—it is the same as in the char of loose samples. It is suggested that a decrease in emission of hydrocarbons in the composition of volatile products can be connected with the morphology of inorganic components and the location of their agglomerates in the tablet.

The changes in composition and structural-chemical parameters of pyrolysis products cannot be caused by a hindered removal of volatile products of pyrolysis from the briquetted sample only; an interaction is powered by briquetting both between these products and basic biomass components directly in the sample and the volatile products on the surface of formed char. The results of the investigation of influence of the briquetting on the changes in structural-chemical parameters of liquid products of pyrolysis of wheat straw, soft wood, and sunflower husk samples prove that different types of lignocellulose biomass respond differently to this technological stimulus. That is why it is very difficult (or even probably impossible) to propose a single mechanism in order to explain the differences in nature of transformations taking place in briquetted and loose biomass during pyrolysis.

Taking into account that the pyrolysis process analyzed in this paper is the first stage of combustion, it can be suggested that wheat straw and soft wood should be burned in the loose form, and sunflower husk in the briquetted form. This would avoid incomplete combustion of wheat straw briquettes caused by formation of greater slag agglomerates and help prevent the emission of environmentally hazardous compounds.

Supplementary Materials: The following supporting information can be downloaded at: https://www.mdpi.com/article/10.3390/en15124257/s1, Figure S1. Results of microanalysis of SWL sample. Figure S2. Results of microanalysis of the surface of SWD sample. Figure S3. Results of microanalysis of the inside of SWD sample. Figure S4. Results of microanalysis of SHL sample. Figure S5. Results of microanalysis of the surface of SHD sample. Figure S6. Results of microanalysis of the inside of SHD sample.

Author Contributions: Conceptualization, V.Z. and M.B.; methodology, V.Z. and A.S.; validation, V.Z., M.B., and A.S.; investigation, A.S and M.B.; data curation, A.S.; writing—original draft preparation, V.Z. and M.B.; writing—review and editing, V.Z. and M.B.; visualization, A.S. and V.Z; project administration, V.Z.; funding acquisition, V.Z. All authors have read and agreed to the published version of the manuscript.

Funding: This research was funded by the Rector of Jan Kochanowski University in Kielce within the framework of Project SUPB. RN 21.191.

Institutional Review Board Statement: Not applicable.

Informed Consent Statement: Not applicable.

Data Availability Statement: Not applicable.

Acknowledgments: The authors would like to thank Lina Kieush for making biomass samples available for analysis.

Conflicts of Interest: The authors declare no conflict of interest.

References

1. Bajwa, D.S.; Peterson, T.; Sharma, N.; Shojaeiarani, J.; Bajwa, S.G. A Review of Densified Solid Biomass for Energy Production. *Renew. Sust. Energy Rev.* **2018**, *96*, 296–305. [CrossRef]
2. Liu, Z.; Zhang, F.; Yan, S.; Tian, L.; Wang, H.; Liu, H.; Wang, H.; Hu, J. Effects of Temperature and Low-Concentration Oxygen on Pine Wood Sawdust Briquettes Pyrolysis: Gas Yields and Biochar Briquettes Physical Properties. *Fuel Process. Technol.* **2018**, *177*, 228–236. [CrossRef]
3. Albashabsheh, N.T.; Heier Stamm, J.L. Optimization of Lignocellulosic Biomass-to-Biofuel Supply Chains with Densification: Literature Review. *Biomass Bioenergy* **2021**, *144*, 105888. [CrossRef]
4. Halim, S.A.; Swithenbank, J. Characterisation of Malaysian Wood Pellets and Rubberwood Using Slow Pyrolysis and Microwave Technology. *J. Anal. Appl. Pyrolysis* **2016**, *122*, 64–75. [CrossRef]
5. Yang, Z.; Sarkar, M.; Kumar, A.; Tumuluru, J.S.; Huhnke, R.L. Effects of Torrefaction and Densification on Switchgrass Pyrolysis Products. *Bioresour. Technol.* **2014**, *174*, 266–273. [CrossRef] [PubMed]

6. Florentino-Madiedo, L.; Díaz-Faes, E.; García, R.; Barriocanal, C. Influence of Binder Type on Greenhouse Gases and PAHs from the Pyrolysis of Biomass Briquettes. *Fuel Process. Technol.* **2018**, *171*, 330–338. [CrossRef]
7. Muazu, R.I.; Stegemann, J.A. Biosolids and Microalgae as Alternative Binders for Biomass Fuel Briquetting. *Fuel* **2017**, *194*, 339–347. [CrossRef]
8. Zhu, Y.; Yang, W.; Fan, J.; Kan, T.; Zhang, W.; Liu, H.; Cheng, W.; Yang, H.; Wu, X.; Chen, H. Effect of Sodium Carboxymethyl Cellulose Addition on Particulate Matter Emissions during Biomass Pellet Combustion. *Appl. Energy* **2018**, *230*, 925–934. [CrossRef]
9. Si, Y.; Hu, J.; Wang, X.; Yang, H.; Chen, Y.; Shao, J.; Chen, H. Effect of Carboxymethyl Cellulose Binder on the Quality of Biomass Pellets. *Energy Fuels* **2016**, *30*, 5799–5808. [CrossRef]
10. Manouchehrinejad, M.; Mani, S. Torrefaction after Pelletization (TAP): Analysis of Torrefied Pellet Quality and Co-Products. *Biomass Bioenergy* **2018**, *118*, 93–104. [CrossRef]
11. Wu, S.; Zhang, S.; Wang, C.; Mu, C.; Huang, X. High-Strength Charcoal Briquette Preparation from Hydrothermal Pretreated Biomass Wastes. *Fuel Process. Technol.* **2018**, *171*, 293–300. [CrossRef]
12. Araújo, S.; Vilas Boas, M.A.; Neiva, D.M.; de Cassia Carneiro, A.; Vital, B.; Breguez, M.; Pereira, H. Effect of a Mild Torrefaction for Production of Eucalypt Wood Briquettes under Different Compression Pressures. *Biomass Bioenergy* **2016**, *90*, 181–186. [CrossRef]
13. Pradhan, P.; Mahajani, S.M.; Arora, A. Production and Utilization of Fuel Pellets from Biomass: A Review. *Fuel Process. Technol.* **2018**, *181*, 215–232. [CrossRef]
14. Amarasekara, A.; Tanzim, F.S.; Asmatulu, E. Briquetting and Carbonization of Naturally Grown Algae Biomass for Low-Cost Fuel and Activated Carbon Production. *Fuel* **2017**, *208*, 612–617. [CrossRef]
15. Zvicevičius, E.; Raila, A.; Čiplienė, A.; Černiauskienė, Ž.; Kadžiulienė, Ž.; Tilvikienė, V. Effects of Moisture and Pressure on Densification Process of Raw Material from *Artemisia dubia*. *Wall. Renew. Energy* **2018**, *119*, 185–192. [CrossRef]
16. Tag, A.T.; Duman, G.; Yanik, J. Influences of Feedstock Type and Process Variables on Hydrochar Properties. *Bioresour. Technol.* **2018**, *250*, 337–344.
17. Wang, Y.; Wu, K.; Sun, Y. Pelletizing Properties of Wheat Straw Blending with Rice Straw. *Energy Fuels* **2017**, *31*, 5126–5134. [CrossRef]
18. Gendek, A.; Aniszewska, M.; Malat'ák, J.; Velebil, J. Evaluation of Selected Physical and Mechanical Properties of Briquettes Produced from Cones of Three Coniferous Tree Species. *Biomass Bioenergy* **2018**, *117*, 173–179. [CrossRef]
19. Rahaman, S.A.; Salam, P.A. Characterization of Cold Densified Rice Straw Briquettes and the Potential Use of Sawdust as Binder. *Fuel Process. Technol.* **2017**, *158*, 9–19. [CrossRef]
20. Gilvari, H.; de Jong, W.; Schott, D.L. Quality Parameters Relevant for Densification of Bio-Materials: Measuring Methods and Affecting Factors—A Review. *Biomass Bioenergy* **2019**, *120*, 117–134. [CrossRef]
21. Jiang, L.; Liang, J.; Yuan, X.; Li, H.; Li, C.; Xiao, Z.; Huang, H.; Wang, H.; Zeng, G. Co-pelletization of Sewage Sludge and Biomass: The Density and Hardness of Pellet. *Bioresour. Technol.* **2014**, *166*, 435–443. [CrossRef] [PubMed]
22. Yank, A.; Ngadi, M.; Kok, R. Physical Properties of Rice Husk and Bran Briquettes under Low Pressure Densification for Rural Applications. *Biomass Bioenergy* **2016**, *84*, 22–30. [CrossRef]
23. Riva, L.; Nielsen, H.K.; Skreiberg, Ø.; Wang, L.; Bartocci, P.; Barbanera, M.; Bidini, G.; Fantozzi, F. Analysis of Optimal Temperature, Pressure and Binder Quantity for the Production of Biocarbon Pellet to Be Used as a Substitute for Coke. *Appl. Energy* **2019**, *256*, 113933. [CrossRef]
24. Sun, J.; Shen, Z.; Zhang, Y.; Zhang, Q.; Wang, F.; Wang, T.; Chang, X.; Lei, Y.; Xu, H.; Cao, J.; et al. Effects of Biomass Briquetting and Carbonization on $PM_{2.5}$ Emission from Residential Burning in Guanzhong Plain, China. *Fuel* **2019**, *244*, 379–387. [CrossRef]
25. Gao, A.; Wang, Y.; Lin, G.; Li, B.; Hu, X.; Huang, Y.; Zhang, S.; Zhang, H. Volatile-Char Interactions during Biomass Pyrolysis: Reactor Design toward Product Control. *Renew. Energy* **2022**, *185*, 1–7. [CrossRef]
26. Song, Y.; Wang, Y.; Hu, X.; Hu, S.; Xiang, J.; Zhang, L.; Zhang, S.; Min, Z.; Li, C.Z. Effects of Volatile–Char Interactions on in Situ Destruction of Nascent Tar during the Pyrolysis and Gasification of Biomass. Part I. Roles of Nascent Char. *Fuel* **2014**, *122*, 60–66. [CrossRef]
27. Huang, Y.; Liu, S.; Akhtar, M.A.; Li, B.; Zhou, J.; Zhang, S.; Zhang, H. Volatile–Char Interactions during Biomass Pyrolysis: Understanding the Potential Origin of Char Activity. *Bioresour. Technol.* **2020**, *316*, 123938. [CrossRef]
28. Ding, K.; Wang, Y.; Liu, S.; Lin, G.; Syed-Hassan, S.S.A.; Li, B.; Hu, X.; Huang, Y.; Zhang, S.; Zhang, H. Volatile-Char Interactions during Biomass Pyrolysis: Insight into the Activity of Chars Derived from Three Major Components. *J. Anal. Appl. Pyrolysis* **2021**, *159*, 105320. [CrossRef]
29. Guan, G.; Kaewpanha, M.; Hao, X.; Wang, Z.; Cheng, Y.; Kasai, Y.; Abudula, A. Promoting Effect of Potassium Addition to Calcined Scallop Shell Supported Catalysts for the Decomposition of Tar Derived from Different Biomass Resources. *Fuel* **2013**, *109*, 241–247. [CrossRef]
30. Hosoya, T.; Kawamoto, H.; Saka, S. Solid/Liquid- and Vapor-Phase Interactions between Cellulose- and Lignin-Derived Pyrolysis Products. *J. Anal. Appl. Pyrolysis* **2009**, *85*, 237–246. [CrossRef]
31. Chen, Y.; Fang, Y.; Yang, H.; Xin, S.; Zhang, X.; Wang, X.; Chen, H. Effect of Volatiles Interaction during Pyrolysis of Cellulose, Hemicellulose, and Lignin at Different Temperatures. *Fuel* **2019**, *248*, 1–7. [CrossRef]
32. Penzik, M.V.; Kozlov, A.N.; Zhang, S.; Badenko, V.V.; Sosnovsky, I.K.; Shamansky, V.A. A Segmental Analysis of Pyrolysis of Woody Biomass. *Thermochim. Acta* **2022**, *711*, 179209. [CrossRef]

33. Osman, A.I.; Young, T.J.; Farrell, C.; Harrison, J.; Al-Muhtaseb, A.H.; Rooney, D.W. Physicochemical Characterization and Kinetic Modeling Concerning Combustion of Waste Berry Pomace. *ACS Sustain. Chem. Eng.* **2020**, *8*, 17573–17586. [CrossRef]
34. Ojha, D.K.; Kumar, V.S.P.; Vinu, R. Analytical Pyrolysis of Bagasse and Groundnut Shell Briquettes: Kinetics and Pyrolysate Composition Studies. *Bioresour. Technol. Rep.* **2021**, *15*, 100784. [CrossRef]
35. Chen, D.; Cen, K.; Zhuang, X.; Gan, Z.; Zhou, J.; Zhang, Y.; Zhang, H. Insight into Biomass Pyrolysis Mechanism Based on Cellulose, Hemicellulose, and Lignin: Evolution of Volatiles and Kinetics, Elucidation of Reaction Pathways, and Characterization of Gas, Biochar and Bio-Oil. *Combust. Flame* **2022**, *242*, 112142. [CrossRef]
36. Osman, A.I. Mass Spectrometry Study of Lignocellulosic Biomass Combustion and Pyrolysis with NOx Removal. *Renew. Energ.* **2020**, *146*, 484–496. [CrossRef]
37. Feng, D.; Guo, D.; Shang, Q.; Zhao, Y.; Zhang, L.; Guo, X.; Cheng, J.; Chang, G.; Guo, Q.; Sun, S. Mechanism of Biochar-Gas-Tar-Soot Formation during Pyrolysis of Different Biomass Feedstocks: Effect of Inherent Metal Species. *Fuel* **2021**, *293*, 120409. [CrossRef]
38. Couhert, C.; Commandre, J.M.; Salvador, S. Is It Possible to Predict Gas Yields of Any Biomass after Rapid Pyrolysis at High Temperature from Its Composition in Cellulose, Hemicellulose and Lignin? *Fuel* **2009**, *88*, 408–417. [CrossRef]
39. Giudicianni, P.; Cardone, G.; Sorrentino, G.; Ragucci, R. Hemicellulose, Cellulose and Lignin Interactions on *Arundo Donax* Steam Assisted Pyrolysis. *J. Anal. Appl. Pyrolysis* **2014**, *110*, 138–146. [CrossRef]
40. Zevenhoven-Onderwater, M.; Backman, R.; Skrifvars, B.J.; Hupa, M. The Ash Chemistry in Fluidised Bed Gasification of Biomass Fuels. Part I: Predicting the Chemistry of Melting Ashes and Ash–Bed Material Interaction. *Fuel* **2001**, *80*, 1489–1502. [CrossRef]
41. Eom, I.Y.; Kim, J.Y.; Kim, T.S.; Lee, S.M.; Choi, D.; Choi, I.G.; Choi, J.W. Effect of Essential Inorganic Metals on Primary Thermal Degradation of Lignocellulosic Biomass. *Bioresour. Technol.* **2012**, *104*, 687–694. [CrossRef]
42. Khelfa, A.; Bensakhria, A.; Weber, J.V. Investigations into the Pyrolytic Behaviour of Birch Wood and Its Main Components: Primary Degradation Mechanisms, Additivity and Metallic Salt Effects. *J. Anal. Appl. Pyrolysis* **2013**, *101*, 111–121. [CrossRef]
43. Zhao, D.; Dai, Y.; Chen, K.; Sun, Y.; Yang, F.; Chen, K. Effect of Potassium Inorganic and Organic Salts on the Pyrolysis Kinetics of Cigarette Paper. *J. Anal. Appl. Pyrolysis* **2013**, *102*, 114–123. [CrossRef]
44. Fan, H.; Gu, J.; Wang, Y.; Yuan, H.; Chen, Y.; Luo, B. Effect of Potassium on the Pyrolysis of Biomass Components: Pyrolysis Behaviors, Product Distribution and Kinetic Characteristics. *Waste Manag.* **2021**, *121*, 255–264. [CrossRef] [PubMed]
45. Zhu, H.; Yi, B.; Hu, H.; Fan, Q.; Wang, H.; Yao, H. The Effects of Char and Potassium on the Fast Pyrolysis Behaviors of Biomass in an Infrared-Heating Condition. *Energy* **2021**, *214*, 119065. [CrossRef]
46. Lu, Q.; Yuan, S.; Liu, C.; Zhang, T.; Xie, X.; Deng, X.; He, R. A Fe-Ca/SiO_2 Catalyst for Efficient Production of Light Aromatics from Catalytic Pyrolysis of Biomass. *Fuel* **2020**, *279*, 118500. [CrossRef]
47. Shen, Y.; Yu, S.; Yuan, R.; Wang, P. Biomass Pyrolysis with Alkaline-Earth-Metal Additive for Co-Production of Bio-Oil and Biochar-Based Soil Amendment. *Sci. Total Environ.* **2020**, *743*, 140760. [CrossRef] [PubMed]
48. Sarkar, M.; Kumar, A.; Tumuluru, J.S.; Patil, K.N.; Bellmer, D.D. Gasification Performance of Switchgrass Pretreated with Torrefaction and Densification. *Appl. Energy* **2014**, *127*, 194–201. [CrossRef]
49. Guo, F.; Liu, Y.; Wang, Y.; Li, X.; Li, T.; Guo, C. Pyrolysis Kinetics and Behavior of Potassium-Impregnated Pine Wood in TGA and a Fixed-Bed Reactor. *Energy Convers. Manag.* **2016**, *130*, 184–191. [CrossRef]
50. Zhang, L.; Yang, Z.; Li, S.; Wang, X.; Lin, R. Comparative Study on the Two-Step Pyrolysis of Different Lignocellulosic Biomass: Effects of Components. *J. Anal. Appl. Pyrolysis* **2020**, *152*, 104966. [CrossRef]
51. Zubkova, V.; Strojwas, A.; Bielecki, M.; Kieush, L.; Koverya, A. Comparative Study of Pyrolytic Behavior of the Biomass Wastes Originating in the Ukraine and Potential Application of Such Biomass. Part 1. Analysis of the Course of Pyrolysis Process and the Composition of Formed Products. *Fuel* **2019**, *254*, 115688. [CrossRef]
52. Yang, H.; Yan, R.; Chen, H.; Lee, D.H.; Zheng, C. Characteristics of Hemicellulose, Cellulose and Lignin Pyrolysis. *Fuel* **2007**, *86*, 1781–1788. [CrossRef]
53. Zong, P.; Jiang, Y.; Tian, Y.; Li, J.; Yuan, M.; Ji, Y.; Chen, M.; Li, D.; Qiao, Y. Pyrolysis Behavior and Product Distributions of Biomass Six Group Components: Starch, Cellulose, Hemicellulose, Lignin, Protein and Oil. *Energy Convers. Manag.* **2020**, *216*, 112777. [CrossRef]
54. Gong, C.; Thomsen, S.T.; Meng, X.; Pu, Y.; Puig-Arnavat, M.; Bryant, N.; Bhagia, S.; Felby, C.; Ragauskas, A.J.; Thygesen, L.G. Effects of Different Pelleting Technologies and Parameters on Pretreatment and Enzymatic Saccharification of Lignocellulosic Biomass. *Renew. Energy* **2021**, *179*, 2147–2157. [CrossRef]
55. Cao, Z.; Zhang, S.; Huang, X.; Liu, H.; Sun, M.; Lyu, J. Correlations between the Compressive Strength of the Hydrochar Pellets and the Chemical Components: Evolution and Densification Mechanism. *J. Anal. Appl. Pyrolysis* **2020**, *152*, 104956. [CrossRef]
56. Kawamoto, H.; Watanabe, T.; Saka, S. Strong Interactions during Lignin Pyrolysis in Wood—A Study by in Situ Probing of the Radical Chain Reactions Using Model Dimers. *J. Anal. Appl. Pyrolysis* **2015**, *113*, 630–637. [CrossRef]
57. Gargiulo, V.; Giudicianni, P.; Alfè, M.; Ragucci, R. Influence of Possible Interactions between Biomass Organic Components and Alkali Metal Ions on Steam Assisted Pyrolysis: A Case Study On *Arundo donax*. *J. Anal. Appl. Pyrolysis* **2015**, *112*, 244–252. [CrossRef]
58. Fushimi, C.; Katayama, S.; Tsutsumi, A. Elucidation of Interaction among Cellulose, Lignin and Xylan during Tar and Gas Evolution in Steam Gasification. *J. Anal. Appl. Pyrolysis* **2009**, *86*, 82–89. [CrossRef]

59. Wu, S.; Shen, D.; Hu, J.; Zhang, H.; Xiao, R. Cellulose-Lignin Interactions during Fast Pyrolysis with Different Temperatures and Mixing Methods. *Biomass Bioenergy* **2016**, *90*, 209–217. [CrossRef]
60. Kawamoto, H.; Morisaki, H.; Saka, S. Secondary Decomposition of Levoglucosan in Pyrolytic Production from Cellulosic Biomass. *J. Anal. Appl. Pyrolysis* **2009**, *85*, 247–251. [CrossRef]
61. Xia, S.; Li, K.; Xiao, H.; Cai, N.; Dong, Z.; Chen, X.; Chen, Y.; Yang, H.; Tu, X.; Chen, H. Pyrolysis of Chinese Chestnut Shells: Effects of Temperature and Fe Presence on Product Composition. *Bioresour. Technol.* **2019**, *287*, 121444. [CrossRef] [PubMed]
62. Usino, D.O.; Ylitervo, P.; Moreno, A.; Sipponen, M.H.; Richards, T. Primary Interactions of Biomass Components during Fast Pyrolysis. *J. Anal. Appl. Pyrolysis* **2021**, *159*, 105297. [CrossRef]
63. Herrera, R.; Erdocia, X.; Llano-Ponte, R.; Labidi, J. Characterization of Hydrothermally Treated Wood in Relation to Changes on Its Chemical Composition and Physical Properties. *J. Anal. Appl. Pyrolysis* **2014**, *107*, 256–266. [CrossRef]
64. Traoré, M.; Kaal, J.; Cortizas, A.M. Application of FTIR Spectroscopy to the Characterization of Archeological Wood. *Spectrochim. Acta A* **2016**, *153*, 63–70. [CrossRef]
65. Zubkova, V.; Strojwas, A.; Bielecki, M. Analysis of the Pyrolytic Behaviour of Birch, Maple, and Rowan Leaves. *Energies* **2021**, *14*, 2091. [CrossRef]
66. Siddiqui, M.N.; Ali, M.F.; Shirokoff, J. Use of X-ray Diffraction in Assessing the Aging Pattern of Asphalt Fractions. *Fuel* **2002**, *81*, 51–58. [CrossRef]
67. AlHumaidan, F.S.; Hauser, A.; Rana, M.S.; Lababidi, H.M.S.; Behbehani, M. Changes in Asphaltene Structure during Thermal Cracking of Residual Oils: XRD Study. *Fuel* **2015**, *150*, 558–564. [CrossRef]
68. He, P.; Liu, Y.; Shao, L.; Zhang, H.; Lü, F. Particle Size Dependence of the Physicochemical Properties of Biochar. *Chemosphere* **2018**, *212*, 385–392. [CrossRef]
69. Li, K.; Khanna, R.; Zhang, J.; Barati, M.; Liu, Z.; Xu, T.; Yang, T.; Sahajwalla, V. Comprehensive Investigation of Various Structural Features of Bituminous Coals Using Advanced Analytical Techniques. *Energy Fuels* **2015**, *29*, 7178–7189. [CrossRef]
70. Tsaneva, V.N.; Kwapinski, W.; Teng, X.; Glowacki, B.A. Assessment of the Structural Evolution of Carbons from Microwave Plasma Natural Gas Reforming and Biomass Pyrolysis Using Raman Spectroscopy. *Carbon* **2014**, *80*, 617–628. [CrossRef]
71. Wang, Z.; Cao, J.; Wang, J. Pyrolytic Characteristics of Pine Wood in a Slowly Heating and Gas Sweeping Fixed-Bed Reactor. *J. Anal. Appl. Pyrolysis* **2009**, *84*, 179–184. [CrossRef]
72. Azargohar, R.; Nanda, S.; Kozinski, J.A.; Dalai, A.K.; Sutarto, R. Effects of Temperature on the Physicochemical Characteristics of Fast Pyrolysis Bio-Chars Derived from Canadian Waste Biomass. *Fuel* **2014**, *125*, 90–100. [CrossRef]
73. Hwang, H.; Oh, S.; Cho, T.S.; Choi, I.G.; Choi, J.W. Fast Pyrolysis of Potassium Impregnated Poplar Wood and Characterization of Its Influence on the Formation as Well as Properties of Pyrolytic Products. *Bioresour. Technol.* **2013**, *150*, 359–366. [CrossRef] [PubMed]
74. Hu, J.; Jiang, B.; Liu, J.; Sun, Y.; Jiang, X. Influence of Interactions between Biomass Components on Physicochemical Characteristics of Char. *J. Anal. Appl. Pyrolysis* **2019**, *144*, 104704. [CrossRef]
75. Hu, S.; Jiang, L.; Wang, Y.; Su, S.; Sun, L.; Xu, B.; He, L.; Xiang, J. Effects of Inherent Alkali and Alkaline Earth Metallic Species on Biomass Pyrolysis at Different Temperatures. *Bioresour. Technol.* **2015**, *192*, 23–30. [CrossRef]

energies

MDPI

Review

Stannous Tungstate Semiconductor for Photocatalytic Degradation and Photoelectrochemical Water Splitting: A Review

Weixin Qiu and Yang Liu *

School of Chemistry and Chemical Engineering, Central South University, Changsha 410083, China
* Correspondence: yangliu_csu@csu.edu.cn

Abstract: The use of photocatalysis and photoelectrocatalysis is expected to achieve the efficient utilization of solar energy to alleviate and even solve the problems of energy depletion and environmental pfollution. At present, stannous tungstate materials have attracted extensive attention in the fields of photocatalysis and photoelectrocatalysis as favorable candidates for such utilization because of their narrow band gap energy (which is ~1.7 eV for the α phase and ~2.7 eV for the β phase, respectively) and unique band structure (which covers the oxidation and reduction potential of water). However, their practical application is still limited by excessive electron–hole recombination and poor stability. In this review, basic information (crystal and electronic structures) related to photocatalysis and photoelectrocatalysis is presented. Additionally, various strategies to enhance the photocatalytic and photoelectrochemical properties of stannous tungstate materials, such as morphological modification, crystal facet engineering, doping modification, and multicomponent compositing, are summarized. Furthermore, the achievements and difficulties of the relevant studies are discussed. The information presented in this review can provide a reference for subsequent research on the photocatalytic and photoelectrochemical performance of tungstate-based materials.

Keywords: stannous tungstate; photocatalysis; photoelectrocatalysis; water splitting; degradation

1. Introduction

With the rapid development of technology, energy has become a necessity for social operation, and the human consumption of non-renewable traditional energy (from sources such as oil and coal) is decreasing. The energy crisis has become an urgent problem that needs to be solved. Therefore, researchers are enthusiastically looking for alternative renewable clean energy sources. Among the many potential energy sources, solar energy has attracted significant attention due to its large reserves and reproducibility. Although solar energy exhibits a variety of excellent properties, its dispersion, instability and intermittency have led to its low utilization. Therefore, exploring efficient ways to better use solar energy is necessary for the generation of alternative power. Photocatalytic and photoelectrochemical methods show promise for degrading pollutants and generating new kinds of energy (such as hydrogen and chemical) via the use of solar energy. All the relevant reactions are completed by the photo-generated holes and electrons; thus, they follow a similar principle. Since the first report of TiO_2 in 1972 [1], different kinds of materials with a smaller band gap than that of TiO_2 (3.0–3.2 eV) have been explored [2].

One of these materials is stannous tungstate ($SnWO_4$), which is a potential catalyst due to its narrow band gap, suitable band position, and nontoxicity [3]. However, due to its low charge separation efficiency and poor stability [4], its photocatalytic and photoelectrochemical performance has not met theoretical expectations. To solve this problem, many researchers are committed to obtaining a better catalytic performance and higher solar energy efficiency by means of morphology control, doping, and multi-component compositing. In this paper, we describe the synthesis method, physical characteristics, and activities regarding the photocatalysis and photoelectrocatalysis of stannous tungstate. In addition, the challenges and future outlook regarding the use of stannous tungstate in photoelectrochemical water splitting are summarized.

2. Structure, Properties, and Synthesis of $SnWO_4$

In the 1970s, Jeitschko and Sleight reported the crystalline structure and morphology of the low-temperature (α) and high-temperature (β) phases of stannous tungstate materials [5,6]. Orthorhombic α-$SnWO_4$(Figure 1a) exhibits a stable structure below 670 °C, and it tends to transform into cubic β-$SnWO_4$ (Figure 1b) above 670 °C [7]. A reverse transition from the β-phase to the α-phase may occur under negative pressure, and the W is replaced by Mo [8]. In the crystal of α-$SnWO_4$, W and O form WO_6 octahedral structures that are interconnected in four corners, while O binds to Sn in a regular octahedral form [5]. For β-$SnWO_4$, the W atoms of β- form WO_4^{2-} tetrahedrons with O and disperse within the crystal, whereas near the Sn atoms, O is bound to it as a twisted octahedron [6].

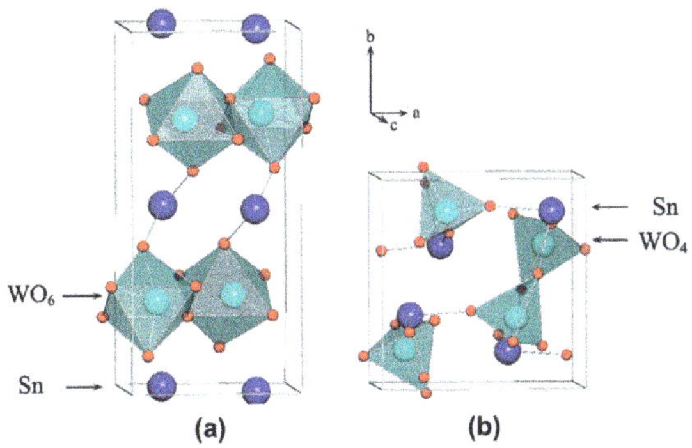

Figure 1. Crystal structures of $SnWO_4$ polymorphs: (**a**) α-$SnWO_4$ (orthorhombic); (**b**) β-$SnWO_4$ (cubic).

This difference in structure leads to different properties for the α and β phases. The W-O bond in the WO_4^{2-} tetrahedron (~1.75 Å) [5] is shorter than that in the WO_6 octahedrons (~1.80–2.15 Å) [6]. It leads to a larger crystal field-splitting energy and a higher orbital energy of W in β-$SnWO_4$. The conduction band edge of the α phase is mainly composed of a W 5d orbital, while the β phase contains an Sn 5p orbital, in addition to the W 5d orbital (Figure 2). The valence bands of both $SnWO_4$ are formed by hybridizing the O 2p and Sn 5s orbitals, and the Sn-O bonds are almost identical in length (~2.20–2.80 Å) [5,6]. Thus, the positions of the valence bands are similar. These characteristics result in a narrower band gap for α-$SnWO_4$ (~1.7 eV) than for β-$SnWO_4$ (~2.7 eV) [3], and α-$SnWO_4$ has a wider range of light absorption and a higher theoretical photocurrent density (~17 mA cm^{-2} [9]) than β-$SnWO_4$. In addition, the β phase is a direct band gap and the α phase is an indirect band gap [3], which leads to a higher carrier separation efficiency for the α phase than the β phase [10]. Due to the wider band gap, the photo-generated electrons (holes) of β-$SnWO_4$ show a more significant reduction (oxidation) activity. Thus, β-$SnWO_4$ was used in photocatalysis, and only α-$SnWO_4$ has been used in photoelectrochemical water splitting and sulfite oxidation. However, in addition to the excellent optical properties mentioned above, the use of stannous tungstate exhibits problems that urgently need to be solved. Since Sn is divalent and oxidizable in $SnWO_4$, it is susceptible to photocorrosion, which is attributed to stannous, and the following reaction occurs: $SnWO_4 + H_2O + 2h^+ \rightarrow SnO_2 + WO_3 + 2H^+$. Therefore, $SnWO_4$ has poor stability, which limits its performance in practical applications [4]. At the same time, a recombination of carriers occurs in the diffusion process of $SnWO_4$. Its diffusion length is short, which limits the number of photo-generated carriers that can arrive at the surface of the crystal, leading to unsatisfactory photocatalytic and photoelectrochemical performance.

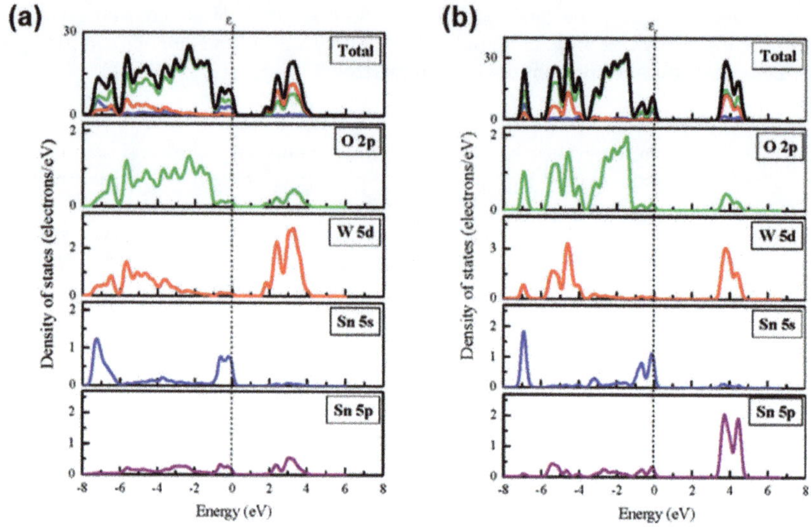

Figure 2. Partial density states of (**a**) α-$SnWO_4$ and (**b**) β-$SnWO_4$ [3].

Various methods have been developed to synthesize $SnWO_4$ materials, and their differences in thermodynamic stability lead to changes in the synthesis of α-$SnWO_4$ and β-$SnWO_4$. Because the α phase is a kind of crystalline phase at low temperatures, its synthesis conditions are milder. Many methods can be used to synthesize the α phase, e.g., hydrothermal methods, solid–solid reaction methods [3,11–13], and magnetron sputtering methods [10,14,15]. Each synthesis method has its own advantages. The hydrothermal method can be used to obtain crystals with a complete crystal phase, small particle size, and uniform distribution. A magnetron sputtering method can obtain dense and uniform high-quality thin films. The solid–solid reaction method possesses a simple process and a large output. However, the synthesis of the high-temperature crystalline β-phase is relatively harsh because it requires higher temperature conditions (which should be obtained by rapid cooling at more than 940 K). It has been reported that most β-$SnWO_4$ powders are synthesized with a high-temperature solid phase method [11,12,16,17]. β-$SnWO_4$ can also be obtained by using $NaWO_4$ as a raw material through kinetic control under more mild conditions [18]. Jan Ungelenk and Claus Feldmann synthesized β-$SnWO_4$ using a microemulsion method at low temperatures [19]. When W and O combine to form a discrete WO_4^{2-} tetrahedron in $NaWO_4$, the structure is similar to that of β-$SnWO_4$, so Na_2WO_4 is likely to be directly converted to β-$SnWO_4$. Additionally, assisted by hydrothermal [20,21] and microwave-assisted hydrothermal methods [22], α-$SnWO_4$ can also be synthesized using $NaWO_4$ and $SnCl_2$ as the precursors.

To synthesize the film for photoelectrochemical water spliting, an indirect or direct coating method can be used. In indirect coating, as-prepared $SnWO_4$ powders are drop-cast onto conductive substrates. This is an easy-to-perform method, but the performance of the photoanode is weak due to the poor charge transfer caused by bad connections between the particles. For direct coating, Moritz Kölbach et al. employed a pulsed laser-deposited method that used an α-$SnWO_4$ target formed by annealing WO_3 (99.99%) and SnO (99.99%) powders [23]. Bozheyev et al. reported on the use of a magnetron co-sputtering method in which Sn (99.99%) and W (99.95%) were the targets, and the deposition was performed with the existence of O_2 [24]. Recently, Gottesman used a new method—rapid thermal processing—to synthesize an α-$SnWO_4$ film. The technique could be used to treat a sample at higher temperatures, without destroying the glass-based F:SnO_2 (FTO), and a desired crystallinity with few defects was obtained [25]. Liu et al. used a chemical vapor deposition to convert WO_3 into α-$SnWO_4$, and the reaction is shown in Equation (1) [26].

An in situ hydrothermal conversion method was also reported, in which WO_3 was used as a precursor to react with Sn^{2+} in the solution [27]. The possible reactions occurring in this method are illustrated as Equations (2)–(3) [28]. As a result, the α-$SnWO_4$ film exhibits similar nanostructured array morphology to that of WO_3 films due to the inherited behavior (Figure 3) [28]. The anion in hydrothermal conversion solutions may also affect the morphology of α-$SnWO_4$ films [29].

$$4WO_3 + 3SnCl_2 \rightleftharpoons 3SnWO_4 + WCl_6 \quad (1)$$

$$WO_3 + H_2O \rightleftharpoons 2H^+ + WO_4{}^{2-} \quad (2)$$

$$WO_4{}^{2-} + Sn^{2+} \rightleftharpoons SnWO_4 \quad (3)$$

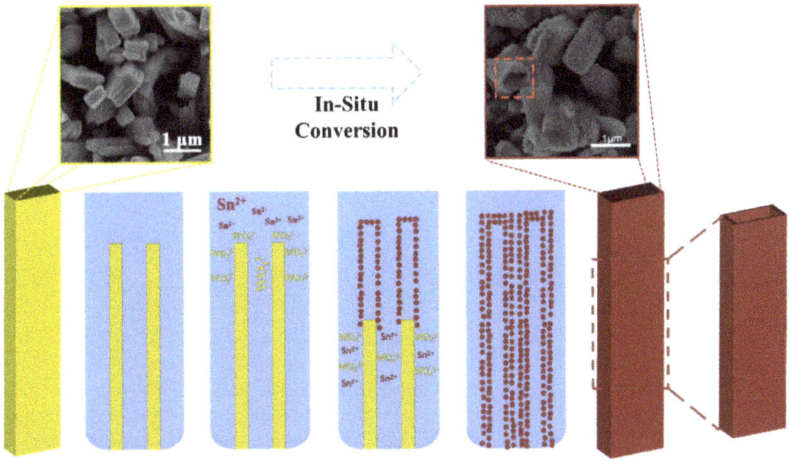

Figure 3. Schematic for in situ conversion from a rod-like WO_3 array to a rod-like $SnWO_4$ [28].

Generally speaking, $SnWO_4$ has good optical properties, such as a narrow band gap and a special band structure, which results in excellent application prospects for the fields of photocatalysis and photoelectrocatalysis (especially in the field of water splitting). However, it is necessary to pay attention to the instability and short carrier diffusion length of the material itself. Furthermore, because of the differences in crystal structure and band structure between the two crystal phases (α phase and β phase), their properties and synthesis methods are different. For instance, the photo-generated electrons (holes) of the β phase have more significant reduction (oxidation) activity than that of the α phase, so the β phase is suitable for photocatalytic degradation. The α phase has a wider light absorption range and higher carrier separation efficiency, so it has potential for use in photoelectrocatalysis.

3. Research on Modification of $SnWO_4$

Since Cho et al. first reported its optical properties in 2009 [3], $SnWO_4$ has been widely used in photocatalytic degradation [30] and photoelectrochemical water splitting [31]. Researchers have been working to modify $SnWO_4$ via morphological modification, facet regulations, doping, and the construction of multicomponent composites in order to obtain better photocatalytic degradation rates of organic pollutants (Table 1), as well as improved photoelectrochemical water splitting and sulfite oxidation performance (Table 2).

Table 1. Recently reported $SnWO_4$-based materials for photocatalysis.

Type of Catalyst	Organic Pollutant	Light	Degradation Efficiency	Ref.
Nest-like α-$SnWO_4$	methyl orange (MO)	300 W Xe arc lamp > 420 nm	96.1% (60 min)	[32]
Flower-like α-$SnWO_4$	MO	300 W tungsten-halogen lamp > 420 nm	95% (80 min)	[33]
α-$SnWO_4$ nanostrips + small nanosheets	MO	500 W Xe lamp	90.4% (30 min)	[34]
α-$SnWO_4$ synthesized with different solvents	MO	300 W Xe lamp > 420 nm	98% (80 min)	[35]
leaf-like β-$SnWO_4$	MO	250 W tungsten-halogen lamp	60% (2 h)	[36]
Bi^{3+}-doped α-$SnWO_4$	MO	500 W Xe lamp > 420 nm	95% (2 h)	[37]
α-$SnWO_4$/Ag-NPs	MO	500 W Xe lamp > 420 nm	97% (70 min)	[38]
α-$SnWO_4$/ZnFe-LDH	MO	500 W Xe lamp	95.1% (40 min)	[39]
SnS/α-$SnWO_4$	MO	500 W Xe lamp	95.6% (90 min)	[40]
CTAB-α-$SnWO_4$	MO	500 W Xe lamp	58.7% (30 min)	[20]
h-BN/α-$SnWO_4$	MO	300 W Xe lamp > 420 nm	94.7% (90 min)	[41]
α-$SnWO_4$/SnO_2	MO	500 W Xe lamp > 420 nm	97% (40 min)	[42]
β-$SnWO_4$-GO	MO	tungsten-halogen lamp > 420 nm	90% (25 min)	[12]
α-$SnWO_4$-GO	MO	300 W tungsten-halogen lamp > 420 nm	41.2% (6 h)	[43]
Zn^{2+} doped α-$SnWO_4$ nanowires	MO	300 W Hg lamp > 420 nm	~100% (90 min)	[44]
β-$SnWO_4$ with different morphology	rhodamine B (RhB)	AM 1.5 G solar light (100 mW cm^{-2})	~22% (2 h)	[45]
Cube β-$SnWO_4$	RhB	8 W UV-lamp with a monowavelength of 366 nm	TOF:1.14	[46]
Spike-cube β-$SnWO_4$	RhB	8 W UV-lamp with a monowavelength of 366 nm	TOF:2.77	[46]
SnS/α-$SnWO_4$	RhB	500 W Xe lamp	97.62% (2 h)	[40]
α-$SnWO_4$/g-C_3N_4	RhB	500 W Xe arc lamp > 420 nm	91% (80 min)	[47]
1D $ZnWO_4$@β-$SnWO_4$	RhB	300 W Xe lamp > 420 nm	~100% (2 h)	[48]
β-$SnWO_4$-GO	RhB	under visible light	91% (25 min)	[12]
α-$SnWO_4$	RhB	100 W tungsten-halogen lamp > 420 nm	~82% (4 h)	[3]
β-$SnWO_4$	RhB	100 W tungsten-halogen lamp > 420 nm	~97% (4 h)	[3]
BiOBr/α-$SnWO_4$	RhB	sunlight	97.9% (1 h)	[49]
spike-cube β-$SnWO_4$	methylene blue (MB)	8 W UV-lamp with a monowavelength of 366 nm	TOF:2.65	[43]
β-$SnWO_4$-rGO nanocomposite	MB	tungsten-halogen lamp > 420 nm	94% (2 h)	[50]
β-$SnWO_4$/ZnO	MB	70 W sodium lamps	82.9% (2 h)	[51]
β-$SnWO_4$ truncated rhombic dodecahedrons	MB	150 W halogen lamp	100% (20 min)	[19]
β-$Sn_{1-n}WO_4$·nα-Sn	MB	halogen bulb (3300 K)	~90% (1 h)	[52]
h-BN/α-$SnWO_4$	tetracycline (TC)	300 W Xe lamp > 420 nm	82.2% (4 h)	[41]
SnS/α-$SnWO_4$	TC	500 W Xe lamp	57.0% (3 h)	[40]
β-$SnWO_4$-rGO nanocomposite	rose bengal (RB)	tungsten-halogen lamp > 420 nm	98% (2 h)	[50]
β-$SnWO_4$ NPs	RB	300 W tungsten bulb	94.6% (2.5 h)	[53]
α-$SnWO_4$/UiO-66(NH_2)/g-C_3N_4	ibuprofen (IPF)	high pressure Xe lamp source simulated sunlight	95.5% (2 h)	[54]
β-$Sn_{1-n}WO_4$·nα-Sn	basic green 4 (BG)	halogen bulb (3300 K)	~63% (1 h)	[52]
BiOBr/α-$SnWO_4$	BG	sunlight	95.5% (45 min)	[49]
β-$Sn_{1-n}WO_4$·nα-Sn	methyl red (MR)	halogen bulb (3300 K)	~71% (1 h)	[52]

Table 2. Recent report on the performance of α-$SnWO_4$ based photoanodes in PEC water splitting.

Photoanode	Photocurrent Density (at 1.23 V vs. RHE)	Electrolyte	Ref.
α-$SnWO_4$ nanowires	0.032 mA cm^{-2}	0.1 M KPi buffer (pH ≈ 5)	[27]
α-$SnWO_4$ porous nanostructure	0.080 mA cm^{-2}	0.5 M KPi buffer (pH ≈ 7)	[31]
α-$SnWO_4$ nanocrystalline particles	~0.750 mA cm^{-2}	0.5 M KPi buffer and 0.5 M Na_2SO_3 (pH ≈ 7)	[4]
α-$SnWO_4$ microcrystalline particles	0.375 mA cm^{-2}	0.5 M Na_2SO_4 (pH ≈ 7)	[24]
α-$SnWO_4$ coral-like morphology	0.420 mA cm^{-2}	KOH/H_3BO_3 buffer and 0.2 M Na_2SO_3 (pH ≈ 9)	[55]
2D nanosheets α-$SnWO_4$	0.411 mA cm^{-2}	0.2 M KPi buffer (pH ≈ 7)	[56]
Nail-like α-$SnWO_4$	0.300 mA cm^{-2}	0.2 M KPi buffer (pH ≈ 7)	[29]
2D long-plate α-$SnWO_4$	0.790 mA cm^{-2}	0.2 M KPi buffer (pH ≈ 7)	[28]
NiO_x coated RTP-α-$SnWO_4$	0.950 mA cm^{-2}	0.5 M KPi buffer with 0.5 M Na_2SO_3 (pH ≈ 7)	[25]

3.1. Morphological Modification

By regulating the synthesis technique, which is a universal method for enhancing the catalytic performance of a material, a target material with a specific morphology can be obtained. The morphology of a material is usually designed for the following purposes [57]: (1) to increase the specific surface area and reduce the nanometer size, as a high specific surface area can provide more active sites, and a small nanometer size can provide a shorter charge transfer path that improves carrier separation efficiency and enhances light absorption efficiency; (2) to improve the crystallinity of materials, as a high crystallinity can reduce the recombination center and improve carrier separation efficiency, but it is difficult to simultaneously improve crystallinity and specific surface area; (3) to obtain special structural shapes (such as 0D, 1D, 2D, and 3D materials), as different dimensions have diverse characteristics. Thus, directional design can improve the performance of the materials.

In the study of $SnWO_4$ for photocatalysis, there have been many attempts to obtain improved properties by adjusting the morphologies of the substrate materials. Zhu et al. obtained flower-like α-$SnWO_4$ powders under neutral conditions with a DTAB-assisted hydrothermal method, degrading 95% of methyl orange (MO) within 80 min [33]. Want et al. also discussed the solvent-dependent morphology of α-$SnWO_4$ powders [35], and Liu et al. synthesized nanostrips and nanosheets by changing the water/ethylene glycol ratio through a solvothermal method [34]. Meanwhile, the layered multi-hollow spherical α-$SnWO_4$ powders [37] prepared by Zhu et al. and the nested nanostructure α-$SnWO_4$ [32] prepared by Zhang et al. showed an improved catalytic performance after controlling their morphologies. The increased specific surface area enabled by morphology controlling exposes the active sites and shortens the length needed for photo-generated carriers to arrive at a surface. For a β-$SnWO_4$ photocatalyst, Chen et al. prepared cubes and spike-cube shaped particles, with the latter showing strong photocatalytic activity [46]. Warmuth et al. formed tetrahedral β-$SnWO_4$, which demonstrated better photodegradation activity than a β-$SnWO_4$ material, with truncated octahedra and a spike-cube morphology (Figure 4a) [45]. Ungelenk et al. synthesized β-$SnWO_4$ powders with a microemulsion method. The powders had a rhombohedral dodecahedron shape, and they could complete the degradation of methylene blue (MB) in 20 min [19]. Raj et al. also noted the better

MO degradation performance of leaf-like β-$SnWO_4$ compared with that of sphere-like and irregularly structured β-$SnWO_4$ [36]. Alharthi et al. synthesized honeycomb-like β-$SnWO_4$ with a hydrothermal method and calcined it at 700 °C, which enabled the photodegradation of rose bengal in 150 min [53]. For photoelectrochemical water splitting, it is difficult to control the morphology of the powder present on an electrode. Accordingly, Zhu et al. synthesized a porous α-$SnWO_4$ film through the hydrothermal conversion of WO_3 films [31], and they obtained a photocurrent density of 0.08 mA cm^{-2} at a potential of 1.23 V_{RHE}. In contrast, topography optimized nails (Figure 4b) [27,29], nanosheets (Figure 4c) [56], and long-plates [28] have presented better PEC activity than α-$SnWO_4$, with a powder stacked morphology because the array structures provide pathways for charge transfer, with less recombination.

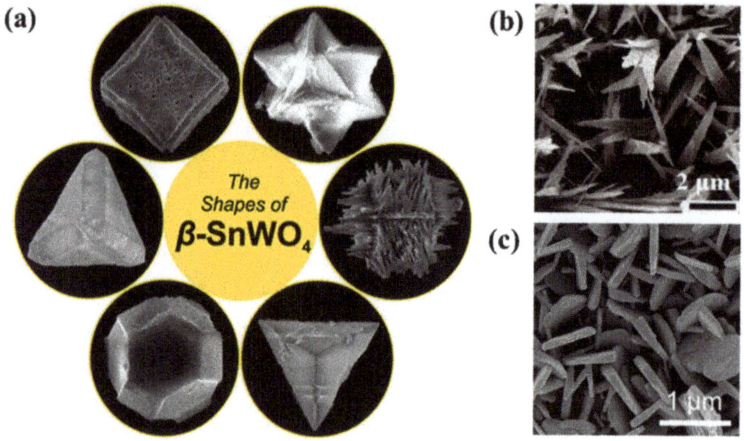

Figure 4. (**a**) Different shapes of β-$SnWO_4$ [45]; α-$SnWO_4$ with (**b**) dense nanorods [27] and (**c**) nanosheet [56] morphology.

3.2. Crystal Facet Engineering

During morphology regulation, the ratio of exposure facets may change as well [58]. It is noted that reaction activities vary for different facets [59,60]. Thus, charge transfer ability can be modified by controlling the anisotropic growth of crystals in a certain direction. The β-$SnWO_4$ enclosed by (100) and (110) facets obtained the strong photocatalytic degradation of methylene blue [19]. Harb et al. found that the (001) facet of α-$SnWO_4$ and the (100) facet of β-$SnWO_4$ showed strong charge separation and transport capabilities via calculations with the HSE06 code [10]. They also discussed the (121), (210), (111), (200), and (040) exposed facets of α-$SnWO_4$ using DFT, and the (210) and (121) facets were shown to perform oxygen evolution reaction due to the positive position of their valence band (Figure 5a) [61]. They also noted that the (001) facet had a high surface energy, but it may be thermodynamically less stable than the other five facets. Wang et al. discussed the oxygen evolution performance of α-$SnWO_4$ (010) facets with different kinds of termination, and O-Sn termination showed better OER activity than did the others [62]. These DFT researchers have demonstrated that crystal facet engineering is another viable method for obtaining a good $SnWO_4$ substrate material. Inspired by these projects, our group synthesized an α-$SnWO_4$ array and added F ions to its precursor solution to increase the ratio of the (001) facet (Figure 5b). The facet-controlled film had a higher photocurrent than the uncontrolled film due to the better OER performance (Figure 5c) [29], and a ~1.9 times greater photocurrent was obtained in an unbiased PEC cell.

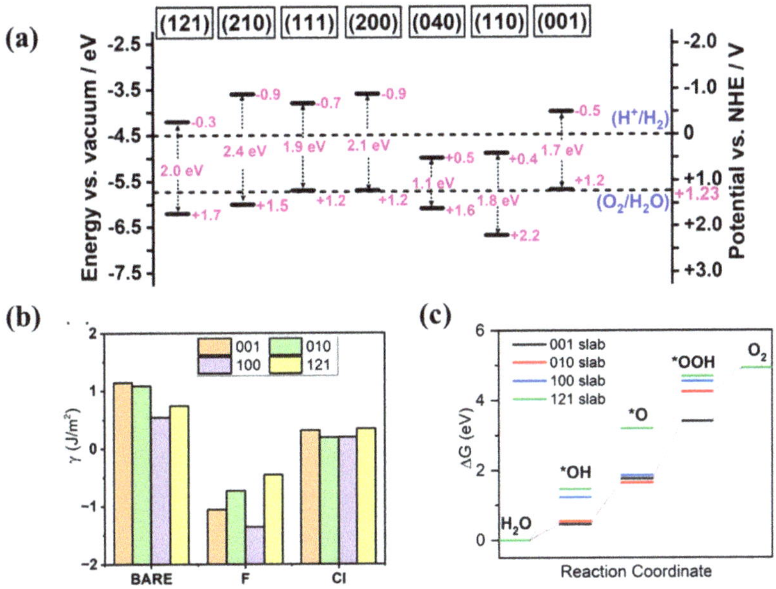

Figure 5. (**a**) VBM and CBM energy levels for the optimized (121)−, (210)−, (111)−, (200)−, (040)−, (110), and (001) oriented α-$SnWO_4$ slabs [61,63]; (**b**) surface free energy (γ) of facets before and after the termination with F and Cl atoms; (**c**) free-energy profiles of OER on (001)−, (010)−, (100)−, and (121) slabs at 0 V (the * means adsorbed state) [29].

3.3. Doping Modification

Doping can change the position of an energy band, reduce the width of a band gap, and improve the efficiency of carrier separation and transfer, thus improving the catalytic performance of the materials. For example, Zn^{2+} doping was introduced into α-$SnWO_4$ by Su et al. [44]. Because Zn and Sn have similar electronegativity values and suitable sizes, Zn^{2+} is used to replace Sn^{2+} in $SnWO_4$ to form $Sn_{1-x}Zn_xWO_4$. This ensures that the Zn 4s and Zn 4p orbitals participate in the hybridization of the valence band and the conduction band, respectively, thus changing the position of the energy band and reducing the band gap. The authors also found that Zn^{2+} doping changed the morphology of the material, and the resulting $Sn_{0.955}Zn_{0.045}WO_4$ could degrade nearly 90% of MO in 20 min. Zhu et al. introduced defects into α-$SnWO_4$ through Bi^{3+} doping [37]. Unlike Zn^{2+} doping, Bi does not participate in the formation of energy band hybridization, but cation vacancies are formed due to changes in the charges. These defects form effective energy levels in the band gap which enhance light absorption, inhibit electron-hole pairing, and improve the photocatalytic properties of the material. Although the mechanisms of Zn^2 and Bi^3 doping are not exactly the same, both of them can improve the performance of $SnWO_4$ by changing its morphology and band structure. At the same time, because β-$SnWO_4$ has a wider band and a lower carrier separation efficiency than α-$SnWO_4$, the doping of the two ions may lead to better performance improvements for β-$SnWO_4$. In addition to these two kinds of hetero ions, there may be more suitable ions for $SnWO_4$ modification (such as Mo and Co) that need to be further studied. In theory, Azofra et al. replaced Sn with Ge and substituted W with Mo in DFT calculations, and they obtained the enhanced orbital hybridization of the VBM/CBM electronic states, which amplified the amount of generated holes/electrons on top of the (110)/(100) facets [63]. However, these result have not been experimentally confirmed.

3.4. Multicomponent Composite

The multicomponent composite strategy includes coating a sedimentary protection layer and building heterojunction. Constructed multicomponent catalysts generally have the following advantages over single catalysts: (1) they expand the light absorption spectra, (2) they promote carrier separation, (3) they inhibit carrier recombination, and (4) they prevent photocorrosion.

In photocatalysis, the compositing of reduced graphene oxide (RGO) and β-$SnWO_4$ can be achieved via microwave heating [12] and hydrothermal methods [50]. In such a composite, RGO, as a supporting material, provides a larger surface area and more active sites than β-$SnWO_4$, thereby improving the specific surface area and catalytic activity of the material. At the same time, the high conductivity of RGO makes the electrons generated by β-$SnWO_4$ move towards the RGO, while the holes remain inside β-$SnWO_4$, which improves the carrier separation efficiency of the materials. The compositing of RGO and α-$SnWO_4$ was reported by Huang et al. [43], and Wang et al. [41] combined hexagonal boron nitride (2D materials analogous to graphite) with α-$SnWO_4$. Both groups improved the photocatalytic activity of α-$SnWO_4$ for degrading methyl orange (MO) and tetracycline. In a composite of Ag-NPs and α-$SnWO_4$ [38], the photo-absorption efficiency can be increased through the surface plasmon resonance (SPR) of Ag-NPs. Because Ag-NPs have lower Fermi levels than the bottom of the conduction band of α-$SnWO_4$, the photoelectrons can migrate to Ag, increasing the efficiency of carrier separation. Moreover, Ag can more easily transfer electrons to a solution when it is in direct contact with the solution, which accelerates the transfer of interfacial carriers. The construction of heterojunctions is widely studied in $SnWO_4$ material research. The formation of heterojunctions is mainly based on the combination of different energy band positions of different materials. Accordingly, photogenerated electrons are easier to transfer to materials with lower conduction bands, while photogenerated holes tend to transfer to materials with higher valence bands. It is possible to enhance the carrier separation efficiency of materials by selecting suitable materials for compositing. For example, the heterojunctions of α-$SnWO_4$ and g-C_3N_4 were shown to effectively inhibit carrier recombination, thereby increasing the catalytic activity of the material [47]. $SnWO_4$/ZnO [51], $SnWO_4$/BiOBr [49], $SnWO_4$/UiO-66/g-C_3N_4 ternary heterojunction [54], one dimensional $ZnWO_4$@$SnWO_4$ core-shell heterojunction [48], and SnS/α-$SnWO_4$ [40] have also been reported. In addition, α-$SnWO_4$ and SnO_2 [42], along with β-$SnWO_4$ [52], can form heterojunctions to improve the efficiency of carrier separation.

In photoelectrocatalysis, Kölbach and Schnell et al. deposited NiO_x on α-$SnWO_4$ to protect the substrate from photocorrosion [4], which improved material stability and maintained a stable current for sulfite oxidation at 1.23 V vs. RHE in 30 min. Furthermore, α-$SnWO_4$ is susceptible to photo-corrosion and oxidation during photoelectrochemical water splitting, which results in a lower photocurrent value (<10 μA cm^{-2}). However, the photocurrent of the material is greatly increased after protection by NiO_x (~0.75 mA cm^{-2}). Another study found that a thin oxide layer (SnO_2) might form at the interface of α-$SnWO_4$ and NiO_x, which explains the reason for the decreasing photovoltage after the loading of NiO_x [9]. Thus, the exploration of alternative techniques for loading protection layers, such as vacuum evaporation and atomic layer deposition, is recommended.

3.5. Other Techniques

In addition to the aforementioned approaches, photocatalysis and photoelectrocatalysis performance may be influenced by the external environment [64], as has been studied in recent years. Liu et al. discussed the PEC performance of $SnWO_4$/Sn electrodes at different operating temperatures, and they found that higher photocurrents could be obtained at 70 °C (Figure 6a) [55]. Schnell studied the pH-dependent stability of α-$SnWO_4$, and pH = 7 (Figure 6b) was suggested to be used for testing due to the formation of the surface oxide layer as a protection layer [65]. Besides, α-$SnWO_4$, with higher crystallinity, was formed by Cottesman, presenting ~0.95 mA/cm^2 after loading with NiO_x for sulfite oxidation.

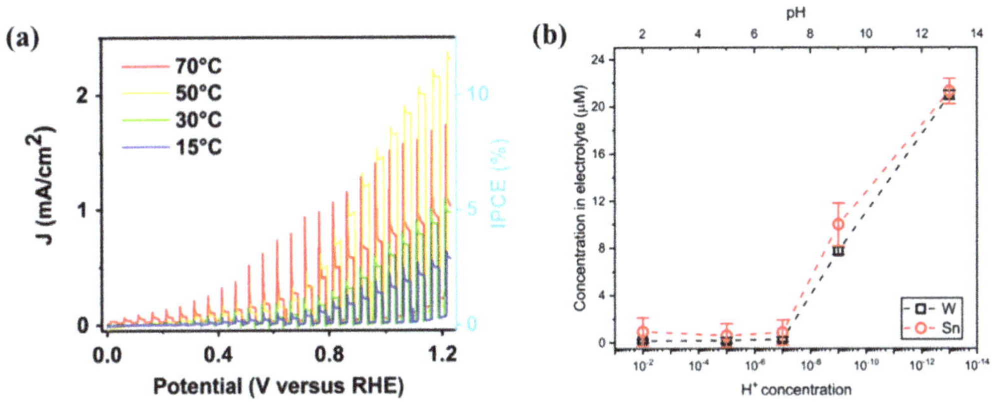

Figure 6. (**a**) LSV curves of the $SnWO_4$ (500 °C H_2) photoanode measured in KOH/H_3BO_3 buffer (pH = 9) with 0.2 M Na_2SO_3 at different temperatures under the chopped 455 nm LED illumination [55]; (**b**) the concentrations of Sn and W dissolved from α-$SnWO_4$ films after photoelectrochemically treated at various pH levels at a potential of 1.23 V vs. RHE, for a total of 1 h [65].

4. Conclusions and Outlook

In general, $SnWO_4$ is considered a type of potential catalytic material for photocatalysis and photoelectrocatalysis, especially for water splitting, because of its optical properties and special band structure (which covers the oxidation and reduction potential of water). β-$SnWO_4$ is always used in photocatalytic degradation due to the high reduction (oxidation) activity of its photo-generated electrons (holes), while α-$SnWO_4$ is employed in both photocatalysis and photoelectrocatalysis. In recent years, many researchers have studied the application of $SnWO_4$ in photocatalysis and photoelectrocatalysis, and they have improved the properties of materials by means of morphology control, doping, and multicomponent combinations. Compared with photocatalysis, photoelectrocatalysis has the advantage of discrete oxidation and reduction processes. This makes it easier to separate gaseous products, study reaction mechanisms and kinetics, and adjust reaction selectivity. Therefore, the application of α-$SnWO_4$ in photoelectrocatalysis needs to be further explored. However, it is worth noting that the catalytic performance of α-$SnWO_4$ and its applications in solar energy have not reached their expected values. For example, the highest reported experimental current for PEC water oxidation is only ~0.79 mA cm^{-2} at 1.23 V_{RHE}, which is far from the theoretical value. Other problems include low efficiency and poor stability. To solve these problems, obtaining a high-performing substrate through morphology control, crystal surface control, and doping may prove to be an effective research route. Then, the PEC performance of α-$SnWO_4$ can be further upgraded by constructing heterojunctions, depositing protective layers, and depositing cocatalysts. Of these methods, loading cocatalysts is the most promising because the catalysts can also act as protection layers. However, a suitable loading method to avoid oxidizing α-$SnWO_4$ is required, and pinholes should be excluded to prevent α-$SnWO_4$ from attaching to the electrolytes.

Author Contributions: W.Q.: writing—original draft and visualization; Y.L.: writing—review and editing, supervision, funding acquisition, and conceptualization. All authors have read and agreed to the published version of the manuscript.

Funding: This study was supported by the National Nature Science Foundation of China (51904356).

Data Availability Statement: Not applicable.

Acknowledgments: W.Q. and Y.L. appreciate GaoShuang He, Keke Wang, Wenzhang Li, and Jie Li's suggestions on this paper.

Conflicts of Interest: The authors declare no conflict of interest.

References

1. Fujishima, A.; Honda, K. Electrochemical Photolysis of Water at a Semiconductor Electrode. *Nature* **1972**, *238*, 37–38. [CrossRef] [PubMed]
2. Tao, X.; Zhao, Y.; Wang, S.; Li, C.; Li, R. Recent advances and perspectives for solar-driven water splitting using particulate photocatalysts. *Chem. Soc. Rev.* **2022**, *51*, 3561–3608. [CrossRef] [PubMed]
3. Cho, I.-S.; Kwak, C.H.; Kim, D.W.; Lee, S.; Hong, K.S. Photophysical, Photoelectrochemical, and Photocatalytic Properties of Novel $SnWO_4$ Oxide Semiconductors with Narrow Band Gaps. *J. Phys. Chem. C* **2009**, *113*, 10647–10653. [CrossRef]
4. Kölbach, M.; Pereira, I.J.; Harbauer, K.; Plate, P.; Höflich, K.; Berglund, S.P.; Friedrich, D.; van de Krol, R.; Abdi, F.F. Revealing the performance limiting factors in α-$SnWO_4$ photoanodes for solar water splitting. *Chem. Mater.* **2018**, *30*, 8322–8331. [CrossRef]
5. Jeitschko, W.; Sleight, A. Stannous tungstate: Properties, crystal structure and relationship to ferroelectric $SbTaO_4$ type compounds. *Acta Crystallogr. Sect. B Struct. Crystallogr. Cryst. Chem.* **1974**, *30*, 2088–2094. [CrossRef]
6. Jeitschko, W.; Sleight, A.W. Synthesis, properties and crystal structure of β-$SnWO_4$. *Acta Crystallogr. Sect. B Struct. Crystallogr. Cryst. Chem.* **1972**, *28*, 3174–3178. [CrossRef]
7. Ke, J.; Younis, M.A.; Kong, Y.; Zhou, H.; Liu, J.; Lei, L.; Hou, Y. Nanostructured Ternary Metal Tungstate-Based Photocatalysts for Environmental Purification and Solar Water Splitting: A Review. *Nano-Micro Lett.* **2018**, *10*, 69. [CrossRef]
8. Gomes, E.O.; Gouveia, A.F.; Gracia, L.; Lobato, A.; Recio, J.M.; Andres, J. A Chemical-Pressure-Induced Phase Transition Controlled by Lone Electron Pair Activity. *J. Phys. Chem. Lett.* **2022**, *13*, 9883–9888. [CrossRef]
9. Schnell, P.; Kolbach, M.; Schleuning, M.; Obata, K.; Irani, R.; Ahmet, I.Y.; Harb, M.; Starr, D.E.; van de Krol, R.; Abdi, F.F. Interfacial Oxide Formation Limits the Photovoltage of α-$SnWO_4/NiO_x$ Photoanodes Prepared by Pulsed Laser Deposition. *Adv. Energy Mater.* **2021**, *11*, 2003183. [CrossRef]
10. Harb, M.; Ziani, A.; Takanabe, K. Critical difference between optoelectronic properties of α- and β-$SnWO_4$ semiconductors: A DFT/HSE06 and experimental investigation. *Phys. Status Solidi B* **2016**, *253*, 1115–1119. [CrossRef]
11. Kuzmin, A.; Anspoks, A.; Kalinko, A.; Timoshenko, J.; Kalendarev, R. Extended x-ray absorption fine structure spectroscopy and first-principles study of $SnWO_4$. *Phys. Scr.* **2014**, *89*, 044005. [CrossRef]
12. Thangavel, S.; Venugopal, G.; Kim, S.-J. Enhanced photocatalytic efficacy of organic dyes using β-tin tungstate–reduced graphene oxide nanocomposites. *Mater. Chem. Phys.* **2014**, *145*, 108–115. [CrossRef]
13. Alexei, K.; Andris, A.; Aleksandr, K.; Janis, T.; Robert, K.; Lucie, N.; François, B.; Tetsuo, I.; Pascale, R. Pressure-induced insulator-to-metal transition in α-$SnWO_4$. *J. Phys. Conf. Ser.* **2016**, *712*, 012122.
14. Kuzmin, A.; Zubkins, M.; Kalendarev, R. Preparation and Characterization of Tin Tungstate Thin Films. *Ferroelectrics* **2015**, *484*, 49–54. [CrossRef]
15. Ziani, A.; Harb, M.; Noureldine, D.; Takanabe, K. UV-Vis optoelectronic properties of α-$SnWO_4$: A comparative experimental and density functional theory based study. *APL Mater.* **2015**, *3*, 096101. [CrossRef]
16. Stoltzfus, M.W.; Woodward, P.M.; Seshadri, R.; Klepeis, J.-H.; Bursten, B. Structure and Bonding in $SnWO_4$, $PbWO_4$, and $BiVO_4$: Lone Pairs vs. Inert Pairs. *Inorg. Chem.* **2007**, *46*, 3839–3850. [CrossRef]
17. Wojcik, J.; Calvayrac, F.; Goutenoire, F.; Mhadhbi, N.; Corbel, G.; Lacorre, P.; Bulou, A. Lattice Dynamics of β-$SnWO_4$: Experimental and Ab Initio Calculations. *J. Phys. Chem. C* **2013**, *117*, 5301–5313. [CrossRef]
18. Pavithra, N.S.; Patil, S.B.; Kumar, S.R.K.; Alharthi, F.A.; Nagaraju, G. Facile synthesis of nanocrystalline β-$SnWO_4$: As a photocatalyst, biosensor and anode for Li-ion battery. *SN Appl. Sci.* **2019**, *1*, 1123. [CrossRef]
19. Ungelenk, J.; Feldmann, C. Synthesis of faceted β-$SnWO_4$ microcrystals with enhanced visible-light photocatalytic properties. *Chem. Commun.* **2012**, *48*, 7838–7840. [CrossRef]
20. Huang, J.; Liu, H.; Zhong, J.; Li, J. Enhanced simulated sunlight-driven photocatalytic performance of $SnWO_4$ prepared in the presence of cetyltrimethylammonium bromide. *Environ. Prog. Sustain. Energy* **2020**, *39*, e13314. [CrossRef]
21. Alharthi, F.A.; Shashank, M.; Shashikanth, J.; Viswantha, R.; Alghamdi, A.A.; Algethami, J.; Alsaiari, M.A.; Jalalah, M.S.; Ganganagappa, N. Hydrothermal synthesis of α-$SnWO_4$: Application to lithium-ion battery and photocatalytic activity. *Ceram. Int.* **2021**, *47*, 10242–10249. [CrossRef]
22. Barros, M.M.P.; Almeida, K.C.; Silva, S.A.; Botelho, G. Synthesis and characterization of α-$SnWO_4$ powders obtained by microwave-assisted hydrothermal method. *Cerâmica* **2022**, *68*, 236–241. [CrossRef]
23. Kölbach, M.; Hempel, H.; Harbauer, K.; Schleuning, M.; Petsiuk, A.; Höflich, K.; Deinhart, V.; Friedrich, D.; Eichberger, R.; Abdi, F.F.; et al. Grain Boundaries Limit the Charge Carrier Transport in Pulsed Laser Deposited α-$SnWO_4$ Thin Film Photoabsorbers. *ACS Appl. Energy Mater.* **2020**, *3*, 4320–4330. [CrossRef]
24. Bozheyev, F.; Akinoglu, E.M.; Wu, L.; Lu, H.; Nemkayeva, R.; Xue, Y.; Jin, M.; Giersig, M. Band gap optimization of tin tungstate thin films for solar water oxidation. *Int. J. Hydrogen Energy* **2020**, *45*, 8676–8685. [CrossRef]
25. Gottesman, R.; Peracchi, I.; Gerke, J.L.; Irani, R.; Abdi, F.F.; van de Krol, R. Shining a Hot Light on Emerging Photoabsorber Materials: The Power of Rapid Radiative Heating in Developing Oxide Thin-Film Photoelectrodes. *ACS Energy Lett.* **2022**, *7*, 514–522. [CrossRef]
26. Zhu, S.; Liu, D.; Li, J.; Kuang, Y. Chemical Vapor Deposition of Crystalized Nanoscale α-$SnWO_4$ Thin Films and Their Photoelectrocatalytic Properties. *ACS Appl. Energy Mater.* **2022**, *5*, 14372–14380. [CrossRef]

27. Pyper, K.J.; Evans, T.C.; Bartlett, B.M. Synthesis of α-$SnWO_4$ thin-film electrodes by hydrothermal conversion from crystalline WO_3. *Chin. Chem. Lett.* **2015**, *26*, 474–478. [CrossRef]
28. Qiu, W.; Zhang, Y.; He, G.; Chen, L.; Wang, K.; Wang, Q.; Li, W.; Liu, Y.; Li, J. Two-Dimensional Long-Plate $SnWO_4$ Photoanode Exposed Active Facets for Enhanced Solar Water Splitting. *ACS Appl. Energy Mater.* **2022**, *5*, 11732–11739. [CrossRef]
29. Liu, Y.; Qiu, W.; He, G.; Wang, K.; Wang, Y.; Chen, L.; Wu, Q.; Li, W.; Li, J. Nail-like α-$SnWO_4$ Array Film with Increased Reactive Facets for Photoelectrochemical Water Splitting. *J. Phys. Chem. C* **2022**, *126*, 15596–15605. [CrossRef]
30. Mirsadeghi, S.; Zandavar, H.; Tooski, H.F.; Rahimi, M.; Sohouli, E.; Rahimi-Nasrabadi, M.; Ganjali, M.R.; Pourmortazavi, S.M. Rapid photodegradation and detection of zolpidem over β-$SnWO_4$ and α-$SnWO_4$ nanoparticles: Optimization and mechanism. *Environ. Sci. Pollut. Res.* **2020**, *28*, 5430–5442. [CrossRef]
31. Zhu, Z.; Sarker, P.; Zhao, C.; Zhou, L.; Grimm, R.L.; Huda, M.N.; Rao, P.M. Photoelectrochemical Properties and Behavior of α-$SnWO_4$ Photoanodes Synthesized by Hydrothermal Conversion of WO_3 Films. *ACS Appl. Mater. Interfaces* **2017**, *9*, 1459–1470. [CrossRef] [PubMed]
32. Zhang, X.; You, X.; Wang, X.; Yu, C.; Xu, L.; Wang, C.; Song, Y.; Zhang, F. Nest-like α-$SnWO_4$ nanostructures assembled by nanowires: Facile synthesis and their superior photocatalytic performance. *J. Alloys Compd.* **2019**, *802*, 502–510. [CrossRef]
33. Zhu, G.; Que, W.; Zhang, J.; Zhong, P. Photocatalytic activity of $SnWO_4$ and SnW_3O_9 nanostructures prepared by a surfactant-assisted hydrothermal process. *Mater. Sci. Eng. B* **2011**, *176*, 1448–1455. [CrossRef]
34. Liu, X.; Liang, B.; Yang, J.; Li, W. Solvent effect on morphological evolution and photocatalytic property of α-$SnWO_4$. *J. Taiwan Inst. Chem. Eng.* **2019**, *95*, 575–582. [CrossRef]
35. Wang, Q.L.; Li, H.B.; Jiang, H.Y.; Ding, S.T.; Song, Z.W.; Shi, J.S. Effect of solvent on α-$SnWO_4$ photocatalyst for degradation of methyl orange under visible light irradiation. *Adv. Perform. Mater.* **2015**, *30*, 288–293. [CrossRef]
36. Raj, A.T.; Thangavel, S.; Rose, A.; Jipsa, C.V.; Jose, M.; Nallamuthu, G.; Kim, S.-J.; Venugopal, G. Influence of Morphology and Common Oxidants on the Photocatalytic Property of β-$SnWO_4$ Nanoparticles. *J. Nanosci. Nanotechnol.* **2016**, *16*, 2541–2547. [CrossRef]
37. Zhu, Z.; Tian, H.; Zhang, M.; Liang, B.; Li, W. Preparation of α-$SnWO_4$ hierarchical spheres by Bi^{3+}-doping and their enhanced photocatalytic activity under visible light. *Ceram. Int.* **2016**, *42*, 14743–14748. [CrossRef]
38. Liu, X.; Liang, B.; Zhang, M.; Long, Y.; Li, W. Enhanced photocatalytic properties of α-$SnWO_4$ nanosheets modified by Ag nanoparticles. *J. Colloid Interface Sci.* **2017**, *490*, 46–52. [CrossRef]
39. Wang, Y.; Zhou, S.; Zhao, G.; Li, C.; Liu, L.; Jiao, F. Fabrication of $SnWO_4$/ZnFe-layered double hydroxide composites with enhanced photocatalytic degradation of methyl orange. *J. Mater. Sci.-Mater. Electron.* **2020**, *31*, 12269–12281. [CrossRef]
40. Liu, X.; Liang, B.; Li, W. In situ decoration of SnS quantum dots on the α-$SnWO_4$ nanosheets for superior visible-light photocatalytic performance. *Appl. Surf. Sci.* **2020**, *531*, 147379. [CrossRef]
41. Wang, J.; Yan, H.; Long, Y.; Li, W. Enhanced photocatalytic property of α-$SnWO_4$ nanoplates by h-BN decorating. *J. Mater. Sci. Mater. Electron.* **2021**, *32*, 21858–21868. [CrossRef]
42. Yao, S.; Zhang, M.; Di, J.; Wang, Z.; Long, Y.; Li, W. Preparation of α-$SnWO_4$/SnO_2 heterostructure with enhanced visible-light-driven photocatalytic activity. *Appl. Surf. Sci.* **2015**, *357*, 1528–1535. [CrossRef]
43. Huang, R.K.; Ge, H.; Lin, X.J.; Guo, Y.L.; Yuan, R.S.; Fu, X.Z.; Li, Z.H. Facile one-pot preparation of α-$SnWO_4$/reduced graphene oxide (RGO) nanocomposite with improved visible light photocatalytic activity and anode performance for Li-ion batteries. *RSC Adv.* **2013**, *3*, 1235–1242. [CrossRef]
44. Su, Y.G.; Hou, L.C.; Du, C.F.; Peng, L.M.; Guan, K.; Wang, X.J. Rapid synthesis of Zn^{2+} doped $SnWO_4$ nanowires with the aim of exploring doping effects on highly enhanced visible photocatalytic activities. *RSC Adv.* **2012**, *2*, 6266–6273. [CrossRef]
45. Warmuth, L.; Feldmann, C. β-$SnWO_4$ with Morphology-Controlled Synthesis and Facet-Depending Photocatalysis. *ACS Omega* **2019**, *4*, 13400–13407. [CrossRef]
46. Chen, Y.-C.; Lin, Y.-G.; Hsu, L.-C.; Tarasov, A.; Chen, P.-T.; Hayashi, M.; Ungelenk, J.; Hsu, Y.-K.; Feldmann, C. β-$SnWO_4$ Photocatalyst with Controlled Morphological Transition of Cubes to Spikecubes. *ACS Catal.* **2016**, *6*, 2357–2367. [CrossRef]
47. Liang, Q.; Jin, J.; Liu, C.; Xu, S.; Yao, C.; Chen, Z.; Li, Z. Hydrothermal fabrication of α-$SnWO_4$/g-C_3N_4 heterostructure with enhanced visible-light photocatalytic activity. *J. Mater. Sci. Mater. Electron.* **2017**, *28*, 11279–11283. [CrossRef]
48. Zhuang, H.; Xu, W.; Lin, L.; Huang, M.; Xu, M.; Chen, S.; Cai, Z. Construction of one dimensional $ZnWO_4$@$SnWO_4$ core-shell heterostructure for boosted photocatalytic performance. *J. Mater. Sci. Technol.* **2019**, *35*, 2312–2318. [CrossRef]
49. Chowdhury, A.P.; Shambharkar, B.H. Fabrication and characterization of BiOBr-$SnWO_4$ heterojunction nanocomposites with boosted photodegradation capability. *Chem. Eng. J. Adv.* **2020**, *4*, 100040. [CrossRef]
50. Alharthi, F.A.; Alsaiari, M.A.; Jalalah, M.S.; Shashank, M.; Shashikanth; Alghamdi, A.A.; Algethami, J.S.; Ganganagappa, N. Combustion synthesis of β-$SnWO_4$-rGO: Anode material for Li-ion battery and photocatalytic dye degradation. *Ceram. Int.* **2021**, *47*, 10291–10300. [CrossRef]
51. Elviera; Yulizar, Y.; Apriandanu, D.O.B.; Surya, R.M. Fabrication of novel $SnWO_4$/ZnO using Muntingia calabura L. leaf extract with enhanced photocatalytic methylene blue degradation under visible light irradiation. *Ceram. Int.* **2022**, *48*, 3564–3577. [CrossRef]
52. Ungelenk, J.; Feldmann, C. Nanoscale β-$Sn_{1-n}WO_4$·nα-Sn-A highly efficient photocatalyst for daylight-driven degradation of organic dyes and its real "green" synthesis. *Appl. Catal. B-Environ.* **2011**, *102*, 515–520. [CrossRef]

53. Alharthi, F.A.; AlFawaz, A.; Ahmad, N. Photocatalytic degradation of anionic dye using well-crystalline $SnWO_4$ polyoxometalate. *Phys. Scr.* **2022**, *97*, 085813. [CrossRef]
54. Wei, Q.; Xiong, S.; Li, W.; Jin, C.; Chen, Y.; Hou, L.; Wu, Z.; Pan, Z.; He, Q.; Wang, Y.; et al. Double Z-scheme system of α-$SnWO_4$/UiO-66(NH_2)/g-C_3N_4 ternary heterojunction with enhanced photocatalytic performance for ibuprofen degradation and H_2 evolution. *J. Alloys Compd.* **2021**, *885*, 160984. [CrossRef]
55. Liu, D.; Chen, X.; Qiao, Y.; Zhou, Y.; Kuang, Y. Awakening the Photoelectrochemical Activity of α-$SnWO_4$ Photoanodes with extraordinary Crystallinity Induced by Reductive Annealing. *Adv. Energy Sustain. Res.* **2021**, *3*, 2100146. [CrossRef]
56. He, G.; Li, W.; Qiu, W.; Xu, C.; Wang, K.; Chen, L.; Wang, Y.; Liu, Y.; Li, J. Constructing a Two-Dimensional $SnWO_4$ Nanosheet Array Film for Enhanced Photoelectrochemical Performance. *ACS Appl. Energy Mater.* **2022**, *5*, 11883–11891. [CrossRef]
57. Wu, H.; Tan, H.L.; Toe, C.Y.; Scott, J.; Wang, L.; Amal, R.; Ng, Y.H. Photocatalytic and Photoelectrochemical Systems: Similarities and Differences. *Adv. Mater.* **2019**, *32*, 1904717. [CrossRef] [PubMed]
58. Liu, Y.; Wygant, B.R.; Kawashima, K.; Mabayoje, O.; Hong, T.E.; Lee, S.-G.; Lin, J.; Kim, J.-H.; Yubuta, K.; Li, W.; et al. Facet effect on the photoelectrochemical performance of a WO_3/$BiVO_4$ heterojunction photoanode. *Appl. Catal. B* **2019**, *245*, 227–239. [CrossRef]
59. Chen, L.; Li, W.; Qiu, W.; He, G.; Wang, K.; Liu, Y.; Wu, Q.; Li, J. Oriented $CuWO_4$ Films for Improved Photoelectrochemical Water Splitting. *ACS Appl. Energy Mater.* **2022**, *14*, 47737–47746. [CrossRef]
60. Liu, C.; Chen, L.; Su, X.; Chen, S.; Zhang, J.; Yang, H.; Pei, Y. Activating a TiO_2/$BiVO_4$ Film for Photoelectrochemical Water Splitting by Constructing a Heterojunction Interface with a Uniform Crystal Plane Orientation. *ACS Appl. Energy Mater.* **2022**, *14*, 2316–2325. [CrossRef]
61. Harb, M.; Cavallo, L.; Basset, J.-M. Remarkable Influence of α-$SnWO_4$ Exposed Facets on Their Photocatalytic Performance for H_2 and O_2 Evolution Reactions. *J. Phys. Chem. C* **2020**, *124*, 18684–18689. [CrossRef]
62. Wang, W.; Wu, Y.; Chen, D.-L.; Liu, H.; Xu, M.; Liu, X.; Xin, L. The surface reconstruction induced enhancement of the oxygen evolution reaction on α-$SnWO_4$ (010) based on a density functional theory study. *Phys. Chem. Chem. Phys.* **2022**, *24*, 19382–19392. [CrossRef] [PubMed]
63. Azofra, L.M.; Cavallo, L.; Basset, J.-M.; Harb, M. Need for Rationally Designed $SnWO_4$ Photo(electro)catalysts to Overcome the Performance Limitations for O_2 and H_2 Evolution Reactions. *J. Phys. Chem. C* **2021**, *125*, 8488–8496. [CrossRef]
64. Liu, Y.; Chen, L.; Zhu, X.; Qiu, H.; Wang, K.; Li, W.; Cao, S.; Zhang, T.; Cai, Y.; Wu, Q.; et al. Effects of operating temperature on photoelectrochemical performance of $CuWO_4$ film photoanode. *J. Electroanal. Chem.* **2022**, *924*, 116859. [CrossRef]
65. Schnell, P.; Dela Cruz, J.M.C.M.; Kölbach, M.; van de Krol, R.; Abdi, F.F. pH-Dependent Stability of α-$SnWO_4$ Photoelectrodes. *Chem. Mater.* **2022**, *34*, 1590–1598. [CrossRef]

energies

MDPI

Review

Gel Fuels: Preparing, Rheology, Atomization, Combustion

Dmitrii Glushkov *, Kristina Paushkina and Andrei Pleshko

Heat Mass Transfer Laboratory, National Research Tomsk Polytechnic University, Tomsk 634050, Russia
* Correspondence: dmitriyog@tpu.ru; Tel.: +7-(3822)701-777 (ext. 1953)

Abstract: The review presents the results of experimental and theoretical studies obtained in recent years within the framework of the main areas of research of gel fuels: choice of component composition and substantiation of fuel preparation technologies and their rheological characteristics; fuel transportation and atomization processes; and the processes of their ignition and combustion. The main advantages of gel fuels in comparison with widely used liquid and solid fuels are considered. The advantages and disadvantages of known experimental approaches to the study of the combustion processes of gel fuels are analyzed. The well-known physical and mathematical models of gel fuels ignition are given, including those under conditions of melt droplets dispersion. The tendencies of further development of gel fuels within the framework of the combustion theory are formulated.

Keywords: gel fuel; dispersion; combustion; rheology; atomization

1. Introduction

At present, three main types of fuels are widely used in practice: liquid [1,2], solid [3,4] and gaseous [5,6]. Liquid fuels (gasoline, diesel fuel, kerosene, hydrazine, heptyl, alcohols, naphtha, benzene-gasoline mixtures, fuel oil, kerosene and methoxydiethylborane/tetrahydrofuran and others), as a rule, are used in internal combustion engines (motor transport, shipbuilding) [7], jet engines (aviation) [8], liquid-propellant rocket engines [9] and technological installations in thermal power engineering [10]. Solid fuels (coals, combustible shale, metallized composite solid fuels and others) are used as energy resources in coal-fired steam and water-heating boilers [11], blast furnaces [12] and solid rocket engines [13]. Gaseous fuels (methane, propane, dimethyl ether, coke oven gas and others) are widely used in various applications, but the most efficient energy potential of these fuels is used in gas turbines [14].

Along with the aforementioned fuels, in recent years, in transport technology and in the energy sector, a promising direction of development is the design of effective compositions and methods for obtaining various types of composite fuels: synthesis gas [15], coal-water slurries [16], coal-water slurries with petrochemicals [17], suspension fuels [18], gel fuels [19,20] and fuel briquettes [21]. This is mainly attributable to the tasks of improving the environmental, energy, economic and operational characteristics of both technological systems of devices in transport and energy, and the processes of storage, transportation and combustion of new fuels types.

In recent years, the prospects for space exploration have been the main incentives for developments in the rocket and space industry. The solution for this problem requires an integrated approach to conducting fundamental research and using their results in practice to make rational, technical decisions when conducting development work. One of the main problems is the need to develop new fuels, study their properties, as well as study the physicochemical processes that occur during their ignition and combustion.

Solid and liquid propellants were widely used in rocket and space technology at the end of the XX and the beginning of the XXI century. Each of them has its own advantages and disadvantages compared to one other. To date, the potential of such fuels is almost completely exhausted, according to many experts [22–24]. Gel fuels can become an alternative energy resource within the framework of the implementation of the program for developments in rocket and space technology. These propellants combine the advantages of solid and liquid rocket propellants [24], of which the main ones are related to safety aspects (a small amount of vapor is released during leaks compared to liquid fuels; indifference to shock, friction and electrostatic discharge; minimal risk of accidental ignition, because the combustion process is sufficiently well controlled and can be stopped; the elastic-deformable or viscous-plastic state prevents the formation of cracks in the fuel charge, so their development does not affect the increase in the combustion area, uncontrolled combustion or explosion), energy efficiency (relatively high specific thrust impulse compared to composite solid propellants) and storage (long-term storage without maintaining special conditions; flexible packaging; relatively low proportion of solid fine particle settling in high-viscosity formulations compared to liquid fuels). Gel fuels have higher energy characteristics than composite solid propellants; in particular, a specific thrust impulse is about 3000–3500 m/s [25]. Replacing a composite solid propellant with a gel fuel will make it relatively easy to implement dynamic control of engine thrust during the time in practice [24]. Furthermore, gel fuels, compared to liquid fuels, have lower fire hazard rates because of the minimization of evaporation losses and leakage during storage. Gel fuel components and their combustion products tend to have a lower environmental impact than typical liquid propellants [26].

The aggregate state of gel fuel can differ significantly depending on its purpose and environmental characteristics—from liquid (with high viscosity) to solid (elastically deformable). Within the framework of ignition theories of solid and liquid condensed substances [27–31], the properties of composite solid propellants and liquid fuels have been studied in full, and mathematical models have been developed to predict the characteristics of physicochemical transformations that occur when they are heated. The results of experimental studies [22,23] indicate a fairly significant difference in the regularities and characteristics of the physicochemical processes occurring during the ignition and combustion of gel fuels, compared with typical solid and liquid fuels. Therefore, an urgent task is a comprehensive review of studies on gel fuels. An analysis of publications on this topic in international periodicals made it possible to identify the following main areas of research conducted to date:

1. Preparation of fuel compositions based on various components [32–37];
2. Study of the rheological characteristics of fuels [35,36,38–41];
3. Study of the processes of gel fuels atomization [41–44];
4. Study of the processes of fuel composition, ignition and combustion [22,23,45–48];
5. Application of fuels in practice [32,49–53].

Below are the results of the analysis of the main achievements in recent years for each of the abovementioned research areas.

This review is focused on a brief coverage of modern achievements in the field of gel fuels. Only the most important results of leading experts in the field of gel fuels are presented here, with an emphasis on the published research results of recent years, including the results of studies that have not yet been published. Within the framework of the topic of gel fuels, this review is primarily focused on the study of elastically deformable fuel compositions, whereas most other studies are devoted to plastically deformable gel fuels.

2. Fuel Preparation

Physically, gel fuels are rheologically modified liquids, the properties of which are changed by adding various thickeners (gelling agents) [24,54–56]. Depending on the type of thickeners, it is currently conventionally accepted [24,54,55] in the scientific periodical literature as the distinguishing of two fundamentally different (in terms of rheological properties and physical and mechanical characteristics) types of gel fuels. On the one hand, these are "soft substances" (Figure 1a) that exhibit the physical and mechanical characteristics of both a solid and a liquid, depending on the level of shear stress [24,54].

On the other hand, these are materials with a three-dimensional elastically deformable solid framework (Figure 1b) containing finely dispersed liquid drops in the matrix cells [55].

Figure 1. Appearance of gel fuels: (**a**) Liquid and plastically deformable fuel compositions based on kerosene: K-100—100 % kerosene, K-99—99 wt% kerosene, 1 wt% silicon dioxide, K-95—95 wt% kerosene, 5 wt% silicon dioxide, K-90—90 wt% kerosene, 10 wt% silicon dioxide, K-85—85 wt% kerosene, 15 wt% silicon dioxide; (**b**) Elastically deformable: 50 vol% oil + 48 vol% aqueous solution of PVA (10 wt%) + 2 vol% emulsifier.

The first type of fuel is obtained, as a rule, by adding organic and inorganic thickeners to combustible liquids, and the second is obtained by adding polymeric thickeners. Rheological properties, physico-mechanical and physico-chemical characteristics of combustible liquids change quite significantly after thickening. After obtaining the gel fuels and for a certain period of time during storage, their viscosity increases by several times compared to combustible liquids in the initial state, and the density increases significantly, especially if finely dispersed metal particles are added to the fuel composition [56]. These components also contribute to the growth of the thermal effect of the combustion process [57] and the specific energy density [24]. Thus, the production and practical application of gel fuels can provide a high level of control over operational characteristics, uniformity and reproducibility of fuel compositions, as well as energy characteristics.

Mainly potential technologies for the use of gel fuels, for example, in rocket and ramjet engines, afterburners, involve the implementation of the processes of their transportation through pipelines, supply to the combustion chamber and organization of direct combustion, similar to the processes characteristic of typical liquid fuels in the initial state [24,54–57]. It is worth noting that in addition to thickening flammable liquids, gel oxidizers are also used in practice to solve some problems [58,59]. In the early stages of development of gel fuels in the middle of the XX century, they were considered solely as a replacement for mixed solid fuels [58–63]. Accordingly, the corresponding formulations of the first gel fuels and oxidizers were developed [58–63].

However, later, it was proposed to use gel propellants instead of liquid propellants [54,64], since the thickening of the latter made it possible to minimize their main operational disadvantages, as well as to improve the energy characteristics mainly by adding metal powders, the particles of which can be fairly evenly distributed in the volume of the thickened combustible liquid. This has led to gel fuels being predominantly characterized as non-Newtonian fluids. The compositions most widely used in experimental studies are given in references Table 1.

Table 1. Compositions of typical gel fuels.

Rocket Fuel	Thickener [a]	Additives [a,b,c]	Notes	Ref.
Water (98.5%)	Carbopol 941 (0.5%) (2–7 μm)	NaOH (1%)	Mechanical mixing	[23]
JP-5 JP-8 IRFNA HP	AO (2–5%) SiO_2 (4–8%) SiO_2 (3–5%) SiO_2 (5–7%)	–	Mechanical mixing	[65]
Jet A1 (85%)	Thixatrol 289 (7.5%)	MIAK (7.5%)	Mechanical mixing	[66]
Jet A1 (NA)	Thixcin R (4–7%)	NA	Mixing at 40–60 °C for 2.5 h at 1000 rpm.	[67]
MMH	HPC (3%) Aerosil (6%) Hybrid Gel (2–3%)	–	Mechanical/acoustic mixing at 60 °C	[68]
Jet A1 (85%)	Thixatrol ST (7.5%) (23 μm)	Xylene (7.5%)	Mixing at 63 °C for 1 h at 1200 rpm.	[34]
ISROsene (85–65%)	Thixatrol ST (7.5%) (23 μm)	Xylene (7.5%) Nano aluminum oxide (50–150 nm) Oleic acid (10%)	Mixing at 63 °C for 1 h at 1200 rpm.	[69,70]
RP-1	Carbosil (5%) (44 μm)	0–55% Alex (100 nm) Tween 85 (0–1.3%)	–	[71]

Notes. [a]—relative ratios of components in %wt. [b]—weight percent surfactant dependent on metal content. [c]—liquid additives are used as solvents for thickeners.

As a rule, the main component of these fuel compositions are flammable liquids used as fuels in aerospace technology, such as liquefied hydrogen [72–74]; hypergolic fuels, including hydrazine [36,75–77], monomethylhydrazine (MMH) [36,63,64,75,77–79], unsymmetrical dimethylhydrazine (UDMH) [22,36,63], tetramethylethylenediamine (TMEDA) [80]; various grades of kerosene-based jet fuel JP (JP-1, JP-5, JP-8 and JP-10) [63,75–77] and Jet A1 [23,34,50,81,82], RP-1 [54,71]; and alcohol-containing fuels based on methanol [58] and ethanol [58,83,84]. Other common liquid components of gel fuels are isopropanol [58], oils of petroleum origin [61], spent organic solvents (benzene, acetone and chlorinated solvents) [61], nitromethane [85], ethanolamine [86], aniline [63], nitropropane [63], isooctane [87], hexane [73,74], propane [74], ethane [74], paraffin [50], glycerol [88] and ammonium ethyl acetate [89].

Oxidizers used in gel form are mainly used in conjunction with hypergolic fuels. Among such oxidants, red fuming nitric acid (RFNA) [62,87] inhibited red fuming nitric acid (IRFNA) [62,64,75,77] and hydrogen peroxide [75–77,87,89], as well as cryogenic substances such as substituted bromine trifluoride [62], bromine pentafluoride-chlorine trifluoride [62], liquefied oxygen [89] and oxygen difluoride; a large number of nitroalkanes and nitrogen tetroxide [59] can be noted as well.

A significant amount of research related to the development of fuel formulations is aimed at identifying suitable types of thickeners for specific combustible liquids and oxidizers [24,77]. As a rule, most thickeners are commercially available manufactured compounds. Both inorganic [90] and organic thickeners are used, many of which are polymeric [91]. For thickeners, typical selection criteria apply [90–92]: they must be compounds that are applicable at low concentrations; they must be combustible, provide a reversible system; and thickeners should be non-aggressive, environmentally friendly and economically affordable.

Inorganic thickeners, such as silica, work well with fuels and oxidizers [91]. However, they are inert and, therefore, their use obviously leads to losses and a decrease in the energy efficiency of the fuel during combustion. In addition, their tendency to form covalent bonds reduces the reversibility of the gel fuel and negatively affects its flow characteristics.

Polymeric thickeners such as Carbopol, cellulose compounds, Xanthan and others are used as aqueous solutions to thicken combustible liquids. These compounds are straight or branched chain polymers and form a strong entangled or cross-linked network through hydrogen bonds when dissolved in water [91]. Cellulosic compounds and agar-agar can only thicken fuels. Although the rheology of these polymer gels exhibits shear thinning characteristics, the molecular bonding of the polymer chains to water prevents liquid breakdown and subsequent aerosol formation, which impairs atomization. However, atomization can be achieved for fuels containing agar-agar as a thickener, which absorbs water and forms a network around the liquid rather than binding to its molecules [92]. These examples illustrate the importance of the thickener–liquid interaction in fuel atomization and highlight the need for proper thickener selection.

Unlike polymer gels, low molecular weight thickeners are used to obtain some organic gels based on organic liquids [93]. Organogels resemble polymer gels in the structure of molecular bonds that form a matrix, in the cells of which finely dispersed liquid drops are located. They are easier to spray compared to polymer gels because of a component with reduced tensile viscosity. Moreover, they burn together with liquid fuel and oxidizers. For example, the castor oil derivative Thixcin R is used to thicken a suspension of coal and oil [61], as well as Jet A1 [94]. Other castor oil derivatives are widely used as a non-polymeric organic thickener—Thixatrol [23,34,50,69,70,82,94], Thixatrol ST [23,34,50,82,91,95], MPA 60 [96], Thixatrol 289 [97,98], Thixcin R [61,94], Thixatrol Plus [99].

Among inorganic thickeners, and most likely among all types of thickeners used for the preparation of gel fuels, various forms of silicon dioxide are most widely used [72,75,79,86,87,99]; for example Cabosil and Aerosil. It is worth noting that the results of numerous studies indicate the need for caution when choosing these commercially available silica gel thickeners for the preparation of gel fuels, since the latter do not always satisfy the basic typical requirements for rheological properties and physical and mechanical characteristics. To date, a large number of combinations have been proposed between widely used liquid fuels and thickeners, but only a small part of these fuel compositions has ever been used in experimental studies of the gel fuels spraying and combustion.

Equally important components of gel fuels are energy additives, for example, metal powders. The concentration of the metal is the determining factor if it is used as an energy additive. The typical range of metal content in fuel by mass is 5–60% [91,98,100]. The justification for the advantage of metallized fuel compositions is to obtain suspensions that have a greater value of thrust per kilogram of oxidizer (or air). Preference is given to metals with the highest volume and mass values of combustion heat [101–104]. It is known that Mg is characterized by the highest thrust per kilogram of air at stoichiometric mixture ratios, followed by Al [101]. In practice, Al_2O_3 is sticky and melts at the flame temperature of a typical gel fuel [100,101], whereas boron combustion is difficult to initiate [102]. Magnesium oxides do not melt and, therefore, Mg is a promising component for metallized gel fuels. The efficient use of boron requires the organization of a more complex fuel combustion mode [103,104].

The components range and gel fuels compositions are diverse, so the method of their preparation in each specific case can be quite different, especially when using thickeners of various origins (polymeric, organic, and inorganic). However, most techniques for preparing gel fuels rely on the mixing of thickener, solvent (optional), metal particles, catalysts, stabilizing surfactants (if metals are added) and liquid fuel at room conditions or elevated temperatures using mechanical or acoustic methods. The order of adding the listed components to the mixture, their concentrations and mixing times are quite significantly different in each specific case [34,75,94,105–108]. The main stages and conditions for the preparation of gel fuels are four possible combinations of combustible liquid component/oxidizers and thickeners: fuel–organic (polymeric) thickener, oxidizer–inorganic thickener, fuel–inorganic thickener and fuel–organic (non-polymeric) dispersed thickener. It should be noted that compositions based on organic thickeners and oxidizing agents are of no practical interest, since the vast majority of oxidizing agents are highly reactive with organic thickeners [105].

In [19,109,110], methods for the preparation of gel fuels are presented by thickening waste oils of petroleum origin with a polymeric thickener (without additional components and with the addition of finely dispersed solid combustible particles), as well as by thickening kerosene with an inorganic thickener (silicon dioxide).

Based on the analysis of references [19,34,75,94,105–110], it was concluded that the component composition of gel fuels is very diverse and, in each specific case, is determined by the availability of components, such as a thickener (organic, inorganic or polymeric) and the necessary rheological properties or physical and mechanical characteristics of the final product, as a rule. The latter, in turn, have a direct impact on all subsequent stages of the gel fuel life cycle, in particular, the processes of its storage, transportation through fuel lines, spraying in combustion chambers and direct combustion.

3. Rheology

Gel fuels are characterized by non-Newtonian properties. According to their rheological properties, they can be conditionally divided into viscoelastic and viscoplastic substances [111]. After preparation, the gel fuel is viscoelastic, but under conditions of exceeding the limiting stress (yield stress), it exhibits viscoplastic properties. Using the approaches and methods of rheology and physical chemistry, the deformation under the action of stress is studied, including flow and elastic deformation. Structural and mechanical properties of gel fuels are studied within the framework of colloid chemistry models using research methods adopted in rheology. The rheological characteristics given in this paper are relevant for plastically deformable fuels and for elastically deformable ones in a molten (liquefied) state suitable for transportation and atomization.

3.1. Shear Strain

The shear of gel fuel occurs at all stages of the technological process of its use (from preparation to combustion). Under shear conditions, the geometric shape of a unit of mass changes at a constant volume.

A typical process of burning gel fuel (Figure 2) in practice consists of six main stages [45,112–115], which successively replace each other: (1) storage; (2) moving from the fuel tank to the transport channel; (3) movement along the fuel pipeline (mainly through pipelines); (4) atomization (or dispersion); (5) movement of fuel fragments (particles or drops) in the combustion chamber; (6) melting, evaporation and combustion of fuel. Storage is the longest period during which the gel fuel causes the least intake due to gravity.

At the second stage, under the conditions of moving gel fuel from the tank to the transport channel, the shear rate is 2.4 s^{-1} at a fuel flow rate of 0.015 m/s and a characteristic time of about 53 s. Stages three to six are fast-paced, and each is completed in less than 1 s. The shear rate at each of these stages varies over a wide range and depends on the design of technological elements and the speed of fuel movement in them.

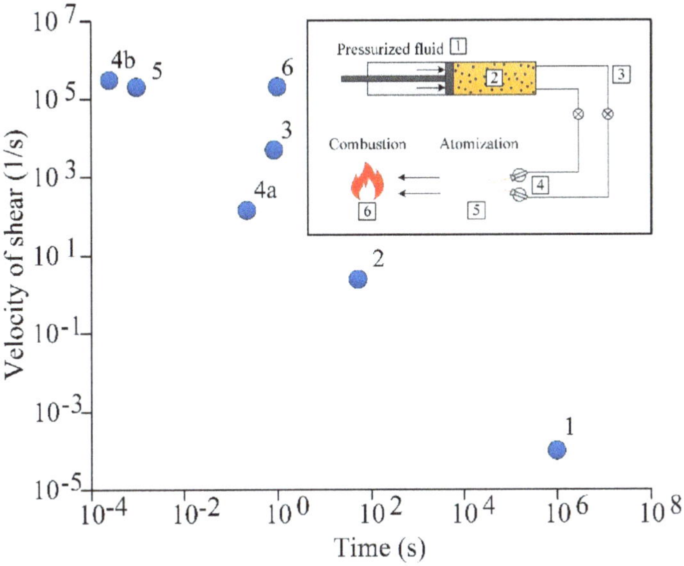

Figure 2. Shear rate and characteristic time of the stages implementation of the typical gel fuel life cycle [45,112–115]: 1—storage; 2—moving from the fuel tank to the transport channel; 3—moving in fuel line; 4—atomization with nozzle (4a) or hydraulic nozzle (4b); 5—movement of fuel fragments (particles or droplets) in the combustion chamber; 6—melting, evaporation and combustion of fuel.

At the third stage, the shear rate recorded under the conditions of the movement of gel fuel through the pipeline at a speed of 2.4 m/s for 0.84 s is 4.8×10^3 s^{-1}.

When dispersing fuel at the fourth stage, the speed of movement of gel fuel fragments depends on the method of atomization and the design of the corresponding spray device. It is known [45,112–115] that when organizing atomization with the help of nozzles, the speed is about 0.23 m/s and the time of movement is 0.23 s. Under such conditions, the shear rate is close to 141 s^{-1} (Figure 2, point 4a). When organizing atomization by means of a hydraulic nozzle (Figure 2, point 4b), the shear rate reaches 3×10^5 s^{-1} at a fuel particle velocity of 38.2 m/s and a time of 2.6×10^{-4} s.

At the fifth and sixth stages, the velocities of the fuel particles are, as a rule, the same. It was registered [45,112–115] that under the conditions of evaporation and combustion of fuel particles when they move through the combustion chamber at a speed of 20 m/s for 1 s, the shear rate is 2×10^5 s^{-1}.

The shear rate of the gel fuel reaches its maximum values in the combustion chamber because the viscosity of the fuel decreases significantly with increasing temperature. When particles of gel fuel move in a stream of high-temperature gases, a combination of the following interrelated physical and chemical processes occurs: (1) fuel melting; (2) free surface evaporation; (3) formation of a gas–vapor mixture in the vicinity of a melt droplet (it should be noted that the temperature of the vapor entering the oxidizer medium does not exceed the boiling temperature of the combustible liquid, on the basis of which the gel fuel is prepared); (4) in the presence of solid inclusions in the composition of the gel fuel, relatively cold metal particles enter, together with vapors, into the oxidizer medium; (5) as a result of the heating of combustible vapors and solid combustible components, they ignite and burn out. A more detailed description of the combustion stages of gels will be given in Section 5.

The shear rate at each of these stages varies over a wide range and depends on the design of the technological elements and the speed of fuel movement in them. Shear rate is an important characteristic of gel fuel, which determines not only its shape, but also rheological properties throughout the life cycle of existence from preparation to combustion [45,112–115].

3.2. Rheological and Physical–Mechanical Characteristics

Gel fuels are rheologically complex non-Newtonian fluids. Viscosity, yield strength, thixotropy and viscoelasticity are the most important rheological characteristics that determine not only the shape of the gel fuel, but also the stability of properties at all stages of its life cycle. During this period, it is especially important to control the viscosity of the gel fuel. The yield stress is important to control at the stage of storage, spraying and combustion. Control of thixotropy parameters is important at the stage of preparation and storage, and viscoelasticity at the stage of combustion.

Structural-mechanical properties of gel fuels can be estimated by constructing rheological curves (flow and viscosity curves). The construction of rheological curves of gel fuels is carried out according to the results of measurements obtained using a viscometer or rheometer. The flow equations of Herschel-Bulkley [116], Casson [117], Bingham [118], Ostwald [119], Carro-Yashida [120,121], the modified Herschel-Bulkley equation [65,122,123], three-parameter equation [124] and the modified Ostwald equation [42] are most often used to describe the rheological behavior of gel fuels. The wide possibilities of varying the composition of gel fuels do not allow for deriving a universal expression for describing the rheological and physical–mechanical characteristics. Experimental points of fuels with similar compositions can be approximated by different equations of flow or viscosity. The choice of equation depends on the researcher.

3.2.1. Viscosity

Shear viscosity (or dynamic viscosity, effective viscosity, apparent viscosity) is a physical quantity that describes the resistance of a substance to a viscous shear flow or energy loss in a viscous shear flow, and characterizes the internal friction of a gel fuel [111]. This friction occurs between the layers of fuel as a result of its movement. With an increase in shear rate, the viscosity of gel fuels decreases, tending to a constant value corresponding to the complete liquefaction (degradation) of the fuel.

During their life cycle, gel fuels are subject to different shear rates. During storage, when the fuel is exposed to low-intensity vibration, the shear rate does not exceed 10^{-2} s^{-1} [45,112–115]. The shear rate reaches up to 10^3 s^{-1} [45,112–115] during loading/unloading, moving gel fuel from the tank to the transport channel, as well as when it moves along the fuel line. When a significant momentum is transferred to the gelled fuel under dispersing conditions, for example, through a hydraulic nozzle, the shear rate exceeds 10^3 s^{-1}. The shear rate exceeds 10^5 s^{-1} [45,112–115] under conditions of movement of fine fuel particles in the volume of the combustion chamber, as well as during the evaporation and combustion of fuel for a relatively short period of time. Conventionally, three ranges of shear rate changes can be distinguished for typical gel fuels corresponding to different stages of their life cycle:

1. Low (less than 1 s^{-1});
2. Medium (from 1 to 10^5 s^{-1});
3. High (more than 10^5 s^{-1}).

It is impossible to establish a strict dependence on the viscosity of the gel fuel on the shear rate, since the viscosity, in addition to the shear rate, is significantly affected by the component composition [125]. Experimental results [69,70,113] are known on the study of the change in viscosity (determined by pressure drop) on the shear rate of xylene gel (Jet A1 jet fuel, Thixatrol ST organic thickener) [113] and gel propellant with aluminum oxide particles (ISROsene kerosene, Thixatrol ST organic thickener, xylene) [69,70] in the conditions of movement along the fuel line to the spray device and during their atomization. Detailed information on methods and means of measuring tensile viscosity is presented in the reference [126].

Experimental results [69,70,113] are known on the study of the change in viscosity (determined from the pressure drop) on the shear rate of xylene gel (Jet A1 liquid jet fuel, Thixatrol ST organic thickener) [113] and gel propellant with aluminum oxide particles (ISROsene kerosene, Thixatrol ST organic thickener, xylene) [69,70] while moving along the fuel line to the spray device and during their atomization (Figure 3).

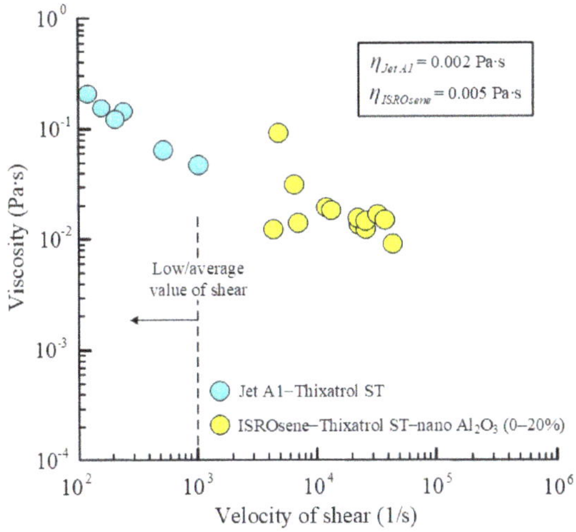

Figure 3. Viscosity change with shear rate under conditions of xylene gel (Jet A1 jet fuel, Thixatrol ST organic thickener) [113] and gel propellant with Al_2O_3 particles (ISROsene kerosene, Thixatrol ST organic thickener, xylene) [69,70] to spray device and during atomization.

The results [69,70] (Figure 3) illustrate that with increasing shear rate, the viscosity of xylene gel (Jet A1 jet fuel, Thixatrol ST organic thickener) decreases, and the viscosity of gel propellant (ISROsene kerosene, Thixatrol ST organic thickener, xylene) is quasi-constant. The spread of values according to the data (kerosene ISROsene, organic thickener Thixatrol ST, xylene) is due to different concentrations of 0–20% alumina particles in the fuel composition [69,70].

The shear rate arising from the movement of small fragments of fuel through the combustion chamber, as well as from the evaporation and combustion of fuel is very high (more than 10^5 s^{-1}). The viscosity of a gel fuel under such conditions is several orders of magnitude greater than that of a typical Newtonian fluid, such as water [69,70].

3.2.2. Tensile Strength

The flow of gel fuels begins only when the limiting shear stress (tensile strength) is exceeded [113]. This means that gelled fuels liquefy under conditions where the shear stress exceeds the threshold value of tensile strength. It is known [126] that the tensile strength depends on the structure of the substance. Gel fuels, characterized by a high tensile strength compared to liquid fuels, are sprayed at the dispersion stage without sedimentation, which significantly increases the energy potential of the fuel. The results of direct measurements of the tensile strength, for example, using paddle mixers [127,128], depend on the sensitivity of the measuring device.

The tensile strength of gel fuels is not a constant value [129], as it depends on the registration time. There are studies [77,96,130], in which the tensile strength was determined using a conical penetrometer and a rheometer with an ascending sphere [130], as well as the minimum drop in capillary pressure at the moment of the start of liquefaction [77] and the maximum torque required to spin the rotor [96,130].

Using a Heavy Duty Shop Press 20 Ton hydraulic press, the tensile strength of gel fuels pellets (obtained after 15 cycles of freezing/thawing oil emulsions) with a characteristic size of 20 mm was measured for the following compositions group: 100–20 vol% PVA aqueous solution (5, 10 wt%) + 0–80 vol% oil. It has been established [109] that an increase in the oil concentration in fuel composition leads to a decrease in the tensile strength (Figure 4). At identical oil concentrations, the difference in the tensile strength of pellets based on 5% and 10% PVA solutions is 73–78% [109].

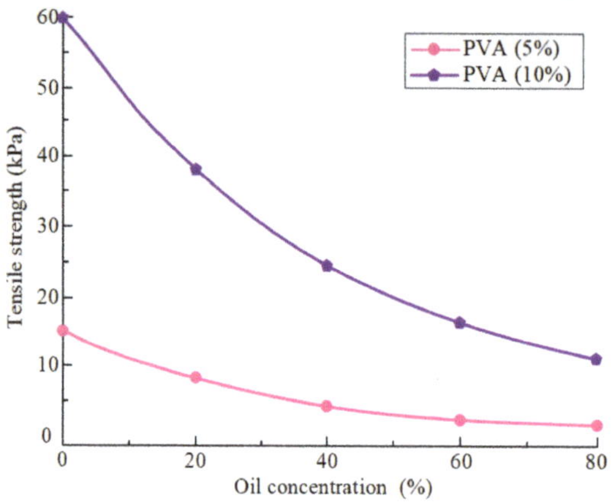

Figure 4. The dependence of the tensile strength of oil-filled cryogels fuel pellets on the oil concentration at different concentrations of PVA aqueous solutions [109].

The influence of elastic deformation along the axis of symmetry at the application of 6.5 kPa on the loss of moisture (oil) was established [109] for the same compositions. The authors found that the lower the density of the polymer matrix in the gel fuel, the greater the mass loss of the sample after deformation (Figure 5). The moisture loss of compositions based on a 10% PVA solution is on average 1% less than that of compositions with a 5% PVA solution. At higher oil concentrations, the density of the polymer matrix is lower. During deformation of oil-filled cryogels with an oil concentration of 80%, the loss of moisture due to deformation is 2–2.5 times greater than for cryogels containing 20% oil.

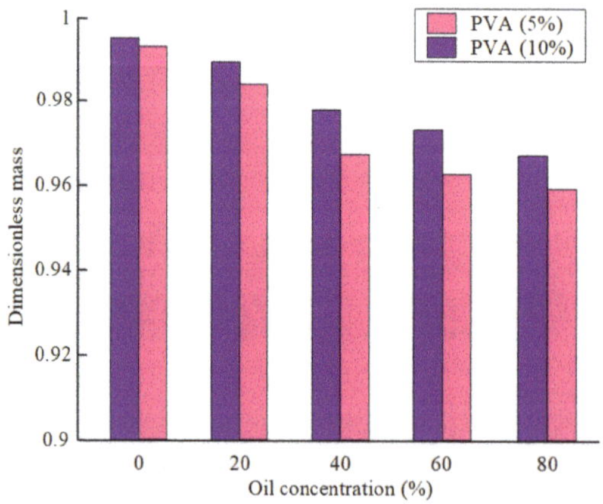

Figure 5. Masses of fuel pellets after deformation, reduced to their initial mass, at different concentrations of an aqueous solution of PVA [109].

In accordance with the technique described in detail in [110], the tensile strength of six compositions of oil-filled cryogels containing particles with different concentrations of coal dust with a dispersion of no more than 140 μm, as well as cryogels containing oil and coal particles, was studied. It has been established [110] that the adding of finely dispersed solid particles (up to 50%) into the cryogel composition leads to an increase in the tensile strength of fuel pellets by 1.7–2.3 times. Gel fuel pellets based on oil-filled cryogels with finely dispersed coal particles are characterized by lower elasticity and, as a result, higher tensile strength values compared to oil-free cryogel pellets.

3.2.3. Viscoelastic Properties

Gel fuels are viscoelastic substances that exhibit both viscosity and elasticity upon deformation [131]. Viscoelastic properties prevent leakage or spillage of gel fuels during storage and transportation, if the applied shear stress does not exceed the tensile strength. After deformation, the gel fuel partially or completely restores its geometric shape, if the tensile strength is not exceeded. Conventionally, gel fuels can be classified according to their viscoelastic properties into three groups: viscoelastic liquids, viscoelastic solids and elastic solids [131–134].

The Table 2 presents the results of studies of the viscoelastic properties of gel fuels and their analogues based on experimental studies of creep or oscillating shear.

Table 2. Viscoelastic properties of gel fuels and their analogues.

Gel	Method	Notes	Ref.
0.25–0.35% aqueous solution of Carbopol 934, stainless steel particles SUS304 (0–3 vol%)	resilience test after unloading; vibration shear test	similar viscoelastic properties of non-metallized and metallized gels; highly metallized gels are brittle	[91]
Liquid fuel Jet A1, organic thickener Thixatrol ST, xylene (85:7.5:7.5)	vibration shear test	dynamic modulus depends on frequency	[113]
Jet fuel JP-5, aluminum octoate (2–5%)	resilience test after unloading	viscoelastic solids; yield strength was not used	[77]
Nitromethane, 4–8% nanosized silica particles	vibration shear test	modulus of elasticity does not depend on frequency; the modulus of elasticity increases with increasing thickener content	[135]
2% aqueous solution of polysaccharide	resilience test after unloading; vibration shear test	δ = 15–33° (ω = 0.1–10 Hz); verification of the Cox–Mertz relation was performed	[122]
Water, 4–6% hydroxypropylcellulose solution (monomethylhydrazine + hydroxypropylcellulose simulant)	vibration shear test	physical function does not depend on frequency; significant difference between G' and G'' at ω = 0.2 10 Hz; the concentration of hydroxypropylcellulose does not affect G'	[136]
0.1–4% aqueous solution of Carbopol 941, triethanolamine	vibration shear test	compliance of microstructures with the frequency and range of voltage; triethanolamine contributes to the formation of fragile and rigid structures; the resulting gel (without triethanolamine) is Maxwell's fluid	[131]
Monomethylhydrazine, 6% aerosil; Monomethylhydrazine, 3% hydroxypropyl cellulose; Monomethylhydrazine, 2% hydroxypropyl cellulose + 3% aerosil	vibration shear test	hydroxypropylcellulose and hybrid gels are viscoelastic liquids and perfectly elastic substances; Aerosil based gels are viscoelastic solids	[133]
Water, 1% polysaccharide; water, 3% hydroxypropyl cellulose; water, Cekol 2000; water, Cekol 3000	vibration shear test	type and concentration of the thickener do not affect G'; physical function dependent on frequency for hydroxypropyl cellulose and Cekol	[137]

When conducting experimental studies on the creep of gel fuel, the shear stress is controlled (set) and the shear rate is changed in steps or in a continuous mode [138]. As a result, information is obtained not only about the viscoelastic properties of the gel fuel, but also about the creep yield strength of the gel fuel at a long-term constant shear stress [50]. The type of typical dependences obtained in the study of the creep of gel fuel [138] is shown in Figure 6.

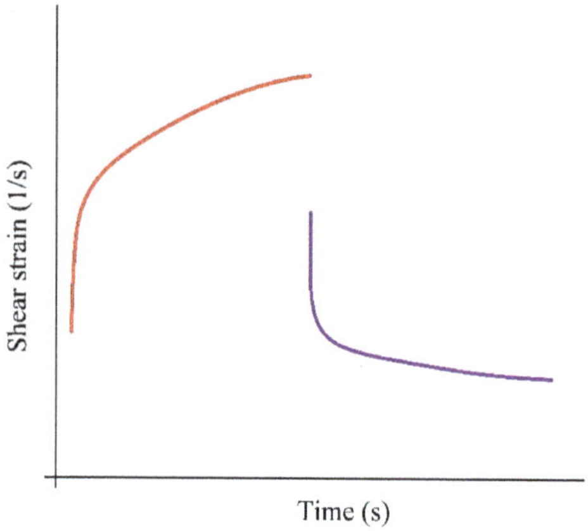

Figure 6. Type of typical dependences in the study of creep of gel fuel [138].

Experimental studies on oscillatory shear are carried out using rheometers or rheogoniometers. During dynamic (cyclic) tests, the gel fuel is subjected to a deformation that varies according to a sinusoidal law. Sinusoidal shear oscillations with a small amplitude are implemented using cone (or cylinder) oscillations, and the value of the deformation amplitude is fixed. The magnitude of stresses arising in the fuel is measured. Gel fuel is characterized by linear viscoelasticity if the resulting stresses change according to a sinusoidal law, but are out of phase with the deformation. The type of typical dependencies in the study of oscillatory shear is shown in Figure 7 [126,138].

The authors of [109] proposed a technique for studying the characteristics of elastic deformation of oil-filled cryogels based on 5% and 10% PVA aqueous solutions. It has been established that the higher the concentration of PVA in an aqueous solution and the lower the oil content in the fuel, the higher the elasticity of the pellets under otherwise identical conditions (Figure 8). The difference between the elastic modulus of pellets based on 5% and 10% PVA solutions with different oil concentrations is 27–71%. The results obtained qualitatively agree with the results in [139].

In accordance with the technique described in detail by the authors [109], the elastic deformation of heterogeneous gel fuels based on oil-free and oil-filled cryogels containing particles of coal dust with a dispersion of no more than 140 μm was studied [110]. It was concluded [110] that in the practical application of fuel pellets based on oil-filled cryogels, the necessary physical and mechanical characteristics (elasticity, tensile strength) can be achieved by varying both the concentration of PVA in an aqueous solution and oil in the primary emulsion.

It should be noted that currently, there are no known results of changing the viscoelastic properties of gel fuels or their analogues under conditions of ultra-high values of the cyclic frequency of shear deformation achieved during atomization in dispersing devices.

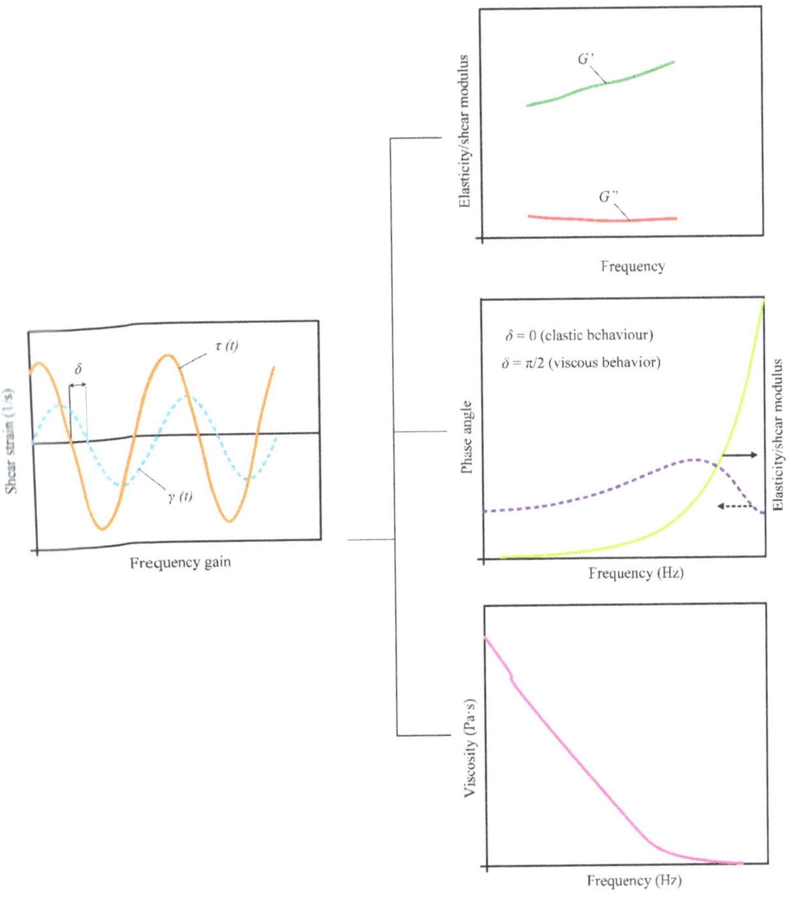

Figure 7. Typical trends obtained in the study of gel fuel for oscillatory shear [126,138].

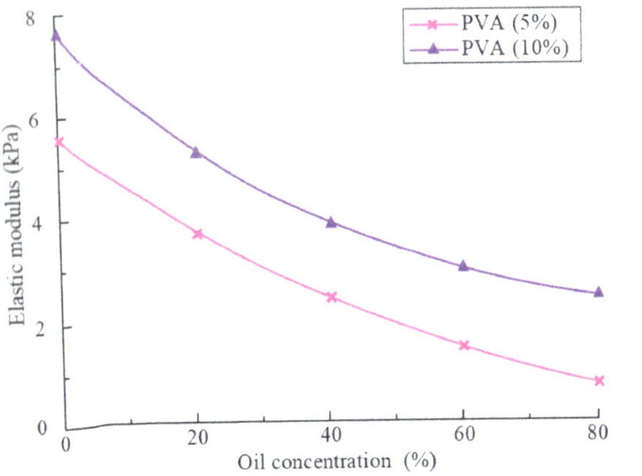

Figure 8. Elastic modulus of fuel pellets of oil-filled cryogels vs. oil concentration at various concentrations of aqueous solution of PVA.

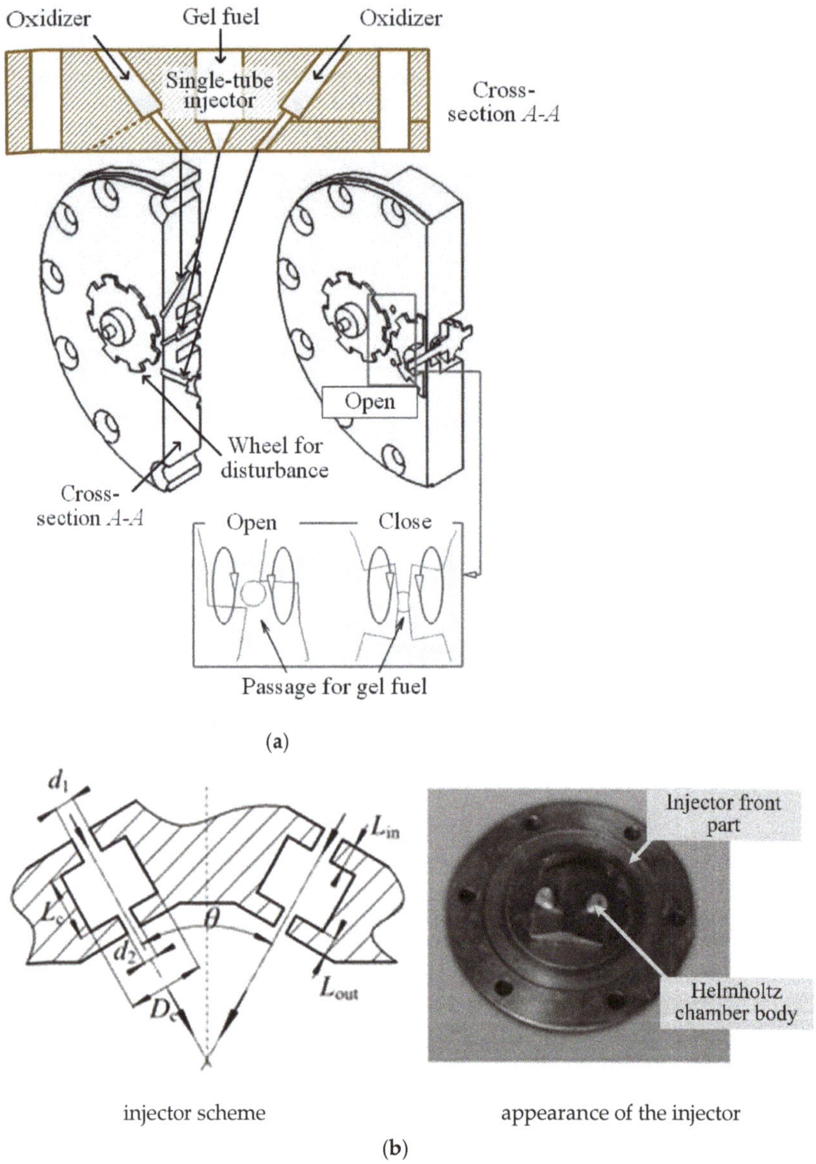

Figure 11. Schematic representation of an improved three-jet spray atomization (**a**) and a two-jet spray atomization with a Helmholtz chamber (**b**) [97].

4.1. Main Characteristics of Atomization

The analysis of video frames of high-speed video recording makes it possible to conduct a qualitative description of the mechanisms of destruction of gel fuel flows after their atomization from spray devices. Predictive assessment of the process of destruction of gel fuels is possible based on the results of an analysis of spray characteristics, which includes the breakup length (L_b), droplet size relative to known estimates of the average size (D_{32}) and spray cone angle (α). Under conditions of shock wave formation, the frequency (f), length (λ), and wave propagation velocity (u_p) are evaluated.

4.1.1. Breakup Length

The breakup length of the gel fuel spray membrane is the most important characteristic in determining the geometric dimensions of the combustion chambers. The results of studies of the breakup length of gel fuel and their simulants when flowing through a hydraulic nozzle [69,70,122,143,146,149,150,167,168] during vortex spraying [43,169] and two-jet spraying [69,70,143,146,149,150,167,168,170] have been established. Figure 12 shows the known results of flow breakup under the conditions of shock wave formation. It can be seen from Figure 12 that gels containing Thixatrol ST, as well as micro- and nanoparticles of aluminum and its oxide (Jet A1–7.5% Thixatrol ST–7.5% MIAK–35.4% m-Al and ISROsene–7.5% Thixatrol ST–7.5% xylene–(0–20% n-Al_2O_3)), are characterized by a significantly shorter breakup length than gels containing polymeric thickeners, carbopol, ethanol and xanthan gum. When spraying aqueous solutions containing more than 0.5% of carbopol thickeners and xanthan gum, the breakup length is longer compared to the breakup length of water. The addition of micro- and nanoparticles of metals to the composition of gel fuels reduces the breakup length, and an increase in the concentration of the thickener in the composition increases it.

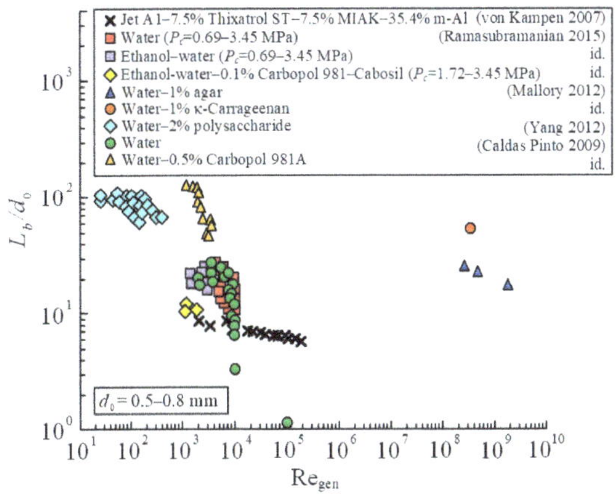

(a)

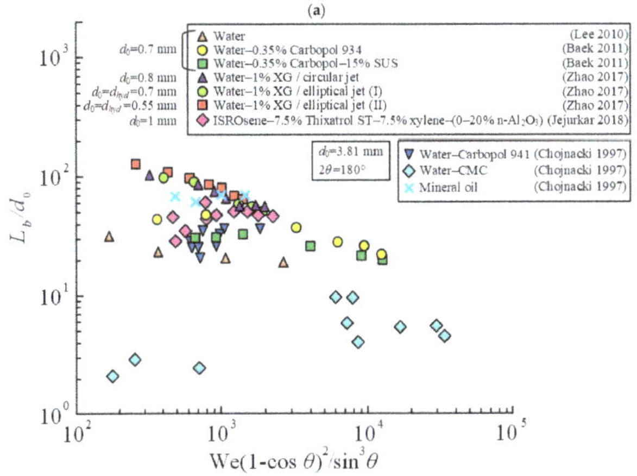

(b)

Figure 12. The breakup length of gel fuels and their simulants when flowing through a two-jet atomizer, depending on the: (**a**) Reynolds number [65,143,144,150,168,171]; (**b**) Weber number [68,69,149,167,170].

Under the conditions of vortex spraying of gel fuels, four typical regions were identified [172] along the flow breakup length (Figure 13): I—intact hollow conical sheet; II—the formation of holes connected by ligaments; III—turbulent flow of ligaments; and IV—breakup of ligaments into droplets.

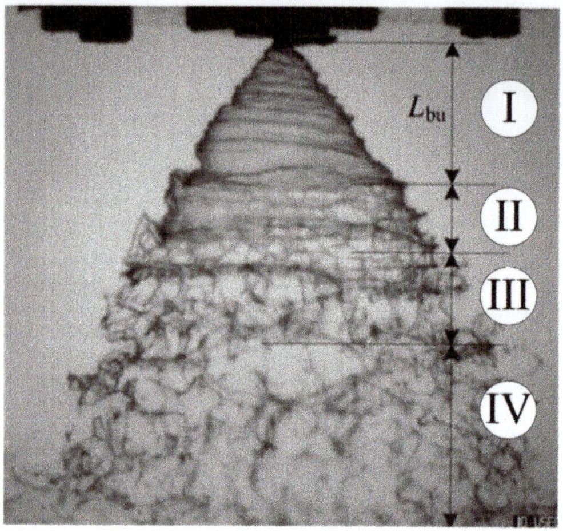

Figure 13. Typical regions of flow destruction along the breakup length [172].

In region I, the flow under the centrifugal force is transformed into a conical hollow thin layer. In region II, holes of a thin layer are formed, connected by ligaments. In area III, the thin layer is completely destroyed into ligaments. The latter, colliding with each other, form the structure of the "fluid web" [172]. In area IV, the ligaments form a structure in the form of "beads on a string" or partially collapse into droplets of irregular geometric shape. The length of the breakup of the gel fuel is determined by the length of the region I (Figure 13).

4.1.2. Spray Angle

Figure 14 shows the results of determining the dependences of the spray angle (α) on the momentum flux ratio (M) [95,99], injection pressure (P_{inj}) [148], volume flow (Q) [173] and Weber number (We_{gel}) [68].

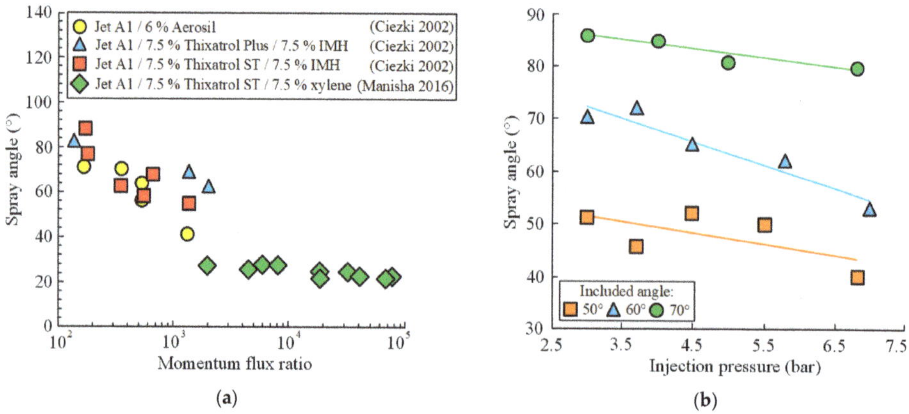

Figure 14. *Cont.*

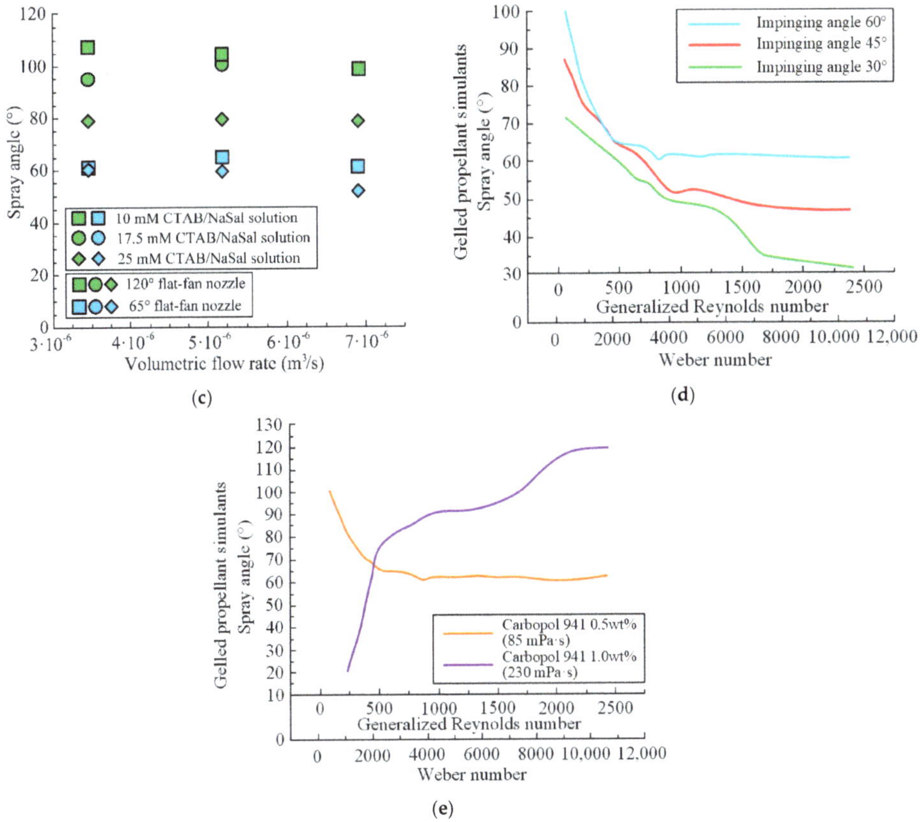

Figure 14. Spray angle versus: (**a**) Momentum flux ratio [95,99]; (**b**) Injection pressure [148]; (**c**) Volumetric flow rate [173]; (**d**,**e**) Weber and Reynolds numbers [68].

It is known [95] that the spray angle of the gel fuel decreases with an increase in the ratio of the mass flow rate of the gel fuel to a similar characteristic of the air flow, as well as with an increase in the viscosity of the fuel itself. In this case, the spray angle increases under conditions of a low ratio of the flow of gel fuel to air (2.7–4.0%). As this ratio increases (up to 8%), the spray angle decreases. The latter is due to the fact that at a ratio of the flow of gel fuel to air in the range from 4% to 8%, the viscosity increases, and the volume of the flow decreases [95]. At a low mass flow rate of the gel fuel, the cross-sectional area of the air flow increases, reducing the cross-sectional area of the fuel flow and increasing its speed. As a result, the layer decomposes into ligaments and further into droplets mainly in the axial direction. Consequently, the angle of the spray cone decreases. An increase in the ratio of mass flow rates of gel fuel to air flow by more than 8% does not affect the spray angle and remains practically constant within the high range of this ratio. The mass flow rate of the gel fuel is the dominant factor influencing the reduction in viscosity and thus reduces the spray angle as it increases. It can be seen (Figure 14a) that the composition of the gel fuel thickener has little effect on the spray angle [95,99].

Figure 14b shows the results of study [148] on the influence of pressure (from 3 to 7 bar, which corresponds to an average jet velocity of 10–20 m/s) on the spray angle. The experiments were carried out [148] with two-jet spraying of a gel fuel consisting of liquid kerosene (60%), aluminum particles 15 μm (30%), organophilic clay (8%) and propylene glycol (2%) at collision angles of 50°, 60° and 70°. It can be seen (Figure 14b) that the spray angle decreases as the injection pressure increases. As the injection pressure increases, the impulse of the shock wave force increases and, consequently, the breakup length decreases [148].

The effect of volumetric flow rate on the spray angle (Figure 14c) was studied [106] by spraying solutions containing 10 to 25 mmol of cetyltrimethylammonium bromide (CTAB) and sodium salicylate (NaSal) in distilled deionized water through a stainless steel swirl nozzle with angle openings of 65° and 120°. It was found [106] that an increase in volume flow had little effect on the spray angle (Figure 14c). However, it was found [106] that an increase in elasticity greatly reduces the spray angle.

Figure 14d,e show the results [68] of determining the dependences of the spray angle on the generalized Reynolds number and Weber number under two-jet spray conditions. It was established [68] that the spray angle of gel fuels depends on the injection rate, jet collision angle and viscosity. With an increase in the injection rate of gel fuel, the spray angle decreases due to a decrease in the viscosity of the fuel due to stretching. As the Reynolds number increases, the spray angle approaches the jet collision angle (Figure 14d). Figure 14e shows that the curves characterizing the dependences of the spray angles of 0.5% and 1.0% aqueous solution of Carbopol with a viscosity of 85 mPa·s and 230 mPa·s, respectively, have different shapes. Under conditions of high-viscosity gel fuel atomization, the spray angle increases with an increase in the Weber number. Under conditions of low viscosity, the spray angle decreases and, at a Reynolds number of 570, tends to the jet collision angle [68].

4.1.3. Droplet Size Distribution

At present, a fairly extensive experimental database of statistical data on droplet sizes during the spraying of gel fuels and their simulants has been accumulated [69,144,148,149,174–176]. Figure 15 shows dependences of the ratio of Sauter mean droplet diameter to the injector outlet diameter (D_{32}/d_0) on the parameter (d_0/u_j) for gel fuels and their simulants [69,146,148,149,151,153,168,177] as well as water [146,148,149,168]. The parameter (d_0/u_j) characterizes the operation of the spray device, where u_j is the velocity at the injector outlet. Based on the results of the analysis of Figure 15, it can be concluded that the diameter of droplets of gel fuels decreases to the size of droplets of a Newtonian liquid with a decrease in the diameter of the outlet of the atomizer and an increase in speed.

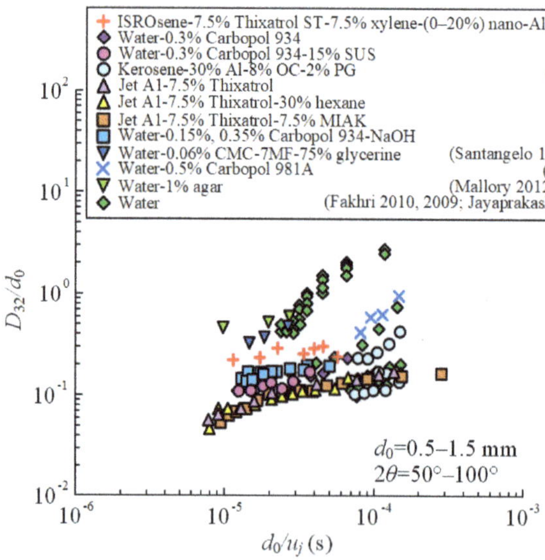

Figure 15. Dimensionless Sauter mean droplet diameters from the parameter of the spraying device (d_0/u_j) [69,146,148,149,151,153,168,171,178].

The droplet size can be reduced to 50 μm by changing the design of spraying devices [97]. Figure 16 shows the results of an experimental determination of the dimensionless Sauter mean droplet diameters, obtained with two-jet atomization (Figure 16) followed by a jet collision with the teeth of a pair of wheels (Figure 16a), using structures that initiate flow pulsations of three-jet spraying (Figure 16a) when using a porous insert (Figure 10) at the outlet of the "flash-boiling atomization" nozzle (Figure 16b) and atomizers with an internal impact on the flow (Figure 16c).

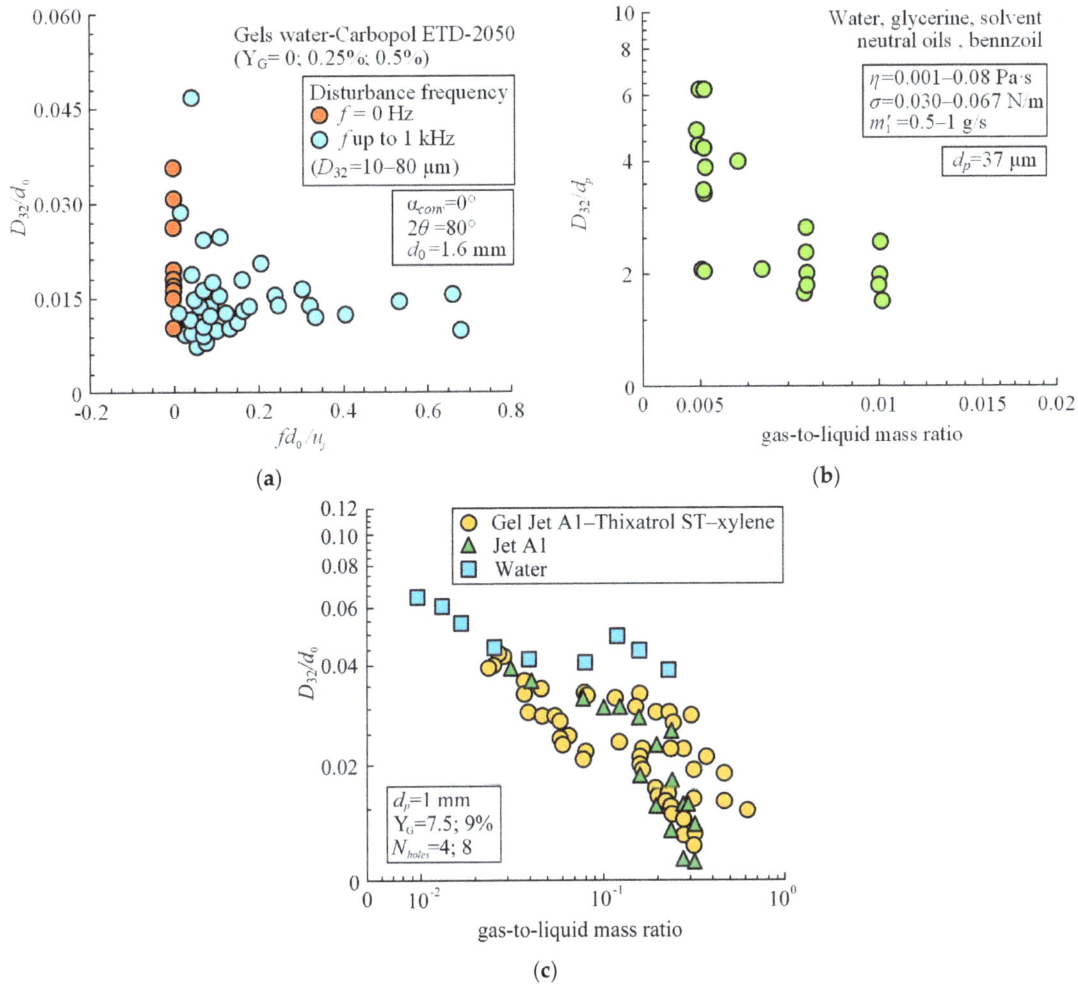

Figure 16. Dimensionless Sauter mean diameters of droplets. Designs of spraying devices: (**a**) Two-jet spraying followed by collision of the jets with the teeth of a pair of wheels and three-jet spraying with pulsating flows [97]; (**b**) "Flash-boiling atomizer" with a porous insert (pore diameter d_p) at the nozzle outlet [166,176]; (**c**) Atomizer with internal impact on flow [81,95,179].

It can be seen from Figure 16a that as the pulsation frequency of the three-jet spray and the wheel speed increase, the average droplet size decreases to values from 10 to 80 μm. The porous insert at the outlet of the "flash-boiling atomization" nozzle also significantly reduces the size of the resulting droplets (Figure 16b) while maintaining a low mass air flow. When using atomizer designs with an internal impact, the average size of the formed droplets is in the range from 10 to 70 μm (Figure 16c).

Atomization efficiency mainly depends on the way the atomizing gas collides with the atomized liquid and the angle of their collision. If the jet collision angle is large, then the average droplet size does not depend on the diameter (thickness) of the gel fuel flow and the ambient pressure. If the collision angle is small (the air and fuel flows practically move in parallel), then after their collision, the average size of the droplets depends on the thickness of the fuel flow and the ambient air pressure. Atomizer designs that maintain prolonged contact between air and fuel, such as wave atomization and "flash-boiling atomization", perform more efficiently than designs based on two- and three-jet atomization.

5. Combustion

With the further development of the topic of gel fuels, such fuel can become an alternative to mixed solid fuel [180]. When using gel fuel, the same scheme of engine operation can be realized as when using high-power charges on an open surface. The aggregation state of gel fuel (depending on the surrounding conditions and on the purpose of the fuel) can change quite significantly – from liquid to solid elastically deformable The main difference between gel fuel and mixed solid fuel is the presence of the most complex physical and chemical processes occurring during its heating [180]. For example, when a mixed solid fuel containing ammonium perchlorate and butyl rubber is heated, the exothermic interaction of the fuel and oxidizer occurs in the near-surface layer of the heated region. Processes occurring during the induction period are well described in terms of the mathematical model of solid-phase ignition of condensed matter [181]. Gel fuels are characterized by gas-phase ignition. During fuel heating, a gas mixture is formed as a result of the evaporation of the fuel and mixing with an external oxidizer [48]. To predict the characteristics of gel fuel gas-phase ignition processes, appropriate mathematical models are needed, which differ, for example, from the models of ignition of solid [181] and liquid [182] fuels' condensed matter. The development of such predictive mathematical models is impossible without experimental investigations of ignition processes, which are a rather difficult task in the development of new fuels.

5.1. Experimental Investigation of Ignition and Combustion of Gel Fuel

Fuel particles were ignited under conditions of radiative-convective heating occurring in a high-temperature stationary air medium, which was generated in the volume of a ceramic tube (inner diameter 50 mm, length 450 mm) of a muffle furnace [183–185]. The range of air temperature variation in the furnace was T_g = 700–1000 °C. After its stabilization, the fuel particle located on the holder was introduced into the cavity of the ceramic tube of the furnace using a minirobotic arm. The processes occurring during the induction period, t_d, were recorded by a color high-speed video camera.

Experimental studies [186] were carried out for two compositions of gel fuel based on an aqueous solution (10 wt%) of polyvinyl alcohol (PVA). The first composition (No. 1) was prepared without the addition of fine solid particles as follows: 50 vol% oil and 48 vol% aqueous solution of PVA and 2 vol% emulsifier. The second composition (No. 2) was prepared on the basis of the first by adding 30 wt% coal dust (140 μm) to the primary oil emulsion (50 vol% oil and 48 vol% PVA aqueous solution and 2 vol% emulsifier).

A fundamental difference was established between the mechanisms of ignition of single particles (melt droplets) of two compositions of gel fuels in a high-temperature stationary air medium [186]. Both fuels, when heated in the temperature range of 700–1000 °C, ignite with dispersion of melt droplets caused by the multicomponent composition of the fuel. During the melting of oil-filled cryogels, the liquid components of the fuel are separated. A shell of molten thickener is formed on the surface of the droplet. Under it is a flammable liquid. Because of such a structure of a droplet of molten gel fuel, when it is heated, processes occur that are not typical for the process of ignition of a droplet of a single-component combustible liquid, such as:

1. Formation of a large number of bubbles in the near-surface layer of the droplet;
2. An increase in the number and size of the bubbles, accompanied by a significant change in the shape of the drop;
3. Collapse of bubbles and dispersion of the initial droplet with the separation of a group of droplets of a significantly smaller size (1–2 orders smaller than the initial size of the fuel particle);
4. The previous process is accompanied by the release of combustible liquid vapor into the oxidizer medium through the thickener layer (outer shell) when the droplet is destroyed;
5. When the limiting conditions are reached, gas-phase ignition occurs in the vicinity of the droplet.

In contrast to liquid single-component fuels, the initiation of combustion of gel fuels occurs not in a small vicinity of a droplet, but in a rather large area (Figure 17), which has a positive effect on the development of subsequent combustion and intensification of the burnout of the components.

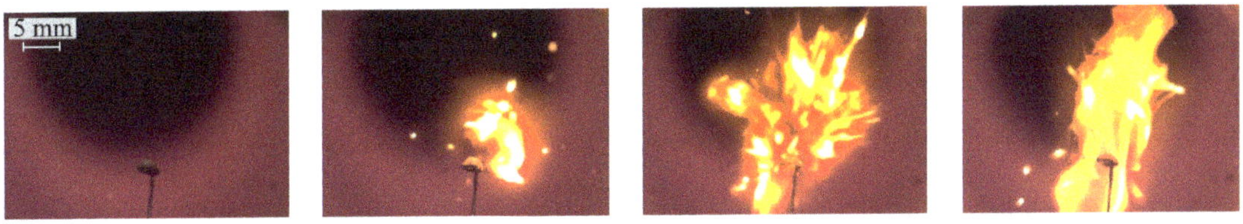

t = 3.0 s t_d = 3.1 s t = 3.2 s t = 3.5 s

Figure 17. Typical videoframes of ignition and combustion of a particle of gel fuel (composition No. 1—without solid particles: 50 vol% oil + 48 vol% aqueous solution of PVA + 2 vol% emulsifier) at T_g = 900 °C [186].

The difference between the ignition mechanisms of these two gel fuel compositions is as follows. As a result of the collapse of bubbles and the dispersion of molten fuel droplet No. 2, vapors of a combustible liquid and fine coal particles, which are impregnated with this liquid, are blown into the high-temperature oxidizing medium. Therefore, gas-phase ignition of the gel fuel containing finely dispersed solid particles occurs more intensively and rather evenly in a large vapor-gas region (Figure 18).

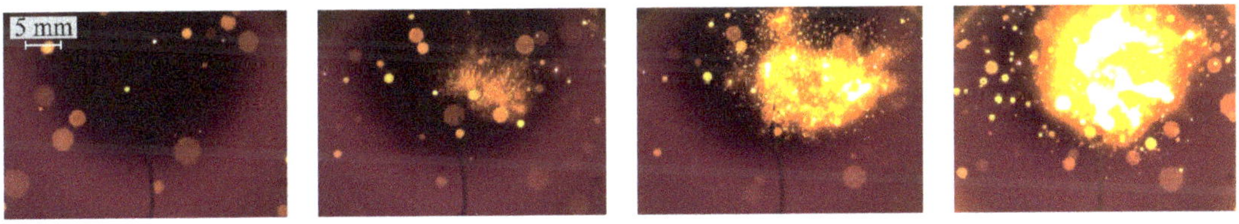

t = 1.6 s t_d = 1.7 s t = 1.8 s t = 1.9 s

Figure 18. Typical videoframes of ignition and combustion of a particle of gel fuel (composition No. 2—with finely dispersed carbonaceous particles: 70 wt% (50 vol% oil + 48 vol% aqueous solution of PVA + 2 vol% emulsifier) + 30 wt% coal dust) at T_g = 900 °C [186].

By varying the temperature of the heated air in the range of 700–1000 °C and the velocities of entry into the combustion chamber of 0.06–0.10 m/s, the velocity (Figure 19a) and the size of the burnout area of fine fragments and the vapor-gas mixture after the dispersion of droplets of molten fuels were established (Figure 19b).

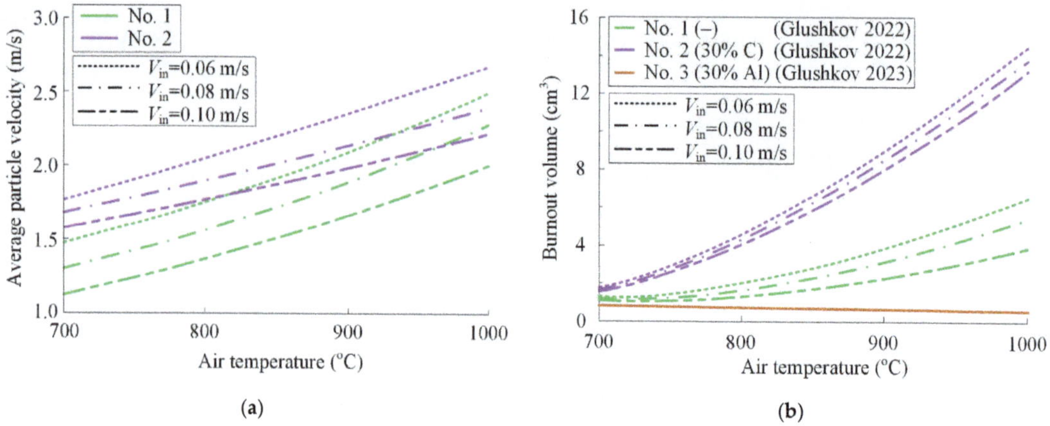

Figure 19. Characteristics of the dispersion of melts of gel fuels No. 1 and No. 2 at different rates of injection of fuel particles into the combustion chamber: (**a**) Speeds of movement of finely dispersed fragments during droplet dispersion [186]; (**b**) Dimensions of areas in which the process of burnout occurs [186,187].

Under such conditions, for an oil-filled cryogel without solid inclusions, the average velocities of fine particles after dispersion change from 1.1 to 2.2 m/s (by 90–100%), for a fuel composition, with addition of carbonaceous particles, ranging from 1.6 to 2.5 m/s (by 45–55%). The authors found (Figure 19b) two different trends in the dependence of the size of the burnout area on the temperature of the energy source. Compositions with a conspicuous gas-phase combustion without dispersion of a droplet melt (Figure 19b—No. 3: with the addition of 30 wt% V-ALEX 100 aluminum nanopowder to the primary oil emulsion) are characterized by a decrease in the size of the burnout region with increasing ambient temperature. Under radiant heating, after the droplet melts, the fuel evaporates monotonically according to the d^2 law. In the vicinity of the droplet, a spherical combustible gas-vapor mixture is formed, which ignites when the ignition temperature is reached. The higher the ambient temperature, the faster the necessary conditions for the ignition of the gas-vapor mixture are reached, the less time the combustible liquid has to evaporate and the smaller the size of the area in which the combustible gas-vapor mixture is formed. For fuel compositions that disperse during heating and combustion, the opposite trend is characteristic—the size of the burnout area increases with increasing ambient temperatures.

In this case, the energy release power of gel fuels is about 190 W in a volume of about 8 cm^3 with a burn-up of 10 mg of gel fuel (against about 25 W in a volume of about 1 cm^3 with a longer burning time of a single-component liquid fuel droplet).

Within the method of the two-color pyrometry [186], the temperature trends of the flame during the combustion of single particles of gel fuels were established. Fuel composition No. 1 is characterized by a series of episodes of dispersion after ignition of a fuel particle. The extremes on the temperature trends (Figure 20) correspond to the mechanism of ignition and combustion of the oil-filled cryogel described above, and illustrate a series of episodes of dispersion, proceeding until the complete burnout of the fuel components is achieved. The addition of fine combustible particles to the fuel composition not only increases the intensity of its ignition, but also ensures the occurrence of a fairly stable fuel burnout without significant temperature fluctuations during the combustion process (Figure 20). The result of experimental studies indicates that the regularities and characteristics of the ignition of gel fuels (both with the addition of solid particles and without them) can vary greatly. Such a wide variety of combustion mechanisms and characteristics is one of the advantages of using gel fuels compared to liquid and solid fuels.

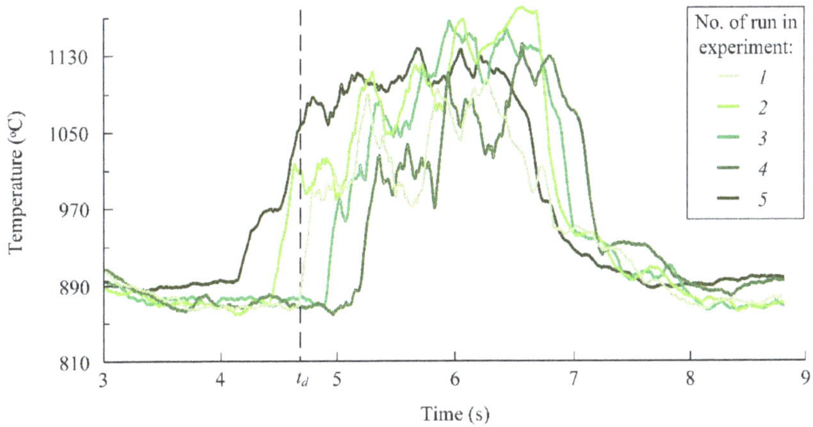

(a)

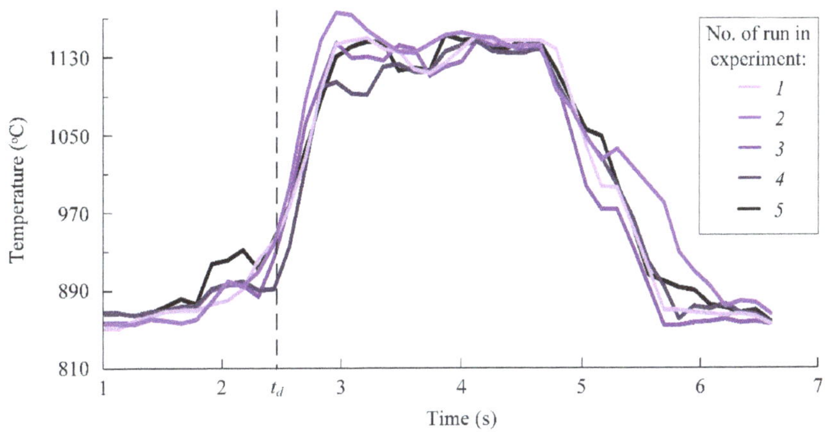

(b)

Figure 20. Flame temperature during the combustion of gel fuel at T_g = 850 °C: (**a**) No. 1 and (**b**) No. 2; *1*–*5*—number of run in a series of experiments.

5.2. Mathematical Models of Ignition Processes

Based on the results of experimental studies [109,188,189], a scheme was developed (Figure 21) for the ignition of a typical gel fuel particle based on an organic polymer thickener in a high-temperature air medium. It is assumed [109,188,189] that a cold fuel particle ($T = T_0$) of a spherical shape with a radius r_p at the initial time ($t = 0$) is introduced into the region of still air heated to high temperatures ($T = T_a$) (Figure 21a). When the particle is heated (mainly due to the radiant heat supply), its surface temperature reaches the melting temperature of the gel fuel (T_{melt} = 330 K) and a melting front is formed, the boundary of which moves towards the center of the particle (Figure 21b). After the gel fuel is melted, its components are almost instantly separated. A thin thickener layer is formed on the surface of the melt, under which a combustible liquid is located (Figure 21c). The thickness of the thickener layer decreases until it completely evaporates (Figure 21d). After that, the combustible liquid begins to evaporate from the free surface (Figure 21d). In the vicinity of a fuel particle, a combustible gas-vapor mixture is formed as a result of diffusion (Figure 21e). When limiting conditions are reached, gas-phase ignition occurs (Figure 21e).

The event of excess temperature in the zone of exothermic reaction of the temperature of heated air ($T > T_a$) is accepted as the ignition criterion [109,188,189].

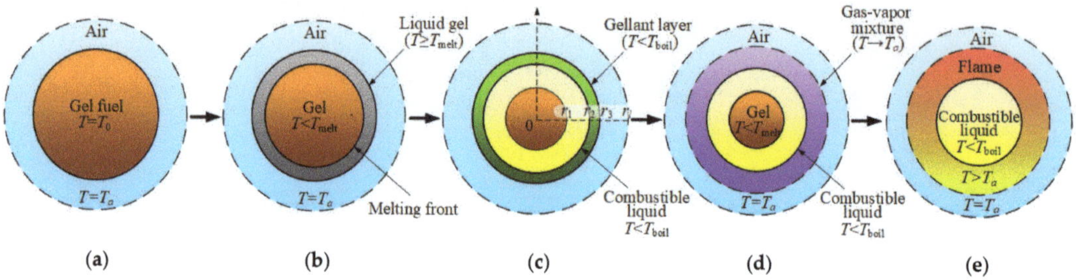

Figure 21. Scheme of the ignition process of a gel fuel particle [109,188,189]: (**a**) initial moment ($t = 0$); (**b**) gel fuel melting; (**c**) separation of gel components; (**d**) formation of combustible gas-vapor mixture; (**e**) ignition moment ($t = t_d$).

As a result of numerical simulation of the ignition process [109,188,189], temperature profiles in the fuel particle/air system (typical ones are shown in Figure 22) at different times during the induction period t_d with the initial particle radius $r_p = 1.25$ mm and heated air temperature $T_a = 1200$ K. The inflection points on the temperature profiles at $0 < r \leq r_p$ correspond to the coordinate of the fuel melt droplet surface. The concentration of the thickener (water solution of PVA) in the composition of the gel fuel was $\mu_4 = 48\%$.

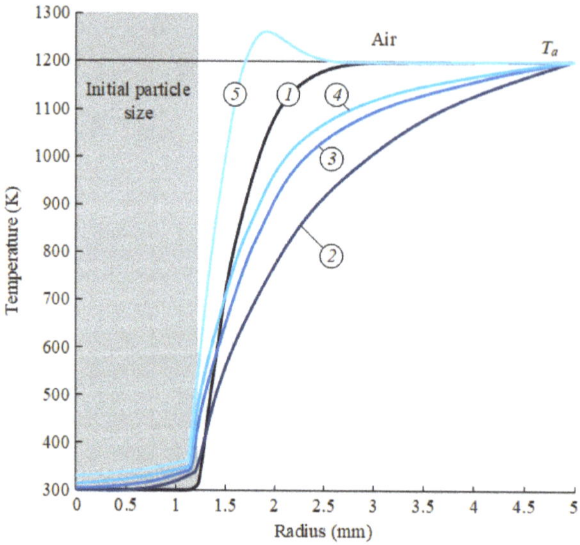

Figure 22. Temperature profiles in the fuel particle/air system at $T_a = 1200$ K, $r_p = 1.25$ mm, $\mu_4 = 48\%$ at different time points [189]: *1*—$t = 0.1$ s; *2*—$t = 0.386$ s; *3*—$t = 0.77$ s; *4*—$t = 1.16$ s; *5*—$t = 1.55$ s.

The induction period t_d can be conditionally divided into two stages. The first stage is the predominance of heat and mass transfer processes during the heating of a fuel particle, the formation of a combustible gas-vapor mixture and its heating. The second stage is the predominance of chemical reaction processes during the interaction of the components of the gas mixture. The characteristic times of these stages differ significantly. The duration of the first stage is 70–90% of the time t_d (depending on the temperature of the heated air). The second stage is much more intense and its duration depends mainly on the kinetic parameters of the process of chemical reaction of the combustible component and the oxidizer. Therefore, the processes of heat and mass transfer have a rather significant effect on the intensity of ignition of the fuel as a whole.

It has been established [189] that the temperature T_a = 800 K is the minimum required for ignition of a gel fuel particle. At lower air temperatures, the fuel does not burn. Under such conditions, the energy of the heating source is not enough to ignite the resulting gas-vapor mixture. The results of numerical simulation are in good agreement with the experimental data [109] obtained under identical conditions. The difference in ignition delay times does not exceed the random error (20%) of recording t_d in experiments. [109]. The theoretically established fuel ignition delay times in the heated air temperature range of 800–1473 K [189] are reduced by more than six times (Figure 23). When close to the limiting conditions of ignition (when t_d = 8.15 s), the duration of the induction period, t_d, asymptotically tends to infinity at $T_a \rightarrow 800$ K (Figure 23). Under conditions of high ambient temperatures (at T_a > 1150 K), when physicochemical processes reach their maximum intensity, the fuel ignition delay time does not decrease to less than 1.3 s (Figure 23). Thus, for ignition conditions, the preparation time for the fuel particle/air system for the development of an intense exothermic reaction process is about 1.2 s at an oxidizer temperature exceeding 1150 K. As a rule, in practice, the fuel enters the combustion chamber in the form of a flow of particles which sizes may vary within a certain range. Therefore, for the possible practical application of the results of studies [189], a more detailed analysis of the influence of this factor on the ignition characteristics of gel fuel is required.

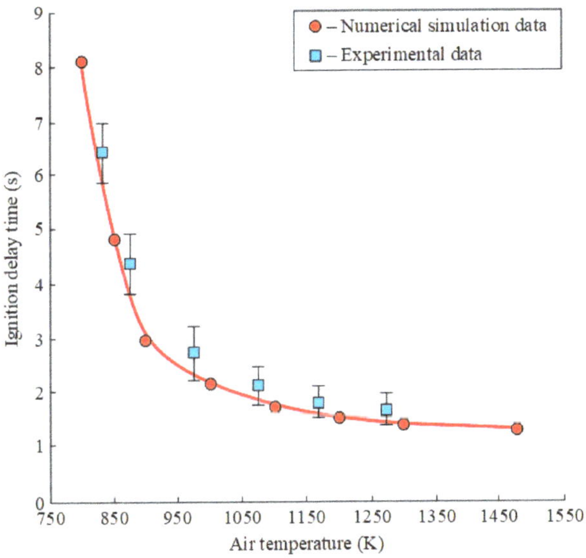

Figure 23. Dependence of the ignition delay time of gel fuel particles on the temperature of heated air at r_p = 1.25 mm, μ_4 = 48% [189].

6. Conclusions

Over the past two decades, a fairly large amount of new knowledge about gel fuels has been obtained within the previously identified areas of research on fuel component composition, rheological characteristics, atomization and combustion processes [24,50,55,64,65,70,81,91,94,100,115,144,149–151,190–192]. In addition to laboratory studies, full-scale tests of the first rocket engines operating on gel fuels were carried out. Future advances in this area contributing to the emergence of mass-produced engines will depend on success in the implementation of research and development work in the following areas, consistent with the global trend in the development of fundamental and applied research on gel fuels.

6.1. Fuel Composition

The difficulties in optimizing high metal content (up to 60%) in gel fuels require the increased attention of the scientific community. Studies have shown that promising gel fuels with a high content of fine metal particles behave like unstable slurry fuels. Experience with the use of liquid fuels shows that the metal content can reach 60%. This result should be replicated for gel propellants as part of a large-scale research cycle.

Considering the direct connection between the component composition and rheology (as well as the atomization process), it is advisable to optimize fuel compositions iteratively, including rheology and atomization data in the corresponding search algorithm.

In the future, the plan is to apply systematic multi-criteria methods for searching for optimal gel fuel compositions.

Previously, large-scale studies of compositions in the selected areas in the framework of the study of gel fuels, including metallized ones, were not carried out. Thus, there is no information on fuels' stability and the influence of the processes of components settling and separation on their uniformity of composition, rheological characteristics, atomization and combustion. Most of the results were obtained for fuel samples weighing less than 1 kg, or in some cases for samples weighing 1–10 kg. The influence of the scale factor on the quality of the prepared gel fuel is unknown.

6.2. Rheology

The effect of tensile strength, thixotropy and viscoelasticity on failing structures remains to be seen. This area of research should be expanded by considering secondary decay events.

The rheological properties of "brittle" or "extremely fluid" gels can play a decisive role in their spray characteristics. Since gels with a high concentration of the metal component can be "brittle", it is important to study the correlation between the destruction processes of such gels and their rheological properties.

Research should be carried out on methods of viscosity reduction in laboratory and in-line measurements to increase viscosity reduction beyond what is currently achieved by shifting fuel delivery and injection systems. The results of such studies should be useful for the effective design of appropriate systems for aerospace industry objects with limited weight and size characteristics.

6.3. Atomization

The destruction modes of gel fuels during spraying should be systematically checked to identify the influence of rheological properties.

Original approaches to the development of the design of advanced injectors should consider the rheological behavior of gel propellants. This will not only effectively organize fuel atomization in the combustion chamber, but also maintain its combustion characteristics at a predetermined level.

The feature of gel fuels containing particles of thickeners (especially silica) and metal particles has been reliably established—to form fragments of irregular shape during spraying. Further dynamics of changes in the shape and size of these fragments and their effect on the flame should be studied in more detail. Approaches to the formation of a component composition that combine two types of thickeners, one of which is polymeric, should prove useful in balancing the tendency to form irregular fragments and relatively easy fuel atomization.

It is known that the secondary dispersion of metalized liquid fuel droplets is critical for the ignition and combustion of metal particles. It is assumed that similar problems will arise for metallized gel fuels, which needs to be confirmed.

6.4. Combustion

Promising mathematical models of the combustion of gel fuel single droplets based on a polymer thickener should describe the physical processes that lead to the oscillatory nature of the injection of combustible liquid vapor into the oxidizer medium and the formation of a vapor-gas mixture and its burnout. To develop the corresponding mathematical models, a theoretical basis is needed in the form of the results of experimental studies cycle using high-speed video recording and non-contact diagnostics of gas flows.

A detailed understanding of the regularities and characteristics of physicochemical processes occurring during flaring of gel fuels is necessary to overcome the gap between theoretical and empirical knowledge in order to increase the reliability of known predictive models describing the corresponding processes.

Experimental data on the characteristics of metallized gel fuel flare combustion have not yet been published. The results of fundamental research are important for carrying out development work in the design of a ramjet engine.

Research is required to determine effective solutions to reduce fuel viscosity in a short period of time, reduce the overall dimensions of the nozzle while maintaining its performance, increase the degree of fuel metallization and control the dynamics of solid metal oxide particles and the processes of their interaction with the combustion chamber walls and nozzle.

Author Contributions: D.G.—writing, review and editing; K.P.—writing, review and editing and A.P.—writing, review and editing. All authors have read and agreed to the published version of the manuscript.

Funding: The reported study was funded by the Russian Science Foundation grant number [18-13-00031], URL (accessed on 26.12.2022): https://rscf.ru/project/21-13-28043/.

Conflicts of Interest: The authors declare no conflict of interest.

References

1. Carasillo, D.A. *Liquid Fuels: Types, Properties and Production*; Nova Science: New York, NY, USA, 2012; ISBN 9781614704355.
2. Houghton-Alico, D. *Alcohol Fuels: Policies, Production, and Potential*; Taylor & Francis: Abingdon, UK, 2019; ISBN 9780429704642.
3. De Souza-Santos, M.L. *Solid Fuels Combustion and Gasification: Modeling, Simulation, and Equipment Operations*, 2nd ed.; CRC Press: Boca Raton, FL, USA, 2010; ISBN 9781420047509.
4. Tillman, D.; Duong, D.; Harding, N.S. *Solid Fuel Blending*; Butterworth-Heinemann: Oxford, UK, 2012; ISBN 9780123809322.
5. Müller, H.; Kiefer, J. Fuels | Gaseous. In *Encyclopedia of Analytical Science*; Elsevier: Amsterdam, The Netherlands, 2019; ISBN 9780081019832.
6. Shah, Y.T. *Chemical Energy from Natural and Synthetic Gas*; CRC Press: Boca Raton, FL, USA, 2017; ISBN 9781315302348.
7. Verhelst, S.; Turner, J.W.; Sileghem, L.; Vancoillie, J. Methanol as a Fuel for Internal Combustion Engines. *Prog. Energy Combust. Sci.* **2019**, *70*, 43–88. [CrossRef]
8. Li, X.; Qin, S.; Huang, X.; Liu, H. Multi-Component Effect and Reaction Mechanism for Low-Temperature Ignition of Kerosene with Composite Enhancer. *Combust. Flame* **2019**, *199*, 401–410. [CrossRef]
9. Sakaki, K.; Kakudo, H.; Nakaya, S.; Tsue, M.; Suzuki, K.; Kanai, R.; Inagawa, T.; Hiraiwa, T. Combustion Characteristics of Ethanol/Liquid-Oxygen Rocket-Engine Combustor with Planar Pintle Injector. *J. Propuls. Power* **2017**, *33*, 514–521. [CrossRef]
10. García-Contreras, R.; Martínez, J.D.; Armas, O.; Murillo, R.; García, T. Study of a Residential Boiler under Start-Transient Conditions Using a Tire Pyrolysis Liquid (TPL)/Diesel Fuel Blend. *Fuel* **2015**, *158*, 744–752. [CrossRef]
11. Glushkov, D.O.; Kuznetsov, G.V.; Chebochakova, D.A.; Lyakhovskaya, O.E.; Shlegel, N.E.; Anufriev, I.S.; Shadrin, E.Y. Experimental Study of Coal Dust Ignition Characteristics at Oil-Free Start-up of Coal-Fired Boilers. *Appl. Therm. Eng.* **2018**, *142*, 371–379. [CrossRef]
12. Li, Y.; Zhang, X.; Zhang, J.; Zhou, J.; Yan, H. Numerical Simulation and Optimization of Pulverized Coal Injection with Enriched Oxygen into Blast Furnace. *Appl. Therm. Eng.* **2014**, *67*, 72–79. [CrossRef]
13. Padwal, M.B.; Varma, M. Thermal Decomposition and Combustion Characteristics of HTPB-Coarse AP Composite Solid Propellants Catalyzed with Fe2O3. *Combust. Sci. Technol.* **2018**, *190*, 1614–1629. [CrossRef]
14. Emami, M.D.; Shahbazian, H.; Sunden, B. Effect of Operational Parameters on Combustion and Emissions in an Industrial Gas Turbine Combustor. *J. Energy Resour. Technol. Trans. ASME* **2019**, *141*, 012202. [CrossRef]
15. Dubinin, A.M.; Shcheklein, S.E. Mini Coal-Fired CHP Plant on the Basis of Synthesis Gas Generator (CO + H2) and Electrochemical Current Generator. *Int. J. Hydrogen Energy* **2017**, *42*, 26048–26058. [CrossRef]

16. Vershinina, K.Y.; Glushkov, D.O.; Kuznetsov, G.V.; Strizhak, P.A. Differences in the Ignition Characteristics of Coal–Water Slurries and Composite Liquid Fuel. *Solid Fuel Chem.* **2016**, *50*, 88–101. [CrossRef]
17. Miccio, F.; Raganati, F.; Ammendola, P.; Okasha, F.; Miccio, M. Fluidized Bed Combustion and Gasification of Fossil and Renewable Slurry Fuels. *Energies* **2021**, *14*, 7766. [CrossRef]
18. Strizhak, P.A.; Piskunov, M.V.; Volkov, R.S.; Legros, J.C. Evaporation, Boiling and Explosive Breakup of Oil–Water Emulsion Drops under Intense Radiant Heating. *Chem. Eng. Res. Des.* **2017**, *127*, 72–80. [CrossRef]
19. Glushkov, D.O.; Kuznetsov, G.V.; Nigay, A.G.; Yashutina, O.S. Heat and Mass Transfer Induced by the Ignition of Single Gel Propellant Droplets. *J. Energy Inst.* **2019**, *92*, 1944–1955. [CrossRef]
20. Padwal, M.B.; Natan, B.; Mishra, D.P. Gel Propellants. *Prog. Energy Combust. Sci.* **2021**, *83*, 100885. [CrossRef]
21. Tabakaev, R.; Shanenkov, I.; Kazakov, A.; Zavorin, A. Thermal Processing of Biomass into High-Calorific Solid Composite Fuel. *J. Anal. Appl. Pyrolysis* **2017**, *124*, 94–102. [CrossRef]
22. Feng, S.; He, B.; He, H.; Su, L.; Hou, Z.; Nie, W.; Guo, X. Experimental Studies the Burning Process of Gelled Unsymmetrical Dimethylhydrazine Droplets under Oxidant Convective Conditions. *Fuel* **2013**, *111*, 367–373. [CrossRef]
23. Mishra, D.P.; Patyal, A.; Padhwal, M. Effects of Gellant Concentration on the Burning and Flame Structure of Organic Gel Propellant Droplets. *Fuel* **2011**, *90*, 1805–1810. [CrossRef]
24. Natan, B.; Rahimi, S. The Status of Gel Propellants in Year 2000. *Int. J. Energetic Mater. Chem. Propuls.* **2001**, *5*, 172–194. [CrossRef]
25. Tsutsuran, V.I.; Petrukhin, N.V.; Gusev, S.A. *Military and Technical Analysis of a State and Prospect of Development of Rocket Fuels*; MO RF: Moscow, Russia, 1999.
26. Klapötke, T.M. *Chemistry of High-Energy Materials*; Walter de Gruyter GmbH: Berlin, Germany, 2015.
27. Zeldovich, Y.B.; Barenblatt, G.I.; Librovich, V.B.; Makhviladze, G.M. *Mathematical Theory of Combustion and Explosions*; Plenum: New York, NY, USA, 1985.
28. Williams, F.A. *Combustion Theory*; Westview Press: Boulder, CO, USA, 1985.
29. Merzhanov, A.G.; Khaikin, B.I. Theory of Combustion Waves in Homogeneous Media. *Prog. Energy Combust. Sci.* **1988**, *14*, 1–98. [CrossRef]
30. Vilyunov, V.N.; Zarko, V.E. *Ignition of Solids*; Elsevier: Amsterdam, The Netherlands, 1989.
31. Okajima, S.; Kumagai, S. Further Investigations of Combustion of Free Droplets in a Freely Falling Chamber Including Moving Droplets. *Symp. Combust.* **1975**, *15*, 401–407. [CrossRef]
32. Manzhai, V.N.; Fufaeva, M.S. Polyvinyl Alcohol Cryogels as an Efficient Spent-Oil Utilization Method. *Chem. Technol. Fuels Oils* **2015**, *51*, 487–492. [CrossRef]
33. Dove, M.F.A.; Logan, N.; Mauger, J.P.; Allan, B.D.; Arndt, R.E.; Hawk, C.W. Aluminum Alloy Compatibility with Gelled Inhibited Red Fuming Nitric Acid. *J. Propuls. Power* **1996**, *12*, 585–590. [CrossRef]
34. Padwal, M.B.; Mishra, D.P. Synthesis of Jet A1 Gel Fuel and Its Characterization for Propulsion Applications. *Fuel Process. Technol.* **2013**, *106*, 359–365. [CrossRef]
35. Gupta, B.L.; Varma, M.; Munjal, N.L. Rheological Studies on Virgin and Metallized Unsymmetrical Dimethyl Hydrazine Gelled Systems. *Propellants, Explos. Pyrotech.* **1986**, *11*, 45–52. [CrossRef]
36. Verma, M.; Gupta, B.L.; Pandey, M. Formulation & Storage Studies on Hydrazine-Based Gelled Propellants. *Def. Sci. J.* **1996**, *46*, 435–442. [CrossRef]
37. Varghese, T.L.; Gaindhar, S.C.; David, J.; Jose, J.; Muthiah, R.; Rao, S.S.; Ninan, K.N.; Krishnamurthy, V.N. Developmental Studies on Metallised UDMH and Kerosene Gels. *Def. Sci. J.* **1995**, *45*, 25–30. [CrossRef]
38. Arnold, R.; Santos, P.H.S.; Campanella, O.H.; Anderson, W.E. Rheological and Thermal Behavior of Gelled Hydrocarbon Fuels. *J. Propuls. Power* **2011**, *27*, 151–161. [CrossRef]
39. Martinez-Pastor, J.; Franco, P.; Oton-Martinez, R.A. Rheology of Double-Base Gelled Propellants as the Basis for Extrusion Process Modelling: Influence of Normal Force on Slip Layer and Flow Curves. *Int. J. Mater. Form.* **2020**, *13*, 219–233. [CrossRef]
40. Jyoti, B.V.S.; Baek, S.W. Rheological Characterization of Metalized and Non-Metalized Ethanol Gel Propellants. *Propellants, Explos. Pyrotech.* **2014**, *39*, 866–873. [CrossRef]
41. Rahimi, S.; Natan, B. The Flow of Gel Fuels in Tapered Injectors. *J. Propuls. Power* **2000**, *16*, 458–464. [CrossRef]
42. Mansour, A.; Chigier, N. Air-Blast Atomization of Non-Newtonian Liquids. *J. Non-Newton. Fluid Mech.* **1995**, *58*, 161–194. [CrossRef]
43. Guan, H.-S.; Li, G.-X.; Zhang, N.-Y. Experimental Investigation of Atomization Characteristics of Swirling Spray by ADN Gelled Propellant. *Acta Astronaut.* **2018**, *144*, 119–125. [CrossRef]
44. Fu, Q.-F.; Duan, R.-Z.; Cui, K.-D.; Yang, L.-J. Spray of Gelled Propellants from an Impinging-Jet Injector under Different Temperatures. *Aerosp. Sci. Technol.* **2014**, *39*, 552–558. [CrossRef]
45. Nachmoni, G.; Natan, B. Combustion Characteristics of Gel Fuels. *Combust. Sci. Technol.* **2000**, *156*, 139–157. [CrossRef]
46. Galecki, D.L. Ignition and Combustion of Metallized Propellants. In Proceedings of the 25th Joint Propulsion Conference, Monterey, CA, USA, 12–16 July 1989; p. AIAA 1989-2883.
47. He, B.; Nie, W.; He, H. Unsteady Combustion Model of Nonmetalized Organic Gel Fuel Droplet. *Energy Fuels* **2012**, *26*, 6627–6639. [CrossRef]
48. Mishra, D.P.; Patyal, A. Effects of Initial Droplet Diameter and Pressure on Burning of ATF Gel Propellant Droplets. *Fuel* **2012**, *95*, 226–233. [CrossRef]

49. Manzhai, V.N.; Fufaeva, M.S.; Egorova, L.A. Fuel Briquettes Based on Finely Dispersed Coke Particles and Polyvinyl Alcohol Cryogels. *Solid Fuel Chem.* **2013**, *47*, 43–46. [CrossRef]
50. Ciezki, H.K.; Naumann, K.W. Some Aspects on Safety and Environmental Impact of the German Green Gel Propulsion Technology. *Propellants Explos. Pyrotech.* **2016**, *41*, 539–547. [CrossRef]
51. Kukushkin, V.; Ivanchenko, A. The Pasty Propellant Rocket Engines Development. In Proceedings of the 29th Joint Propulsion Conference and Exhibit, Monterey, CA, USA, 28–30 June 1993; p. AIAA 1993-1754.
52. Palaszewski, B.; Powell, R. Launch Vehicle Performance Using Metallized Propellants. *J. Propuls. Power* **1994**, *10*, 828–833. [CrossRef]
53. Moghaddam, A.S.; Rezaei, M.R.; Tavangar, S. Experimental Investigation of Characteristic Length Influence on a Combustion Chamber Performance with Liquid and Gelled UDMH/IRFNA Bi-Propellants. *Propellants Explos. Pyrotech.* **2019**, *44*, 1154–1159. [CrossRef]
54. Rapp, D.C.; Zurawski, R.L. Characterization of Aluminum/RP-1 Gel Propellant Properties. In Proceedings of the 24th Joint Propulsion Conference, Boston, MA, USA, 11–13 July 1988; American Institute of Aeronautics and Astronautics: Reston, VA, USA; p. 25.
55. Kajiwara, K. Structure of Gels. In *Gels Handbook*; Osada, Y., Kajiwara, K., Fushimi, T., Irasa, O., Hirokawa, Y., Matsunaga, T., Shimomura, T., Wang, L., Ishida, H., Eds.; Academic Press: London, UK, 2001; Volume 1, pp. 122–171.
56. Tepper, F.; Lerner, M.I.; Ginley, D.S. Metallic Nanopowders: An Overview. In *Dekker Encyclopedia of Nanoscience and Nanotechnology*; CRC Press: Boca Raton, FL, USA, 2004; pp. 1921–1934.
57. Kaledin, L.; Tepper, F. Metallic Nanopowders: Rocket Propulsion. In *Dekker Encyclopedia of Nanoscience and Nanotechnology*; CRC Press: Boca Raton, FL, USA, 2011; pp. 2165–2175.
58. Ray, A.B. Process of Producing Solidified Fuel. U.S. Patent 2,046,209, 30 June 1936.
59. Long, C.G. Reinforced Gelled Propellants. U.S. Patent 3,035,950, 22 May 1962.
60. Longwell, J.P.; Bieber, H. Gelled Solid Rocket Propellant Reinforced with Unoriented Microfibers. U.S. Patent 3,369,943, 20 February 1968.
61. Tarpley, W.B. Thixotropic Gel Fuels. U.S. Patent 4,156,594, 29 May 1979.
62. Tarpley, W.B. Thixotropic Oxidizer Propellant Mixtures. U.S. Patent 3,449,178, 10 June 1969.
63. Tarpley, W.B. Thixotropic Organic Liquid Propellant Compositions with Solid Storage Characteristics. U.S. Patent 3,471,344, 7 October 1969.
64. Hodge, K.F.; Crofoot, T.A.; Nelson, S. Gelled Propellants for Tactical Missile Applications. In Proceedings of the 35th Joint Propulsion Conference and Exhibit, Los Angeles, CA, USA, 20–24 June 1999; American Institute of Aeronautics and Astronautics: Reston, Virigina, 1999; pp. 1–7.
65. Ramasubramanian, C.; Notaro, V.; Lee, J.G. Characterization of Near-Field Spray of Nongelled- and Gelled-Impinging Doublets at High Pressure. *J. Propuls. Power* **2015**, *31*, 1642–1652. [CrossRef]
66. Rahimi, S.; Natan, B. Thixotropic Effect of Inorganic Gel Fuels. *J. Propuls. Power* **2000**, *16*, 1182–1184. [CrossRef]
67. Yoon, C.; Heister, S.D.; Campanella, O.E. Plain-Orifice Gelled Propellant Flow Characteristics with Rheological Hysteresis. In Proceedings of the 48th AIAA/ASME/SAE/ASEE Joint Propulsion Conference and Exhibit, Atlanta, GA, USA, 30 July–1 August 2012; p. AIAA-2012-4135.
68. Lee, I.; Koo, J. Break-up Characteristics of Gelled Propellant Simulants with Various Gelling Agent Contents. *J. Therm. Sci.* **2010**, *19*, 545–552. [CrossRef]
69. Jejurkar, S.Y.; Yadav, G.; Mishra, D.P. Characterization of Impinging Jet Sprays of Gelled Propellants Loaded with Nanoparticles in the Impact Wave Regime. *Fuel* **2018**, *228*, 10–22. [CrossRef]
70. Jejurkar, S.Y.; Yadav, G.; Mishra, D.P. Visualizations of Sheet Breakup of Non-Newtonian Gels Loaded with Nanoparticles. *Int. J. Multiph. Flow* **2018**, *100*, 57–76. [CrossRef]
71. Mordosky, J.W.; Zhang, B.Q.; Kuo, K.K.; Tepper, F.; Kaledin, L.A. Spray Combustion of Gelled RP-1 Propellants Containing Nano-Sized Aluminum Particles in Rocket Engine Conditions. In Proceedings of the 37th Joint Propulsion Conference and Exhibit, Salt Lake City, UT, USA, 8–11 July 2001; p. AIAA-2001-3274.
72. McKinney, C.D.; Tarpley, W.B. *Gelling of Liquid Hydrogen*; NASA-CR-54967; NASA: Washington, DC, USA, 1966.
73. Staikovich, J.; Adams, S.; Palaszewski, B. Nanoparticulate Gellants for Metallized Gelled Liquid Hydrogen with Aluminum. In Proceedings of the 32nd Joint Propulsion Conference and Exhibit, Lake Buena Vista, FL, USA, 1–3 July 1996; p. AIAA-96-3234.
74. Wong, W.; Starkovich, J.; Adams, S.; Palaszewski, B. Cryogenic Gellant and Fuel Formulation for Metallized Gelled Propellants: Hydrocarbons and Hydrogen with Aluminum. In Proceedings of the 30th Joint Propulsion Conference cosponsored by AIAA, ASME, SAE, and ASEE, Indianapolis, IN, USA, 27–29 June 1994; p. AIAA-94-3175.
75. Rahimi, S.; Hasan, D.; Peretz, A. Development of Laboratory-Scale Gel-Propulsion Technology. *J. Propuls. Power* **2004**, *20*, 93–100. [CrossRef]
76. Rahimi, S.; Peretz, A.; Natan, B. Rheological Matching of Gel Propellants. *J. Propuls. Power* **2010**, *26*, 376–378. [CrossRef]
77. Rahimi, S.; Peretz, A.; Natan, B. On Shear Rheology of Gel Propellants. *Propellants, Explos. Pyrotech.* **2007**, *32*, 165–174. [CrossRef]
78. Solomon, Y.; DeFini, S.J.; Pourpoint, T.L.; Anderson, W.E. Gelled Monomethyl Hydrazine Hypergolic Droplet Investigation. *J. Propuls. Power* **2013**, *29*, 79–86. [CrossRef]
79. Cho, K.Y.; Pourpoint, T.L.; Son, S.F.; Lucht, R.P. Microexplosion Investigation of Monomethylhydrazine Gelled Droplet with OH Planar Laser-Induced Fluorescence. *J. Propuls. Power* **2013**, *29*, 1303–1310. [CrossRef]

80. Thynell, S.T.; Adair, J.H.; Goddard, I.W.A.; Hanson, R.K.; Law, C.K.; Lee, J.; Rabitz, H.A.; Risha, G.A.; Williams, F.A.; Yang, V.; et al. *Spray and Combustion of Gelled Hypergolic Propellants (Final Report)*; Pennsylvania State University: University Park, PA, USA, 2014; pp. 1–347.
81. Padwal, M.B.; Mishra, D.P. Effect of Air Injection Configuration on the Atomization of Gelled Jet A1 Fuel in an Air-Assist Internally Mixed Atomizer. *At. Sprays* **2013**, *23*, 327–341. [CrossRef]
82. Padwal, M.B.; Mishra, D.P. Experimental Characterization of Gelled Jet A1 Spray Flames. *Flow Turbul. Combust.* **2016**, *97*, 295–337. [CrossRef]
83. Jyoti, B.V.S.; Baek, S.W. Formulation and Comparative Study of Rheological Properties of Loaded and Unloaded Ethanol-Based Gel Propellants. *J. Energ. Mater.* **2015**, *33*, 125–139. [CrossRef]
84. Jyoti, B.V.S.; Naseem, M.S.; Baek, S.W. Hypergolicity and Ignition Delay Study of Pure and Energized Ethanol Gel Fuel with Hydrogen Peroxide. *Combust. Flame* **2017**, *176*, 318–325. [CrossRef]
85. Von Kampen, J.; Madlener, K.; Ciezki, H.K. Characteristic Flow and Spray Properties of Gelled Fuels with Regard to the Impinging Jet Injector Type. In Proceedings of the Collection of Technical Papers—AIAA/ASME/SAE/ASEE 42nd Joint Propulsion Conference, Sacramento, CA, USA, 9–12 July 2006; Volume 4, pp. 2639–2650.
86. Jyoti, B.V.S.; Naseem, M.S.; Baek, S.W.; Lee, H.J.; Cho, S.J. Hypergolicity and Ignition Delay Study of Gelled Ethanolamine Fuel. *Combust. Flame* **2017**, *183*, 102–112. [CrossRef]
87. Weiser, V.; Gläser, S.; Kelzenberg, S.; Eisenreich, N.; Roth, E. Investigations on the Droplet Combustion of Gelled Mono-and Bipropellants. In Proceedings of the 41st AIAA/ASME/SAE/ASEE Joint Propulsion Conference and Exhibit, Tucson, AZ, USA, 10–13 July 2005; p. AIAA-2005-4474.
88. Boyer, E.; Risha, G.A.; Kuo, K.K.; Devendorf, T.E. Combustion Characteristics of Non-Toxic Non-Hypergolic Bipropellants. In Proceedings of the 38th AIAA/ASME/SAE/ASEE Joint Propulsion Conference and Exhibit, Indianapolis, IN, USA, 7–10 July 2002; p. AIAA-2002-4297.
89. Tarpley, W.B., Jr. Thixotropic Liquid Propellant Compositions with Solid Storage Characteristics. U.S. Patent 3,470,040, 30 September 1969.
90. Coguill, S.L. *Synthesis of Highly Loaded Gelled Propellants*; Technical Report; Resodyn Corp.: Butte, MT, USA, 2009.
91. Baek, G.; Kim, C. Rheological Properties of Carbopol Containing Nanoparticles. *J. Rheol.* **2011**, *55*, 313–330. [CrossRef]
92. Green, J.; Rapp, D.; Roncace, J. Flow Visualization of a Rocket Injector Spray Using Gelled Propellant Simulants. In Proceedings of the 27th Joint Propulsion Conference, Sacramento, CA, USA, 24–26 June 1991; pp. 1–16.
93. Terech, P.; Weiss, R.G. Low Molecular Mass Gelators of Organic Liquids and the Properties of Their Gels. *Chem. Rev.* **1997**, *97*, 3133–3160. [CrossRef]
94. Varma, M.; Pein, R. Optimisation of Processing Conditions for Gel Propellant Production. *Int. J. Energ. Mater. Chem. Propuls.* **2009**, *8*, 501–513. [CrossRef]
95. Padwal, M.B.; Mishra, D.P. Characteristics of Gelled Jet A1 Sprays Formed by Internal Impingement of Micro Air Jets. *Fuel* **2016**, *185*, 599–611. [CrossRef]
96. Goldin, M.; Pfeffer, R.; Shinnar, R. Break-up of a Capillary Jet of a Non-Newtonian Fluid Having a Yield Stress. *Chem. Eng. J.* **1972**, *4*, 8–20. [CrossRef]
97. Chernov, V.; Natan, B. Effect of Periodic Disturbances on Non-Newtonian Fluid Sprays. *At. Sprays* **2008**, *18*, 723–738.
98. Arnold, R.; Anderson, W.E. Droplet Burning of JP-8/Silica Gels. In Proceedings of the 48th AIAA Aerospace Sciences Meeting Including the New Horizons Forum and Aerospace Exposition, Orlando, FL, USA, 4–7 January 2010. [CrossRef]
99. Ciezki, H.K.; Robers, A.; Schneider, G. Investigation of the Spray Behavior of Gelled Jet A-1 Fuels Using an Air Blast and an Impinging Jet Atomizer. In Proceedings of the 38th AIAA/ASME/SAE/ASEE Joint Propulsion Conference and Exhibit, Indianapolis, IN, USA, 7–10 July 2002.
100. Negri, M.; Ciezki, H.K. Combustion of Gelled Propellants Containing Microsized and Nanosized Aluminum Particles. *J. Propuls. Power* **2015**, *31*, 400–407. [CrossRef]
101. Olson, W.T.; Breitwieser, R. *NACA Research on Slurry Fuels through 1954*; NACA-RM-E55B14 (21 April 1955); NACA: Washington, DC, USA, 1955; pp. 1–31.
102. Gany, A. Comprehensive Consideration of Boron Combustion in Air Breathing Combustion. In Proceedings of the 42nd AIAA/ASME/SAE/ASEE Joint Propulsion Conference & Exhibit, Sacramento, CA, USA, 9–12 July 2006; p. AIAA-2006-4567.
103. Haddad, A.; Natan, B.; Arieli, R. The Performance of a Boron-Loaded Gel-Fuel Ramjet. *Prog. Propuls. Phys.* **2011**, *2*, 499–518. [CrossRef]
104. Natan, B.; Gany, A. Combustion Characteristics of a Boron-Fueled Solid Fuel Ramjet with Aft-Burner. *J. Propuls. Power* **1993**, *9*, 694–701. [CrossRef]
105. Munjal, N.L.; Gupta, B.L.; Varma, M. Preparative and Mechanistic Studies on Unsymmetrical Dimethyl Hydrazine-Red Fuming Nitric Acid Liquid Propellant Gels. *Propellants, Explos. Pyrotech.* **1985**, *10*, 111–117. [CrossRef]
106. Santos, P.H.S.; Arnold, R.; Anderson, W.E.; Carignano, M.A.; Campanella, O.H. Characterization of JP-8/SiO 2 and RP-1/SiO 2 Gels. *Eng. Lett.* **2010**, *18*, 41–48.
107. Brändle, R.; Liebl, J.; Pein, R. Preparation and Characterization of Gelled Fuels. *Int. J. Energ. Mater. Chem. Propuls.* **2007**, *6*, 307–321. [CrossRef]
108. Kim, H.; Ko, T.; Kim, S.; Yoon, W. Spray Characteristics of Aluminized-Gel Fuels Sprayed Using Pressure-Swirl Atomizer. *J. Non-Newton. Fluid Mech.* **2017**, *249*, 36–47. [CrossRef]

109. Vershinina, K.Y.; Glushkov, D.O.; Nigay, A.G.; Yanovsky, V.A.; Yashutina, O.S. Oil-Filled Cryogels: New Approach for Storage and Utilization of Liquid Combustible Wastes. *Ind. Eng. Chem. Res.* **2019**, *58*, 6830–6840. [CrossRef]
110. Glushkov, D.O.; Kuznetsov, G.V.; Nigay, A.G.; Yanovsky, V.A.; Yashutina, O.S. Ignition Mechanism and Characteristics of Gel Fuels Based on Oil-Free and Oil-Filled Cryogels with Fine Coal Particles. *Powder Technol.* **2020**, *360*, 65–79. [CrossRef]
111. Kirsanov, E.A.; Matveenko, V.N. *Non-Newtonian Flow of Dispersed, Polymer and Liquid Crystal Systems. Structural Approach*; Technosphere: Moscow, Russia, 2016.
112. Solomon, Y.; Natan, B. Experimental Investigation of the Combustion of Organic-Gellant-Based Gel Fuel Droplets. *Combust. Sci. Technol.* **2006**, *178*, 1185–1199. [CrossRef]
113. Padwal, M.B.; Mishra, D.P. Interactions among Synthesis, Rheology, and Atomization of a Gelled Propellant. *Rheol. Acta* **2016**, *55*, 177–186. [CrossRef]
114. Rahimi, S.; Natan, B. Atomization of Gel Propellants through an Air-Blast Triplet Atomizer. *At. Sprays* **2006**, *16*, 379–400. [CrossRef]
115. Solomon, Y.; Natan, B.; Cohen, Y. Combustion of Gel Fuels Based on Organic Gellants. *Combust. Flame* **2009**, *156*, 261–268. [CrossRef]
116. Herschel, W.H.; Bulkley, R. Konsistenzmessungen von Gummi-Benzollösungen. *Kolloid-Zeitschrift* **1926**, *39*, 291–300. [CrossRef]
117. Casson, N. A Flow Equation for Pigment-Oil Suspensions of the Printing Ink Type. In *Rheology of Disperse Systems*; Mill, C.C., Ed.; Pergamon Press: London, UK, 1959; pp. 84–104.
118. Bingham, E.C. *Fluidity and Plasticity*; McGraw-Hill: New York, NY, USA, 1922; 503p.
119. Ostwald, W.; Aueibach, R. Ueber Die Viscositat Kolloider Losungen Im Struktur-, Laminar- Und Turbulenz-Gebiet. *Kolloid Zeitschrift* **1926**, *38*, 261–264. [CrossRef]
120. Carreau, P.J. Rheological Equations from Molecular Network Theories. *Trans. Soc. Rheol.* **1972**, *16*, 99–127. [CrossRef]
121. Yasuda, K.; Armstrong, R.C.; Cohen, R.E. Shear Flow Properties of Concentrated Solutions of Linear and Star Branched Polystyrenes. *Rheol. Acta* **1981**, *20*, 163–178. [CrossRef]
122. Yang, L.J.; Fu, Q.F.; Zhang, W.; Du, M.L.; Tong, M.X. Spray Characteristics of Gelled Propellants in Novel Impinging Jet Injector. *J. Propuls. Power* **2013**, *29*, 104–113. [CrossRef]
123. Madlener, K.; Frey, B.; Ciezki, H.K. Generalized Reynolds Number for Non-Newtonian Fluids. *Prog. Propuls. Phys.* **2009**, *1*, 237–250. [CrossRef]
124. Yarin, A.L.; Zussman, E.; Theron, A.; Rahimi, S.; Sobe, Z.; Hasan, D. Elongational Behavior of Gelled Propellant Simulants. *J. Rheol.* **2004**, *48*, 101–116. [CrossRef]
125. Macosko, C.W. *Rheology: Principles, Measurements and Applications*; Willey-VCH, Inc.: New York, NY, USA, 1994.
126. Nijenhuis, K.; McKinley, G.; Spiegelberg, S.; Barnes, H.; Aksel, N.; Heymann, L.; Odell, J. Non-Newtonian Flows. In *Springer Handbook of Experimental Fluid Mechanics*; Springer: Berlin/Heidelberg, Germany, 2007; pp. 619–743. [CrossRef]
127. Dzuy, N.Q.; Boger, D.V. Yield Stress Measurements for Concentrated Suspension. *J. Rheol.* **1983**, *27*, 321–349. [CrossRef]
128. Schram, G. *Fundamentals of Practical Rheology and Rheometry*; KolosS: Moscow, Russia, 2003.
129. Cheng, D.C.H. Yield Stress: A Time-Dependent Property and How to Measure It. *Rheol. Acta* **1986**, *25*, 542–554. [CrossRef]
130. Office of Advanced Research and Technology. *System Analysis of Gelled Space-Storable Propellants*; Aerojet Liquid Rocket Company: Washington, DC, USA, Contract NAS7-473 SA-1. *Summ. Rep.* 1038-04F, July 1970.
131. Kim, J.Y.; Song, J.Y.; Lee, E.J.; Park, S.K. Rheological Properties and Microstructures of Carbopol Gel Network System. *Colloid Polym. Sci.* **2003**, *281*, 614–623. [CrossRef]
132. Gupta, B.L.; Varma, M.; Goel, S.B. Rheological Characterization of Fuming Nitric Acid Gel. *Propellants, Explos. Pyrotech.* **1986**, *11*, 85–90. [CrossRef]
133. Dennis, J.D.; Kubal, T.D.; Campanella, O.; Son, S.F.; Pourpoint, T.L. Rheological Characterization of Monomethylhydrazine Gels. *J. Propuls. Power* **2013**, *29*, 313–320. [CrossRef]
134. Jyoti, B.V.S.; Baek, S.W. Rheological Characterization of Ethanolamine Gel Propellants. *J. Energ. Mater.* **2016**, *34*, 260–278. [CrossRef]
135. Teipel, U.; Forter-Barth, U. Rheological Behavior of Nitromethane Gelled with Nanoparticles. *J. Propuls. Power* **2005**, *21*, 40–43. [CrossRef]
136. Mallory, J.; DeFini, S.J.; Sojka, P. Formulation of Gelled Propellant Simulants. In Proceedings of the 46th AIAA/ASME/SAE/ASEE Joint Propulsion Conference & Exhibit, Nashville, TN, USA, 25–28 July 2010.
137. Mallory, J.A.; Sojka, P.E. Impinging JET Structure and Breakup Using Gelled Propellant Simulant. In Proceedings of the 24th European Conference on Liquid Atomization and Spray Systems (ILASS-Europe 2011), Estoril, Portugal, 5–7 September 2011.
138. Altunina, L.K.; Manzhai, V.N.; Fufaeva, M.S. Mechanical, Thermal, and Physical Properties of Cryogels and Foamed Cryogels Produced from Aqueous Solutions of Polyvinyl Alcohol. *J. Appl. Chem.* **2006**, *79*, 1689–1692.
139. Santos, P.H.S.; Carignano, M.A.; Campanella, O.H. Investigation of Thixotropy in Gelled Jet Propellan. In Proceedings of the World Congress on Engineering and Computer Science, San Francisco, CA, USA, 20–22 October2010.
140. Santos, P.H.S.; Carignano, M.A.; Campanella, O.H. Qualitative Study of Thixotropy in Gelled Hydrocarbon Fuels. *Eng. Lett.* **2011**, *19*, 13–19.
141. Benmouffok-Benbelkacem, G.; Caton, F.; Baravian, C.; Skali-Lami, S. Non-Linear Viscoelasticity and Temporal Behavior of Typical Yield Stress Fluids: Carbopol, Xanthan and Ketchup. *Rheol. Acta* **2010**, *49*, 305–314. [CrossRef]
142. Chen, A.; Guan, X.; Li, X.; Zhang, B.; Zhang, B.; Song, J. Preparation and Characterization of Metalized JP-10 Gel Propellants with Excellent Thixotropic Performance. *Propellants, Explos. Pyrotech.* **2017**, *42*, 1007–1013. [CrossRef]

143. von Kampen, J.; Alberio, F.; Ciezki, H.K. Spray and Combustion Characteristics of Aluminized Gelled Fuels with an Impinging Jet Injector. *Aerosp. Sci. Technol.* **2007**, *11*, 77–83. [CrossRef]
144. Caldas Pinto, P.; Hopfe, N.; Ramsel, J.; Naumann, W.; Thumann, A.; Kurth, G. Scalability of Gelled Propellant Rocket Motors. In Proceedings of the 7th European Conference for Aeronautics and Space Sciences (EUCASS), Milano, Italy, 3–7 July 2017; p. 9.
145. Natan, B.; Solomon, Y.; Perteghella, V. Hypergolic Ignition by Fuel Gellation and Suspension of Reactive or Catalyst Particles. *J. Propuls. Power* **2011**, *27*, 1145–1148. [CrossRef]
146. Fakhri, S.; Lee, J.G.; Yetter, R.A. Effect of Nozzle Geometry on the Atomization and Spray Characteristics of Gelled-Propellant Simulants Formed by Two Impinging Jets. *At. Sprays* **2010**, *20*, 1033–1046. [CrossRef]
147. Dennis, J.D.; Yoon, C.; Santos, P.H.; Mallory, J.A.; Fineman, C.N.; Pourpoint, T.L.; Son, S.F.; Heister, S.D.; Sojka, P.E.; Campanella, O.H. Characterization of Gelling Systems for Development of Hypergolic Gels. In Proceedings of the 4th European Conference for Aerospace Sciences (EUCASS), Saint Petersburg, Russia, 4–8 July 2011; pp. 1–15.
148. Jayaprakash, N.; Chakravarthy, S.R. Impingement Atomization of Gel Fuels. In Proceedings of the 41st Aerospace Sciences Meeting and Exhibit, Reno, NV, USA, 6–9 January 2003.
149. Baek, G.; Kim, S.; Han, J.; Kim, C. Atomization Characteristics of Impinging Jets of Gel Material Containing Nanoparticles. *J. Non-Newton. Fluid Mech.* **2011**, *166*, 1272–1285. [CrossRef]
150. Yang, L.J.; Fu, Q.F.; Qu, Y.Y.; Gu, B.; Zhang, M.Z. Breakup of a Power-Law Liquid Sheet Formed by an Impinging Jet Injector. *Int. J. Multiph. Flow* **2012**, *39*, 37–44. [CrossRef]
151. Ma, Y.C.; Bai, F.Q.; Chang, Q.; Yi, J.M.; Jiao, K.; Du, Q. An Experimental Study on the Atomization Characteristics of Impinging Jets of Power Law Fluid. *J. Non-Newton. Fluid Mech.* **2015**, *217*, 49–57. [CrossRef]
152. Ma, D.J.; Chen, X.D.; Khare, P.; Yang, V. Atomization Patterns and Breakup Characteristics of Liquid Sheets Formed by Two Impinging Jets. In Proceedings of the 49th AIAA Aerospace Sciences Meeting including the New Horizons Forum and Aerospace Exposition, Orlando, FL, USA, 4–7 January 2011. [CrossRef]
153. Desyatkov, A.; Madlener, K.; Ciezki, H.K.; Natan, B. Experimental Investigation of Gelled Fuel with High Vapor Pressure Species and Metal Additives. In Proceedings of the 44th AIAA/ASME/SAE/ASEE Joint Propulsion Conference & Exhibit, Hartford, CT, USA, 21–23 July 2008. [CrossRef]
154. Dennis, J.D.; Pourpoint, T.L.; Son, S.F. Ignition of Gelled Monomethylhydrazine and Red Fuming Nitric Acid in an Impinging Jet Apparatus. In Proceedings of the 47th AIAA/ASME/SAE/ASEE Joint Propulsion Conference & Exhibit, San Diego, CA, USA, 31 July–3 August 2011. [CrossRef]
155. James, M.D.; Kubal, T.D.; Son, S.F.; Anderson, W.E.; Pourpoint, T.L. Calibration of an Impinging Jet Injector Suitable for Liquid and Gelled Hypergolic Propellants. In Proceedings of the 45th AIAA/ASME/SAE/ASEE Joint Propulsion Conference & Exhibit, Denver, CO, USA, 2–5 August 2009. [CrossRef]
156. Kampen, J.; Ciezki, H.K.; Tiedt, T.; Madlener, K. Some Aspects of the Atomization Behavior of Newtonian and of Shear-Thinning Gelled Non-Newtonian Fluids with an Impinging Jet Injector. In Proceedings of the 42nd AIAA Joint Propulsion Conference and Exhibition, Sacramento, CA, USA, 9–12 July 2006.
157. Palaszewski, B.; Zakany, J.S. Metallized Gelled Propellants: Oxygen /RP-1 /Aluminum Rocket Combustion Experiments. In Proceedings of the 31st Joint Propulsion Conference and Exhibit, San Diego, CA, USA, 10–12 July 1995.
158. Negri, M.; Redaelli, M.; Ciezki, H. Recent Results on Thread Formation with an Impinging Jet Injector. In Proceedings of the 48th AIAA/ASME/SAE/ASEE Joint Propulsion Conference & Exhibit, Atlanta, GA, USA, 30 July–1 August 2012.
159. Connell, T.L.; Risha, G.A.; Yetter, R.A.; Natan, B. Hypergolic Ignition of Hydrogen Peroxide/Gel Fuel Impinging Jets. *J. Propuls. Power* **2018**, *34*, 182–188. [CrossRef]
160. Bailardi, G.; Negri, M.; Ciezki, H.K. Several Aspects of the Atomization Behavior of Various Newtonian Fluids with a Like-on-like Impinging Jet Injector. In Proceedings of the 23rd Annual Conference on Liquid atomization and Spray Systems (ILASS-Europe 2010), Brno, Czech Republic, 6–9 September 2010; pp. 1–9.
161. James, M.D. Liquid and Gelled Sprays for Mixing Hypergolic Propellants Using an Impinging Jet Injection System. Master's Thesis, Purdue University, West Lafayette, IN, USA, 2010.
162. Ryan, H.M.; Anderson, W.E.; Pal, S.; Santoro, R.J. Atomization Characteristics of Impinging Liquid Jets. *J. Propuls. Power* **1995**, *11*, 135–145. [CrossRef]
163. Rodrigues, N.S.; Sojka, P.E. Spatially Resolved Gelled Propellant Spray Characteristics of Impinging Jets. In Proceedings of the Proceedings of the ASME 2015 International Mechanical Engineering Congress and Exposition. Volume 1: Advances in Aerospace Technology, Houston, TX, USA, 13–19 November 2015; American Society of Mechanical Engineers: New York, NY, USA, 2015; p. V001T01A036.
164. Rahimi, S.; Natan, B. Atomization Characteristics of Gel Fuels. In Proceedings of the 34th AIAA/ASME/SAE/ASEE Joint Propulsion Conference and Exhibit, Cleveland, OH, USA, 3–15 July 1998; p. 3830.
165. Connell, T.L.; Risha, G.A.; Yetter, R.A.; Natan, B. Ignition of Hydrogen Peroxide with Gel Hydrocarbon Fuels. *J. Propuls. Power* **2018**, *34*, 170–181. [CrossRef]
166. Sutherland, J.J.; Sojka, P.E.; Plesniak, M.W. Ligament-Controlled Effervescent Atomization. *At. Sprays* **1997**, *7*, 383–406. [CrossRef]
167. Zhao, F.; Qin, L.Z.; Fu, Q.F.; Mo, C.J.; Yang, L.J. Spray Characteristics of Elliptical Power-Law Fluid-Impinging Jets. *J. Fluids Eng. Trans. ASME* **2017**, *139*, 40. [CrossRef]

168. Fakhri, S.; Lee, J.G.; Yetter, R.A. Atomization and Spray Characteristics of Gelled-Propellant Simulants Formed by Two Impinging Jets. In Proceedings of the 45th AIAA/ASME/SAE/ASEE Joint Propulsion Conference and Exhibit, Denver, CL, USA, 2–5 August 2009; pp. 1–11.
169. Yang, L.-J.; Fu, Q.-F.; Zhang, W.; Du, M.-L.; Tong, M.-X. Atomization of Gelled Propellants from Swirl Injectors with Leaf Spring in Swirl Chamber. *At. Sprays* **2011**, *21*, 949–969. [CrossRef]
170. Chojnacki, K.; Feikema, D.; Chojnacki, K.; Feikema, D. Study of Non-Newtonian Liquid Sheets Formed by Impinging Jets. In Proceedings of the 33rd Joint Propulsion Conference and Exhibit, Seattle, WA, USA, 6–9 July 1997; pp. 1–11.
171. Mallory, J.A.; Sojka, P.E. A Study of Gelled Propellant Simulants Using Impinging Jet Injectors. In Proceedings of the 12th Triennial International Conference on Liquid Atomization and Spray Systems (ICLASS 2012), Heidelberg, Germany, 2–6 September 2012; pp. 1–8.
172. Yang, L.J.; Fu, Q.F.; Qu, Y.Y.; Zhang, W.; Du, M.L.; Xu, B.R. Spray Characteristics of Gelled Propellants in Swirl Injectors. *Fuel* **2012**, *97*, 253–261. [CrossRef]
173. Thompson, J.C.; Rothstein, J.P. The Atomization of Viscoelastic Fluids in Flat-Fan and Hollow-Cone Spray Nozzles. *J. Non-Newton. Fluid Mech.* **2007**, *147*, 11–22. [CrossRef]
174. McCabe, J.D.; Coil, M.A. A Graphical Spray Analysis Method for Gel Spray Characterization. In Proceedings of the 46th AIAA/ASME/SAE/ASEE Joint Propulsion Conference & Exhibit, Nashville, TN, USA, 25–28 July 2010. [CrossRef]
175. Buckner, H.N.; Sojka, P.E. Effervescent Atomization of High Viscosity Fluids: Part I. Newtonian Liquids. *At. Sprays* **1991**, *1*, 239–252. [CrossRef]
176. Sutherland, J.J.; Sojka, P.E.; Plesniak, M.W. Entrainment by Ligament-Controlled Effervescent Atomizer-Produced Sprays. *Int. J. Multiph. Flow* **1997**, *23*, 865–884. [CrossRef]
177. Mallory, J.A. Jet Impingement and Primary Atomization of Non-Newtonian Liquids. Ph.D. Thesis, Purdue University, West Lafayette, IN, USA, 2012.
178. Santangelo, P.J.; Sojka, P.E. A Holographic Investigation of the Near-Nozzle Structure of an Effervescent Atomizer-Produced Spray. *At. Sprays* **1995**, *5*, 137–155. [CrossRef]
179. Padwal, M.B.; Mishra, D.P. Severity of Operating Parameters and Breakup Modes in the Internally Impinging Atomizer. In Proceedings of the 27th European Conference on Liquid Atomization and Spray Systems (ILASS-Europe 2016), Brighton, UK, 4–7 September 2016; p. AA-07.
180. Glushkov, D.O.; Nigay, A.G.; Yashutina, O.S. The Gel Fuel Ignition at Local Conductive Heating. *Int. J. Heat Mass Transf.* **2018**, *127*, 1203–1214. [CrossRef]
181. Gallier, S.; Ferrand, A.; Plaud, M. Three-Dimensional Simulations of Ignition of Composite Solid Propellants. *Combust. Flame* **2016**, *173*, 2–15. [CrossRef]
182. Sazhin, S.S.; Shchepakina, E.; Sobolev, V. Order Reduction in Models of Spray Ignition and Combustion. *Combust. Flame* **2018**, *187*, 122–128. [CrossRef]
183. Yang, D.; Xia, Z.; Huang, L.; Ma, L.; Feng, Y.; Xiao, Y. Exprimental Study on the Evaporation Characteristics of the Kerosene Gel Droplet. *Exp. Therm. Fluid Sci.* **2018**, *93*, 171–177. [CrossRef]
184. Ojha, P.K.; Prabhudeva, P.; Karmakar, S.; Maurya, D.; Sivaramakrishna, G. Combustion Characteristics of JP-10 Droplet Loaded with Sub-Micron Boron Particles. *Exp. Therm. Fluid Sci.* **2019**, *109*, 109900. [CrossRef]
185. Javed, I.; Baek, S.W.; Waheed, K. Effects of Dense Concentrations of Aluminum Nanoparticles on the Evaporation Behavior of Kerosene Droplet at Elevated Temperatures: The Phenomenon of Microexplosion. *Exp. Therm. Fluid Sci.* **2014**, *56*, 33–44. [CrossRef]
186. Glushkov, D.O.; Paushkina, K.K.; Pleshko, A.O.; Vysokomorny, V.S. Characteristics of Micro-Explosive Dispersion of Gel Fuel Particles Ignited in a High-Temperature Air Medium. *Fuel* **2022**, *313*, 123024. [CrossRef]
187. Glushkov, D.O.; Paushkina, K.K.; Pleshko, A.O.; Yanovsky, V.A. Ignition and Combustion Behavior of Gel Fuel Particles with Metal and Non-Metal Additives. *Acta Astronaut.* **2023**, *202*, 637–652. [CrossRef]
188. Glushkov, D.O.; Kosintsev, A.G.; Kuznetsov, G.V.; Vysokomorny, V.S. Experimental Research and Numerical Simulation of Gel Fuel Ignition by a Hot Particle. *Fuel* **2021**, *291*, 120172. [CrossRef]
189. Glushkov, D.O.; Kosintsev, A.G.; Kuznetsov, G.V.; Vysokomorny, V.S. Numerical Simulation of Ignition of a Typical Gel Fuel Particle, Based on Organic Polymer Thickener, in a High-Temperature Air Medium. *Acta Astronaut.* **2021**, *178*, 272–284. [CrossRef]
190. Tepper, F.; Lerner, M.I.; Ginley, D.S. Metallic Nanopowders. In *Dekker Encyclopedia of Nanoscience and Nanotechnology*, 3rd ed.; CRC Press: Boca Raton, FL, USA, 2014; pp. 2491–2502, ISBN 9781351238175.
191. Bulushev, D.A.; Yuranov, I.; Lieber, C.M. Metallic Nanopowders: Rocket Propulsion. In *Dekker Encyclopedia of Nanoscience and Nanotechnology*, 3rd ed.; Putyera, K., Schwarz, J.A., Eds.; CRC Press: Boca Raton, FL, USA, 2014; Volume 7, pp. 2503–2512, ISBN 9781351238175.
192. Haddad, B.; Liazid, A.; Ferreira, P. A Multi-Criteria Approach to Rank Renewables for the Algerian Electricity System. *Renew. Energy* **2017**, *107*, 462–472. [CrossRef]

energies

MDPI

Perspective

Circular Economy and Green Chemistry: The Need for Radical Innovative Approaches in the Design for New Products

Amos Ncube [1], Sandile Mtetwa [2], Mahak Bukhari [3], Gabriella Fiorentino [4,*] and Renato Passaro [5]

1 International PhD Programme "Environment, Resources and Sustainable Development", Department of Science and Technology, Parthenope University of Naples, Centro Direzionale–Isola C4, 80143 Naples, Italy
2 Yusuf Hamied Department of Chemistry, University of Cambridge, Lensfield Road, Cambridge CB2 1EW, UK
3 School of Environment, Beijing Normal University, 19 Xinjiekouwai St., Haidian District, Beijing 100875, China
4 ENEA (Italian National Agency for New Technologies, Energy and Sustainable Economic Development), Department for Sustainability, Division Resource Efficiency, Research Centre of Portici, P.le E. Fermi 1, Portici, 80055 Naples, Italy
5 Department of Science and Technology, Parthenope University of Naples, Centro Direzionale–Isola C4, 80143 Naples, Italy
* Correspondence: gabriella.fiorentino@enea.it

Abstract: The idea of a circular economy (CE) has gained ground over the past ten years as a means of addressing sustainable development and getting around the limitations of the current and linear dominant production and consumption patterns. The primary goal of a CE is to encourage the adoption of closing-the-loop production methods to improve resource use efficiency, modify chemical processes, and increase product and material lifespan. According to the 2030 Agenda for Sustainable Development, which focuses on 17 Sustainable Development Goals, 14 of which call for the appropriate application of green chemistry (GC) concepts and patterns, the role that chemistry may play in the shift toward more sustainable models is critical. By serving as the foundation for novel products made from renewable feedstocks and designed to be reused, recycled, or recovered with the associated minimum energy requirements, green and sustainable chemistry could be the key to unlocking the economic potential of the CE toward new product design and ultimately solving waste management problems. The aim of this perspective paper, while using a variety of literature sources, is to essentially capture the main issues associated with the CE and GC paradigms and how these two approaches can merge toward sustainable business models and the production of new materials. This integration focuses on reducing waste, conserving resources, and minimizing negative environmental impacts, while also considering economic viability. However, the obstacles to achieving implementation of the CE and GC principles are investment, environmental education, and legislation. To advance toward the circular economy and green chemistry, international agreements should be reconsidered to provide an appropriate framework, including the creation of incentives for businesses and individuals to adopt circular practices, the establishment of education programs to promote the benefits of circular practices, and the development of regulations to support the transition to sustainable production and consumption patterns.

Keywords: circular economy; green chemistry; energy efficient chemical production; waste management

1. Introduction

To address sustainable development and get around the limitations of the current linear production and consumption patterns, the circular economy (CE) concept has gained traction in recent years [1]. The primary goal of the CE, as stated by the Ellen Macarthur foundation [2], is to promote the adoption of closing-the-loop production patterns to improve resource usage efficiency and longevity. Increased circularity is thought to be a key driver of the Sustainable Development Goals (SDGs) of the United Nations 2030 Agenda for Sustainable Development [3] (Mancini and Raggi, 2021) and the broader objectives of sustainability [4]. Transforming the linear economy, which has been the dominant model since the onset of the industrial revolution, into a circular one is by no means an easy task, as most products we use are designed for a make–dispose culture [5]. This inadvertently results in the depletion of resources and the loss of wealth from waste and, nowadays, the widely accepted solution to dispose products is waste recycling [6]. In the context of our current environmental challenges, recycling will not adequately overcome the sheer amount of waste that the world produces today [7]. The CEs point of departure from the recycling economy is through the incorporation of many closed loops and preventive design mechanisms before the waste is generated and disposed [8]. In fact, the CE concept goes beyond recycling, through waste prevention and reduction, but also technological and organizational innovations across and within value chains [4].

As noted by the United Nations in the 2030 Agenda, 14 out of 17 SDGs call for the appropriate application of green chemistry (GC) concepts and patterns that are considered vital for allowing the shift toward more sustainable approaches [9]. GC is an approach that focuses on product designs and procedures that eliminate or minimize the impact of hazardous chemicals on the environment [10]. GC possesses the potential to reduce the hazardous impact of chemicals on the environment and human health, thereby improving quality of life and the state of the environment [11]. Besides the perceived environmental and social benefits of GC, its adaptation also provides economic benefits [9], such as the reduced use of chemicals, reduced capital used in waste storage, and treatments and environmental compensation payments for the damage caused [12]. The choice of safer and more sustainable chemistries has a significant impact on the product lifecycle, including the potential sustainability of these materials' production, usage, reusing, recycling, and ultimately, the end of its life [13]. Overall, the goal of GC is to establish molecular sustainability by going beyond research in laboratories and expanding the focus toward industry and communities, as a way of responding to current environmental, economic, and societal challenges [14].

Generally, the CE and GC are expected to present new and innovative business opportunities, through (1) better planning of resource use, (2) replacing fossil energies and material resources with renewables, (3) reusing and recycling and, finally, (4) circular governance. The present study focuses on the role played by the circular economy and green chemistry in the needed search for radically innovative approaches to design new products from various sources, including from waste fractions. The aim of this perspective paper, while based on a variety of literature sources, is to essentially capture the main issues associated with the CE and GC paradigms and how these two approaches can merge toward sustainable business models and the production of new materials. Integrating the circular economy and green chemistry into research is important because it leads to the development of sustainable and environmentally friendly production methods. This integration focuses on reducing waste, conserving resources, and minimizing negative environmental impacts, while also considering economic viability. It promotes the use of renewable resources, closed-loop systems, and the development of biodegradable products, which in turn helps to address issues related to resource depletion, pollution, and climate change. Ultimately, the integration of the circular economy and green chemistry within research can help to create a more sustainable future for both the environment and the economy.

1.1. The Circular Economy

The rise of the circular economy concept in Europe is credited to Walter Stahel, who sketched an economic model aimed at closing loops in industrial processes with a visionary outlook on creating jobs, improving economic performance, preserving resources, and preventing waste [15]. Stahel was a member of the Club of Rome, an elite group founded by the entrepreneur Aurelio Peccei (1908–1984), comprising intellectuals, politicians, and business people who hold influential and powerful positions in global affairs and environmental protection. The most known output of the Club of Rome was the support to and the publication of the seminal book "The Limits to Growth". Around the year 2010, Ellen MacArthur was incorporated into the same group and became a member [2], followed by Anders Wijkman, a former member of the European Parliament and the President of the Club of Rome (2012–2018). The latter eventually delivered a report based on models that predicted positive effects on climate, environment, and economy, which resulted in the first EU Circular Economy Package. Since then, the Circular Economy (CE) concept has started receiving increased attention in Europe and has attracted many disciplines, from environmental to economics and social sciences [16,17]. However, much earlier, around the 2000s, China was already playing a central role in theorizing, implementing, and upscaling, embedding CE principles in its vision of an "ecological civilisation" with a focus on industrial innovation [18]. Albeit in parallel, China and Europe have continued to share common perspectives on the CE discourse relating to waste reduction and resource efficiency coupled with increased material circularity [19]. Many national governments and international economic policy bodies are now attempting to develop strategies for the implementation of CE practices at micro, meso, and macro levels [20]. As a result, the CE is virtually appearing everywhere from the Chinese 5-year plans to the European Green Deals, to the formation of non-governmental circular economy networks and in many corporates such as Ikea, iPhone, and Renault, among others [21]. In Europe, the transition to the circular economy follows a bottom-up approach which includes initiatives of environmental organizations, civil society and non-governmental organizations [22]. In contrast, China promotes a top-down approach based on national policies toward environmental responsibility and development strategies [22]. According to Genovese and Pansera [23], bottom-up initiatives can only become effective if they are complemented and coupled with top-down approaches, such as government support through incentives and rewards for positive externalities by companies and organizations. This complementary approach can be considered of great importance for many developing countries in Asia and Africa, which are yet to develop strong policies toward the transition to a circular economy.

1.2. Green Chemistry

The emergence of green chemistry dates to the 1980s, when the negative effects of industrial development became apparent. Concerns grew over environmental damage and occupational hazards that were heavily tied to industrial activities [24]. Especially in the chemical sector, the rise in harmful chemical products at the time demanded a more structural approach to implementing policies targeted at their reduction. This led to the convening of multiple global meetings that resulted in policies and initiatives such as the "Brundtland Report 1985", the "Alternative Synthetic Routes for Pollution Prevention" Initiative in 1991, and the "Agenda 21" of the 1990s. These, together with the reports that followed, had the collective consensus of prioritizing environmental, social awareness, and sustainable development [25]. At its core, green chemistry focuses on using chemical expertise and knowledge to reduce or eliminate the formation and use of hazardous substances throughout the planning, manufacturing, and application of chemicals [26]. For example, traditionally, chemical industry residues are pre-treated to reduce toxicity before disposal. However, such measures are extremely cost extensive. Applying a green chemistry approach when dealing with industrial residues requires the utilization of multi-dimensional aspects, such as greener process design and techniques to limit waste generation during production, being also mindful of the cost and energy involved in the entire process and, lastly, considering the processing and disposal of any left-over residues [27]. In 1998, Paul Anastas and John C. Warner provided an excellent breakdown of the "12 principles of green chemistry" [28], which included elements such as atom efficiency, energy efficiency, degradation, and waste reduction, providing an essential toolbox for sustainable chemical design and synthesis, which halts pollution at a molecular level [29].

Over the years, scientists have been reporting on key trends that align with GC principles. Such areas of research include the following: (i) studying catalytic and biocatalytic processes to produce highly selective, pure chemicals without producing hazardous by-products [30,31], (ii) selection for new, safe, and sustainable source materials, such as biomass [32,33], (iii) creating chemicals that are eco-friendly and less hazardous, (iv) innovative alternatives, toward non-toxic and renewable solvents, such as supercritical fluids, water, and ionic liquids [34], (v) innovative reaction conditions, such as light reacting, ultrasound, and microwave [35] and (vi) novel alternatives for the decontamination of polluted water and air for quality improvement, such as photocatalytic reactions. While it is not possible to use all 12 principles simultaneously in constructing a green chemistry sustainable process, it is possible to apply as many as is feasible at various points of a product's lifecycle [36,37].

2. The Nexus between Green Chemistry and the Circular Economy

It is increasingly clear that green chemistry plays a crucial role in shifting from linear to circular product and material use by allowing the continuing sustainable cycle of resources up to the disposal [38]. Green chemistry can be considered as a strategy to move toward the circular economy, by modifying and making the manufacturing processes of products efficient in preventing and reducing waste. Much research has been dedicated toward the development of synthetic procedures that can align with the 12 guiding principles of GC with a further step to also include concepts of circularity [39,40]. Some of the examples include keeping resources in use for as long as possible [41] as well as the usage of chemicals made from resources that are usually considered as waste to reduce the dependence on depleting non-renewable resources [42]. For the successful implementation of circularity, durable raw materials are an important component; therefore, advances in chemistry are required to increase a product's lifespan for use, reuse, and recycling [43]. Lastly, an important element that helps provide consistency, maintain quality, and offer safety to prevent potentially harmful impacts on the environment and human health is an appropriate level of purity of all materials and products [2]. In this regard, there is a need to create product chemicals and materials that are green and sustainable [44]. Most chemicals are made without keeping toxicity levels in mind and the impact they will have on humans and the environment [45], whereas GC emphasizes that environmental integration and safety measures in chemical systems are insufficient to address societal challenges and promote a transition toward sustainability. Rather, economic, social, political, and technological factors must also be considered [46].

Green chemistry ultimately forms the foundation of a safe and sustainable circular economy [47]. Both GC and the CE adhere to the notion of rethinking the design of chemicals, materials, and products [48], and can be applied together to build a value chain and business model that describes the full range of activities needed to create a product or service [49]. Since GC focuses on molecular synthesis for the design of new products and materials, it can play an essential role in securing circularity in two essential sustainability aspects: (i) designing safer, less toxic chemicals and chemical processes and (ii) using renewable feedstocks. More importantly, sustainable GC could be the key to unlocking the economic potential of the CE in waste management and new product design, by providing the foundation for cutting-edge products that employ renewable feedstocks, are made to be reused, recycled, or recovered, and have minimal energy requirements connected with them. For this reason, the above-mentioned aspects will be explored in deeper detail to showcase the nexus between GC and the CE, whilst demonstrating that GC can be regarded as a necessary tool toward the transition to a CE and a bioeconomy. Figure 1 shows a schematic representation of a system that can integrate the CE and GE in a strategic sustainability framework. The main goal of the CE and GC is, firstly, to correct the linear system that leads to the production of the main product (blue flows) and by-products (residues that need further inputs to be upgraded and become usable, yellow flows) and secondly, to design new and innovative products (green flows feeding back) based on the application of CE and GC approaches to produce value-added products back into the market.

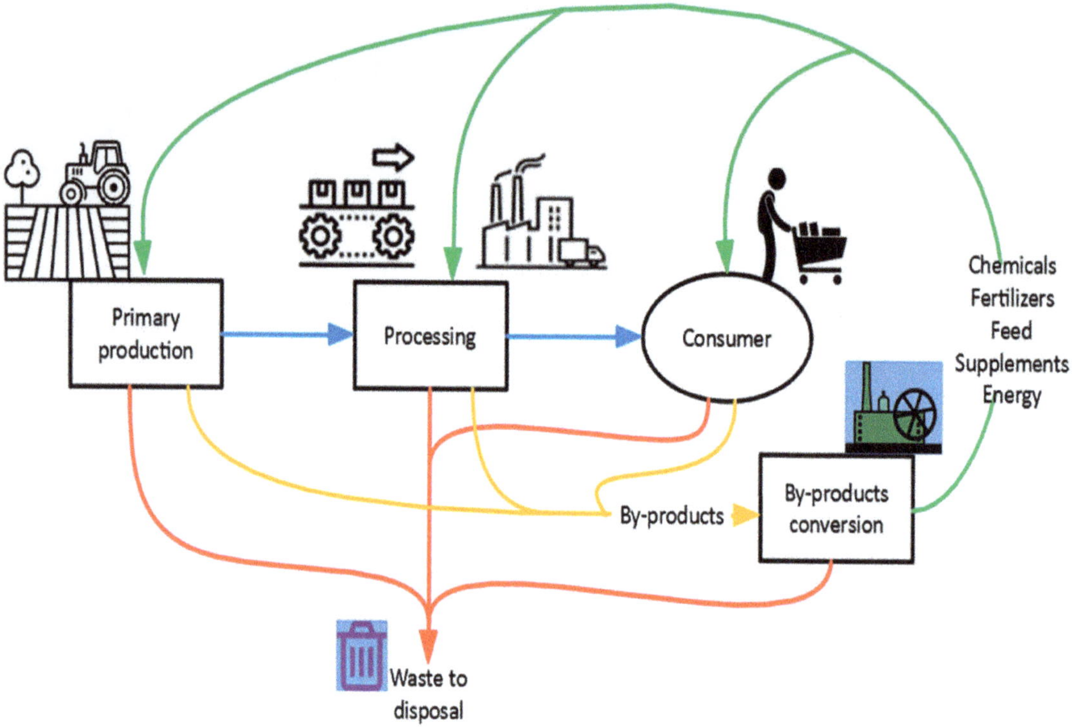

Figure 1. Schematic representation of a system that can integrate the circular economy and green chemistry in a strategic sustainability framework.

2.1. Green Chemistry in Practice

2.1.1. Waste Management Using a GC Approach

Over the next 30 years, the global annual waste is anticipated to increase from 2.01 billion tons in 2016 to 3.4 billion tons [50]. Carelessly discarding materials or products that were of no further use was not a cause of concern in past decades, as natural resources were seemingly unlimited [51]. Such a practice is commonly referred to as a linear process, where entities change or progress straight from one stage to another, i.e., used to discard [52]. He et al. [53] conducted a review of municipal solid waste (MSW) management across 219 countries and concluded that the dominant MSW management practice remains to be landfill disposal in both developed and underdeveloped countries. The decomposition of solid waste in landfilling, if not properly managed, is known to have a significant negative influence on the soil, air, water quality, and human health [54]. In particular, hazardous waste streams have resulted in undesirable bioactivity and ecotoxicity in cases like that of the thalidomide-inducing birth defects in the city of Bhopal (India) or the sinking shallow tube wells being drilled into groundwater, which contaminated the drinking water supplies with arsenic in Bangladesh [55]. Furthermore, MSW is the fourth largest contributor to global greenhouse gas (GHG) emissions, being responsible for 5.5–6.4% of global methane (CH_4) emissions annually [56]. The CE uses the waste hierarchy system to deal with waste management (EU Directive 2008/98/EC). The hierarchy prioritizes handling waste by (1) prevention, (2) preparation for reuse, (3) recycling, (4) energy recovery, and (5) disposal. This strategy helps promote environmental preservation and returns valuable materials to the economy [57]. On the other hand, the first principle of GC is waste minimization, with consideration to each of the three main stages of a material's life, namely production, use, and disposal. While GC emphasizes how waste needs to be prevented at the production stage, there are effective interventions that can be performed at the usage and disposal stages. In this framework, a useful tool is the product life cycle analysis (LCA), which can be used to assess the environmental impacts of a new product from its raw material sourcing (residual feedstock) up to its end-of-life disposal. The application of the LCA can help monitor the progress of the circular economy and green chemistry intervention by providing a comprehensive view of the lifecycle of innovative products, identifying a product's hotspots and by providing strategies for closing the loop and reducing waste. For example, the LCA of new products may reveal that the packaging is the most significant contributor to its environmental impact and this information can further inform the development of more sustainable packaging solutions, such as biodegradable or compostable packaging. Overall, a LCA provides a comprehensive view of a product's environmental impact and provides information for the development of more sustainable solutions within a circular economy and green chemistry context.

2.1.2. Waste Management–Production Stage (Chemical Synthesis Techniques)

From being a major contributor to waste production, solvents and auxiliaries frequently pose health and environmental risks. Using safer chemicals still stands as a core pillar of GC [58]. Principle five of GC considers two main options: the search for synthesis techniques that minimize the use of solvents or do not require them (mechanochemistry), and the replacement of dangerous solvents with safe solvents, such as water, supercritical carbon dioxide, or benign solvents, which are specially created for a particular procedure [59,60]. In 1992, Roger Sheldon first proposed the idea of the E-Factor, or environmental impact factor. This measure aids in calculating the waste produced per kilogram of a product. A prominent and early example of this is the synthesis of ethylene oxide, which was made via a chlorohydrin intermediate. The E-Factor for the entire synthesis was calculated as five, meaning that for every kilogram of the product, 5 kg of waste was to be disposed. This value ignores wastewater that has been contaminated with chlorine by-products, hence the E-Factor is likely to be higher. By swapping chlorine with molecular oxygen in the synthetic procedure, the E-Factor decreased to 0.3 (0.3 kg of waste per 1 kg of product). The formation of wastewater was also eliminated as the non-chlorinated alternative method produced ~16 times less waste than the traditional method. Since then, numerous publications by green chemists' have reported instances where novel procedures or products have allowed many synthetic methods to be compliant with the aspects of sustainability. Of wide interest is the use of water, a benign solvent, which is risk-free and safe from hazards. It can be an effective solvent for large-scale chemical processing. Due to the hydrophobic effect and the fact that many organic compounds do not dissolve in water, the characteristics of water have even improved reaction speeds [61]. However, water is not a universal solvent, and other alternatives need to be considered [62]. Supercritical fluids (SCFs) are an option for conventional (mostly toxic) organic solvents. SCFs are substances that can co-exist as both a gas and liquid above a certain temperature and due to this, they exhibit properties that are desirable for solvents [63]. The toxic organic solvents, which are traditionally used for the decaffeination of green coffee beans, and the perchloroethylene used in dry cleaning have now been replaced by supercritical CO_2. This strategy has resulted in reduced waste generation and cost-effectiveness, since large amounts of solvents were required for the conventional processes [64].

Beyond laboratory research, eliminating waste at the production stage using GC is a widely utilized technique in the industry. The GC approach has brought positive results, as demonstrated by several examples. The varnish and paint industry is already producing solvent-free paints and lacquers. The detractor industry has eliminated all phosphorus-containing detergents [65]. Recently, a non-toxic, vegetable-based hair dye called "Hairprint" from the Warner Babcock Institute for Green Chemistry has been created as an alternative to the poisonous, skin-irritating, and carcinogenic colors [66]. Companies, such as Merck, BMS, and Solutia, among others, have demonstrated how safer chemicals or procedures have aided them in more sustainable manufacturing, which lessens the amount of waste generated, but more so can have the impact of improving the yields of synthetic products. This, in turn, leads to greater economic viability of operations. Elevance CleanTM 1200 is a volatile organic compound (VOC)-free bio-based solvent that is another powerful, environmentally friendly degreasing solvent created by Elevance Renewable Sciences. For its exceptional cleaning capabilities, Elevance CleanTM 1200 won the bio-based product innovation of the year award at the 2015 WBM bio business awards. This non-flammable solvent complies with the numerous stringent environmental laws as it is made from natural oils. The company has since then ventured into personal care products, coatings, and surfactants, which are formulated from non-toxic sources with a smaller environmental footprint [67]. The above practical examples allude that GC aims to lessen, and possibly eliminate, the risk of waste rather than restricting it by regulating exposure to dangerous substances.

2.1.3. Waste Management–Usage Stage (Recycling)

A popular strategy of waste handling at the usage stage is recycling. Several materials successfully undergo recycling, but recycling is particularly important for plastics, since they are a major component of solid waste, being not biodegradable and taking more than a thousand years to degrade into the soil. Moreover, if burned, some plastics are toxic. Therefore, it is essential to recycle them as much as possible. The four ways of recycling plastics are primary, secondary, tertiary, and quaternary. Primary and secondary recycling typically follow a closed-loop system, where mechanical means are used to shred existing material and the pelletized form is remolded to produce the same material again. Quaternary recycling involves the incineration of "hard to recycle" materials, which are usually contaminated and not adequate for primary and secondary recycling. Recovery of energy from incineration is recommended [68]. Although primary and secondary recycling are the current conventional forms of preserving plastic material in circulation, their closed-loop nature limits their potential for creating a sustainable economic value of the process. Additionally, the continued mechanical processing of plastic reduces its robustness with time, hence a material cannot be recycled forever [69]. At the moment, tertiary recycling, namely the process by which a waste plastic material is converted into chemicals and fuels, is on the rise. This aspect of recycling is gaining increased attention worldwide and is often referred to as "chemical recycling" [70].

Chemical recycling is a universal term that describes the breakdown of plastic into its original monomers, sub-polymer fragments, or other small species using heat or chemicals. As the products formed are precursor materials, these can be useful in a myriad of applications, either to re-form the plastic again or to make other polymers for different products [71]. In chemical degradation, the breakdown process can be assisted by GC-derived techniques, such as glycolysis, hydrolysis, methanolysis, and aminolysis, where certain solvents are employed to "cleave" bonds between monomers, thereby fragmenting them. This can be regarded as the reverse of condensation-type reactions. Unfortunately, due to the nature of the chemical recycling process, not all polymers can be recycled in such a way. Only materials that are initially formed through condensation reactions (such as polyesters, polyamides, or polyurethanes) can be chemically depolymerized [72]. However, since most thermosetting plastics, which make up ~87% of the plastics in circulation, are formed from addition polymerization (the opposite of condensation), the technique of using solvents is currently limited in its application [73]. However, it should be noted that polymers from addition polymerization can be broken down using gasification, pyrolysis, and thermal cracking, where heat and/or mechanical force are used.

Despite these limits, there is huge potential for chemical recycling as a tool for sustainability. The most easy-to-handle plastics are polyethylene terephthalate (PET) and high-density polyethylene (HDPE). PET is regarded as one of the most recycled plastics by volume, globally. Although PET can be recycled through primary or secondary means, there are some drawbacks. Firstly, the PET needs to be clean enough or considered as pure waste for suitability in the extrusion and melt-recovery process. This is considered a hurdle, due to inconsistencies in consumer-led waste separation and cleaning before discarding or sorting at the municipal waste management level [74]. Another significant problem with the recycling of PET by the re-extrusion method is that the mechanical properties of the recycled material are greatly reduced with each reuse [75]. For these reasons, chemical recycling emerges as a more attractive route to manage PET waste and the extension of such technologies should be amplified in industry.

There are few companies implementing chemical recycling technologies at a commercial scale. A prominent example is a patented technology by the start-up "Worn Again Technologies", which recycles PET into pellets. During the process, benzaldehyde, benzyl acetate, benzyl benzoate, or other suitable solvents are used to dissolve the polyesters in a solvent system, followed by the chilling of the solvent system and precipitation before filtration to separate the materials. After separation, the plastic is washed, dried, and molded into pellets and/or converted into fibers. Cotton textiles, post-consumer PET bottles, and PET-containing plastic containers can all be processed using this method. Worn Again has also partnered with companies such as Himes and Kering to promote the reduction of textile waste [76]. In Germany, to create polyamides and polyesters from multilayer plastic post-industrial waste, APK AG is demonstrating its new cycling method on a plant that can produce 8000 tons of recycled materials annually. The technology dissolves plastic using a mixture of solvents from a group of cycloalkanes, alkanes, or isooctane [77]. Additionally, CreaSolv, a solvent-based procedure developed by The Fraunhofer Institute, creates plastics with qualities similar to those of virgin materials, while successfully eliminating the impurities and additives [78].

Solvent selection for plastic mixes, which are common in actual plastic waste, remains a challenge in chemical recycling. Two or more plastics are often combined to form multi-component polymeric materials, with each plastic chosen based on its useful properties. For example, the multilayer plastic films used to make food packaging boxes are extremely difficult to recycle due to their intricate compositions and the incompatibility of their various polymer layers. Additionally, the inherent contaminant nature of colored PET deters its suitability to undergo chemical recycling [79]. To address this challenge, innovations such as STRAP have been developed to generate solvent systems for recycling multilayer polymers through selective dissolution. STRAP includes experimentation, computer modelling, and process design tools to achieve its desired targets. Selective dissolution ultimately enables the separation of different plastics and is tolerant to the additives and impurities present in waste (https://polyloop.fr/strap-recycling/?lang=en, accessed on 11 January 2023). Techno-economic analysis shows that selective dissolution can produce plastics at costs similar to virgin resins, making the process economically viable [80].

Although chemical recycling is currently controversial and is a widely debated topic in the CE community, there is increased awareness of its benefits, as acknowledged by evidence-based research [81].

2.1.4. Waste Management–Disposal Stage (Waste to Energy and Value-Added Products)

The combined GC and CE approach to handling waste and avoiding landfilling or incineration mainly involves the prevention of waste by using it as a resource. As far as biowaste is concerned, the concept of waste to energy is simpler to apply than reusing or recycling, thus biowaste is a commonly used feedstock for recovering bioenergy. Biowaste is biodegradable and transformable waste, which is primarily produced from industrial and agricultural processes, or from residential activities that become municipal waste [82]. Bioconversion is a critical step in dealing with the rising demand for raw materials, manufacturing costs, environmental pollution, and waste management [83]. Over time, the focus on waste management has shifted to the circular economy notion, in which resources, energy, and materials are continuously recycled [84]. The development of biorefineries, which turn biowaste into bioenergy and high-value chemicals, has been encouraged in response to the pressures of environmental depletion and resource scarcity [85], and is considered an important step toward implementing a successful circular bioeconomy [86].

Research has shown that biowaste produces high-value materials, such as biohydrogen, biogas, and biohythane, which could help with the current energy crisis, and, consequently, with emissions reduction [87]. Given that the only by-product of hydrogen combustion is water, it is one of the cleanest and pollution-free fuels available today [88]. As a result, hydrogen is essential for the advancement of low-carbon economies and technology, as it is projected to play a dominant role within the global energy landscape [89]. A range of biotechnologies are used to create biohydrogen, including light-dependent processes, such as photo-fermentation and bio-photolysis, as well as light-independent ones, such as dark fermentation and microbial electrolysis cells [90]. Equally clean as renewable sources of energy, biogas can be produced from food waste, manure, straw, or sewage, through the Anaerobic Digestion (AD) process [91]. AD is regarded as one of the most cost-effective biological treatments for biowaste treatment. It lessens the environmental effects of waste disposal, while enabling energy recovery and the production of nutrient-rich digestate materials [92]. AD converts organic waste into biogas that is used in combined heat and power (CHP) units for the production of heat and electricity, while the digestate by-product is used as a biofertilizer on agricultural lands [93]. The EU has seen a rise in the number of AD plants, from 244 plants in 2010 to 688 in 2016 [94]. Recently, there has been a shift toward small-scale AD plants because they have several advantages over the traditional centralized management system, including fewer transportation needs, a higher community involvement, and the chance to build regional nutrient and energy loops. Anaerobic digestion can also result in the production of biohythane, which is composed of gases similar to those found in biogas, but in different ratios, consisting of 5–10% H_2, 50–60% CH_4, and 35–45% CO_2 [95]. Biohythane considerably increases the efficiency of traditional compression ignition engines and noticeably reduces the emissions of hydrocarbons and nitrogen oxides [96].

Together with waste to energy technologies, the minimization of waste can be achieved by employing biowaste conversion into value added products [97]. Examples reported in the literature include fermentation processes, where solid biowaste can be converted into bio-based products, such as biobutanol, biodiesel, citric acid, ethanol, hydrogen, and lactic acid. Lactic acid (LA) can act as the source of numerous different molecules, which makes it a versatile organic acid. LA is one of the US Department of Energy's "Top Value-Added chemicals from Biomass", due to its importance in the bio-based industry [98]. The manufacture of polylactic acid (PLA), a biopolymer with characteristics similar to those of polyester, has sparked attention, and, in 2018, the PLA segment accounted for 28.3% of the revenue generated by the LA market as a whole. According to recent projections, the market value of LA is expected to increase from USD 2.64 billion in 2018 to over USD 9 billion by 2025 [98]. Recent developments in nanotechnology have allowed researchers to transform liquid, solid, and even gaseous biomass waste into either value-added materials or products with minimum adverse effects [99]. Biomass-derived carbon nanomaterials are widely used for the manufacture of batteries, fuel cells, electro-catalysis, water purification, etc. [100]. For example, nanomaterials can be utilized as electrode modification materials in electrochemical sensors. The synthesis of these materials using plant-based methods has been evidenced using green chemistry technology and is much regarded as a less hazardous route to their manufacture [101]. Moreover, the synthesis of novel biowaste-derived carbon nanomaterials (e.g., graphene quantum dots, carbon nanotubes, and graphene) is widely explored due to their sustainable and cost-effective production on a large-scale [102].

Overall, the various biorefinery techniques that have emerged so far have demonstrated to be sustainable methods for integrated bioproducts, such as biochemicals, biopolymers, bioplastics, biofuels, and biofertilizers that are further used for commercial, agricultural, and industrial applications [103]. The global market value of biowaste-to-energy technology is approximately USD 25.32 billion and is expected to increase to USD 40 billion in 2023 [104].

2.1.5. Utilizing Minimum Energy Requirements

Energy that is not used in a synthetic or manufacturing process can likewise be viewed as waste. It is highly desirable to develop chemical processes or systems that consume minimal energy. An example of what green chemists have accomplished to reduce energetic requirements is to lower the energy barrier of a chemical reaction or select suitable reactants, so that the transformation can happen at room temperature, with all the direct and indirect benefits associated with it [30]. The reduction of energy requirements is achieved by employing catalysts, so that a process can be carried out at lower temperatures and pressures. With new catalytic reactions and catalyst types, catalysis offers a range of advantages in terms of process utilization, selectivity, the use of alternative reaction media, and, most importantly, energy savings [105]. An alternative to catalysis is microwave-assisted organic synthesis, which has developed rapidly since the 1980s. The use of microwave energy allows for the reduced energy requirement from conventional fossil fuel-based sources. To increase reproducibility, researchers increasingly employ specialized commercial equipment that measures and regulates the power input, temperature, and pressure [106]. Energy savings, when compared to conventional heating, rely on how well electric energy is converted to microwave energy, as well as on the reactor's properties, the volume of the reaction mixture, and the reaction component's capability to absorb microwave energy. However, the use of microwave heating remains rather unexplored on an industrial scale [107].

Increasing a chemical system's energy efficiency is just one aspect of the solution. Alternative forms of energy are also required. A myriad of sustainable fuels is suitable for powering industrial plants, which target the reduction of carbon-based energy sources [108]. Biofuels, solar energy, wind, hydropower, geothermal energy, and hydrogen fuel cells are a few of these examples. Although the status quo has not currently made use of most of these alternative energies, due to their low capacity and economic non-viability, they are expected to have huge potential in the next decade to power industries [109]. As such, green chemists will play a crucial role in addressing this challenge. One way of doing this is by understanding and developing chemical systems that can transform solar radiation into voltaic energy [110]. Although interest has been shown in organic, inorganic, and hybrid solar cells, organic solar cells have attracted the most attention due to their higher efficiency. The development of materials and polymers that can effectively convert light into current is still a difficult task, but it is essential for the success of this strategy. Another option for addressing the impending rise in energy demands would be proton-exchange membrane (PEM) fuel cells that use hydrogen and oxygen gases. By splitting water, hydrogen is formed and can be used as an energy carrier. A key point is to ensure the that hydrogen utilized is green hydrogen, produced from non-carbon-emitting renewable resources [111].

Another example of GC supporting the reduction of energy requirements comes from the textile industry, which is a key contributor to both water contamination and usage. According to a shocking document published in 2017 titled the Ellen MacArthur Foundation New Textiles Economy report, ca. 50 L of water is required to dye a kilogram of material when using conventional dyeing methods. Drying the dyed material also uses up a lot of energy. To circumvent this energy-intensive practice, supercritical carbon dioxide was discovered by a Dutch start-up (DyeCoo) as an alternative solvent, introducing water-free dyeing. By eliminating the water in this process, less energy is needed for drying, rendering the new process energy economical [112].

2.1.6. Use of Renewable Feedstocks

The use of renewable feedstock is regarded as sustainable as the industrial overuse of finite natural mineral resources is leading to their depletion. Nature generates 180 billion metric tons of renewable biomass each year, but only approximately 4% is currently used by humans. It consists of about 75% carbohydrates, 20% lignin, and the remaining portion is made up of lipids, proteins, and terpenes [113]. The chemical transformations of mono and disaccharides, the reactions of oils, fatty acids, and terpenes, as well as the chemistry of glucose fermentation products, are all methods for turning biomass into usable commodities. The production of ethanol, lactic acid, surfactants, furfural, d-sorbitol sweetener, and medicines, such as penicillin, are the main non-food uses of carbohydrates. It is possible to crosslink defatted soy flour, a cheap commercial product that is mostly made of soy protein and carbs, and use it in biodegradable composite materials with plant fibers, such as flax, hemp, or bamboo. The eco-friendly composites hold promise as a substitute for non-biodegradable composites, such as polypropylene or glass in the construction industry [114,115]. To duplicate the adhesive proteins found in mussels, soy flour has been chemically modified; the resulting product has been commercialized as a wood glue, taking the place of hazardous urea-formaldehyde resins that are often used in plywood and particleboard [116,117].

As most renewable raw materials or feedstock are plant-based, adapting to them is challenging. To avoid burdening the food production sector with the production of feedstock for industrial uses, numerous studies in green chemistry suggest the use of agricultural and food wastes as raw materials for a range of industrial processes. This option is economically advantageous and reduces competition with food production. In addition, using waste helps to prevent pollution and waste management problems [11,118]. Cellulose, lignin, suberin, and other wood components, as well as lactic acid, chitin, starch, glycerol, and oil, are examples of renewable materials. For instance, lignin is a significant waste product of the pulp and paper sector. It has long been burnt as a fuel at industrial sites. Recently, it has been discovered to have use in producing items such as dispersants, additives, and raw materials, for the manufacturing of other compounds, such as vanillin or humic acid [119,120]. Another prevalent natural polymer, which makes up the exoskeleton of arthropods, is chitin (e.g., crustaceans). It is a significant by-product of the seafood industry and can undergo a chemical process, known as deacetylation, to form chitosan. Chitosan has been used for a wide range of industrial purposes, including water filtration and biomedical applications. It should be possible to replace the current petroleum feedstock with a significant number of raw materials by recycling this bioindustrial waste [121].

Research today is mostly focused on creating biodegradable plastics from renewable resources. The production of bioplastics is seen as a more sustainable alternative to traditional petroleum-based plastics, which have a significant impact on the environment. Bioplastics have been designed, in particular, to be biodegradable and compostable, reducing the amount of waste that goes into landfills and the ocean. In addition, the utilization of renewable raw materials in the production of bioplastics has a positive impact on energy use and CO_2 emissions [122].

The American business NatureWorks was founded in the late 1980s on the premise of making bottles made from lactic acid polymers, the lactic acid derived from the fermentation of dextrose generated from starch, most typically corn. The company has expanded to include the manufacturing of raw materials for 3D printing, clothing, cleaning wipes, trash bags, and toys [123]. The popular brand Coca-Cola created the first recyclable bottle made of 30% polyethylene blends, where the ethylene is derived from plant sources and not petroleum products. In 2021, a decade later, the company debuted a 100% plant-based bottle, made of corn-derived b-terephthalic acid [124]. To replace polyols made from petroleum, BioBased Technologies (previously Cargill) created a commercial technology for turning vegetable oils into polyols. The moderate process conditions also reduce the reliance on non-renewable feedstocks and save energy [125]. These are some examples showing how bioplastics are increasingly becoming part of our everyday life, but many improvements are still needed. Despite the benefits of bioplastics, they still have to face some environmental challenges, such as high production costs, limited availability of raw materials, and the management of bioplastics waste [126]. To overcome these issues, researchers and scientists have been exploring new sources of raw materials that can be used to produce bioplastics such as algae, cellulose, microbes, and food waste, through bioconversion processes to create a more sustainable future for the plastic industry. Additionally, in the food business, biodegradable packaging has a bright future, though its evolution is influenced by a wide range of variables, including politics, legal changes, and the increased worldwide need for food and energy resources.

A step further from utilizing renewable feedstocks for product formation, biocatalysis has shown to be a significant green chemistry-related technology that aids sustainable manufacturing. Biocatalysis makes use of enzymes, the natural catalysts, which are not only simple to manufacture but also biodegradable and renewable [127]. More and more businesses are focusing on creating and utilizing enzymes as biocatalysts to meet sustainability goals. The 2003-founded California start-up "Newlight Technologies" received USD 9.2 million in funding to develop a carbon-negative technology that creates AircarbonTM (a thermoplastic), by mixing air and methane emissions. About 40% of AircarbonTM is oxygen from the air, and 60% of it is carbon and hydrogen from methane emissions. The technology itself was nothing new, but Newlight Technologies' use of a unique biocatalyst increased the yield by nine times and cut the cost by a factor of three, making AircarbonTM more affordable than plastics generated from oil. Following its commercial scale-up in 2013, AircarbonTM was used for product manufacturing by several well-known companies, including Dell, IKEA, Sprint, and Vinmar [128].

2.2. *How Do We Move from Theory to Practice?*

While the principles of GC have aided the process of optimization for linear processes since the 1980s, the framework of GC alone no longer fits the status quo of the need for ensuring *true* sustainability. The work by Keijer et al. [129] provided an excellent commentary on the differences between GC and what they term circular chemistry (CC), the latter being a direct interlinking of the GC approach and the CE framework. The authors postulated a new set of guiding principles that hopefully can allow a more holistic consideration when designing processes or products. According to the 12 principles of GC, the key takeaway points are that waste should be prevented, products should be manufactured in a way that allows them to be degraded, safer synthesis materials and chemicals should be produced, and renewability, from an energy source and feedstock point of view, should be incorporated. Going beyond that, CC proposes that waste should not only be prevented at the production stage, but it must be used as a resource and ultimately a "renewable feedstock". This subject has been highlighted earlier and fits well with the CE agenda. In addition, the new framework of CC requires due diligence when evaluating the nature of certain bio-based catalysts, which are purported as "green" in their action, yet the catalysts themselves may be manufactured using linear processes. Lastly, one of GCs principles is manufacturing products that are designed for degradability. The CE, on the other hand, values ensuring the longevity of a material by keeping it in circulation for a long time whilst maintaining its stability. CC considers the raw materials (chemicals) and energy stored in a material as long-term investment. Therefore, by promoting the reusability of a product, CC conserves the stored energy rather than requiring additional input. Ultimately, the principles of CC, as influenced by the circular economy, lie in considering not only the environmental viability, but economic gains too. At best, green chemistry serves as a transition to fulfilling a circular economy and the concept of "circular chemistry" should be emphasized when moving forward.

3. Policy and Structural Issues Related to CE and GC Integration and Adoption

The concept of the circular economy is receiving an increase in visibility and awareness through its incorporation in discussion forums at workshops, international events, and initiatives on trade opportunities that are linked to recycling and waste management. Unfortunately, the concept is still misrepresented, as its ideas are often reduced to mere recycling and waste management, which is not representative of its innovative new product design with the incorporation of green chemistry. From an international trade context, the circular economy is much more than the trade-in of recycled goods and waste management, rather it is tackling the fundamentals that create and maintain value in the economic system through innovation aided by green chemistry. The opportunity to debunk these misconceptions is seldom allowed in public discourse or otherwise not deemed a priority. The role of governments in promoting a circular economy should be adopted toward supporting the rethinking and shifting away from market failure fixers toward being market shapers who set the tone and direction of economic growth and social equity. The scale and complexity of our societal problems require global coordination and collaboration and a global platform for reflection. At their core, the circular economy, green chemistry, and any other innovation should be models where growth is achieved through reduced consumption and sustainability is appropriately achieved through material substitution, cleaner production, or offsets.

Product Service Systems (PSSs) and waste management (WM) value chains may be appealing business models for stakeholders within the circular economy [130]. The PSS is an integrated combination of products and services, and feasibility is the normal stakeholder requirement for business. Waste plans are governed by laws, policies, rules, local conditions, agreements, and stakeholders, which have specific prerequisites and regional bounds [131]. Subsequently, viable regional PSSs are unique. Importantly, WMs practical actions have been founded on the rising knowledge and the steady formation of public opinion in favor of measures that encourage reuse and recycling [132]. It has also been required to restructure society, which includes changing national norms and laws as well as providing incentives and motivation through financial support for investments and other obligations, for a successful circular economy promotion and implementation [133]. Palmer and Truong [134] report that GC initiatives can be profitable in the long term. According to the available data, between 2012 and 2017, US facilities reported 2226 GC activities for 147 Toxics Release Inventory (TRI) chemicals and chemical categories. Most GC activities were reported by the chemical production, fabricated metals, computer, and electronics industries [135]. As per the projections, the US market for chemical end-use products would grow from USD 149.9 billion in 2016 to USD 884.1 billion in 2026 [136]. According to a recent study, GC has gone global, and when bio-based and renewable products replace conventional products and offer new revenue sources to businesses and local economies, the global chemical industry will increase to more than USD 1.5 trillion annually [137].

Due to cost savings, regulations, and consumer demands, GC has experienced substantial expansion. However, this growth remains slow and uneven as large companies in developed markets have implemented GC strategies, but smaller companies in developing countries still have a long way to GC implications [136]. There is a two-way link between businesses and consumers, and businesses' attitudes toward sustainability may have an impact on how their clients and investors behave [138]. Consumer product companies have benefited greatly from the increased consumer awareness and demand for green products. For some industries, including the pharmaceutical sector, cost reductions from GC have taken the number one spot as a motivator [139].

Even though GC has experienced a rapid expansion in recent years, implementation nuances remain between various worldwide marketplaces and even amongst businesses that are operating within the same industry. Despite its potential for cost savings, GC still necessitates initial expenditures and a transition away from extremely capital-intensive infrastructures [140]. According to US rules, manufacturers are required to re-certify with the Food and Drug Administration (FDA) whenever they alter their production methods. Since this procedure is both expensive and time-consuming, it serves to deter businesses from making the necessary investments to create waste-reducing, atom-efficient chemistry [107]. Since most industries are driven by financial gains, the voluntary adoption of sustainable practices seems less likely. To impose greener behaviors, a powerful, appealing, and balanced regulation is needed. The REACH (Registration, Evaluation, Authorization, and Restriction of Chemical Substances), developed and introduced by the European Union in 2007, is the most promising and significant regulation. On the one hand, REACH requires chemical companies to disclose more information about the risks that their products pose to the environment and human health; on the other hand, it offers the possibility of registration exemptions for a process that supports new, sustainable innovation for five years. To establish a sustainable chemical industry, other nations have been inspired by the European Union to develop rules along similar lines [141].

Green chemistry education is also another factor that may dampen the progress of wider GC and CE adoption in industry. Despite initiatives such as "beyondbenign" to enhance GC teaching (https://www.beyondbenign.org/, accessed on 11 January 2023), the majority of chemistry departments worldwide do not include GC in their curricula, which results in a lack of knowledge about the methods and approaches that are now available [29]. It takes an understanding of green engineering, biotechnology, economics, and, most importantly, toxicity to create a successful green process. The lack of training in these fields among chemists generally makes it more difficult to use green chemistry on an industrial scale [142].

The lack of green solutions remains a point of concern for some processes. The complexity of supply chains, resistance to change, perception of green products being more expensive and less effective, and the technical difficulties in identifying greener technologies and materials are some of the major obstacles reported by researchers [143]. If a green technique is not economically appealing, it may still be rejected on a large scale even if all other criteria are in its favor. An example is the work of Martyn Poliakoff from Nottingham University, who developed an innovative approach to employ supercritical CO_2 as a solvent in what at that time was the first continuous flow reactor of this nature [144]. The process was implemented by Thomas Swan and Company, UK. Without any by-product production, the system resulted in the removal of an expensive and energy-intensive separation that the traditional technique required. The factory, however, was unable to offer chemicals at a lower cost than those produced using conventional non-green processes because of the lack of government subsidies. So, after operating commercially from 2002 to 2009, this plant's production was stopped [145]. Focusing on the supply chain dynamics, it is advised that the extensive and international supply chain must undergo several modifications because of the commercialization of green processes. However, such alterations come at a price. For instance, a BPA-free (bisphenol-A-free) covering for food packaging created by Eden Organic Foods was found to be compatible with some foods, such as beans, but not with extremely acidic tomato sauce. To ensure the compatibility of all product types, there is a need to change the coating types for various food types. This unfortunately suggests a smaller market and a change in manufacturing equipment, both of which would raise the cost of production, eventually affecting market price [107].

The above alludes to the fact that an environmentally friendly technique does not guarantee financial success. The industrial use of green processes is frequently hampered by regulatory, economic, political, and technical issues.

4. Conclusions

The circularity of products plays a vital role in the sustainability of the environment. The circular economy focuses on using the product for as long as possible. The use of chemicals is a key driving force toward product development or manufacturing processes. Hence, advancements in chemistry, where the enhancement of chemical designs to achieve an increase in lifespan along with reduced toxicity in a product, should be prioritized. In consideration of the relationship between chemicals and circularity, green chemistry provides a foundation for secure and lasting circular economy practices. With the implementation of green chemistry and circular economy practices, countries can use their resources to their maximum potential with less input in raw material procurements and waste management costs, as well as limiting the rate of depletion of non-renewable resources. GC–CE implementations are not only important for developed countries, but are even more important for developing countries. Developing countries are facing waste management issues and most of their industries are still using linear waste management techniques, which puts pressure on already failing economies. As per the international monetary fund (IMF) world economic output report of 2022, the global economy will likely shrink by one-third by 2023 and will worsen by 2024. Economic growth in the USA is estimated to drop from 5.7 in 2021 to 1.0 in 2023, and in the European area from 5.2 in 2021 to 0.5 in 2023 [146]. During this phase of extreme global economic recession, there is an opportunity presented for adopting GC–CE principles to help sustain vibrant economies. This could be in the form of saving costs for raw material procurement, transportation cost, waste management, and storage costs, along with providing job opportunities for workers in GC–CE-related industries. With the successful adaptation of green chemistry, which is tightly linked with a circular economy approach, the rising issue of waste management is likely to be resolved. In addition, the resource, labor, and energy burden of the disposal of waste will lessen, resulting in a decrease in carbon footprint and health issues caused by waste emissions in the environment.

The main obstacles to achieving a circular economy are investment, environmental education, and legislation. Investing in new technologies and processes for the circular economy is necessary, but it can be difficult to justify the initial costs without clear benefits. Environmental education is important in promoting the circular economy and green chemistry, but it is not widespread enough to reach everyone who needs it. Legislation is needed to support the circular economy, but it can be slow to change, and not all countries have the same level of commitment to the cause. To advance toward the circular economy and green chemistry, international agreements should be reconsidered to provide a framework for investment, education, and legislation. This could include the creation of incentives for businesses and individuals to adopt circular practices, the establishment of education programs to promote the benefits of circular practices, and the development of regulations to support the transition to a circular economy.

In developed countries, there is much awareness of the current environmental crisis, which is likely to increase exponentially over the next few years. However, in developing countries, awareness is comparably very low, especially of the fact that natural resources are depleting and that they should not be over-exploited. Populace ignorance then leads to the over-exploitation of already vulnerable resources. Governments and policy-makers should not only introduce policies for firms and industries, but should also largely invest in education and information dissemination means regarding the best environmental practices. Additionally, countries in the developing world should prioritize local innovations that promote GC–CE whilst providing financial support for such projects, as for many change agents finances remain a major impediment even in developed countries. There should be funding extended to academia, allowing researchers in tertiary institutions to research solutions that are based on GC–CE principles. To ensure a sustainable future and mitigate the effects of climate change, the integration of a circular economy and green chemistry approaches can potentially make sustainable ambitions, such as GHG emissions reductions, possible.

Funding: This research received no external funding.

Data Availability Statement: Not applicable.

Acknowledgments: The authors also gratefully acknowledge the support from the project "Realising the Transition toward the Circular Economy: Models, Methods and Applications (ReTraCE)", funded by the H2020-MSCA ITN-2018 program (Grant Number: 814247), and from the project "A Just Transition to Circular Economy (JUST2CE)", funded by the H2020-SC5-2020-2 scheme (Grant Number: 101003491).

Conflicts of Interest: The authors declare no conflict of interest.

References

1. European Commission. Closing the Loop—An EU Action Plan for the Circular Economy. 2015. Available online: http://eur-lex.europa.eu/legal-content/EN/TXT/?uri=CELEX:52015DC0614 (accessed on 11 January 2023).
2. Ellen MacArthur Foundation. Ellen-MacArthur-Foundation-Towards-the-Circular-Economy—Vol. 1. Ellen MacArthur Foundation. 2013, p. 1. Available online: https://emf.thirdlight.com/link/x8ay372a3r11-k6775n/@/preview/1?o (accessed on 11 January 2023).
3. Mancini, E.; Raggi, A. A review of circularity and sustainability in anaerobic digestion processes. *J. Environ. Manag.* **2021**, *291*, 112695. [CrossRef]
4. Pansera, M. The Origins and Purpose of Eco-Innovation. *Glob. Environ.* **2016**, *4*, 128–155. [CrossRef]
5. Stahel, W.R. The circular economy. *Nature* **2016**, *531*, 435–438. [CrossRef] [PubMed]
6. Webster, K. A Circular Economy is about the Economy. *Circ. Econ. Sustain.* **2021**, *1*, 115–126. [CrossRef]
7. WEF. *The World Needs a Circular Economy. Help Us Make It Happen*; World Economic Forum: Geneva, Switzerland, 2020.
8. D'Amato, D.; Korhonen, J. Integrating the green economy, circular economy and bioeconomy in a strategic sustainability framework. *Ecol. Econ.* **2021**, *188*, 107143. [CrossRef]
9. Kurowska-Susdorf, A.; Zwierżdżyński, M.; Bevanda, A.M.; Talić, S.; Ivanković, A.; Płotka-Wasylka, J. Green analytical chemistry: Social dimension and teaching. *TrAC-Trends Anal. Chem.* **2019**, *111*, 185–196. [CrossRef]
10. Armenta, S.; Esteve-Turrillas, F.A.; Garrigues, S.; de la Guardia, M. Green Analytical Chemistry: Concepts, evolution, and recent developments. *Green Approaches Chem. Anal.* **2023**, 1–37. [CrossRef]
11. Raj, A.; Chowdhury, A.; Ali, S.W. Green chemistry: Its opportunities and challenges in colouration and chemical finishing of textiles. *Sustain. Chem. Pharm.* **2022**, *27*, 100689. [CrossRef]
12. Adam, D.H.; Ende Supriyadi, Y.N.; Siregar, Z.M.E. Green manufacturing, green chemistry and environmental sustainability: A review. *Int. J. Sci. Technol. Res.* **2020**, *9*, 2209–2211.
13. Smith, S.L.; Raynes, D.B.; Howard, K.L. National leadership and cross-sector collaboration could help overcome differences in stakeholder definitions of sustainable chemistry. *Curr. Res. Green Sustain. Chem.* **2022**, *5*, 100222. [CrossRef]
14. Zimmerman, J.B.; Anastas, P.T.; Erythropel, H.C.; Leitner, W. Designing for a green chemistry future. *Science* **2020**, *367*, 397–400. [CrossRef] [PubMed]
15. Stahel, W.R.; Reday, G. *The Potential for Substituting Manpower for Energy-Research Contract No 76/13-V/343/78-EN*; Commission of the European Communities: Brussels, Belgium, 1977.
16. Chizaryfard, A.; Trucco, P.; Nuur, C. The transformation to a circular economy: Framing an evolutionary view. *J. Evol. Econ.* **2021**, *31*, 475–504. [CrossRef]
17. Harris, S.; Martin, M.; Diener, D. Circularity for circularity's sake? Scoping review of assessment methods for environmental performance in the circular economy. *Sustain. Prod. Consum.* **2021**, *26*, 172–186. [CrossRef]
18. Yuan, Z.; Bi, J.; Moriguichi, Y. The circular economy: A new development strategy in China. *J. Ind. Ecol.* **2006**, *10*, 4–8. [CrossRef]
19. McDowall, W.; Geng, Y.; Huang, B.; Barteková, E.; Bleischwitz, R.; Türkeli, S.; Kemp, R.; Doménech, T. Circular Economy Policies in China and Europe. *J. Ind. Ecol.* **2017**, *21*, 651–661. [CrossRef]
20. Calisto Friant, M.; Vermeulen, W.J.V.; Salomone, R. A typology of circular economy discourses: Navigating the diverse visions of a contested paradigm. *Resour. Conserv. Recycl.* **2020**, *161*, 104917. [CrossRef]
21. Hart, J.; Adams, K.; Giesekam, J.; Tingley, D.D.; Pomponi, F. Barriers and drivers in a circular economy: The case of the built environment. *Procedia CIRP* **2019**, *80*, 619–624. [CrossRef]
22. Ghisellini, P.; Cialani, C.; Ulgiati, S. A review on circular economy: The expected transition to a balanced interplay of environmental and economic systems. *J. Clean. Prod.* **2016**, *114*, 11–32. [CrossRef]
23. Genovese, A.; Pansera, M. The Circular Economy at a Crossroads: Technocratic Eco-Modernism or Convivial Technology for Social Revolution? *Capital. Nat. Social.* **2020**, *32*, 95–113. [CrossRef]
24. Clark, J.H. Green chemistry: Challenges and opportunities. *Green Chem.* **1999**, *1*, 1–8. [CrossRef]
25. Linthorst, J.A. An overview: Origins and development of green chemistry. *Found. Chem.* **2019**, *12*, 55–68. [CrossRef]
26. Murphy, M.A. Early Industrial Roots of Green Chemistry and the history of the BHC Ibuprofen process invention and its Quality connection. *Found. Chem.* **2018**, *20*, 121–165. [CrossRef]
27. De Marco, B.A.; Rechelo, B.S.; Tótoli, E.G.; Kogawa, A.C.; Salgado, H.R.N. Evolution of green chemistry and its multidimensional impacts: A review. *Saudi Pharm. J.* **2019**, *27*, 1–8. [CrossRef]
28. Anastas, P.T.; Warner, J.C. *Green Chemistry: Theory and Practice*; Oxford University Press: New York, NY, USA, 1998.
29. Armstrong, L.B.; Rivas, M.C.; Douskey, M.C.; Baranger, A.M. Teaching students the complexity of green chemistry and assessing growth in attitudes and understanding. *Curr. Opin. Green Sustain. Chem.* **2018**, *13*, 61–67. [CrossRef]
30. Anastas, P.T.; Bartlett, L.B.; Kirchhoff, M.M.; Williamson, T.C. The role of catalysis in the design, development, and implementation of green chemistry. *Catal. Today* **2000**, *55*, 11–22. [CrossRef]
31. Rodriguez-padron, A.D.; Puente-santiago, A.; Alina, M.; Munoz-batista, M.; Luque, R. Environmental Catalysis: Present and Future. *ChemCatChem* **2019**, *11*, 18–38. [CrossRef]
32. Abdussalam-Mohammed, W.; Ali, A.Q.; Errayes, A.O. Green Chemistry: Principles, Applications, and Disadvantages. *Chem. Methodol.* **2020**, *4*, 408–423. [CrossRef]

33. Kühlborn, J.; Groß, J.; Opatz, T. Making natural products from renewable feedstocks: Back to the roots? *Nat. Prod. Rep.* **2020**, *37*, 380–424. [CrossRef] [PubMed]
34. Sheldon, R.A. ScienceDirect The greening of solvents: Towards sustainable organic synthesis. *Curr. Opin. Green Sustain.* **2019**, *18*, 13–19. [CrossRef]
35. Chatel, G.; Varma, R.S. Ultrasound and microwave irradiation: Contributions of alternative physicochemical activation methods to Green Chemistry. *Green Chem.* **2019**, *21*, 6043–6050. [CrossRef]
36. Ahmed, S.N.; Haider, W. Heterogeneous photocatalysis and its potential applications in water and wastewater treatment: A review. *Nanotechnology* **2018**, *29*, 342001. [CrossRef]
37. Al-mamun, M.R.; Kader, S.; Islam, M.S.; Khan, M.Z.H. Photocatalytic activity improvement and application of UV-TiO_2 photocatalysis in textile wastewater treatment: A review. *J. Environ. Chem. Eng.* **2019**, *7*, 103248. [CrossRef]
38. Sheldon, R.A.; Woodley, J.M. Role of Biocatalysis in Sustainable Chemistry. *Chem. Rev.* **2018**, *118*, 801–838. [CrossRef] [PubMed]
39. Erythropel, H.C.; Zimmerman, J.B.; de Winter, T.M.; Petitjean, L.; Melnikov, F.; Lam, C.H.; Lounsbury, A.W.; Mellor, K.E.; Janković, N.Z.; Tu, Q.; et al. The Green ChemisTREE: 20 years after taking root with the 12 principles. *Green Chem.* **2018**, *20*, 1929–1961. [CrossRef]
40. Jamarani, R.; Erythropel, H.C.; Nicell, J.A.; Leask, R.L.; Marić, M. How Green is Your Plasticizer? *Polymers* **2018**, *10*, 834. [CrossRef] [PubMed]
41. Wasserbaur, R.; Sakao, T.; Milios, L. Interactions of governmental policies and business models for a circular economy: A systematic literature review. *J. Clean. Prod.* **2022**, *337*, 130329. [CrossRef]
42. Arun, K.B.; Madhavan, A.; Anoopkumar, A.N.; Surendhar, A.; Liz Kuriakose, L.; Tiwari, A.; Sirohi, R.; Kuddus, M.; Rebello, S.; Kumar Awasthi, M.; et al. Integrated biorefinery development for pomegranate peel: Prospects for the production of fuel, chemicals and bioactive molecules. *Bioresour. Technol.* **2022**, *362*, 127833. [CrossRef]
43. Mesa, J.; González-Quiroga, A.; Maury, H. Developing an indicator for material selection based on durability and environmental footprint: A Circular Economy perspective. *Resour. Conserv. Recycl.* **2020**, *160*, 104887. [CrossRef]
44. Pivnenko, K. *Waste Material Recycling: Assessment of Contaminants Limiting Recycling*; Department of Environmental Engineering, Technical University of Denmark (DTU): Lyngby, Denmark, 2016.
45. Hurst, G.A. Systems thinking approaches for international green chemistry education. *Curr. Opin. Green Sustain. Chem.* **2020**, *21*, 93–97. [CrossRef]
46. Yilan, G.; Cordella, M.; Morone, P. Evaluating and managing the sustainability of investments in green and sustainable chemistry: An overview of sustainable finance approaches and tools. *Curr. Opin. Green Sustain. Chem.* **2022**, *36*, 100635. [CrossRef]
47. Guo, Z.; Wang, A.; Wang, W.Y.; Zhao, Y.L.; Chiang, P.C. Implementing Green Chemistry Principles for Circular Economy towards Sustainable Development Goals. *Chem. Eng. Trans.* **2021**, *88*, 955–960. [CrossRef]
48. To, M.H.; Uisan, K.; Ok, Y.S.; Pleissner, D.; Lin, C.S.K. Recent trends in green and sustainable chemistry: Rethinking textile waste in a circular economy. *Curr. Opin. Green Sustain. Chem.* **2019**, *20*, 1–10. [CrossRef]
49. Ali, S.S.; Elsamahy, T.; Abdelkarim, E.A.; Al-tohamy, R.; Kornaros, M.; Ruiz, A.; Zhao, T.; Li, F.; Sun, J. Biowastes for biodegradable bioplastics production and end-of-life scenarios in circular bioeconomy and biorefinery concept. *Bioresour. Technol.* **2022**, *363*, 127869. [CrossRef] [PubMed]
50. Kaza, S.; Yao, L.; Bhada-Tata, P.; van Woerden, F. *What a Waste 2.0*; The World Bank: New York, NY, USA, 2018.
51. Annamalai, S.; Chandrasekaran, K.; Shin, W.S.; Sundaram, M.; Khaleel, T.M. Beyond dumping: New strategies in the separation of preservative salt from tannery waste mixed salt and its reuse for tannery industrial application. *Environ. Res.* **2022**, *214 Pt 2*, 113885. [CrossRef] [PubMed]
52. Sheldon, R.A.; Norton, M. Green chemistry and the plastic pollution challenge: Towards a circular economy. *Green Chem.* **2020**, *22*, 6310–6322. [CrossRef]
53. He, R.; Sandoval-Reyes, M.; Scott, I.; Semeano, R.; Ferrão, P.; Matthews, S.; Small, M.J. Global knowledge base for municipal solid waste management: Framework development and application in waste generation prediction. *J. Clean. Prod.* **2022**, *377*, 134501. [CrossRef]
54. Chavez-Rico, V.S.; Bodelier, P.L.E.; van Eekert, M.; Sechi, V.; Veeken, A.; Buisman, C. Producing organic amendments: Physicochemical changes in biowaste used in anaerobic digestion, composting, and fermentation. *Waste Manag.* **2022**, *149*, 177–185. [CrossRef]
55. Anastas, P.T. Beyond Reductionist Thinking in Chemistry for Sustainability. *Trends Chem.* **2019**, *1*, 145–148. [CrossRef]
56. Maria, C.; Góis, J.; Leitão, A. Challenges and perspectives of greenhouse gases emissions from municipal solid waste management in Angola. *Energy Rep.* **2020**, *6*, 364–369. [CrossRef]
57. Pires, A.; Martinho, G. Waste hierarchy index for circular economy in waste management. *Waste Manag.* **2019**, *95*, 298–305. [CrossRef]
58. Anastas, P.; Eghbali, N. Green Chemistry: Principles and Practice. *Chem. Soc. Rev.* **2010**, *39*, 301–312. [CrossRef] [PubMed]
59. Ardila-fierro, K.J.; Hernández, J.G. Sustainability Assessment of Mechanochemistry by Using the Twelve Principles of Green Chemistry. *ChemSusChem* **2021**, *14*, 2145–2162. [CrossRef]
60. López-Lorente, Á.I.; Pena-Pereira, F.; Pedersen-Bjergaard, S.; Zuin, V.G.; Ozkan, S.A.; Psillakis, E. The ten principles of green sample preparation. *TrAC Trends Anal. Chem.* **2022**, *148*, 116530. [CrossRef]

61. Sadatshojaei, E.; Wood, D.A. Water, the most accessible eco-friendly solvent, and extraction and separation agent. In *Green Sustainable Process for Chemical and Environmental Engineering and Science: Green Solvents for Environmental Remediation*; Elsevier: Amsterdam, The Netherlands, 2021; pp. 283–292. [CrossRef]
62. Kim, D.; Nunes, S.P. Green solvents for membrane manufacture: Recent trends and perspectives. *Curr. Opin. Green Sustain. Chem.* **2021**, *28*, 100427. [CrossRef]
63. Gulzar, T.; Farooq, T.; Kiran, S.; Ahmad, I.; Hameed, A. Green chemistry in the wet processing of textiles. In *The Impact and Prospects of Green Chemistry for Textile Technology*; Elsevier Ltd.: Amsterdam, The Netherlands, 2019. [CrossRef]
64. Vandeponseele, A.; Chatel, G.; Draye, M.; Piot, C. Subcritical water and supercritical carbon dioxide: Efficient and selective eco-compatible solvents for coffee and coffee by-products valorization. *Green Chem.* **2020**, *22*, 8544–8571. [CrossRef]
65. Ivanković, A.; Dronjić, A.; Bevanda, A.M.; Talić, S. Review of 12 Principles of Green Chemistry in Practice. *Int. J. Sustain. Green Energy* **2017**, *6*, 39–48. [CrossRef]
66. Hairprint. *Biological Hair Color*. Hairprint. 2017. Available online: https://www.myhairprint.com/ (accessed on 11 January 2023).
67. Elevance. Elevance Biorefinery Metathesis Technology. 2022. Available online: https://elevance.com/technology/ (accessed on 11 January 2023).
68. Beghetto, V.; Sole, R.; Buranello, C.; Al-abkal, M. Recent Advancements in Plastic Packaging Recycling: A Mini-Review. *Materials* **2021**, *14*, 4782. [CrossRef]
69. Bracquené, E.; Martinez, M.G.; Wagner, E.; Wagner, F.; Boudewijn, A.; Peeters, J.; Duflou, J. Quantifying the environmental impact of clustering strategies in waste management: A case study for plastic recycling from large household appliances. *Waste Manag.* **2021**, *126*, 497–507. [CrossRef]
70. Rahimi, A.R.; Garciá, J.M. Chemical recycling of waste plastics for new materials production. *Nat. Rev. Chem.* **2017**, *1*, 46. [CrossRef]
71. Lee, A.; Liew, M.S. Tertiary recycling of plastics waste: An analysis of feedstock, chemical and biological degradation methods. *J. Mater. Cycles Waste Manag.* **2021**, *23*, 32–43. [CrossRef]
72. Thiounn, T.; Smith, R.C. Advances and approaches for chemical recycling of plastic waste. *J. Polym. Sci.* **2020**, *58*, 1347–1364. [CrossRef]
73. Huang, J.; Veksha, A.; Chan, W.P.; Giannis, A.; Lisak, G. Chemical recycling of plastic waste for sustainable material management: A prospective review on catalysts and processes. *Renew. Sustain. Energy Rev.* **2022**, *154*, 111866. [CrossRef]
74. Achilias, D.S.; Karayannidis, G.P. The Chemical Recycling of Pet in the Framework of Sustainable Development. *Water Air Soil Pollut. Focus* **2004**, *4*, 385–396. [CrossRef]
75. Assadi, R.; Colin, X.; Verdu, J. Irreversible structural changes during PET recycling by extrusion. *Polymer* **2004**, *45*, 4403–4412. [CrossRef]
76. WAT. *A World Where Resources Are Kept in Constant Circulation, DRIVING Economic, Social and Environmental Benefits*. Worn again Technologies. 2021. Available online: https://wornagain.co.uk/ (accessed on 11 January 2023).
77. APK. Newcycling®—An Economic and Ecological Recycling Technology for a Real Circular Economy for Plastics. 2022. Available online: https://www.apk.group/en/newcycling/ (accessed on 11 January 2023).
78. CreaCycle. *The CreaSolv®Process*. CreaCycleGmbH. 2010. Available online: https://www.creacycle.de/en/the-process.html (accessed on 11 January 2023).
79. Li, H.; Marie Kirkelund, G. Pulsed stirring for energy efficiency improvements during electrodialytic extraction of As, Cd, Cr, Cu, Pb, and Zn from municipal solid waste incineration fly ash and air pollution control residue. *Sep. Purif. Technol.* **2022**, *290*, 120835. [CrossRef]
80. Li, H.; Aguirre-Villegas, H.A.; Allen, R.D.; Bai, X.; Benson, C.H.; Beckham, G.T.; Bradshaw, S.L.; Brown, J.L.; Brown, R.C.; Cecon, V.S.; et al. Expanding plastics recycling technologies: Chemical aspects, technology status and challenges. *Green Chem.* **2022**, *24*, 8899–9002. [CrossRef]
81. European Chemical Agency. Annual Report. 2021. Available online: https://echa.europa.eu/documents/10162/11872732/mb_05_2022_2_annual_report_2021_mb65_en.pdf/7688a1e9-5d23-59fb-213c-2bd940c052ff?t=1660039291431 (accessed on 11 January 2023).
82. De Boni, A.; Melucci, F.M.; Acciani, C.; Roma, R. Community composting: A multidisciplinary evaluation of an inclusive, participative, and eco-friendly approach to biowaste management. *Clean. Environ. Syst.* **2022**, *6*, 100092. [CrossRef]
83. Zuin, V.G. Circularity in green chemical products, processes and services: Innovative routes based on integrated eco-design and solution systems. *Curr. Opin. Green Sustain. Chem.* **2016**, *2*, 40–44. [CrossRef]
84. Orejuela-Escobar, L.M.; Landázuri, A.C.; Goodell, B. Second generation biorefining in Ecuador: Circular bioeconomy, zero waste technology, environment and sustainable development: The nexus. *J. Bioresour. Bioprod.* **2021**, *6*, 83–107. [CrossRef]
85. Nair, L.G.; Agrawal, K.; Verma, P. An overview of sustainable approaches for bioenergy production from agro-industrial wastes. *Energy Nexus* **2022**, *6*, 100086. [CrossRef]
86. Kefalew, T.; Lami, M. Biogas and bio-fertilizer production potential of abattoir waste: Implication in sustainable waste management in Shashemene City, Ethiopia. *Heliyon* **2021**, *7*, e08293. [CrossRef] [PubMed]
87. Masilela, P.; Pradhan, A. Systematic literature review of the sustainability and environmental performance of dark fermentative biohydrogen production. *J. Clean. Prod.* **2022**, *372*, 133541. [CrossRef]

88. Zhou, P.; Gao, S.; Wang, B.; Wang, Y.; Li, C.; Wang, Y.; Sun, B. Influence of hydrogen fuel cell temperature safety on bus driving characteristics and stack heating mode. *Int. J. Hydrogen Energy*, 2022, *in press*. [CrossRef]
89. Ozawa, A.; Kudoh, Y.; Murata, A.; Honda, T.; Saita, I.; Takagi, H. Hydrogen in low-carbon energy systems in Japan by 2050: The uncertainties of technology development and implementation. *Int. J. Hydrogen Energy* **2018**, *43*, 18083–18094. [CrossRef]
90. Zhang, Q.; Liu, H.; Shui, X.; Li, Y.; Zhang, Z. Research progress of additives in photobiological hydrogen production system to enhance biohydrogen. *Bioresour. Technol.* **2022**, *362*, 127787. [CrossRef]
91. Ananthi, V.; Ramesh, U.; Balaji, P.; Kumar, P.; Govarthanan, M.; Arun, A. A review on the impact of various factors on biohydrogen production. *Int. J. Hydrogen Energy*, 2022, *in press*. [CrossRef]
92. Ranieri, L.; Mossa, G.; Pellegrino, R.; Digiesi, S. Energy recovery from the organic fraction of municipal solid waste: A real options-based facility assessment. *Sustainability* **2018**, *10*, 368. [CrossRef]
93. Tampio, E.; Ervasti, S.; Rintala, J. Characteristics and agronomic usability of digestates from laboratory digesters treating food waste and autoclaved food waste. *J. Clean. Prod.* **2015**, *94*, 86–92. [CrossRef]
94. Thiriet, P.; Bioteau, T.; Tremier, A. Optimization method to construct micro-anaerobic digesters networks for decentralized biowaste treatment in urban and peri-urban areas. *J. Clean. Prod.* **2020**, *243*, 118478. [CrossRef]
95. Rena; Mohammed Bin Zacharia, K.; Yadav, S.; Machhirake, N.P.; Kim, S.H.; Lee, B.D.; Jeong, H.; Singh, L.; Kumar, S.; Kumar, R. Bio-hydrogen and bio-methane potential analysis for production of bio-hythane using various agricultural residues. *Bioresour. Technol.* **2020**, *309*, 123297. [CrossRef]
96. Prashanth Kumar, C.; Rena Meenakshi, A.; Khapre, A.S.; Kumar, S.; Anshul, A.; Singh, L.; Kim, S.H.; Lee, B.D.; Kumar, R. Bio-Hythane production from organic fraction of municipal solid waste in single and two stage anaerobic digestion processes. *Bioresour. Technol.* **2019**, *294*, 122220. [CrossRef] [PubMed]
97. Fiorentino, G.; Ripa, M.; Ulgiati, S. Chemicals from biomass: Technological versus environmental feasibility. A review. *Biofuels Bioprod. Biorefining* **2017**, *11*, 195–214. [CrossRef]
98. López-Gómez, J.P.; Pérez-Rivero, C.; Venus, J. Valorisation of solid biowastes: The lactic acid alternative. *Process Biochem.* **2020**, *99*, 222–235. [CrossRef]
99. Santana-Mayor, Á.; Rodríguez-Ramos, R.; Herrera-Herrera, A.V.; Socas-Rodríguez, B.; Rodríguez-Delgado, M.Á. Deep eutectic solvents. The new generation of green solvents in analytical chemistry. *TrAC-Trends Anal. Chem.* **2021**, *134*, 116108. [CrossRef]
100. Tiwari, S.K.; Bystrzejewski, M.; De Adhikari, A.; Huczko, A.; Wang, N. Methods for the conversion of biomass waste into value-added carbon nanomaterials: Recent progress and applications. *Prog. Energy Combust. Sci.* **2022**, *92*, 101023. [CrossRef]
101. Kaya, S.I.; Cetinkaya, A.; Ozkan, S.A. Green analytical chemistry approaches on environmental analysis. *Trends Environ. Anal. Chem.* **2022**, *33*, e00157. [CrossRef]
102. Wang, Z.; Shen, D.; Wu, C.; Gu, S. State-of-the-art on the production and application of carbon nanomaterials from biomass. *Green Chem.* **2018**, *20*, 5031–5057. [CrossRef]
103. Goswami, L.; Kayalvizhi, R.; Dikshit, P.K.; Sherpa, K.C.; Roy, S.; Kushwaha, A.; Kim, B.S.; Banerjee, R.; Jacob, S.; Rajak, R.C. A critical review on prospects of bio-refinery products from second and third generation biomasses. *Chem. Eng. J.* **2022**, *448*, 137677. [CrossRef]
104. Awasthi, M.K.; Sarsaiya, S.; Wainaina, S.; Rajendran, K.; Awasthi, S.K.; Liu, T.; Duan, Y.; Jain, A.; Sindhu, R.; Binod, P.; et al. Techno-economics and life-cycle assessment of biological and thermochemical treatment of bio-waste. *Renew. Sustain. Energy Rev.* **2021**, *144*, 110837. [CrossRef]
105. Wen, C.; Yin, A.; Dai, W.L. Recent advances in silver-based heterogeneous catalysts for green chemistry processes. *Appl. Catal. B Environ.* **2014**, *160–161*, 730–741. [CrossRef]
106. Verma, C.; Quraishi, M.A.; Ebenso, E.E. Microwave and ultrasound irradiations for the synthesis of environmentally sustainable corrosion inhibitors: An overview. *Sustain. Chem. Pharm.* **2018**, *10*, 134–147. [CrossRef]
107. Ratti, R. Industrial applications of green chemistry: Status, Challenges and Prospects. *SN Appl. Sci.* **2020**, *2*, 263. [CrossRef]
108. Anastas, P.T.; Beach, E.S. Green chemistry: The emergence of a transformative framework. *Green Chem. Lett. Rev.* **2007**, *1*, 9–24. [CrossRef]
109. Collins, J.; Gourdin, G.; Qu, D. Modern Applications of Green Chemistry: Renewable Energy. *Green Chem. Incl. Approach* **2018**, 771–860. [CrossRef]
110. Leitner, W.; Quadrelli, E.A.; Schlögl, R. Harvesting renewable energy with chemistry. *Green Chem.* **2017**, *3*, 2015–2016. [CrossRef]
111. Çelik, D.; Yıldız, M. Investigation of hydrogen production methods in accordance with green chemistry principles. *Int. J. Hydrogen Energy* **2017**, *42*, 23395–23401. [CrossRef]
112. Dyecoo. *DyeCoo*. 2022. Available online: https://dyecoo.com/ (accessed on 11 January 2023).
113. Melero, J.A.; Iglesias, J.; Garcia, A. Biomass as renewable feedstock in standard refinery units. Feasibility, opportunities and challenges. *Energy Environ. Sci.* **2012**, *5*, 7393–7420. [CrossRef]
114. Bukartyk, M.; Zholobko, O.; Wu, X. Green Synthesis of Soy Protein Nanocomposites: E ff ects of Cross- Linking and Clay Nanoparticles on the Mechanical Performance. *ACS Omega* **2022**, *7*, 5883–5893. [CrossRef]
115. Swain, S.N.; Biswal, S.M.; Nanda, P.K.; Nayak, P.L. Biodegradable Soy-Based Plastics: Opportunities and Challenges. *J. Environ. Polym. Degrad.* **2004**, *12*, 35–42. [CrossRef]
116. Chen, N.; Lin, Q.; Zheng, P.; Rao, J.; Zeng, Q.; Sun, J. A sustainable bio-based adhesive derived from defatted soy flour and epichlorohydrin. *Wood Sci. Technol.* **2019**, *53*, 801–817. [CrossRef]

117. Samson, D.O.; Mat Jafri, M.Z.; Hashim, R.; Sulaiman, O.; Aziz, M.Z.A.; Yusof, M.F.M.; Shukri, A. *Rhizophora* spp. Particleboards incorporating defatted soy flour bonded with NaOH/IA-PAE: Towards a water equivalent phantom material. *Radiat. Phys. Chem.* **2020**, *176*, 109057. [CrossRef]
118. Pfaltzgra, L.A.; De, M.; Cooper, E.C.; Budarin, V.; Clark, H. Food waste biomass: A resource for high-value chemicals. *Green Chem.* **2013**, *15*, 307–314. [CrossRef]
119. Kazzaz, A.E.; Feizi, Z.H.; Fatehi, P. Grafting strategies for hydroxy groups of lignin for producing materials. *Green Chem.* **2019**, *21*, 5714–5752. [CrossRef]
120. Österberg, M.; Sipponen, M.H.; Mattos, B.D.; Rojas, O.J. Spherical lignin particles: A review on their sustainability and applications. *Green Chem.* **2020**, *22*, 2712–2733. [CrossRef]
121. Shirvan, A.R.; Shakeri, M.; Bashari, A. Recent advances in application of chitosan and its derivatives in functional finishing of textiles. In *The Impact and Prospects of Green Chemistry for Textile Technology*; Elsevier: Amsterdam, The Netherlands, 2019; pp. 107–133. [CrossRef]
122. Raj, T.; Chandrasekhar, K.; Morya, R.; Kumar Pandey, A.; Jung, J.H.; Kumar, D.; Singhania, R.R.; Kim, S.H. Critical challenges and technological breakthroughs in food waste hydrolysis and detoxification for fuels and chemicals production. *Bioresour. Technol.* **2022**, *360*, 127512. [CrossRef]
123. NatureWorks LLC. NatureWorks. 2022. Available online: https://www.natureworksllc.com/About-NatureWorks (accessed on 11 January 2023).
124. The Coca-Cola Company. Coca-Cola Collaborates with Tech Partners to Create Bottle Prototype Made from 100% Plant-Based Sources. Thecocacolacompany. 2021. Available online: https://www.coca-colacompany.com/news/100-percent-plant-based-plastic-bottle (accessed on 11 January 2023).
125. Brzeska, J. A Brief Introduction to the Polyurethanes According to the Principles of Green Chemistry. *Processes* **2021**, *9*, 1929. [CrossRef]
126. Preka, R.; Fiorentino, G.; De Carolis, R.; Barberio, G. The challenge of plastics in a circular perspective. *Front. Sustain. Cities* **2022**, *4*, 920242. [CrossRef]
127. Domínguez de María, P. Biocatalysis, sustainability, and industrial applications: Show me the metrics. *Curr. Opin. Green Sustain. Chem.* **2021**, *31*, 100514. [CrossRef]
128. NewLight. From Greenhouse Gas to Regenerative Materials That Improve the World. 2022. Available online: https://www.newlight.com/ (accessed on 11 January 2023).
129. Keijer, T.; Bakker, V.; Slootweg, J.C. Circular chemistry to enable a circular economy. *Nat. Chem.* **2019**, *11*, 190–195. [CrossRef]
130. Scafà, M.; Carbonari, S.; Papetti, A.; Rossi, M.; Germani, M. A new method for Product Service System: The case of urban waste management. *Procedia CIRP* **2018**, *73*, 67–72. [CrossRef]
131. Kurpiela, S.; Teuteberg, F. Strategic planning of product-service systems: A systematic literature review. *J. Clean. Prod.* **2022**, *338*, 130528. [CrossRef]
132. Hsieh, C.C.; Lathifah, A. Ordering and waste reuse decisions in a make-to-order system under demand uncertainty. *Eur. J. Oper. Res.* **2022**, *303*, 1290–1303. [CrossRef]
133. Meglin, R.; Kytzia, P.S.; Habert, P.G. Regional environmental-economic assessment of building materials to promote circular economy: Comparison of three Swiss cantons. *Resour. Conserv. Recycl.* **2022**, *181*, 106247. [CrossRef]
134. Palmer, M.; Truong, Y. The Impact of Technological Green New Product Introductions on Firm Profitability. *Ecol. Econ.* **2017**, *136*, 86–93. [CrossRef]
135. Gaona, S.D. The Utility of the Toxic Release Inventory (TRI) in Tracking Implementation and Environmental Impact of Industrial Green Chemistry Practices in the United States. In *Green Chemistry*; Saleh, H.E.-D.M., Koller, M., Eds.; IntechOpen: Rijeka, Croatia, 2017. [CrossRef]
136. Veleva, V.R.; Cue, B.W. The role of drivers, barriers, and opportunities of green chemistry adoption in the major world markets. *Curr. Opin. Green Sustain. Chem.* **2019**, *19*, 30–36. [CrossRef]
137. Jiahuey, Y.; Liu, Y.; Yu, Y. Measuring green growth performance of China's chemical industry. *Resour. Conserv. Recycl.* **2019**, *149*, 160–167. [CrossRef]
138. Karl, H.; Jim, L.; Jane, C.; Meet the 2020 Consumers Driving Change. IBM Institute for Business Value. 2020. Available online: https://www.ibm.com/downloads/cas/EXK4XKX8 (accessed on 11 January 2023).
139. Veleva, V.R.; Cue, B.W.; Todorova, S.; Thakor, H.; Mehta, N.H.; Padia, K.B. Benchmarking green chemistry adoption by the Indian pharmaceutical supply chain. *Green Chem. Lett. Rev.* **2018**, *11*, 439–456. [CrossRef]
140. Fernandez Rivas, D.; Cintas, P. On an intensification factor for green chemistry and engineering: The value of an operationally simple decision-making tool in process assessment. *Sustain. Chem. Pharm.* **2022**, *27*, 100651. [CrossRef]
141. Borchert, F.; Beronius, A.; Ågerstrand, M. Characterisation and analysis of key studies used to restrict substances under REACH. *Environ. Sci. Eur.* **2022**, *34*, 83. [CrossRef]
142. Goh, H.Y.; Wen, W.; Wong, C.; Ong, Y.Y. A Study To Reduce Chemical Waste Generated in Chemistry Teaching Laboratories. *J. Chem. Educ.* **2020**, *97*, 87–96. [CrossRef]
143. Yadav, R.; Pathak, G.S. Young consumers' intention towards buying green products in a developing nation: Extending the theory of planned behavior. *J. Clean. Prod.* **2016**, *135*, 732–739. [CrossRef]

144. Licence, P.; Ke, J.; Sokolova, M.; Ross, S.K.; Poliakoff, M. Chemical reactions in supercritical carbon dioxide: From laboratory to commercial plant. *Green Chem.* **2003**, *5*, 99–104. [CrossRef]
145. Han, X.; Poliakoff, M. Continuous reactions in supercritical carbon dioxide: Problems, solutions and possible ways forward. *Chem. Soc. Rev.* **2012**, *41*, 1428–1436. [CrossRef] [PubMed]
146. IMF. *World Economic Outlook Countering the Cost of Living Crisis*; IMF: Washington, DC, USA, 2022.

energies

MDPI

Review

Pyrolysis and Extraction of Bark in a Biorefineries Context: A Critical Review

Umut Şen [1,*], Bruno Esteves [2] and Helena Pereira [1]

1 Centro de Estudos Florestais, Instituto Superior de Agronomia, Tapada da Ajuda, 1349-017 Lisbon, Portugal; hpereira@isa.ulisboa.pt
2 CERNAS Research Centre, Polytechnic Institute of Viseu, 3504-510 Viseu, Portugal; bruno@estgv.ipv.pt
* Correspondence: umutsen@isa.ulisboa.pt

Abstract: Bark-based biorefineries are estimated to become a trending topic in the coming years, particularly with their adsorbent applications and antioxidant production. While the potential benefits of bark-based biorefineries are well-recognized, it is not known how to develop a bark-based biorefinery considering different unit operations and the potential end products. The characterization and screening of different barks for chemical composition is the initial step in biorefinery development, along with the selection of the relevant conversion processes. This state-of-the-art review provides background knowledge on the pyrolysis and extraction studies of bark as the key conversion operations. The results of recent (2010–2022) bark pyrolysis and extraction studies were critically analyzed for process conditions, product properties, and product yields, as well as a discussion of energy-saving possibilities. A biorefinery scheme was proposed based on these data. The current knowledge gaps were identified and future directions were evaluated, which include the production of charcoals, platform chemicals, and extract profiling for specific applications. The results indicate that barks are particularly rich in hydrophilic extractives with potential antioxidant properties and pyrolysis operations, resulting in functional chars that may be used in value-added applications as adsorbent materials. A biorefinery scheme allows for the production of platform chemicals, antioxidant extracts and biochars from barks while contributing to the reduction of waste and environmental pollution.

Keywords: biomass; biofuels; extractives; fractionation; platform chemicals

1. Introduction

Bark is defined in plant anatomy as the tissues outside the vascular cambium that externally cover the stem and branches [1,2]. Tree barks are heterogeneous complex solid materials, and they are one of the most important forest by-products available in the field or in the wood-processing industries, potentially reaching an annual amount as high as 400 million cubic meters worldwide [3,4]. The majority of produced bark is not valorized as a resource for value-added products, but rather is disposed of as a waste stream or burned at domestic scale. Only a few barks are converted into value-added products, such as that of the cork oak, where the cork in the outer bark is harvested periodically to produce cork products, namely cork stoppers for the wine industry, valued worldwide [5]. Cork is also used in a number of other applications, including insulation, adsorbents, composites, fillers, cement-replacement materials, etc. [5–8]. Other important industrial barks include birch bark, which is a source of bioactive triterpenoid extractives such as betulin [9], as well as a number of oak and chestnut barks that are utilized for tannin extraction [10]. In former times, willow and poplar barks were extracted to produce salicylates, which are precursors of aspirin [11], which is currently produced from sodium phenolate, a petroleum-based benzene, and CO_2 using the Kolbe–Schmitt process [12] following acetylation with acetic anhydride. The barks of oaks, chestnut, pines, and birch, among other species, are currently being analyzed to produce chemicals and materials. However, many other unused or less known tree barks may have potential economic benefits. According to the latest estimation, probably less than 100 of the approximately 73,000 different tree species on Earth [13] are currently industrially used, and they are mostly by-products of high-value timber production [3].

The biorefineries and circular bioeconomy concepts were developed to reduce waste, produce sustainable products, and increase the overall efficiency and economy of biomass processing [14,15]. The overall aim is to use biomass feedstocks to produce chemicals, energy and materials, as is the case in conventional petroleum refineries [16]. Barks are considered potential resources to produce sustainable and biodegradable products in a biorefinery approach [17–21]. In order to develop efficient bark-based biorefineries, it is important to characterize the raw material and envisage the specific potential conversion paths. The chemical composition of bark is among the most important parameters for indicating the possible conversion paths, because it shows the content and composition of chemical fractions that will be available during bark conversion, namely ash, extractives, lignin, polysaccharides, and suberin.

Slow pyrolysis is considered the key thermochemical process for converting bark into energy products, because it increases the efficiency of fuel applications and also produces functional biochars that may be used as solid amendment, adsorbent, etc. [3]. The combustion of bark does not increase the net CO_2 release and releases less SO_x emissions into the atmosphere than coal because bark is considered carbon-neutral and contains much less sulfur, thereby contributing less to greenhouse gas emissions compared with coal [10]. Emerging technologies such as biomass gasification can further reduce CO_2 emissions if bark is used as fuel [22]. To increase gasification efficiency, barks may be subject to a previous pyrolysis process such as torrefaction [23], slow pyrolysis [24], or hydrothermal carbonization [25].

The extraction of bark may also be an essential step in biorefinery development, since barks contain many more extractives than wood and other lignocellulosic materials [26]. Bark extracts, often present in substantial amounts and with an outstanding chemical diversity, may therefore be considered for a large range of applications depending on extract yield and composition. Thus, extraction plays an important role in the valorization of bark, and the production of chemicals is an important step in bark-based biorefineries.

While barks are potentially important resources for biorefineries, their systematic conversion through cascade processing encompassing different conversion routes has not been previously analyzed, and the existing studies concentrate usually on a single conversion process. However, a previous study has shown that the conversion efficiency of spruce bark was significantly improved by using an integrated hot water extraction, pyrolysis, and anaerobic digestion process [27]. A successful bark-based biorefinery will therefore require the evaluation of different conversion processes. To achieve this objective, the chemical composition, and particularly the extract profiles, of different barks must be known.

This state-of-the-art review analyzes bark-based pyrolysis studies and extract profiles of different barks, mainly focusing on the period between 2010 and 2022, in order to integrate them into a cascade process for biorefineries. The development of bark-based biorefineries is discussed, the knowledge gaps are identified, and future directions in bark-based biorefineries are proposed.

2. Tree Barks and Biorefinery Research

2.1. *Trends in Bark-Based Biorefineries*

The first step in the prospective design of a biorefinery is to evaluate the previous studies and trends on the conversion pathways for the target biomass feedstock. A search in the Web of Science (WOS) on the number of publications and research keywords about the lignocellulosic biomass in a biorefinery context is a preliminary phase and has been a common approach in many studies.

It is difficult to estimate the number of studies targeting bark biorefinery by using only the publication titles, because not all have "biorefinery" in the title. Therefore, the screening for biorefinery studies on barks was conducted using the abstract keywords of "bark" and "waste", since most biorefinery studies are performed with waste materials. The results show that the number of publications on barks has steadily increased, starting from 25 and reaching 92 between 2010 and 2022 (Figure 1). It is therefore reasonable to expect a higher number of publications on bark conversion in the near future. Therefore, to estimate future trends, a forecast was performed (Figure 1). The forecast indicates that the number of bark-based biorefinery studies will most likely increase, and reach between 173 to 208 by 2035, with a 0.95 confidence level.

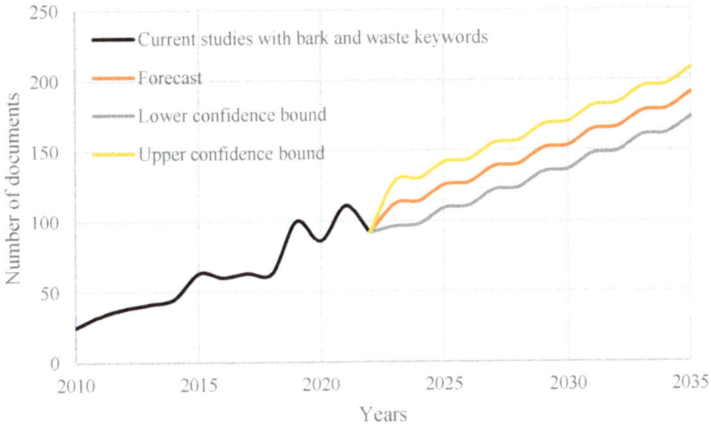

Figure 1. Number of publications in Web of Science using the keywords "bark and waste" in abstracts between 2010 and 2022, and the prognosis of future studies until 2035 with a 0.95 confidence level.

In a second step, using the same approach, abstracts were searched for the terms bark, extraction, and pyrolysis, which resulted in only 28 research items. Considering the low number of co-occurring studies, research titles including bark crossed with pyrolysis or extraction were searched for an extended time frame (1945–2023) and co-occurrence maps were created [28]. These maps show an overview of the studied topics on "bark and pyrolysis" or on "bark and extraction" (Figure 2). The different co-occurring research topics were analyzed, as shown by the different color clusters, and the most-studied topics are represented with larger circles. These maps indicate that studies on bark extraction are more common (310 research items) than on bark pyrolysis (91 research items), with antioxidant production and adsorption applications making up the principal research clusters.

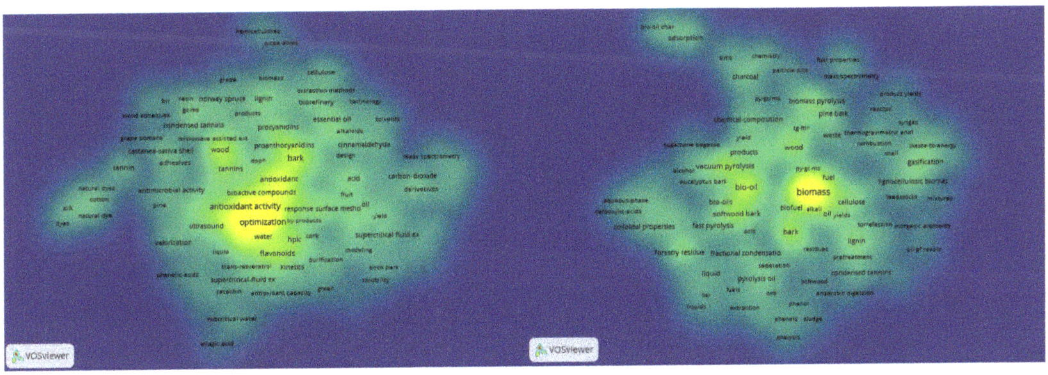

Figure 2. Co-occurrence map of publications with the titles "bark and extraction" on the **left**, and "bark and pyrolysis" on the **right** in WOS between 1945 and 2023.

Based on the results of the co-occurrence maps, we deduced that antioxidant production and adsorption applications from barks will be the most important topics in the near future. To estimate future trends, the same forecasting as carried out in Figure 1 was performed, and Figure 3 was created, which shows estimations of the possible development of these topics in the coming years.

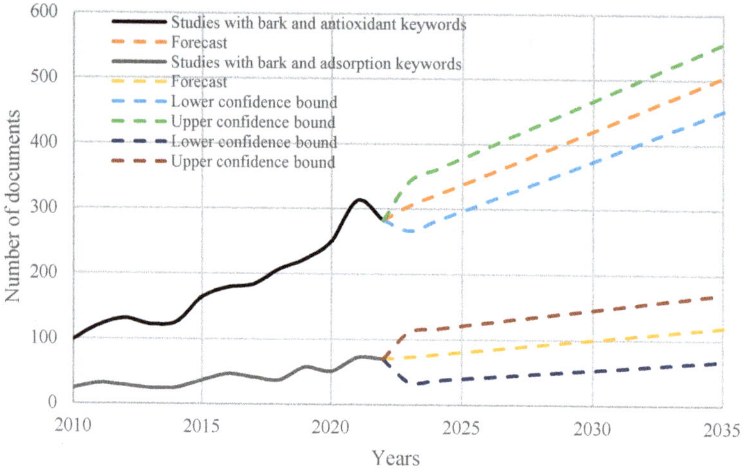

Figure 3. Number of publications in Web of Science using the keywords "bark and adsorption" and "bark and antioxidant" in abstracts between 2010 and 2022.

2.2. *Current Bark-Based Biorefinery Applications*

The previous section gives information on the development and direction of bark-based biorefineries in the near future. Currently, the application of an integrated biorefinery concept for bark is in its infancy, but recent studies show that its use may be increasing. Therefore, it is important to analyze these works. The bark-based biorefinery studies were performed using barks such as Norway spruce, Scots pine, radiata pine, birch, and aspen, based on a previous screening and evaluation of potentially valuable chemical components including extractives, non-cellulosic polysaccharides (mostly pectin and starch), lignin, cellulose, and suberin [27,29–34] (Figures 4–10). Interestingly, although almost all of these studies considered extraction as the first step, they did not consider a further refinement of extracts. Acetone, hot water, and supercritical CO_2 ($scCO_2$) extractions were tested for the screening of bark extracts [29,31,32]. The polysaccharide-containing hot water extracts of spruce bark showed immunostimulating activity, suggesting that the extraction of barks may be further refined to obtain bioactive compounds [35].

The non-cellulosic polysaccharides (NCPs), which are rich sources of glucose, arabinose, and galacturonic acids, were obtained from spruce bark with up to 20% yield by using pressurized hot water extraction [36]. The polysaccharide content of barks is lower than that of woods, agricultural wastes, and energy crops [3], but the extraction of NCPs enables the production of functional materials by the addition of other polysaccharide-derived cellulose nanocrystals (CNC) such as NCP/CNC films [20]. The CNCs may be derived from barks [19] or from other lignocellulosic feedstocks [34].

Suberin extraction is another approach in bark biorefineries when the bark has a high content of cork. The application of ionic liquids seems to be a promising method to obtain highly pure suberin [37], which may be used to produce functional materials [31]. Suberin monomers obtained by alkaline hydrolysis may be integrated into biorefineries after purification and polymerization, such as the production of polyepoxy acid [34] as an additive for NCP/CNC films.

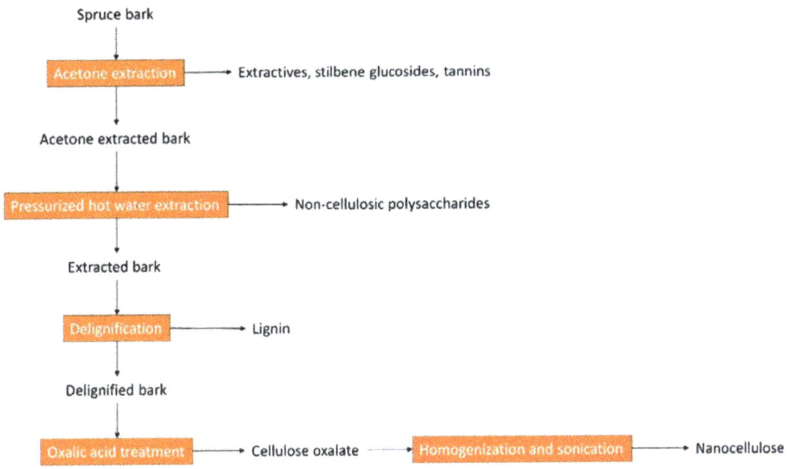

Figure 4. Spruce bark biorefineries [29].

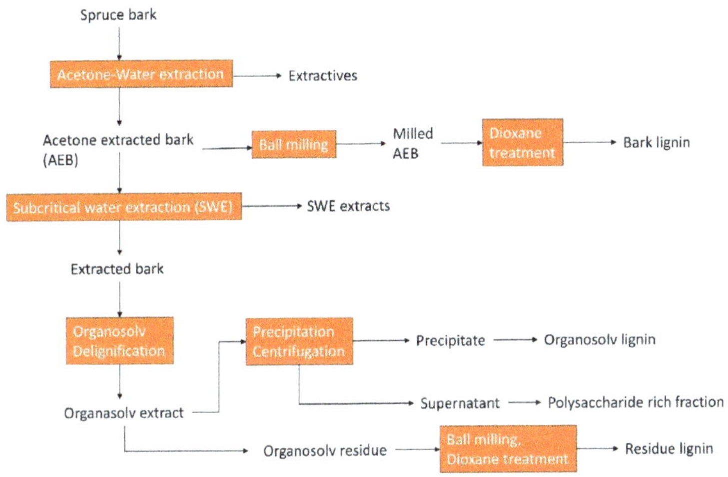

Figure 5. Spruce bark biorefineries [30].

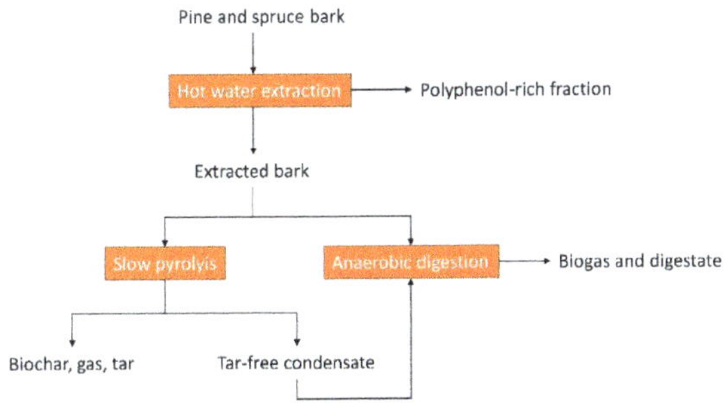

Figure 6. Pine and spruce bark biorefineries [27].

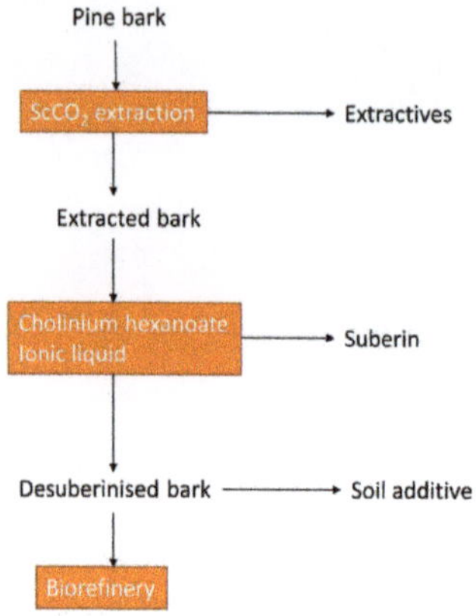

Figure 7. Pine bark biorefineries [31].

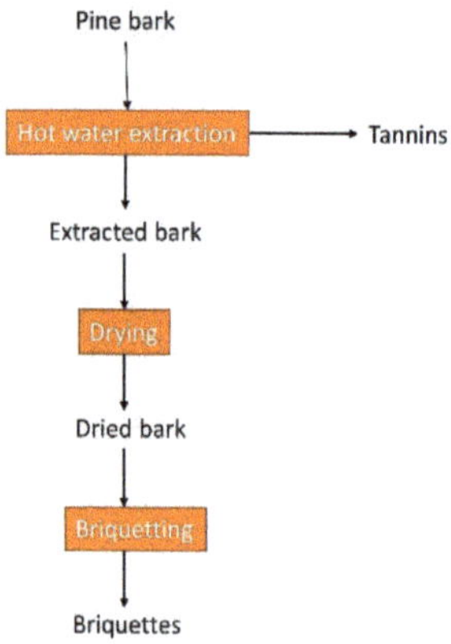

Figure 8. Pine bark biorefineries [32].

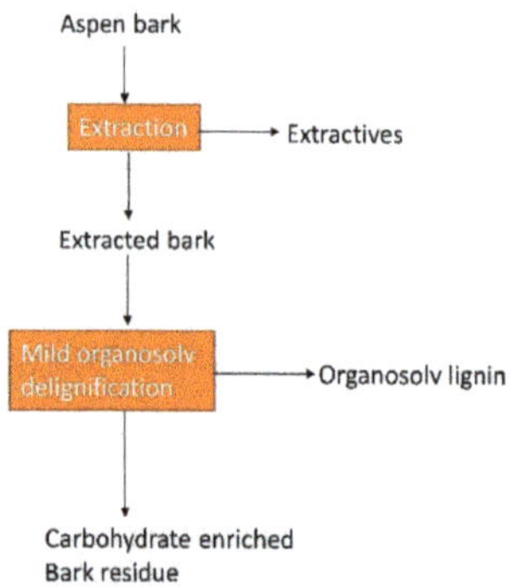

Figure 9. Aspen bark biorefineries [33].

The extraction and fractionation of bark lignin through ball milling and organosolv extraction [30] is possibly the most promising approach in bark biorefineries considering the high lignin content of barks. The bark-based lignins are alternative materials to fossil-based materials in the production of polymers and biofuels [30].

The efficiency of bark-based biorefineries may be increased by integrating different extraction, thermal conversion, and biological processes, such as extraction, slow pyrolysis, and anaerobic digestion [27]. The extraction and slow pyrolysis methods seem to be particularly suitable for the conversion of barks in a biorefinery concept, as explained in the next chapters. Biological processes may not be efficient in a bark biorefinery process considering the recalcitrance of the biomass and low polysaccharide content. However, current information is limited and more studies are required to evaluate the performance of biological processes with bark feedstock.

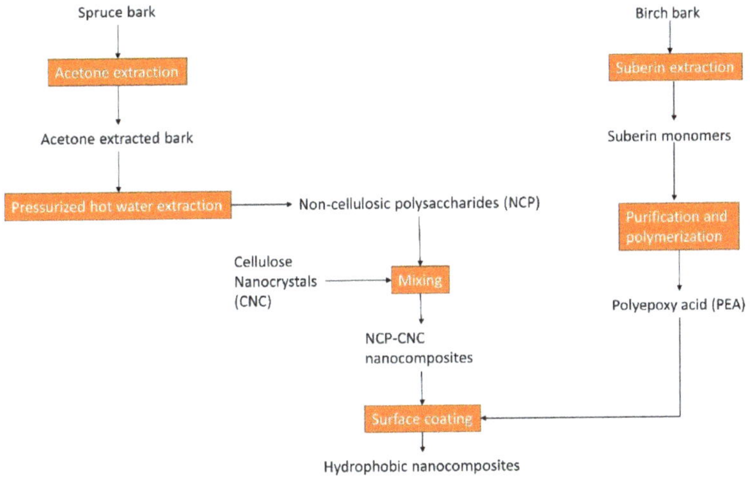

Figure 10. Spruce and birch bark biorefineries [34].

3. Pyrolysis of Bark

Pyrolysis is the thermal degradation of organic matter in the absence of oxygen. The pyrolysis process may be performed conventionally under an inert gas flow or industrially by sealing the pyrolysis chamber to prevent air oxidation [38]. The pyrolysis process may be integrated into bark-based biorefineries in a so-called thermochemical platform to produce biochars with lower moisture content, higher calorific value, and better grindability properties [39]. Different pyrolysis processes have been applied to lignocellulosic biomass, particularly to wood, which is composed mainly of cellulose and hemicellulose polysaccharides (70–75%) and contains low amounts of inorganics (less than 1%) [26]. The chemical composition of wood makes it a promising raw material for fast pyrolysis to produce bio-oils because, under high heating rates, high heat transfer coefficients, a 500 °C final pyrolysis temperature, and fast removal of the pyrolysis vapors and char, liquid yields may reach up to 80% of the wood, where 64% is bio-oil [40]. Under these conditions, the wood polysaccharides undergo an almost total conversion by devolatilization, depolymerization, and condensation (tar formation) reactions [41]. The low ash content of wood and the fast removal of pyrolysis vapors limit catalytic char formation, thereby maximizing the liquid yield in fast pyrolysis [42]. However, the fast pyrolysis bio-oils are reactive materials and require upgrading for use in industrial applications as drop-in fuels [43].

Bark is composed mainly of lignin, polysaccharides and also polyphenols and suberin, and contains a much higher amount of ash (usually over 10%) [10]. The chemical variability of barks is also much higher than that of wood, e.g., barks may be suberin-rich and the range of their extractives content is large. These chemical features of bark increase the char formation and decrease the bio-oil yield. Therefore, the fast pyrolysis of bark is usually not economically feasible due to low bio-oil yields. On the other hand, bark is a potential material for torrefaction, slow pyrolysis, and hydrothermal carbonization where the aim is to produce chars, referred to as torrefied material, biochars, and hydrochars, respectively. These processes are analyzed in this chapter.

3.1. Torrefaction

The torrefaction process is the lowest-temperature (mild) pyrolysis process, usually performed at temperatures between 200 °C and 300 °C [39,44]. This range may be extended to 180–320 °C to include the degradation range of hemicelluloses [45]. Under these conditions, the degradation of macro chemical components in biomass is small and mostly limited to hemicelluloses, and only partially to cellulose [39]. Biomass chemical composition is therefore the determining factor of mass loss in torrefaction. Hardwood hemicelluloses that are mainly composed of xylans are more degraded than softwood hemicelluloses [46] that are mainly composed of glucomannans [10] during torrefaction.

The torrefied biomass has a chemical composition similar to low-grade coals [38,47], contains less oxygen and fewer hemicelluloses, and is more hydrophobic [48]. Therefore, it is easier and less costly to grind [49,50]. The torrefied material is usually used as fuel for gasification or combustion [44]. Torrefaction may also reduce the potassium content in the lignocellulosic torrefied material, which leads to fewer ash-related problems during fuel applications [49].

The torrefaction properties of materials, such as mass loss and the resulting calorific value, depend both on the torrefaction conditions (duration and temperature) and biomass types [51]. These properties may be tuned by using torrefaction indices such as torrefaction severity index (TSI) or torrefaction severity factor (TSF), and the final material properties may be predicted [51].

The energy yield of biomass in torrefaction is calculated as a product of mass yield and the ratio of calorific values of the torrefied material and the raw biomass [49]. The energy yield indicates the amount of energy lost during torrefaction [52]. As the torrefaction temperature increases, the mass yield decreases, while the calorific ratio increases, resulting in an energy yield decrease, thereby suggesting the optimization potential of the process [49]. Possibly the most efficient method to decrease the energy yield and increase the efficiency of torrefaction is the integration of torrefaction with another process, such as co-combustion, pyrolysis, or gasification [53]. Further energy and time optimization can be achieved by using catalysts. For instance, the potassium impregnation of biomass before torrefaction was shown to increase the mass loss rate, providing a time- and energy-saving possibility [54]. Furthermore, performing torrefaction under oxygen-lean conditions, similarly to gasification, was proposed to achieve cost and energy savings [50,55].

One study using life cycle analysis (LCA) of torrefied biomass showed that torrefied biomass allows energy and greenhouse gas savings [53,56]. Under oxidative conditions, torrefaction is faster, and the operation cost is lower, but mass and energy yield are lower due to higher torrefaction temperatures [53].

3.2. Slow Pyrolysis

Slow pyrolysis (carbonization) is the main biochar production process of lignocellulosic biomass and is usually performed at moderate temperatures between 400 and 700 °C [57–59] over a few hours [59]. The produced biochars may be used as fuels [60,61], soil-amendment materials [62–64], adsorbents, or activated carbon precursors [65]. The use of biochars has been growing in recent years, and soil applications of biochars have predominated because of the following facts: biochars are recalcitrant materials in soil, they have high porosity and thus provide shelter for soil microorganisms, and they contain inorganic material that provides a source of nutrients [66]. However, they may also cause ecotoxicity [67] by leaching pyrolysis condensates, which depends on the used biomass type and applied pyrolysis temperature [38]. Biochars are also low-cost alternatives to costly activated carbons in wastewater treatment, and showed high pesticide removal efficiency from aqueous solutions [68]. Wood species may be considered as the best lignocellulosic feedstock for biochar production because wood biochars possess a higher surface area and lower ash content than other biomass types, making wood biochars more suitable for fuel, soil amendment, and adsorption applications [69]. However, wood is a versatile solid material and finds uses in many competing applications from solid wood to composites [70].

and other lignocellulosic feedstocks such as waste barks that are more available and have a lower cost than wood may become of interest. The activated carbons produced from bark biochars may be used as adsorbents or may be integrated into a biological process to enhance the caproate production [71].

Large biomass particles (5 mm to 50 mm) can be used in slow pyrolysis [59], which has a cost advantage compared with fast pyrolysis and gasification. The energy optimization in slow pyrolysis may be achieved by the utilization of pressurized pyrolysis to increase biochar and gas yields [72]. As in torrefaction, pyrolysis intensity may be tuned by using different temperatures for specific applications, e.g., lower [73] or decreasing [74] pyrolysis temperatures were shown to be beneficial in achieving energy savings while producing biochars that have similar properties to high-temperature biochars. Yet another option to decrease cost and increase energy efficiency in slow pyrolysis is to use the CO_2-containing pyrolysis gas in the pyrolysis process, rather than using expensive N_2 gas [72].

The production of biochars has positive environmental impacts, such as reduced fossil depletion, reduced water eutrophication, etc., while the transportation of biochars has a negative environmental impact on their life cycle [75].

3.3. Hydrothermal Carbonization (HTC)

Hydrothermal carbonization is defined as the combined dehydration and decarboxylation of biomass in a suspension with water under saturated pressure for several hours [76]. It is an interesting method of producing chars (referred to as hydrochars) from biomass without an energy-consuming pre-drying step [77]. Hydrothermal carbonization may be considered suitable for the conversion of materials with elevated moisture content [78–80]. The temperature applied is usually between 180 °C and 250 °C [71]. Similar to biochars, hydrochars may be used as fuels, soil amendment material, or activated carbon precursors [78,81–83]. The fuel quality of hydrochars was shown to be comparable to low-temperature biochars [84]. Hydrochars contain less ash than raw biomass and biochars, which is an advantage in fuel applications [85,86]. However, the surface areas of hydrochars are much smaller than those of biochars [78]; therefore, surface activation may be required for adsorption applications.

The process temperature and duration of the HTC process are considered to be the determining factors of its environmental performance [80,87,88] because they determine the hydrochar properties and energy yield. On the other hand, after the production of hydrochars, additional costly steps are required. Compared to dry carbonization biochars, hydrochars may contain higher moisture content, which requires a drying step, and the generated HTC effluents containing dissolved organic and inorganic compounds need to be treated before their discharge into the environment [89,90].

Table 1 summarizes important pyrolysis studies with bark, including the conditions used in the different processes and the bark species, and the obtained char yields and HHV. The bark char yields and calorific value are similar to those of other lignocellulosic biomass types. Fast-growing eucalyptus barks and abundant oak and conifer barks were the main barks analyzed.

Table 1. Conditions, char yields, and char calorific value of different barks in torrefaction, slow pyrolysis, and hydrothermal carbonization.

Process	Bark	Temperature (°C)	Time (min)	Char Yield (%)	HHV (MJ kg^{-1})	Reference
Torrefaction	*Picea abies*	225–300	30–60	61–90	20–24	[91]
	Abies alba	250	60	71	32 *	[92]
	Eucalyptus saligna and *E. grandis*	220–280	60–300	71–91	17–23	[93]
	Eucalyptus globulus and *E. nitens*	250–280	15–30	60–95	19	[94]
	Quercus cerris	200–300	30–60	62–90	-	[38]

Table 1. *Cont.*

Process	Bark	Temperature (°C)	Time (min)	Char Yield (%)	HHV (MJ kg^{-1})	Reference
Slow pyrolysis	*Leucaena leucocephala*	350–550	60–180	47 **	23 *	
	Pine	300–850	60	35–68	-	[95]
	Maesopsis eminii	500	60	32	-	[96]
	Larix sibirica	500–700	30–60	28–63	-	[97]
	Quercus suber	400–700	-	-	32	[98]
	Quercus cerris	325	30–60	52–73	-	[38]
	Quercus petraea	400–500	10	-	20	[99]
HTC	Conifer mixture (80% fir, 15% spruce, 5% pine) and *Robinia pseudoacacia*	150–185	30	71–93	20–22	[100]
	Eucalyptus	220–300	120–600	40–46	20–29	[101]
	Oak and pine	220	300	67–73	-	[102]

* Calculated using an empirical correlation based on ultimate analysis, ** optimum yield.

4. Extraction of Bark

Barks are rich sources of extractives, which are conventionally defined as low-molecular-weight compounds [10]. The amount of extractive in barks is higher than in other lignocellulosic materials and can reach 40% [26]. The bark extractives can be broadly grouped into hydrophilic and lipophilic compounds, depending on the solvent polarity used for their solubilization, and their amount and composition differ in the outer and inner bark and according to the used solvent [10]. Bark may also contain high-molecular-weight polyphenols such as condensed tannins [103] that are structurally similar to low-molecular-weight phenols but have higher molecular weights [10]. Alternatively, extractives may be defined as biomass constituents, which can be extracted by neutral solvents [104].

The most common bark extraction is of condensed tannins [105], which in different barks vary significantly between 5% and 50% [10]. Condensed tannins are used in the manufacture of leather as tanning agents or as phenol replacement in the production of phenol–formaldehyde resins for particle board and plywood [10,105]. However, barks may also contain interesting lipophilic and hydrophilic extractives that have bioactive properties.

Barks should be screened for lipophilic and hydrophilic extract yields to target specific compounds in the biorefinery [106]. There are two different approaches to obtaining extracts from biomass, the first being a conventional sequential solvent extraction, and the second being alternative extraction methods, as analyzed in the following section.

4.1. Sequential Solvent Extraction

Sequential extraction is based on the solubility difference of biomass extractives in solvents with varying polarities. First, lipophilic extractives are removed with nonpolar solvents, and later hydrophilic extractives are removed with polar solvents by increasing solvent polarity.

Interestingly, cork-rich barks contain the highest lipophilic content and the smallest hydrophilic-to-lipophilic extract ratio compared with other barks and wood, as shown in Table 2, where the extraction results were obtained with the same methodology and solvents, since extract composition and yield depend on the extraction process and solvent used. For instance, birch (*Betula pendula*) has the highest amount of lipophilic extractives followed by Turkey oak (*Quercus cerris*) and cork oak (*Quercus suber*) while eucalyptus, chestnut, oak, and some South American tree barks such as *Anadenanthera colubrina* contain a high amount of hydrophilic extracts. Barks contain in general a higher amount of hydrophilic extracts than of lipophilic extracts, as shown by the hydrophilic/lipophilic extract ratios (Table 3).

Table 2. Extraction yields of different barks with different solvents and hydrophilic/lipophilic ratios (RHL).

Bark	Extraction Yield, %				
	DCM	EtOH	Hot Water	RHL	Reference
Quercus cerris	10.9	3.4	2.4	0.53	[107]
Quercus suber—Portugal	5.8	5.9	4.5	1.79	[108]
Quercus suber—Bulgaria	4.4	5.4	2.7	1.84	[109]
Quercus suber—Turkey	7.1	3.6	1.2	0.68	[109]
Quercus variabilis	5.3	2.8	1.1	0.74	[110]
Quercus rotundifolia	1.6	6.4	9.3	9.81	[111]
Quercus faginea	1.9	4.9	6.4	5.95	[112]
Quercus robur	1.1	7.4	14.5	19.91	[113]
Quercus rubra	2.7	2.1	7.3	3.48	[113]
Quercus rubra	7.4	12.7	4.1	2.27	[114]
Albiza niopoides	2.7	3.9	5.2	3.37	[115]
Anadenanthera peregrina	3.0	21.0	4.7	8.57	[116]
Anadenanthera colubrina	2.8	22.5	4.1	9.50	[116]
Betula pendula	32.2	0.6	0.5	0.03	[117]
Betula celtiberica	2.7	4.1	7.5	4.30	[113]
Castanea sativa	2.0	9.5	20.4	14.95	[113]
Copaifera langsdorffii	2.1	17.4	1.8	9.14	[118]
Eucalpytus sideroxylon	1.7	20.1	33.9	31.76	[119]
Fraxinus excelsior	4.3	18.5	6.6	5.84	[113]
Goupia glabra	3.4	15.0	6.2	6.24	[120]
Kiyelmeyera coriacea	7.7	8.2	2.0	1.32	[121]
Myracrodruon urundeuva	4.2	9.3	13.5	5.43	[122]
Plathymenia reticulata	3.4	5.8	3.6	2.76	[123]
Populus x canadensis	3.0	13.7	5.1	6.27	[114]
Robinia pseudoacacia	3.6	7.2	5.5	3.53	[114]
Robinia pseudoacacia	3.8	3.9	5.0	2.34	[113]
Salix	3.3	19.9	5.9	7.82	[114]
Tectona grandis	2.2	2.9	7.3	4.64	[124]
Pinus nigra subsp. laricio	4.0	5.5	4.5	2.50	[114]
Pinus slyvestris	4.2	5.0	9.2	3.38	[125]
Pinus pinea	1.8	9.4	9.4	10.44	[126]
Larix decidua	2.8	19.1	8.2	9.75	[114]
Picea abies	4.8	5.3	11.2	3.44	[125]
Pseudotsuga menziesii	5.6	21.1	2.4	4.20	[127]
Average	4.7	9.5	7.0		
Std.	5.3	6.7	6.3		

Table 3. Comparison of average extractive yields of bark and wood by sequential solvent extraction.

	Average Extractive Content (%)			
	Lipophilic	Hydrophilic	Hydrophilic/Lipophilic Ratio (RHL)	Reference
Barks	4.7	16.5	3.52	This work
Cork-rich barks	10.1	9.2	0.91	[107,108,110,117,121,123,127,128]
Hardwoods	0.7	8.6	12.29	[129,130]
Softwoods	1.5	7.9	5.27	[129,130]

4.2. Alternative Extraction Methods

The conventional sequential solvent extraction method usually uses a Soxhlet apparatus, which has some disadvantages, such as the following:

1. The extraction yield is not optimized, and it is limited because extractions are performed under atmospheric pressure and below 100 °C.
2. The usage of organic solvents renders the method not environmentally friendly.

3. It usually takes a long time, from days to weeks, to obtain the extracts.
4. It has low selectivity to target compounds.
5. The thermally sensitive extracts are lost during the hot water extraction, which may not be appropriate to produce bioactive compounds.

Alternative extraction methods such as accelerated solvent extraction, ultrasound-assisted extraction, microwave-assisted extraction, and maceration offer solutions to these problems.

Accelerated solvent extraction (ASE) is a form of pressurized liquid extraction (PLE) and is perhaps the most well-known technique to overcome the disadvantages of solvent extraction [131]. It is based on using elevated temperatures (up to 200 °C) and pressures (up to 100 bar) and the extraction time and solvent use are reduced significantly [132,133]. The disadvantages of ASE include its higher cost, usage of organic solvents, and possible loss of thermally sensitive extracts (Table 4). Ultrasound-assisted extraction (UAE) is a fast extraction method based on ultrasounds and environmentally friendly solvents such as water and ethanol [134]. As in ASE, it uses less solvent than conventional methods [135]. It is especially suitable for the extraction of hydrophilic and thermally sensitive compounds such as antioxidant compounds [134,135], while its efficiency in the extraction of lipophilic compounds may be limited [136].

Table 4. Evaluation of extraction efficiencies of bark by alternative extraction methods.

	Target Extractive Efficiency		
Extraction Methods	**Hydrophilic Compounds**	**Lipophilic Compounds**	**Limitations**
ASE	Efficient	Efficient	High cost, use of organic solvents, and loss of thermally sensitive compounds
UAE	Efficient	Not efficient	Possible low extract yields, high energy consumption
MAE	Efficient	Efficient	Possible low extract yields, loss of thermally sensitive compounds, and high energy consumption
MAC	Least efficient	Least efficient	Longer extraction time, use of organic solvents, and low extract yields
SFE	Efficient	Most efficient	High cost

Microwave-assisted extraction (MAE) is also a fast method that applies microwaves to extract mainly hydrophilic compounds from biomass, although lipophilic compounds may also be extracted [137]. It is a selective method to extract phenolic compounds [138]. However, it may be considered a highly energy-consuming process, similar to ultrasound-assisted extraction [136]. It is likely that microwave-assisted extraction results in lower extract yields than conventional solvent extraction with bark because the extracts may undergo degradation under elevated extraction times and microwave powers [138]. However, studies with bark and MAE are scarce, and more data are needed to better understand and evaluate the potential of the MAE extraction route for bark.

Maceration (MAC) is the simplest extraction method based on the extraction of biomass with an organic solvent at low temperatures and at atmospheric pressure in a single step. Under these conditions, the extract yield is lower than with the sequential solvent extraction and the extraction time is longer. The use of organic solvents is also a disadvantage of maceration. However, maceration has certain advantages because it is more suitable for the extraction of thermally sensitive compounds and is less energy-consuming (Table 4).

Supercritical fluid extraction (SFE) removes extractives from lignocellulosic biomass with supercritical fluids, i.e., fluids that are above their critical temperature and pressure. The most commonly used supercritical fluid is CO_2, which has a critical temperature and pressure of 31.1 °C and 73.8 bar, respectively [139]. Supercritical CO_2 has interesting properties: it is an environmentally friendly solvent, it is nonpolar, and its density changes at different temperatures and pressures. Compared to water, another environmentally friendly solvent, its critical pressure and temperature are much lower.

Supercritical CO_2 is particularly suitable for selectively extracting lipophilic compounds. It is possible to increase solvent polarity by using a co-solvent such as ethanol [139], and therefore the CO_2-SFE also allows the extraction of hydrophilic compounds from lignocellulosic biomass. Another important advantage of SFE is that the extracts retain their bioactive properties [140]. The main disadvantage of this method is its relatively higher cost than other extraction methods (Table 4).

The alternative extraction methods are relatively new but have been increasingly applied to lignocellulosic biomass and barks. Triterpenoids such as friedelin, lupenone, and betulin from bark are extractives that have been obtained with SFE methods [141–150].

It is necessary to evaluate the cost-effectiveness and the environmental impact of different extraction methods on different barks, and a life cycle analysis (LCA) is also an important step in developing biorefinery routes [151,152]. LCA studies comparing different extraction methods of barks are scarce. An LCA study was performed for Norway spruce (*Picea abies*) bark to compare three different extraction methods, namely hot water extraction (HWE), ultrasound-assisted extraction (UAE), and supercritical fluid extraction (SFE), to extract polyphenols [153]. The hot water extraction (HWE) method showed the lowest environmental impact per unit of extracted phenolic compound [153].

5. Critical Evaluation of Bark-Based Biorefineries

5.1. Screening Barks for Biorefinery

Pyrolysis (slow pyrolysis, torrefaction, and hydrothermal carbonization) and extraction are considered the key conversion routes for bark, given that the technological conditions of pyrolysis processes and the high extractive content of bark favor these two routes. The development of efficient biorefineries requires the screening of barks' properties, namely moisture content, ash content, chemical composition, higher heating value, extractive yield, and composition.

Moisture content varies greatly between different barks and different positions in the tree, as well as between different seasons [3,154]. High-moisture-containing barks should be considered for the HTC process to avoid drying costs in dry pyrolysis. The ash content of barks is usually at least ten times higher than that of wood of temperate zones, which usually contains less than 1% ash [10,155]. The upper limit of ash content of bark may be considered as 20% [2]. The high ash content of bark may become problematic in thermal conversion because certain ash components such as potassium and calcium may cause slagging and fouling problems [156,157]. The high-ash-containing barks (>3%) should be pretreated with hot water to reduce the ash content if they are targeted for fuel use [158,159]. However, the ash removal from bark after hot water treatment may be insufficient [160] and in this case, the hydrothermal carbonization route should be applied to obtain higher biochar yields and to avoid ash-related problems during dry pyrolysis [161,162].

The macromolecular chemical composition of bark includes polysaccharides, lignin, and suberin. Hardwood and softwood barks may differ in their lignin and holocellulose contents as well as in their suberin content [163]. The chemical composition of different barks was reviewed in a previous article that reported 18–44% polysaccharides, 32–40% lignin, and 2–34% suberin [3]. Compared with wood, bark contains a higher amount of ash and extractives, and a lower amount of polysaccharides [10]. It also contains suberin and polyphenols which are absent in wood [3,10]. The cork-rich barks with high suberin content may be directed to obtain suberin monomers and lipophilic extracts to produce polymers and triterpenoids [141,164–168]. The lignin-rich barks should be considered for the production of biochars with higher char yields or to produce polyols and polyurethanes [169–174] or aromatics [175]. Different pretreatments such as organosolvent methods may be applied to increase the conversion efficiency of bark macromolecules such as suberin [176]. Bark polysaccharides may be considered for the production of food delivery packaging because they have a cost advantage as waste materials, they provide adequate strength, and they are biodegradable [177]. The fractionation of the polysaccharide fraction of barks is a complicated task due to the recalcitrance of the biomass and low polysaccharide content, although catalytic fractionation may overcome these problems [27,178].

The heating value of bark is slightly higher than that of wood and ranges between 18.7 MJ/kg and 22.7 MJ/kg [10,179]. Softwood barks generally have higher heating values than hardwood barks [179]. Barks with higher calorific values are potentially better suited for fuel applications. The production of cork pellets with a calorific value of 20 MJ/kg is an example of a fuel use for bark [180]. Fuel pellets produced from softwood barks demonstrated high mechanical durability [181], which is an important parameter in handling and transportation [182].

Extractives of barks may be grouped into hydrophilic compounds, lipophilic compounds, and polyphenols such as condensed tannins. Hydrophilic compounds are the main bark extracts, as shown in the previous section, and have antioxidant and bioactive properties [183]. In the biorefinery, these compounds should be extracted in a first step under low temperatures to avoid thermal decomposition and with a suitable extraction method, considering selectivity, efficiency and environmental impact. The most important lipophilic extracts are terpenes and triterpenoids, which are likely to have therapeutic activities, including anticancer, antimicrobial, and antiviral activities [184,185]. Barks with high amounts of lipophilic extractives such as cork-rich barks should be selected for the production of triterpenoids, while barks containing high amounts of hydrophilic extractives should be selected for phenols to produce antioxidants and aromatic compounds [186,187]. Barks containing high amounts of polyphenols should be selected for the production of tannins [188–190].

5.2. Development of a Biorefinery Scheme

A biorefinery scheme for bark-based biorefineries is proposed in Figure 11, taking into account the discussed bark characteristics and screening factors. A total of three conversion loops are determined. The first loop (L1) is the extraction processes, consisting of different extraction methods to target specific extracts. The second loop (L2) is the pyrolysis processes, which consists of HTC, slow pyrolysis, and torrefaction to produce hydrochars, biochars, and fuel pellets. The third loop (L3) is the macromolecular conversion processes, consisting of the depolymerization of the bark structural components. The threshold limits for selecting conversion routes are set at 30% for moisture and at 3% for ash content. A total of nine different biorefinery products are proposed, of which six can be produced by pyrolysis and extraction methods. The transformation of the macro chemical components in bark is beyond the scope of this review and thus only briefly discussed. The proposed biorefinery products (biochars, hydrochars, fuel pellets, simple phenols, triterpenoids, and condensed tannins) suggest that waste barks can be integrated into biorefinery systems to produce chemicals and solid products. The environmental impact of different conversion routes should be analyzed before the implementation of any biorefinery scheme.

The most abundant and versatile bark biorefinery product seems to be simple phenols, which can be obtained by hydrophilic extraction, hydrothermal carbonization, or slow pyrolysis. Simple phenols are platform chemicals that may be used to produce plastics, pharmaceuticals, or other chemicals.

5.3. Current Knowledge Gaps and Future Directions

The results of this review show that bark and biorefinery studies have experienced an increasing trend in recent years, with biorefineries increasingly being directed toward waste materials including bark. However, knowledge gaps exist, and their identification and surmounting are important for developing bark-based biorefineries.

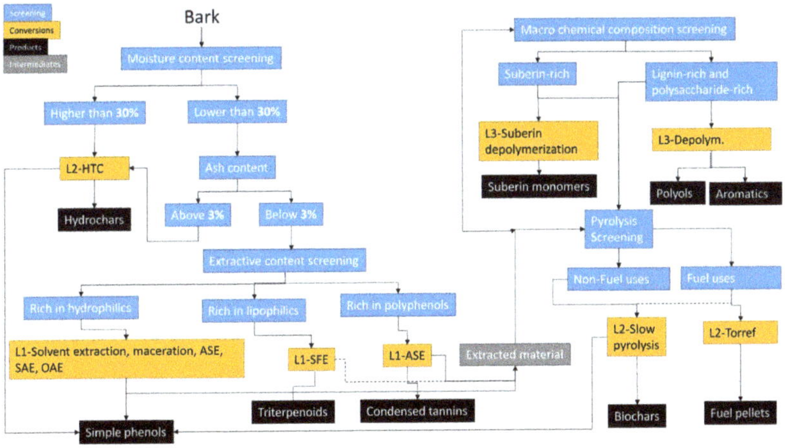

Figure 11. Proposed scheme for bark biorefineries.

The following knowledge gaps on bark conversion were identified in this work:

1. The number of bark characterization studies is insufficient: only a total of 21 articles investigated the extract composition of different barks (Table 2).
2. The applied pyrolysis conditions are non-standardized: a total of 13 articles applied different conditions (Table 1), which complicates the evaluation of the most economic processing conditions.
3. The applied extraction conditions were also variable. Since the extract composition, properties and yield depend on the applied solvent and method, we limited our discussion to cases with similar extraction procedures (Table 2).
4. Studies focused on the screening and application of different bark extracts are scarce: only Jablonsky and co-workers analyzed the potential of softwood bark extracts for the production of chemicals and nutraceuticals [185,191].
5. Energy balance is an important parameter in the evaluation of different processing paths and the scaling up of biorefinery processes. However, the pyrolysis and extraction conditions, as well as chemical properties, of different barks are highly variable which make the energy balance an unhelpful tool. More research is needed to evaluate the energy balance of different bark conversion processes.
6. There is a lack of studies on the environmental impact of bark valorization.
7. Studies combining different processing routes are insufficient: only Rasi and co-workers (2019) studied a cascade processing method consisting of hot water extraction, pyrolysis, and anaerobic digestion for pine and spruce barks [27].
8. There is a lack of studies investigating the production of bark-based platform chemicals such as simple phenols via pyrolysis or extraction.

Considering these limitations, the following recommendations can be made:

1. More research is needed on the chemical and pyrolysis properties of barks. Possibly fewer than 100 barks have been considered for valorization, which is a very small number regarding the huge potential for bark valorization given the existing number of tree species. Future bark valorization studies should apply standardized methods.
2. The bark valorization studies should consider cascade processing, combining different valorization processes instead of a single process.
3. The environmental impact of the different applied conversion processes is largely unknown, as is their economic evaluation. Energy balances should be provided in bark conversion studies.
4. Different bark extracts should be screened for antioxidant or nutraceutical potential, including pharmacokinetic profiles and drug-like properties.

5. The production of phenolic substances and simple phenols should target extractive-rich barks after chemical screening. Barks may be a source of platform chemicals such as simple phenols, as in earlier studies before their replacement with petroleum-based products. It is therefore necessary to re-consider bark for the production of platform chemicals to be used in the food, fragrance, or pharma industries. Efficient and selective production of these chemicals through extraction, pyrolysis, or depolymerization may open up new possibilities for bark valorization, namely using optimized and environmentally friendly conversion methods and improved catalysts [192].

Bark-based biorefineries require a detailed characterization of barks. In general, barks contain a high amount of extractives and inorganic compounds, a low amount of polysaccharides, and also contain suberin and polyphenols, with the cork-rich barks containing the highest amount of lipophilic extracts, e.g., triterpenoids. Therefore, pyrolysis and extraction methods are proposed for bark-based biorefineries, which should be integrated with macromolecular conversion methods.

6. Conclusions

Different pyrolysis and extraction methods were critically reviewed, considering process conditions, energy optimization, and bark as a biorefinery feedstock. The following specific conclusions are derived from this study:

1. The number of bark valorization studies has been increasing in recent years, since barks are widely available feedstocks and may be processed to produce sustainable and environmentally friendly products, including chars, antioxidants, and platform chemicals.
2. Adsorption applications and antioxidant production are predicted to be the most important topics in bark valorization in the near future.
3. More studies are required to screen different barks for their chemical composition, extractives profiles, and drug properties.
4. Bark valorization studies should be designed in the form of cascade processing and should combine different processing paths, including pyrolysis, extraction, and enzymatic digestion or chemical fractioning.

Author Contributions: Conceptualization, U.Ş. and H.P.; investigation, U.Ş. and B.E.; writing—original draft preparation, U.Ş.; writing—review and editing, B.E. and H.P.; visualization, U.Ş.; supervision, H.P. All authors have read and agreed to the published version of the manuscript.

Funding: This research was funded by FCT—Fundação para a Ciência e Tecnologia, grant number UIDB/00239/2020 and UIDB/00681/2020.

Data Availability Statement: Not applicable.

Acknowledgments: U. Şen acknowledges support from FCT through a research contract (DL 57/2016).

Conflicts of Interest: The authors declare no conflict of interest.

References

1. Trockenbrodt, M. Survey and discussion of the terminology used in bark anatomy. *IAWA J.* **1990**, *11*, 141–166. [CrossRef]
2. Harkin, J.M.; Rowe, J.W. *Bark and Its Possible Uses*; Forest Products Laboratory: Madison, WI, USA, 1971; Volume 91.
3. Şen, A.U.; Pereira, H. State-of-the-Art Char Production with a Focus on Bark Feedstocks: Processes, Design, and Applications. *Processes* **2021**, *9*, 87. [CrossRef]
4. Pasztory, Z.; Mohácsiné, I.R.; Gorbacheva, G.; Börcsök, Z. The utilization of tree bark. *BioResources* **2016**, *11*, 7859–7888. [CrossRef]
5. Pereira, H. *Cork: Biology, Production and Uses*; Elsevier: Amsterdam, The Netherlands, 2007; ISBN 978-0-444-52967-1.
6. Silva, S.P.; Sabino, M.A.; Fernandes, E.M.; Correlo, V.M.; Boesel, L.F.; Reis, R.L. Cork: Properties, capabilities and applications. *Int. Mater. Rev.* **2005**, *50*, 345–365. [CrossRef]
7. Menor, M.C.P.; Ros, P.S.; García, A.M.; Caballero, M.J.A. Granulated cork with bark characterised as environment-friendly lightweight aggregate for cement based materials. *J. Clean. Prod.* **2019**, *229*, 358–373. [CrossRef]
8. Lakreb, N.; Sen, U.; Beddiar, A.; Zitoune, R.; Nobre, C.; Gomes, M.G.; Pereira, H. Properties of eco-friendly mortars produced by partial cement replacement with waste cork particles: A feasibility study. *Biomass Convers. Biorefin.* **2022**. [CrossRef]
9. Krasutsky, P.A. Birch bark research and development. *Nat. Prod. Rep.* **2006**, *23*, 919–942. [CrossRef] [PubMed]

10. Fengel, D.; Wegener, G. *Wood: Chemistry, Ultrastructure Reactions*; Walter de Gruyter: Berlin, NY, USA, 1984.
11. Norn, S.; Permin, H.; Kruse, P.R.; Kruse, E. From willow bark to acetylsalicylic acid. *Dan Med. Arb.* **2009**, *37*, 79–98.
12. Luo, J.; Preciado, S.; Xie, P.; Larrosa, I. Carboxylation of phenols with CO_2 at atmospheric pressure. *Chem. Eur. J.* **2016**, *22*, 6798–6802. [CrossRef] [PubMed]
13. Cazzolla Gatti, R.; Reich, P.B.; Gamarra, J.G.P.; Crowther, T.; Hui, C.; Morera, A.; Bastin, J.-F.; De-Miguel, S.; Nabuurs, G.-J.; Svenning, J.-C. The number of tree species on Earth. *Proc. Natl. Acad. Sci. USA* **2022**, *119*, e2115329119. [CrossRef] [PubMed]
14. Atabani, A.E.; Ali, I.; Naqvi, S.R.; Badruddin, I.A.; Aslam, M.; Mahmoud, E.; Almomani, F.; Juchelková, D.; Atelge, M.R.; Khan, T.M.Y. A state-of-the-art review on spent coffee ground (SCG) pyrolysis for future biorefinery. *Chemosphere* **2022**, *286*, 131730. [CrossRef]
15. Velvizhi, G.; Balakumar, K.; Shetti, N.P.; Ahmad, E.; Pant, K.K.; Aminabhavi, T.M. Integrated biorefinery processes for conversion of lignocellulosic biomass to value added materials: Paving a path towards circular economy. *Bioresour. Technol.* **2022**, *343*, 126151. [CrossRef]
16. Fiorentino, G.; Ripa, M.; Ulgiati, S. Chemicals from biomass: Technological versus environmental feasibility: A review. *Biofuels Bioprod. Biorefin.* **2017**, *11*, 195–214. [CrossRef]
17. Şen, A.; Leite, C.; Lima, L.; Lopes, P.; Pereira, H. Industrial valorization of *Quercus cerris* bark: Pilot scale fractionation. *Ind. Crops Prod.* **2016**, *92*, 42–49. [CrossRef]
18. Barbini, S.; Sriranganadane, D.; España Orozco, S.; Kabrelian, A.; Karlström, K.; Rosenau, T.; Potthast, A. Tools for bark biorefineries: Studies toward improved characterization of lipophilic lignocellulosic extractives by combining supercritical fluid and gas chromatography. *ACS Sustain. Chem. Eng.* **2020**, *9*, 1323–1332. [CrossRef]
19. Le Normand, M.; Moriana, R.; Ek, M. Isolation and characterization of cellulose nanocrystals from spruce bark in a biorefinery perspective. *Carbohydr. Polym.* **2014**, *111*, 979–987. [CrossRef] [PubMed]
20. Le Normand, M.; Moriana, R.; Ek, M. The bark biorefinery: A side-stream of the forest industry converted into nanocomposites with high oxygen-barrier properties. *Cellulose* **2014**, *21*, 4583–4594. [CrossRef]
21. Neiva, D.M.; Araujo, S.; Gominho, J.; de Cássia Carneiro, A.; Pereira, H. Potential of *Eucalyptus globulus* industrial bark as a biorefinery feedstock: Chemical and fuel characterization. *Ind. Crops Prod.* **2018**, *123*, 262–270. [CrossRef]
22. Ruiz, J.A.; Juárez, M.C.; Morales, M.P.; Muñoz, P.; Mendívil, M.A. Biomass gasification for electricity generation: Review of current technology barriers. *Renew. Sustain. Energy Rev.* **2013**, *18*, 174–183. [CrossRef]
23. Prins, M.J.; Ptasinski, K.J.; Janssen, F.J.J.G. More efficient biomass gasification via torrefaction. *Energy* **2006**, *31*, 3458–3470. [CrossRef]
24. Parthasarathy, P.; Sheeba, K.N. Combined slow pyrolysis and steam gasification of biomass for hydrogen generation—A review. *Int. J. Energy Res.* **2015**, *39*, 147–164. [CrossRef]
25. Erlach, B.; Harder, B.; Tsatsaronis, G. Combined hydrothermal carbonization and gasification of biomass with carbon capture. *Energy* **2012**, *45*, 329–338. [CrossRef]
26. Sjostrom, E. *Wood Chemistry: Fundamentals and Applications*; Gulf Professional Publishing: Houston, TX, USA; Academic Press: New York, NY, USA, 1993; ISBN 0126474818.
27. Rasi, S.; Kilpeläinen, P.; Rasa, K.; Korpinen, R.; Raitanen, J.-E.; Vainio, M.; Kitunen, V.; Pulkkinen, H.; Jyske, T. Cascade processing of softwood bark with hot water extraction, pyrolysis and anaerobic digestion. *Bioresour. Technol.* **2019**, *292*, 121893. [CrossRef] [PubMed]
28. Van Eck, N.; Waltman, L. Software survey: VOSviewer, a computer program for bibliometric mapping. *Scientometrics* **2010**, *84*, 523–538. [CrossRef]
29. Rietzler, B.; Ek, M. Adding value to spruce bark by the isolation of nanocellulose in a biorefinery concept. *ACS Sustain. Chem. Eng.* **2021**, *9*, 1398–1405. [CrossRef]
30. Rietzler, B.; Karlsson, M.; Kwan, I.; Lawoko, M.; Ek, M. Fundamental Insights on the Physical and Chemical Properties of Organosolv Lignin from Norway Spruce Bark. *Biomacromolecules* **2022**, *23*, 3349–3358. [CrossRef] [PubMed]
31. Bento, A.; Escórcio, R.; Tomé, A.S.; Robertson, M.; Gaugler, E.C.; Malthus, S.J.; Raymond, L.G.; Hill, S.J.; Pereira, C.S. Pinus radiata bark sequentially processed using $scCO_2$ and an ionic liquid catalyst yields plentiful resin acids and alkanoic acids enriched suberin. *Ind. Crops Prod.* **2022**, *185*, 115172. [CrossRef]
32. Wijeyekoon, S.; Suckling, I.; Fahmy, M.; Hall, P.; Bennett, P. Techno-economic analysis of tannin and briquette co-production from bark waste: A case study quantifying symbiosis benefits in biorefinery. *Biofuels Bioprod. Biorefin.* **2021**, *15*, 1332–1344. [CrossRef]
33. Pals, M.; Lauberts, M.; Zijlstra, D.S.; Ponomarenko, J.; Arshanitsa, A.; Deuss, P.J. Mild organosolv delignification of residual aspen bark after extractives isolation as a step in biorefinery processing schemes. *Molecules* **2022**, *27*, 3185. [CrossRef] [PubMed]
34. Li, D.; Moriana, R.; Ek, M. From forest residues to hydrophobic nanocomposites with high oxygen-barrier properties. *Nord. Pulp Pap. Res. J.* **2016**, *31*, 261–269. [CrossRef]
35. Le Normand, M.; Mélida, H.; Holmbom, B.; Michaelsen, T.E.; Inngjerdingen, M.; Bulone, V.; Paulsen, B.S.; Ek, M. Hot-water extracts from the inner bark of Norway spruce with immunomodulating activities. *Carbohydr. Polym.* **2014**, *101*, 699–704. [CrossRef] [PubMed]
36. Le Normand, M.; Edlund, U.; Holmbom, B.; Ek, M. Hot-water extraction and characterization of spruce bark non-cellulosic polysaccharides. *Nord. Pulp Pap. Res. J.* **2012**, *27*, 18–23. [CrossRef]

37. Ferreira, R.; Garcia, H.; Sousa, A.F.; Guerreiro, M.; Duarte, F.J.S.; Freire, C.S.R.; Calhorda, M.J.; Silvestre, A.J.D.; Kunz, W.; Rebelo, L.P.N. Unveiling the dual role of the cholinium hexanoate ionic liquid as solvent and catalyst in suberin depolymerisation. *RSC Adv.* **2014**, *4*, 2993–3002. [CrossRef]
38. Şen, A.U.; Nobre, C.; Durão, L.; Miranda, I.; Pereira, H.; Gonçalves, M. Low-temperature biochars from cork-rich and phloem-rich wastes: Fuel, leaching, and methylene blue adsorption properties. *Biomass Convers. Biorefin.* **2020**, *12*, 3899–3909. [CrossRef]
39. Chen, W.-H.; Peng, J.; Bi, X.T. A state-of-the-art review of biomass torrefaction, densification and applications. *Renew. Sustain. Energy Rev.* **2015**, *44*, 847–866. [CrossRef]
40. Bridgwater, A.V. Principles and practice of biomass fast pyrolysis processes for liquids. *J. Anal. Appl. Pyrolysis* **1999**, *51*, 3–22. [CrossRef]
41. Shoji, T.; Kawamoto, H.; Saka, S. Boiling point of levoglucosan and devolatilization temperatures in cellulose pyrolysis measured at different heating area temperatures. *J. Anal. Appl. Pyrolysis* **2014**, *109*, 185–195. [CrossRef]
42. Bridgwater, A.V. Review of fast pyrolysis of biomass and product upgrading. *Biomass Bioenergy* **2012**, *38*, 68–94. [CrossRef]
43. Bridgwater, A.V. Upgrading fast pyrolysis liquids. In *Thermochemical Processing of Biomass: Conversion into Fuels, Chemicals and Power*; Wiley: Hoboken, NJ, USA, 2011; pp. 157–199.
44. van der Stelt, M.J.C.; Gerhauser, H.; Kiel, J.H.A.; Ptasinski, K.J. Biomass upgrading by torrefaction for the production of biofuels: A review. *Biomass Bioenergy* **2011**, *35*, 3748–3762. [CrossRef]
45. Trinh, T.N.; Jensen, P.A.; Dam-Johansen, K.; Knudsen, N.O.; Sørensen, H.R.; Hvilsted, S. Comparison of lignin, macroalgae, wood, and straw fast pyrolysis. *Energy Fuels* **2013**, *27*, 1399–1409. [CrossRef]
46. Prins, M.J.; Ptasinski, K.J.; Janssen, F.J.J.G. Torrefaction of wood: Part 2. Analysis of products. *J. Anal. Appl. Pyrolysis* **2006**, *77*, 35–40. [CrossRef]
47. Benavente, V.; Fullana, A. Torrefaction of olive mill waste. *Biomass Bioenergy* **2015**, *73*, 186–194. [CrossRef]
48. Ciolkosz, D.; Wallace, R. A review of torrefaction for bioenergy feedstock production. *Biofuels Bioprod. Biorefin.* **2011**, *5*, 317–329. [CrossRef]
49. Niu, Y.; Lv, Y.; Lei, Y.; Liu, S.; Liang, Y.; Wang, D. Biomass torrefaction: Properties, applications, challenges, and economy. *Renew. Sustain. Energy Rev.* **2019**, *115*, 109395. [CrossRef]
50. Kung, K.S.; Shanbhogue, S.; Slocum, A.H.; Ghoniem, A.F. A decentralized biomass torrefaction reactor concept. Part I: Multi-scale analysis and initial experimental validation. *Biomass Bioenergy* **2019**, *125*, 196–203. [CrossRef]
51. Chen, W.-H.; Cheng, C.-L.; Show, P.-L.; Ong, H.C. Torrefaction performance prediction approached by torrefaction severity factor. *Fuel* **2019**, *251*, 126–135. [CrossRef]
52. Chew, J.J.; Doshi, V. Recent advances in biomass pretreatment—Torrefaction fundamentals and technology. *Renew. Sustain. Energy Rev.* **2011**, *15*, 4212–4222. [CrossRef]
53. Chen, W.H.; Lin, B.J.; Lin, Y.Y.; Chu, Y.S.; Ubando, A.T.; Show, P.L.; Ong, H.C.; Chang, J.S.; Ho, S.H.; Culaba, A.B.; et al. Progress in biomass torrefaction: Principles, applications and challenges. *Prog. Energy Combust. Sci.* **2021**, *82*, 100887. [CrossRef]
54. Safar, M.; Lin, B.-J.; Chen, W.-H.; Langauer, D.; Chang, J.-S.; Raclavska, H.; Pétrissans, A.; Rousset, P.; Pétrissans, M. Catalytic effects of potassium on biomass pyrolysis, combustion and torrefaction. *Appl. Energy* **2019**, *235*, 346–355. [CrossRef]
55. Thengane, S.K.; Kung, K.S.; Gupta, A.; Ateia, M.; Sanchez, D.L.; Mahajani, S.M.; Lim, C.J.; Sokhansanj, S.; Ghoniem, A.F. Oxidative torrefaction for cleaner utilization of biomass for soil amendment. *Clean. Eng. Technol.* **2020**, *1*, 100033. [CrossRef]
56. Adams, P.W.R.; Shirley, J.E.J.; McManus, M.C. Comparative cradle-to-gate life cycle assessment of wood pellet production with torrefaction. *Appl. Energy* **2015**, *138*, 367–380. [CrossRef]
57. Brownsort, P.A. *Biomass Pyrolysis Processes: Review of Scope, Control and Variability*; United Kingdom Biochar Reserch Centre: London, UK, 2009; Volume 1, pp. 1–39.
58. Brown, T.R.; Wright, M.M.; Brown, R.C. Estimating profitability of two biochar production scenarios: Slow pyrolysis vs. fast pyrolysis. *Biofuels Bioprod. Biorefin.* **2011**, *5*, 54–68. [CrossRef]
59. Kan, T.; Strezov, V.; Evans, T.J. Lignocellulosic biomass pyrolysis: A review of product properties and effects of pyrolysis parameters. *Renew. Sustain. Energy Rev.* **2016**, *57*, 1126–1140. [CrossRef]
60. Anupam, K.; Sharma, A.K.; Lal, P.S.; Dutta, S.; Maity, S. Preparation, characterization and optimization for upgrading Leucaena leucocephala bark to biochar fuel with high energy yielding. *Energy* **2016**, *106*, 743–756. [CrossRef]
61. Liu, Z.; Han, G. Production of solid fuel biochar from waste biomass by low temperature pyrolysis. *Fuel* **2015**, *158*, 159–165. [CrossRef]
62. Chan, K.Y.; Van Zwieten, L.; Meszaros, I.; Downie, A.; Joseph, S. Agronomic values of greenwaste biochar as a soil amendment. *Soil Res.* **2007**, *45*, 629–634. [CrossRef]
63. Brassard, P.; Godbout, S.; Lévesque, V.; Palacios, J.H.; Raghavan, V.; Ahmed, A.; Hogue, R.; Jeanne, T.; Verma, M. Biochar for soil amendment. In *Char and Carbon Materials Derived from Biomass*; Elsevier: Amsterdam, The Netherlands, 2019; pp. 109–146.
64. Lehmann, J.; Rillig, M.C.; Thies, J.; Masiello, C.A.; Hockaday, W.C.; Crowley, D. Biochar effects on soil biota—A review. *Soil Biol. Biochem.* **2011**, *43*, 1812–1836. [CrossRef]
65. Qian, K.; Kumar, A.; Zhang, H.; Bellmer, D.; Huhnke, R. Recent advances in utilization of biochar. *Renew. Sustain. Energy Rev.* **2015**, *42*, 1055–1064. [CrossRef]
66. Zheng, X.; Xu, W.; Dong, J.; Yang, T.; Shangguan, Z.; Qu, J.; Li, X.; Tan, X. The effects of biochar and its applications in the microbial remediation of contaminated soil: A review. *J. Hazard. Mater.* **2022**, 129557. [CrossRef]

67. Kloss, S.; Zehetner, F.; Dellantonio, A.; Hamid, R.; Ottner, F.; Liedtke, V.; Schwanninger, M.; Gerzabek, M.H.; Soja, G. Characterization of slow pyrolysis biochars: Effects of feedstocks and pyrolysis temperature on biochar properties. *J. Environ. Qual.* **2012**, *41*, 990–1000. [CrossRef]
68. Mandal, A.; Singh, N.; Purakayastha, T.J. Characterization of pesticide sorption behaviour of slow pyrolysis biochars as low cost adsorbent for atrazine and imidacloprid removal. *Sci. Total Environ.* **2017**, *577*, 376–385. [CrossRef] [PubMed]
69. Ronsse, F.; Van Hecke, S.; Dickinson, D.; Prins, W. Production and characterization of slow pyrolysis biochar: Influence of feedstock type and pyrolysis conditions. *Gcb Bioenergy* **2013**, *5*, 104–115. [CrossRef]
70. Tsoumis, G. *Science and Technology of Wood: Structure, Properties, Utilization*; Van Nostrand Reinhold: New York, NY, USA, 1991; Volume 115.
71. Xiang, S.; Wu, Q.; Ren, W.; Guo, W.; Ren, N. Mechanism of powdered activated carbon enhancing caproate production. *Chin. Chem. Lett.* **2023**, *34*, 107714. [CrossRef]
72. Manyà, J.J.; Azuara, M.; Manso, J.A. Biochar production through slow pyrolysis of different biomass materials: Seeking the best operating conditions. *Biomass Bioenergy* **2018**, *117*, 115–123. [CrossRef]
73. Patwa, D.; Bordoloi, U.; Dubey, A.A.; Ravi, K.; Sekharan, S.; Kalita, P. Energy-efficient biochar production for thermal backfill applications. *Sci. Total Environ.* **2022**, *833*, 155253. [CrossRef]
74. Sen, U.; Martins, M.; Santos, E.; Lemos, M.A.; Lemos, F.; Pereira, H. Slow Pyrolysis of *Quercus cerris* Cork: Characterization of Biochars and Pyrolysis Volatiles. *Environments* **2023**, *10*, 4. [CrossRef]
75. Muñoz, E.; Curaqueo, G.; Cea, M.; Vera, L.; Navia, R. Environmental hotspots in the life cycle of a biochar-soil system. *J. Clean. Prod.* **2017**, *158*, 1–7. [CrossRef]
76. Funke, A.; Ziegler, F. Hydrothermal carbonization of biomass: A summary and discussion of chemical mechanisms for process engineering. *Biofuels Bioprod. Biorefin.* **2010**, *4*, 160–177. [CrossRef]
77. Liu, Z.; Quek, A.; Hoekman, S.K.; Balasubramanian, R. Production of solid biochar fuel from waste biomass by hydrothermal carbonization. *Fuel* **2013**, *103*, 943–949. [CrossRef]
78. Kambo, H.S.; Dutta, A. A comparative review of biochar and hydrochar in terms of production, physico-chemical properties and applications. *Renew. Sustain. Energy Rev.* **2015**, *45*, 359–378. [CrossRef]
79. Koprivica, M.; Petrović, J.; Ercegović, M.; Simić, M.; Milojković, J.; Šoštarić, T.; Dimitrijević, J. Improvement of combustible characteristics of *Paulownia* leaves via hydrothermal carbonization. *Biomass Convers. Biorefin.* **2022**. [CrossRef]
80. Gao, Y.; Wang, X.; Wang, J.; Li, X.; Cheng, J.; Yang, H.; Chen, H. Effect of residence time on chemical and structural properties of hydrochar obtained by hydrothermal carbonization of water hyacinth. *Energy* **2013**, *58*, 376–383. [CrossRef]
81. Wang, T.; Zhai, Y.; Zhu, Y.; Li, C.; Zeng, G. A review of the hydrothermal carbonization of biomass waste for hydrochar formation: Process conditions, fundamentals, and physicochemical properties. *Renew. Sustain. Energy Rev.* **2018**, *90*, 223–247. [CrossRef]
82. Libra, J.A.; Ro, K.S.; Kammann, C.; Funke, A.; Berge, N.D.; Neubauer, Y.; Titirici, M.-M.; Fühner, C.; Bens, O.; Kern, J. Hydrothermal carbonization of biomass residuals: A comparative review of the chemistry, processes and applications of wet and dry pyrolysis. *Biofuels* **2011**, *2*, 71–106. [CrossRef]
83. Jain, A.; Balasubramanian, R.; Srinivasan, M.P. Hydrothermal conversion of biomass waste to activated carbon with high porosity: A review. *Chem. Eng. J.* **2016**, *283*, 789–805. [CrossRef]
84. Wang, Y.; Qiu, L.; Zhu, M.; Sun, G.; Zhang, T.; Kang, K. Comparative evaluation of hydrothermal carbonization and low temperature pyrolysis of eucommia ulmoides oliver for the production of solid biofuel. *Sci. Rep.* **2019**, *9*, 5535. [CrossRef] [PubMed]
85. Monedero, E.; Lapuerta, M.; Pazo, A.; Díaz-Robles, L.A.; Pino-Cortés, E.; Campos, V.; Vallejo, F.; Cubillos, F.; Gómez, J. Effect of hydrothermal carbonization on the properties, devolatilization, and combustion kinetics of Chilean biomass residues. *Biomass Bioenergy* **2019**, *130*, 105387. [CrossRef]
86. Zhang, Z.; Zhu, Z.; Shen, B.; Liu, L. Insights into biochar and hydrochar production and applications: A review. *Energy* **2019**, *171*, 581–598. [CrossRef]
87. Mendecka, B.; Lombardi, L.; Micali, F.; De Risi, A. Energy recovery from olive pomace by hydrothermal carbonization on hypothetical industrial scale: A LCA perspective. *Waste Biomass Valorization* **2020**, *11*, 5503–5519. [CrossRef]
88. Wilk, M.; Magdziarz, A.; Kalemba-Rec, I.; Szymańska-Chargot, M. Upgrading of green waste into carbon-rich solid biofuel by hydrothermal carbonization: The effect of process parameters on hydrochar derived from acacia. *Energy* **2020**, *202*, 117717. [CrossRef]
89. Alves, O.; Nobre, C.; Durão, L.; Monteiro, E.; Brito, P.; Gonçalves, M. Effects of dry and hydrothermal carbonisation on the properties of solid recovered fuels from construction and municipal solid wastes. *Energy Convers. Manag.* **2021**, *237*, 114101. [CrossRef]
90. Nobre, C.; Alves, O.; Durão, L.; Şen, A.; Vilarinho, C.; Gonçalves, M. Characterization of hydrochar and process water from the hydrothermal carbonization of Refuse Derived Fuel. *Waste Manag.* **2021**, *120*, 303–313. [CrossRef] [PubMed]
91. Wang, L.; Barta-Rajnai, E.; Skreiberg, Ø.; Khalil, R.; Czégény, Z.; Jakab, E.; Barta, Z.; Grønli, M. Effect of torrefaction on physiochemical characteristics and grindability of stem wood, stump and bark. *Appl. Energy* **2018**, *227*, 137–148. [CrossRef]
92. Butnaru, E.; Brebu, M. The Thermochemical Conversion of Forestry Residues from Silver Fir (*Abies alba* Mill.) by Torrefaction and Pyrolysis. *Energies* **2022**, *15*, 3483. [CrossRef]

93. Almeida, G.; Brito, J.O.; Perré, P. Alterations in energy properties of eucalyptus wood and bark subjected to torrefaction: The potential of mass loss as a synthetic indicator. *Bioresour. Technol.* **2010**, *101*, 9778–9784. [CrossRef] [PubMed]
94. Arteaga-Pérez, L.E.; Segura, C.; Bustamante-García, V.; Cápiro, O.G.; Jiménez, R. Torrefaction of wood and bark from *Eucalyptus globulus* and *Eucalyptus nitens*: Focus on volatile evolution vs. feasible temperatures. *Energy* **2015**, *93*, 1731–1741. [CrossRef]
95. Sessa, F.; Merlin, G.; Canu, P. Pine bark valorization by activated carbons production to be used as VOCs adsorbents. *Fuel* **2022**, *318*, 123346. [CrossRef]
96. Lee, Y.; Park, J.; Ryu, C.; Gang, K.S.; Yang, W.; Park, Y.-K.; Jung, J.; Hyun, S. Comparison of biochar properties from biomass residues produced by slow pyrolysis at 500 °C. *Bioresour. Technol.* **2013**, *148*, 196–201. [CrossRef]
97. Kuznetsov, B.N.; Chesnokov, N.V.; Ivanov, I.P.; Kuznetsova, S.A.; Ivanchenko, N.M. Production of porous carbon materials from bark. *Solid Fuel Chem.* **2015**, *49*, 278–288. [CrossRef]
98. Costa, P.A.; Barreiros, M.A.; Mouquinho, A.I.; Silva, O.E.; Paradela, F.; Oliveira, F.A.C. Slow pyrolysis of cork granules under nitrogen atmosphere: By-products characterization and their potential valorization. *Biofuel Res. J.* **2022**, *9*, 1562–1572. [CrossRef]
99. Saletnik, B.; Saletnik, A.; Zaguła, G.; Bajcar, M.; Puchalski, C. Oak Biomass in the Form of Wood, Bark, Brushwood, Leaves and Acorns in the Production Process of Multifunctional Biochar. *Molecules* **2022**, *27*, 7191. [CrossRef] [PubMed]
100. Paczkowski, S.; Comi, J.; Küttner, I.; Knappe, V.; Russ, M.; Robles, L.A.D.; Jaeger, D.; Pelz, S. Hydrothermal treatment (HTT) for improving the fuel properties of biomass residues. *Biomass Convers. Biorefin.* **2022**, *13*, 6257–6279. [CrossRef]
101. Gao, P.; Zhou, Y.; Meng, F.; Zhang, Y.; Liu, Z.; Zhang, W.; Xue, G. Preparation and characterization of hydrochar from waste eucalyptus bark by hydrothermal carbonization. *Energy* **2016**, *97*, 238–245. [CrossRef]
102. Rodríguez Correa, C.; Stollovsky, M.; Hehr, T.; Rauscher, Y.; Rolli, B.; Kruse, A. Influence of the carbonization process on activated carbon properties from lignin and lignin-rich biomasses. *ACS Sustain. Chem. Eng.* **2017**, *5*, 8222–8233. [CrossRef]
103. Kemppainen, K.; Siika-aho, M.; Pattathil, S.; Giovando, S.; Kruus, K. Spruce bark as an industrial source of condensed tannins and non-cellulosic sugars. *Ind. Crops Prod.* **2014**, *52*, 158–168. [CrossRef]
104. Umezawa, T. Chemistry of extractives. In *Wood and Cellulosic Chemistry*; CRC Press: Boca Raton, FL, USA, 2000; pp. 213–241.
105. Sakai, K. Chemistry of bark. In *Wood and Cellulosic Chemistry*; CRC Press: Boca Raton, FL, USA, 2000; pp. 243–274.
106. Şen, A.; Miranda, I.; Esteves, B.; Pereira, H. Chemical characterization, bioactive and fuel properties of waste cork and phloem fractions from *Quercus cerris* L. bark. *Ind. Crops Prod.* **2020**, *157*, 112909. [CrossRef]
107. Şen, A.; Miranda, I.; Santos, S.; Graça, J.; Pereira, H. The chemical composition of cork and phloem in the rhytidome of *Quercus cerris* bark. *Ind. Crops Prod.* **2010**, *31*, 417–422. [CrossRef]
108. Pereira, H. Variability of the chemical composition of cork. *BioResources* **2013**, *8*, 2246–2256. [CrossRef]
109. Sen, A.; Zhianski, M.; Glushkova, M.; Petkova, K.; Ferreira, J.; Pereira, H. Chemical composition and cellular structure of corks from *Quercus suber* trees planted in Bulgaria and Turkey. *Wood Sci. Technol.* **2016**, *50*, 1261–1276. [CrossRef]
110. Ferreira, J.; Miranda, I.; Şen, U.; Pereira, H. Chemical and cellular features of virgin and reproduction cork from *Quercus variabilis*. *Ind. Crops Prod.* **2016**, *94*, 638–648. [CrossRef]
111. Sousa, V.; Ferreira, J.P.A.; Miranda, I.; Quilhó, T.; Pereira, H. *Quercus rotundifolia* bark as a source of polar extracts: Structural and chemical characterization. *Forests* **2021**, *12*, 1160. [CrossRef]
112. Ferreira, J.P.A.; Miranda, I.; Sousa, V.B.; Pereira, H. Chemical composition of barks from *Quercus faginea* trees and characterization of their lipophilic and polar extracts. *PLoS ONE* **2018**, *13*, e0197135. [CrossRef]
113. Sillero, L.; Prado, R.; Andrés, M.A.; Labidi, J. Characterisation of bark of six species from mixed Atlantic forest. *Ind. Crops Prod.* **2019**, *137*, 276–284. [CrossRef]
114. Vangeel, T.; Neiva, D.M.; Quilho, T.; Costa, R.A.; Sousa, V.; Sels, B.F.; Pereira, H. Tree bark characterization envisioning an integrated use in a biorefinery. *Biomass Convers. Biorefin.* **2021**, *13*, 2029–2043. [CrossRef]
115. Carmo, J.F.; Miranda, I.; Quilhó, T.; Sousa, V.B.; Carmo, F.H.D.J.; Latorraca, J.V.F.; Pereira, H. Chemical and structural characterization of the bark of Albizia niopoides trees from the Amazon. *Wood Sci. Technol.* **2016**, *50*, 677–692. [CrossRef]
116. Mota, G.S.; Sartori, C.J.; Miranda, I.; Quilho, T.; Mori, F.A.; Pereira, H. Bark anatomy, chemical composition and ethanol-water extract composition of Anadenanthera peregrina and Anadenanthera colubrina. *PLoS ONE* **2017**, *12*, e0189263. [CrossRef]
117. Ferreira, J.P.A.; Quilhó, T.; Pereira, H. Characterization of *Betula pendula* outer bark regarding cork and phloem components at chemical and structural levels in view of biorefinery integration. *J. Wood Chem. Technol.* **2017**, *37*, 10–25. [CrossRef]
118. Carmo, J.F.; Miranda, I.; Quilhó, T.; Sousa, V.B.; Cardoso, S.; Carvalho, A.M.; Carmo, F.H.D.J.; Latorraca, J.V.F.; Pereira, H. *Copaifera langsdorffii* bark as a source of chemicals: Structural and chemical characterization. *J. Wood Chem. Technol.* **2016**, *36*, 305–317. [CrossRef]
119. Miranda, I.; Lima, L.; Quilhó, T.; Knapic, S.; Pereira, H. The bark of *Eucalyptus sideroxylon* as a source of phenolic extracts with anti-oxidant properties. *Ind. Crops Prod.* **2016**, *82*, 81–87. [CrossRef]
120. Carmo, J.F.; Miranda, I.; Quilhó, T.; Carvalho, A.M.; Carmo, F.H.D.J.; de Figueiredo Latorraca, J.V.; Pereira, H. Bark characterisation of the Brazilian hardwood *Goupia glabra* in terms of its valorisation. *BioResources* **2016**, *11*, 4794–4807. [CrossRef]
121. Rios, P.; Cabral, V.; Santos, S.; Mori, F.; Graça, J. The chemistry of Kielmeyera coriacea outer bark: A potential source for cork. *Eur. J. Wood Wood Prod.* **2014**, *72*, 509–519. [CrossRef]
122. Sousa, T.B.; da Mota, G.S.; da Araujo, E.S.; Carréra, J.C.; Silva, E.P.; Souza, S.G.; Lorenço, M.S.; Mori, F.A. Chemical and structural characterization of *Myracrodruon urundeuva* barks aiming at their potential use and elaboration of a sustainable management plan. Biomass Convers. *Biorefinery* **2022**, *12*, 1583–1593. [CrossRef]

123. Mota, G.S.; Sartori, C.J.; Ferreira, J.; Miranda, I.; Quilhó, T.; Mori, F.A.; Pereira, H. Cellular structure and chemical composition of cork from *Plathymenia reticulata* occurring in the Brazilian Cerrado. *Ind. Crops Prod.* **2016**, *90*, 65–75. [CrossRef]
124. Baptista, I.; Miranda, I.; Quilhó, T.; Gominho, J.; Pereira, H. Characterisation and fractioning of *Tectona grandis* bark in view of its valorisation as a biorefinery raw-material. *Ind. Crops Prod.* **2013**, *50*, 166–175. [CrossRef]
125. Miranda, I.; Gominho, J.; Mirra, I.; Pereira, H. Chemical characterization of barks from *Picea abies* and *Pinus sylvestris* after fractioning into different particle sizes. *Ind. Crops Prod.* **2012**, *36*, 395–400. [CrossRef]
126. Miranda, I.; Mirra, I.; Gominho, J.; Pereira, H. Fractioning of bark of Pinus pinea by milling and chemical characterization of the different fractions. *Maderas. Cienc. Tecnol.* **2017**, *19*, 185–194. [CrossRef]
127. Ferreira, J.P.A.; Miranda, I.; Gominho, J.; Pereira, H. Chemical characterization of cork and phloem from Douglas fir outer bark. *Holzforschung* **2016**, *70*, 475–483. [CrossRef]
128. Leite, C.; Pereira, H. Cork-containing barks—A review. *Front. Mater.* **2017**, *3*, 63. [CrossRef]
129. Pettersen, R.C. The chemical composition of wood. *Chem. Solid Wood* **1984**, *207*, 57–126.
130. Rowell, R.M. *Handbook of Wood Chemistry and Wood Composites*; CRC press: Boca Raton, LA, USA, 2012; ISBN 0429109091.
131. Gomes, S.V.F.; Portugal, L.A.; dos Anjos, J.P.; de Jesus, O.N.; de Oliveira, E.J.; David, J.P.; David, J.M. Accelerated solvent extraction of phenolic compounds exploiting a Box-Behnken design and quantification of five flavonoids by HPLC-DAD in *Passiflora* species. *Microchem. J.* **2017**, *132*, 28–35. [CrossRef]
132. Richter, B.E.; Jones, B.A.; Ezzell, J.L.; Porter, N.L.; Avdalovic, N.; Pohl, C. Accelerated solvent extraction: A technique for sample preparation. *Anal. Chem.* **1996**, *68*, 1033–1039. [CrossRef]
133. Wang, G.; Lee, A.S.; Lewis, M.; Kamath, B.; Archer, R.K. Accelerated solvent extraction and gas chromatography/mass spectrometry for determination of polycyclic aromatic hydrocarbons in smoked food samples. *J. Agric. Food Chem.* **1999**, *47*, 1062–1066. [CrossRef]
134. Ghitescu, R.-E.; Volf, I.; Carausu, C.; Bühlmann, A.-M.; Gilca, I.A.; Popa, V.I. Optimization of ultrasound-assisted extraction of polyphenols from spruce wood bark. *Ultrason. Sonochem.* **2015**, *22*, 535–541. [CrossRef]
135. Alves, T.P.; Triques, C.C.; Palsikowski, P.A.; da Silva, C.; Fiorese, M.L.; da Silva, E.A.; Fagundes-Klen, M.R. Improved extraction of bioactive compounds from *Monteverdia aquifolia* leaves by pressurized-liquid and ultrasound-assisted extraction: Yield and chemical composition. *J. Supercrit. Fluids* **2022**, *181*, 105468. [CrossRef]
136. Mohammad Azmin, S.N.H.; Abdul Manan, Z.; Wan Alwi, S.R.; Chua, L.S.; Mustaffa, A.A.; Yunus, N.A. Herbal processing and extraction technologies. *Sep. Purif. Rev.* **2016**, *45*, 305–320. [CrossRef]
137. Chupin, L.; Maunu, S.L.; Reynaud, S.; Pizzi, A.; Charrier, B.; Bouhtoury, F.C. Microwave assisted extraction of maritime pine (*Pinus pinaster*) bark: Impact of particle size and characterization. *Ind. Crop. Prod.* **2015**, *65*, 142–149. [CrossRef]
138. Bouras, M.; Chadni, M.; Barba, F.J.; Grimi, N.; Bals, O.; Vorobiev, E. Optimization of microwave-assisted extraction of polyphenols from *Quercus bark. Ind. Crops Prod.* **2015**, *77*, 590–601. [CrossRef]
139. De Melo, M.M.R.; Silvestre, A.J.D.; Silva, C.M. Supercritical fluid extraction of vegetable matrices: Applications, trends and future perspectives of a convincing green technology. *J. Supercrit. Fluids* **2014**, *92*, 115–176. [CrossRef]
140. Rodrigues, V.H.; de Melo, M.M.R.; Portugal, I.; Silva, C.M. Extraction of Eucalyptus leaves using solvents of distinct polarity. Cluster analysis and extracts characterization. *J. Supercrit. Fluids* **2018**, *135*, 263–274. [CrossRef]
141. Şen, A.; De Melo, M.M.R.; Silvestre, A.J.D.; Pereira, H.; Silva, C.M. Prospective pathway for a green and enhanced friedelin production through supercritical fluid extraction of *Quercus cerris* cork. *J. Supercrit. Fluids* **2015**, *97*, 247–255. [CrossRef]
142. de Melo, M.M.R.; Şen, A.; Silvestre, A.J.D.; Pereira, H.; Silva, C.M. Experimental and modeling study of supercritical CO_2 extraction of *Quercus cerris* cork: Influence of ethanol and particle size on extraction kinetics and selectivity to friedelin. *Sep. Purif. Technol.* **2017**, *187*, 34–45. [CrossRef]
143. Vieira, P.G.; de Melo, M.M.R.; Şen, A.; Simões, M.M.Q.; Portugal, I.; Pereira, H.; Silva, C.M. *Quercus cerris* extracts obtained by distinct separation methods and solvents: Total and friedelin extraction yields, and chemical similarity analysis by multidimensional scaling. *Sep. Purif. Technol.* **2020**, *232*, 115924. [CrossRef]
144. de Melo, M.M.R.; Vieira, P.G.; Şen, A.; Pereira, H.; Portugal, I.; Silva, C.M. Optimization of the supercritical fluid extraction of *Quercus cerris* cork towards extraction yield and selectivity to friedelin. *Sep. Purif. Technol.* **2020**, *238*, 116395. [CrossRef]
145. de Melo, M.M.R.; Oliveira, E.L.G.; Silvestre, A.J.D.; Silva, C.M. Supercritical fluid extraction of triterpenic acids from *Eucalyptus globulus* bark. *J. Supercrit. Fluids* **2012**, *70*, 137–145. [CrossRef]
146. Domingues, R.M.A.; de Melo, M.M.R.; Oliveira, E.L.G.; Neto, C.P.; Silvestre, A.J.D.; Silva, C.M. Optimization of the supercritical fluid extraction of triterpenic acids from *Eucalyptus globulus* bark using experimental design. *J. Supercrit. Fluids* **2013**, *74*, 105–114. [CrossRef]
147. Domingues, R.M.A.; Oliveira, E.L.G.; Freire, C.S.R.; Couto, R.M.; Simões, P.C.; Neto, C.P.; Silvestre, A.J.D.; Silva, C.M. Supercritical fluid extraction of *Eucalyptus globulus* Bark—A promising approach for triterpenoid production. *Int. J. Mol. Sci.* **2012**, *13*, 7648–7662. [CrossRef] [PubMed]
148. Rodrigues, V.H.; de Melo, M.M.R.; Portugal, I.; Silva, C.M. Simulation and techno-economic optimization of the supercritical CO_2 extraction of *Eucalyptus globulus* bark at industrial scale. *J. Supercrit. Fluids* **2019**, *145*, 169–180. [CrossRef]
149. Rodrigues, V.H.; de Melo, M.M.R.; Portugal, I.; Silva, C.M. Lupane-type triterpenoids from Acacia dealbata bark extracted by different methods. *Ind. Crops Prod.* **2021**, *170*, 113734. [CrossRef]

150. Yu-hong, Z.; Tao, Y.U.; Yang, W. Extraction of betulin from bark of *Betula platyphylla* by supercritical carbon dioxide extraction. *J. For. Res.* **2003**, *14*, 202–204. [CrossRef]
151. Ubando, A.T.; Ng, E.A.S.; Chen, W.-H.; Culaba, A.B.; Kwon, E.E. Life cycle assessment of microalgal biorefinery: A state-of-the-art review. *Bioresour. Technol.* **2022**, *360*, 127615. [CrossRef]
152. Cherubini, F.; Bargigli, S.; Ulgiati, S. Life cycle assessment (LCA) of waste management strategies: Landfilling, sorting plant and incineration. *Energy* **2009**, *34*, 2116–2123. [CrossRef]
153. Carlqvist, K.; Wallberg, O.; Lidén, G.; Börjesson, P. Life cycle assessment for identification of critical aspects in emerging technologies for the extraction of phenolic compounds from spruce bark. *J. Clean. Prod.* **2022**, *333*, 130093. [CrossRef]
154. Hytönen, J.; Beuker, E.; Viherä-Aarnio, A. Clonal variation in basic density, moisture content and heating value of wood, bark and branches in hybrid aspen. *Silva Fenn.* **2018**, *52*, 9938. [CrossRef]
155. Saka, S. Chemical composition and distribution. In *Wood and Cellulosic Chemistry*; CRC Press: Boca Raton, FL, USA, 2000; pp. 51–81.
156. Priyanto, D.E.; Ueno, S.; Sato, N.; Kasai, H.; Tanoue, T.; Fukushima, H. Ash transformation by co-firing of coal with high ratios of woody biomass and effect on slagging propensity. *Fuel* **2016**, *174*, 172–179. [CrossRef]
157. Teixeira, P.; Lopes, H.; Gulyurtlu, I.; Lapa, N.; Abelha, P. Evaluation of slagging and fouling tendency during biomass co-firing with coal in a fluidized bed. *Biomass Bioenergy* **2012**, *39*, 192–203. [CrossRef]
158. Bandara, Y.W.; Gamage, P.; Gunarathne, D.S. Hot water washing of rice husk for ash removal: The effect of washing temperature, washing time and particle size. *Renew. Energy* **2020**, *153*, 646–652. [CrossRef]
159. Miranda, T.; Montero, I.; Sepúlveda, F.J.; Arranz, J.I.; Rojas, C.V.; Nogales, S. A review of pellets from different sources. *Materials* **2015**, *8*, 1413–1427. [CrossRef] [PubMed]
160. Kemppainen, K.; Inkinen, J.; Uusitalo, J.; Nakari-Setälä, T.; Siika-aho, M. Hot water extraction and steam explosion as pretreatments for ethanol production from spruce bark. *Bioresour. Technol.* **2012**, *117*, 131–139. [CrossRef] [PubMed]
161. Liu, H.; Chen, Y.; Yang, H.; Gentili, F.G.; Söderlind, U.; Wang, X.; Zhang, W.; Chen, H. Hydrothermal carbonization of natural microalgae containing a high ash content. *Fuel* **2019**, *249*, 441–448. [CrossRef]
162. Lynam, J.G.; Reza, M.T.; Yan, W.; Vásquez, V.R.; Coronella, C.J. Hydrothermal carbonization of various lignocellulosic biomass. Biomass Convers. *Biorefinery* **2015**, *5*, 173–181.
163. Harun, J.; Labosky, P. Chemical constituents of five northeastern barks. *Wood Fiber Sci.* **2007**, *17*, 274–280.
164. Ferreira, R.; Garcia, H.; Sousa, A.F.; Freire, C.S.R.; Silvestre, A.J.D.; Rebelo, L.P.N.; Pereira, C.S. Isolation of suberin from birch outer bark and cork using ionic liquids: A new source of macromonomers. *Ind. Crops Prod.* **2013**, *44*, 520–527. [CrossRef]
165. Ekman, R. The Suberin Monomers and Triterpenoids from the Outer Bark of *Betula verrucosa* Ehrh. *Birch Bark Suberin Triterpenoids* **1983**, *37*, 205–211.
166. Gandini, A.; Neto, C.P.; Silvestre, A.J.D. Suberin: A promising renewable resource for novel macromolecular materials. *Prog. Polym. Sci.* **2006**, *31*, 878–892. [CrossRef]
167. Moiteiro, C.; Marcelo Curto, M.J.; Mohamed, N.; Bailén, M.; Martínez-Díaz, R.; González-Coloma, A. Biovalorization of friedelane triterpenes derived from cork processing industry byproducts. *J. Agric. Food Chem.* **2006**, *54*, 3566–3571. [CrossRef]
168. Castola, V.; Marongiu, B.; Bighelli, A.; Floris, C.; Laï, A.; Casanova, J. Extractives of cork (*Quercus suber* L.): Chemical composition of dichloromethane and supercritical CO_2 extracts. *Ind. Crops Prod.* **2005**, *21*, 65–69. [CrossRef]
169. Mahmood, N.; Yuan, Z.; Schmidt, J.; Xu, C.C. Depolymerization of lignins and their applications for the preparation of polyols and rigid polyurethane foams: A review. *Renew. Sustain. Energy Rev.* **2016**, *60*, 317–329. [CrossRef]
170. D'Souza, J.; Yan, N. Producing bark-based polyols through liquefaction: Effect of liquefaction temperature. *ACS Sustain. Chem. Eng.* **2013**, *1*, 534–540. [CrossRef]
171. D'Souza, J.; Camargo, R.; Yan, N. Polyurethane foams made from liquefied bark-based polyols. *J. Appl. Polym. Sci.* **2014**, *131*. [CrossRef]
172. Esteves, B.; Dulyanska, Y.; Costa, C.; Ferreira, J.V.; Domingos, I.; Pereira, H.; de Lemos, L.T.; Cruz-Lopes, L.V. Cork liquefaction for polyurethane foam production. *BioResources* **2017**, *12*, 2339–2353. [CrossRef]
173. Mateus, M.M.; Acero, N.F.; Bordado, J.C.; dos Santos, R.G. Sonication as a foremost tool to improve cork liquefaction. *Ind. Crops Prod.* **2015**, *74*, 9–13. [CrossRef]
174. Dos Santos, R.G.; Bordado, J.C.; Mateus, M.M. Microwave-assisted liquefaction of cork—From an industrial waste to sustainable chemicals. *Ind. Eng. Manag.* **2015**, *4*, 173.
175. Huang, Y.; Duan, Y.; Qiu, S.; Wang, M.; Ju, C.; Cao, H.; Fang, Y.; Tan, T. Lignin-first biorefinery: A reusable catalyst for lignin depolymerization and application of lignin oil to jet fuel aromatics and polyurethane feedstock. *Sustain. Energy Fuels* **2018**, *2*, 637–647. [CrossRef]
176. Karnaouri, A.; Rova, U.; Christakopoulos, P. Effect of different pretreatment methods on birch outer bark: New biorefinery routes. *Molecules* **2016**, *21*, 427. [CrossRef] [PubMed]
177. Ma, J.; He, J.; Kong, X.; Zheng, J.; Han, L.; Liu, Y.; Zhu, Z.; Zhang, Z. From agricultural cellulosic waste to food delivery packaging: A mini-review. *Chin. Chem. Lett.* **2023**, *34*, 107407. [CrossRef]
178. Sun, C.; Wu, W.; Chang, H.; Wang, R.; Wang, K.; Zhong, N.; Zhang, T.; He, X.; Sun, F.; Zhang, E. A tailored bifunctional carbon catalyst for efficient glycosidic bond fracture and selective hemicellulose fractionation. *Bioresour. Technol.* **2022**, *362*, 127861. [CrossRef] [PubMed]

179. Ince, P.J. *Estimating Effective Heating Value of Wood or Bark Fuels at Various Moisture Contents*; US Department of Agriculture, US Forest Service: Madison, WI, USA, 1977; Volume 13.
180. Nunes, L.J.R.; Matias, J.C.O.; Catalão, J.P.S. Energy recovery from cork industrial waste: Production and characterisation of cork pellets. *Fuel* **2013**, *113*, 24–30. [CrossRef]
181. Lehtikangas, P. Quality properties of pelletised sawdust, logging residues and bark. *Biomass Bioenergy* **2001**, *20*, 351–360. [CrossRef]
182. Kaliyan, N.; Morey, R.V. Factors affecting strength and durability of densified biomass products. *Biomass Bioenergy* **2009**, *33*, 337–359. [CrossRef]
183. Tahirović, A.; Bašić, N.; Avdibegović, S. Antioxidant capacity and phenolic content of *Fraxinus ornus* L. and *Fraxinus pennsylvanica* Marsch. leaves and bark extracts. *Rad. Šumar. Fak. Univ. Sarajev.* **2017**, *47*, 1–12. [CrossRef]
184. Dzubak, P.; Hajduch, M.; Vydra, D.; Hustova, A.; Kvasnica, M.; Biedermann, D.; Markova, L.; Urban, M.; Sarek, J. Pharmacological activities of natural triterpenoids and their therapeutic implications. *Nat. Prod. Rep.* **2006**, *23*, 394–411. [CrossRef] [PubMed]
185. Jablonsky, M.; Nosalova, J.; Sladkova, A.; Haz, A.; Kreps, F.; Valka, J.; Miertus, S.; Frecer, V.; Ondrejovic, M.; Sima, J. Valorisation of softwood bark through extraction of utilizable chemicals: A review. *Biotechnol. Adv.* **2017**, *35*, 726–750. [CrossRef]
186. Foti, M.C. Antioxidant properties of phenols. *J. Pharm. Pharmacol.* **2007**, *59*, 1673–1685. [CrossRef]
187. Chen, X.; Zhu, J.; Song, W.; Xiao, L.-P. Integrated cascade biorefinery processes to transform woody biomass into phenolic monomers and carbon quantum dots. *Front. Bioeng. Biotechnol.* **2021**, *9*, 803138. [CrossRef] [PubMed]
188. Li, J.; Maplesden, F. Commercial production of tannins from radiata pine bark for wood adhesives. *Trans. Inst. Prof. Eng. N. Z. Electr. Eng. Sect.* **1998**, *25*, 46–51.
189. Yazaki, Y. Utilization of flavonoid compounds from bark and wood: A review. *Nat. Prod. Commun.* **2015**, *10*, 1934578X1501000333. [CrossRef]
190. Pizzi, A. Condensed tannins for adhesives. *Ind. Eng. Chem. Prod. Res. Dev.* **1982**, *21*, 359–369. [CrossRef]
191. Jablonsky, M.; Haz, A.; Sladkova, A.; Strizincova, P.; Skulcova, A.; Majova, V.; Jablonsky, J. Nutraceuticals as phenolic bioactive compounds analysis of softwood bark and their possibilities of industry applications. *J. Hyg. Eng. Des.* **2019**, *26*, 93–99.
192. Song, Y.; Mobley, J.K.; Motagamwala, A.H.; Isaacs, M.; Dumesic, J.A.; Ralph, J.; Lee, A.F.; Wilson, K.; Crocker, M. Gold-catalyzed conversion of lignin to low molecular weight aromatics. *Chem. Sci.* **2018**, *9*, 8127–8133. [CrossRef]

energies

MDPI

Article

B20 Fuel Compatibility with Steels in Case of Fuel Contamination

Katriina Sirviö *, Jonna Kaivosoja, Carolin Nuortila, Huaying Wang-Alho, Seppo Niemi and Teemu Ovaska

School of Technology and Innovations, University of Vaasa, P.O. Box 700, 65101 Vaasa, Finland; jonna.kaivosoja@uwasa.fi (J.K.); carolin.nuortila@uwasa.fi (C.N.); huaying.wang-alho@uwasa.fi (H.W.-A.); seppo.niemi@uwasa.fi (S.N.); teemu.ovaska@uwasa.fi (T.O.)
* Correspondence: katriina.sirvio@uwasa.fi

Abstract: This study evaluated the compatibility with steels for three B20 fuel samples blended from fossil diesel and used cooking oil methyl ester. One sample was untreated and its concentration of copper was analyzed as <1 ppm. Another sample was doped by adding Cu at a concentration of ≤2 ppm and the third sample by adding Cu at a concentration of ≤4 ppm. Steel samples (carbon steel, stainless steel and a special alloy) were then put into the fuel blends and stored at 50 °C for 692 h. After storing, the metal concentrations of the fuel blends were again analyzed, and signs of corrosion were evaluated visually. The aim of this study was to find out if the fuel already contaminated by copper will affect the corrosion of the chosen steel qualities. Additionally, fuel properties were measured for all three blend samples before the immersion of steels. Visual evaluation of the steels indicated that signs of corrosion were seen in all studied samples, but Cu doping did not increase the signs of corrosion notably. The results also showed that the copper content from 1 to 2 and 4 ppm reduced the oxidation stability and increased the acid number of the fuel samples.

Keywords: biodiesel; diesel fuel; fuel blending; metal contamination; corrosion of steels

1. Introduction

Before achieving the green transition to 100% renewable energy, the transition period may be relieved by blending fossil fuels with sustainable renewable ones. One option for blending components lies in biodiesels, also called fatty acid methyl ester (FAME) fuels, which are mono alkyl esters made of vegetable oils or animal fats. FAME fuels have potential to replace fossil diesel fuel and promote the sustainability in many engine applications. The use of FAME fuels also reduces emissions of greenhouse gases (GHG), if their raw materials are used vegetable oils or nonedible wastewater-based microalgae, and the entire production process is sustainable [1]. When used in internal combustion engines, the combustion properties are quite similar to those of fossil diesel, and FAMEs do not require engine modifications [2].

Biodiesels usually produce fewer emissions of carbon monoxides, hydrocarbons and particulate matter than fossil diesel fuel [3]. Fuel blending results have shown that a relatively high content of biodiesel, 40 vol.%, leads to an optimum increase of brake thermal efficiency and decrease in NOx emissions [4].

Sustainably produced biodiesel is used at large quantities already today. To sustain this positive development, the fuel producers must still optimize the properties of these fuel options. Despite the good properties of biodiesels, using biodiesel has caused a number of practical problems in diesel engines. Among others, important issues to be considered are oxidation stability, chemical composition, ash formation in combustion and the compatibility with metallic structures. Many of the materials used in a diesel engine, especially in the fuel system, might not be compatible with biodiesel.

Most of the mentioned issues are related to the content of some metals in the fuel. Metals can end up in fuel in several ways. Large quantities of sustainable biodiesels are refined from nonedible plants. The elemental composition of seeds varies between different plant species [5] and can also depend on the soil quality in which the feedstock plants have grown [6]. Trace elements from the plant-derived oil can show in biodiesels [7] unless they are removed in the manufacturing process. Heavy metals can be transported from plants to fuel [8]. Trace elements, most of which are metals, can also end up in fuels as a result of fuel contamination. Additionally, metals may be found as a natural, intrinsic part of some lower-quality fossil fuels, which must be taken into account when looking at the blends of bio- and fossil diesels [9]. In biodiesels made of used cooking oils (UCOs), traces of metals may originate from packages. Nevertheless, according to the Cárdenas et al. study [10], for instance, the copper (Cu) contents in different used cooking oils are low, varying from 0.05 to 5 ppm. UCOs are usually pretreated before the biodiesel production process, and metals are removed with other solid impurities. However, metals can also dissolute to fuel from tanks or fuel systems. When in contact with steels, especially carbon steel, biodiesel or its blends may dissolve small traces of metals. In the Sirviö et al. study [11], it was found that a small amount of Cu was dissolved from carbon steel when a carbon steel bar was immersed in B20 fuel.

A fuel contaminated by metals can also threaten the oxidation stability of FAME. Especially high content of Cu seems to have a detrimental effect on the oxidation stability as the metals accelerate the fuel's oxidation [12–14]. A Kugelmeier et al. study [15] stated that metals have a catalytic effect on oxidation stability.

Biodiesels are less compatible with metallic structures and components than fossil diesel fuel. This has to be considered when biodiesels are in contact with the fuel system. The fuel system includes, e.g., fuel tanks, supply pumps and lines, fuel filters plus injection pumps and nozzles which are prone to possible corrosion caused by fuel. Also, the combustion products of biodiesel can affect the combustion chamber walls and valves, as well as the crank mechanism. Several types of steels are largely used in the components prone to these adverse effects [16,17].

The compatibility of steels in contact with biodiesel blends was studied by Kugelmeier et al. [15]. They used biodiesels made of soybean oil, beef tallow and swine lard and formed the blends of B7, B15 and B30 with fossil diesel. In these experiments, carbon steel, stainless steel and aluminum (Al) did not corrode. Cu instead corroded intensively, lost mass and consequently degraded the blends. Fazal et al. [18] found that metals corrode when they were immersed in palm biodiesel. They also stated that, of the studied metals, Cu was causing the highest corrosion rate.

Oni et al. [19] investigated corrosion behaviors of mild steel (MS), Al and Cu immersed in microalgae biodiesel and its blends. Of the studied metals, Cu was the most prone to corrode. They also found that neat biodiesel was more reactive with metals than its blends with fossil diesel fuel.

As seen above, several researchers have studied the material compatibility and corrosion effects of biodiesels. However, these phenomena have rarely been investigated with used cooking oil methyl esters (UCOMEs) and their blends. When assessing the overall compatibility of the fuels, storage and fueling infrastructure, additional knowledge of the stability and corrosivity of UCOMEs should be acquired. The present study increases the scientific knowledge about the UCOME blends for the manufacturers of engines and fuel systems.

The objective of the present study was to investigate the corrosion of carbon steel, stainless steel and a special alloy immersed in metal-doped biodiesel–diesel blends. B20 fuel was chosen to be studied instead of 100% biodiesel, because many engine manufacturers allow B20 fuel as drop-in fuel, while B100 is not generally allowed [20]. The aim of the study was to find out if the fuel, already contaminated by Cu, will affect the corrosion of the chosen steel qualities.

To make B20 fuel, UCOME readily treated with antioxidant and light fuel oil (LFO) were mixed on the volume basis. Three B20 fuel batches were prepared with three different concentrations of Cu (1 ppm, $\leq$2 ppm and $\leq$4 ppm). Steel samples were then put into the fuel blends and stored at 50 °C for more than 600 h. The temperature (of 50 °C) was chosen above the room temperature to accelerate the possible corrosion reactions. However, to make the sample handling convenient, the temperature was chosen not to exceed 50 °C. The Cu concentration was analyzed twice during the storing period to see if metals dissolve from steels to the fuel. Visual evaluations of samples were carried out five times during the immersion. The kinematic viscosity, density, distillation properties, oxidation stability and acid number were measured for the three batches of B20 fuel blends before steel immersion to see if the addition of Cu content from 1 to 2 and 4 ppm has an effect on the mentioned properties.

2. Materials and Methods

The experimental part of the present study was similar to the research by Sirviö et al. [11]. Three B20 fuel batches were prepared from fossil light fuel oil (80 vol.%) and used cooking oil methyl ester or UCOME (20 vol.%). Cu was added to the batches at concentrations of B20-1: 1 ppm, B20-2: ≤2 ppm and B20-3: ≤4 ppm. The metal additive used for doping was Conostan Cu Standard. The metal concentrations were checked by analyzing the samples B20-1, B20-2 and B20-3, being no more than 1 ppm, less than 2 ppm and less than 4 ppm, respectively.

One of the steels was carbon steel (CS; grade Imatra 550), the other stainless steel (SS; 1.4301/1.4307), and the third a special alloy steel called as MoC210M/25CrMo4+SH (CrMo). Ingredient certificates of the steels were available. The concentrations of metals originally found in the steels are listed in Table 1. The amount of Fe was above 70% according to the literature for all the steels [21].

Table 1. The concentration of the elements in the three steel types.

	Carbon Steel	Stainless Steel	MoC210M/25CrMo4+SH
Cu (%)	0.20	0.96	0.20
Fe (%)	>70	>70	>70
Mn (%)	1.13	1.66	0.81
Si (%)	0.24	0.27	0.27
V (%)	0.06		0.01
Pb (%)	0.0003		0.0007
C (%)	0.13	0.023	0.27
Cr (%)	0.20	18.1	0.99
Ni (%)	0.15	8.01	0.19
Mo (%)	0.03	0.29	0.22

Fuel samples of 150 mL were taken, three from each batch, B20-1, B20-2 and B20-3. These samples were filled into borosilicate glass bottles. One round bar of each steel was immersed in one of each fuel sample, B20-1, B20-2 and B20-3, yielding altogether nine samples (Figure 1). The bars were 10 cm long and 2 cm in diameter. Half of the bar was immersed in the fuel blend liquid, and the other half was positioned in the air space above the blend in the glass bottle. The bottles were closed with polypropylene screw caps.

Figure 1. Steel bars immersed in fuel samples. Here, BS (black steel) is referring to CS (carbon steel).

The samples were stored in an oven at 50 °C for 692 h. The oven was a TIN-TN50 Phoenix Instrument, a device manufactured by Phoenix Instrument in Garbsen, Germany.

The Cu concentrations in the fuel blends were analyzed after 308 and 692 h of storing in the oven. The bottles were opened for sampling the fuel. The concentrations of Cu were measured with a PerkinElmer 7000 DV ICP-OES spectrometer, a device manufactured by PerkinElmer, Inc. in Waltham, MA, USA. The analysis procedure entailed diluting a weighed sample with kerosene in a weight ratio of 1:1 and injecting this solution into the spectrometer plasma. Calibration was performed with the known concentrations of the multielement standard which included Cu.

The method evaluates the presence and concentration of these elements by comparing the intensity of the light wavelength emissions with those from the known concentrations of the various elements. The analyses were performed by an in-house method based on Standards EN 14538 [22] and EN 14107 [23] and in accordance with the spectrometer manufacturer's advice.

The samples, fuels and metal bars, were also evaluated visually five times, after 140, 308, 500, 644 and 692 h.

Kinematic viscosity and density were measured with a Stabinger SVM 3000 rotational viscometer, which uses torque and speed measurements to determine viscosity. The device calculates the dynamic viscosity, η (in mPas), from the rotor speed.

The viscometer also has a density-measuring cell which utilizes the oscillating U-tube principle. The kinematic viscosity (mm^2/s), was calculated automatically based on these measurements [24] according to Standard ASTM D7042 [25]. The used device is manufactured by Anton Paar GmbH in Graz, Austria.

The oxidation stability index (OSI) was measured with a Metrohm 873 Biodiesel Rancimat instrument, a device manufactured by Metrohm Ltd. in Herisau Switzerland. This uses an accelerated method to determine the oxidation stability of biodiesel–diesel blends, blowing a stream of air through a heated fuel sample. Vaporizing compounds from the sample drift with air into water, and the consequential change in the water's conductivity is measured. The end point is achieved when the increase in conductivity is at its highest. This method is described in European Standard EN 15751 [26].

The distillation curve was measured with PAC's OptiPMD Micro distillation analyzer, a device manufactured by PAC LP in Houston, TX, USA. This uses a method for the determination of the distillation characteristics of light and middle distillates derived from petroleum and related products, as described in Standard EN ISO 3405 [27].

The acid number was analyzed with a Metrohm Titrando 888 titrator, a device manufactured by Metrohm AG in Herisau, Switzerland. This instrument utilizes the potentiometric titration method. The acid number is measured by a titrimetric method, whereby alkali neutralizes the free fatty acids present in the sample. The sample is diluted with isopropanol and titrated with potassium hydroxide (KOH). The results are expressed as mg KOH/g. The measurement was performed according to Standard EN 14104 [28].

The total ester content and fatty acid profile of neat UCOME were analyzed with a PerkinElmer Clarus 580 gas chromatograph. The method for analyzing total ester content is presented in European Standard EN 14103 [29], and methylheptadecanoate is used as an internal standard. The samples are diluted to heptane. The fatty acid profile is measured using an in-house method that is based on EN 14103.

All the presented results are arithmetic means of at least two replicants. The used methods for analyzing kinematic viscosity, density, oxidation stability and acid number are validated. The relative standard deviations for the measurements were: kinematic viscosity <1%; oxidation stability 4.5%; density <1%; acid number 7.9%; distillation 1.1%.

3. Results

Figure 2 presents the results of the Cu analysis. The results at time 0 h represent the blend batches before immersing the metal bars. As compared with the initial state, the concentration of Cu in the fuel increased in those B20-1 and B20-2 blends where CS was immersed. The increase in B20-1 CS was from 1 to 2 ppm, in B20-2 CS from 2 to 3 ppm and in B20-3 CS from 4 to 5 ppm. The concentration of Cu did not increase in the other samples, but it decreased in the B20-3 SS sample from 4 to 3 ppm. The doping itself did not have any impact on Cu dissolution from CS steel bars into the fuel, as the amount of Cu also increased in the sample which was not doped (B20-1 CS). The results of the B20-3 SS analysis showed slightly smaller concentrations (3 ppm) than the concentration before the steel immersion (≤4 ppm). All the measured changes were small and therefore within the measurement accuracy that was ensured by using the internal standard and a sample with a known concentration. A variation of 1 ppm in the results may still occur between the measurements series. However, the Cu content in all CS samples rose. The dissolution of Cu seems not to depend only on its concentration in the steel as the stainless steel contains more Cu (0.96%) than carbon steel (0.20%) (Table 1), yet Cu concentrations did not rise in fuel samples with stainless steel.

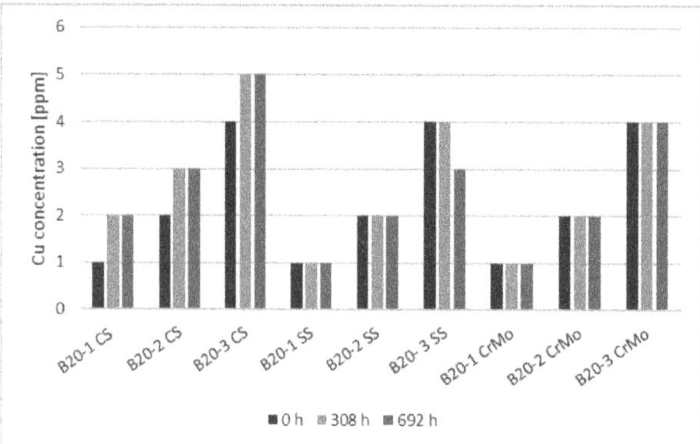

Figure 2. Results of Cu analysis of B20 fuels.

The samples, fuels and metal bars, were also evaluated visually after 140, 308, 500, 644 and 692 h. Notable signs of corrosion were not detected on the steel bars, only minor signs which were seen after 692 h. However, the color of the fuels containing 2 and 4 ppm of Cu changed from yellow to light yellow already after 140 h. The color change was clearer in later evaluations and was also seen after 692 h in B20- CS (which contained Cu less than 1 ppm). After 692 h, some sediment was seen in the bottom of the bottles which contained stainless steel and CrMo steel.

In addition to the immersion and corrosion tests, the present study also covered analyses of UCOME biodiesel and its blends with fossil diesel. These measurements were made before steel immersion. All the results of fuel analyses below are average values of at least two replicates. The total ester content and fatty acid profile of UCOME is given in Table 2.

Table 2. Total ester content and fatty acid profile of UCOME.

Parameter	Method	UCOME	EN 14214 [30]
Total ester content (%)	EN 14103	96.9 ± 0.28	min. 96.5
Linoleic acid (%)	EN 14103	1.85 ± 0.003	max. 12
C14 (%)	EN 14103/in-house method	0.93 ± 0.004	
C16 (%)	EN 14103/in-house method	26.2 ± 0.141	
C18 (%)	EN 14103/in-house method	7.28 ± 3.89	
C18:1 (%)	EN 14103/in-house method	35.3 ± 2.92	
C18:2 n-6 (%)	EN 14103/in-house method	23.8 ± 4.34	
C20:2 n-6 (%)	EN 14103/in-house method	0.1 ± 1.80	
C22 (%)	EN 14103/in-house method	0.21 ± 3.66	

The density, kinematic viscosity and distillation results of the B20-1, B20-2 and B20-3 samples complied with the limits set for diesel fuel in Standard EN 590 (Table 3). Cu doping did not affect the kinematic viscosity and density of the fresh samples. The density of B20-1 was 816 kg/m^3 and that of B20-2 and B20-3 was 815 kg/m^3. The kinematic viscosity of all three samples was 2.57 mm^2/s.

Table 3. Fuel properties measured for the B20 samples and UCOME used for blending. The LFO fulfilled the requirements of EN 590.

Parameter	Method	B20-1	B20-2	B20-3	UCOME	EN 590 [31]
OSI	EN 15751	22	1	1	5	min. 20 (min. 8 *)
Density at 15 °C, kg/m^3	ASTM D7042	816	815	815	878	800–845 (class 1)
Kinematic viscosity at 40 °C, mm^2/s	ASTM D7042	2.57	2.57	2.57	4.29	1.50–4.00 (class 1)
Acid number (mg KOH/g)	EN 14104	0.12	0.13	0.15	0.24	max. 0.50 *
Distillation,	EN ISO 3405					
% recovered at 250 °C, vol. %		35	35	35	0	max. 65
% recovered at 350 °C, vol. %		100	100	100	99	min. 85
95% recovered, °C		340	340	340	351	360

* EN 14214.

The acid number was compared to the limit set in the biodiesel standard EN 14214, and all the results of the B20 samples were clearly below the limit (max. 0.50 mgKOH/g). The acid number for B20-1 was 0.12, for B20-2 0.13 and for B20-3 0.15 mgKOH/g, showing a slightly increasing trend corresponding to increasing amount of Cu (Figure 3). However, it must be borne in mind that EN 14214 applies to 100% biodiesel, whereas the samples were B20 blends.

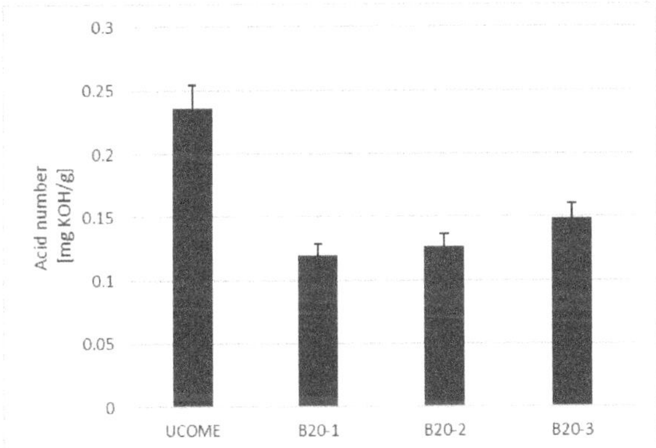

Figure 3. Acid number results of UCOME and the B20 samples.

The distillation results showed that the B20 fuels conformed to the limits set for diesel fuel distillation in Standard EN 590. This stipulates that a maximum of 65 vol.% has to be recovered at 250 °C, a minimum of 85 vol.% at 350 °C and at least 95 vol.% at 360 °C. For B20-1, B20-2 and B20-3, the recovery at 250 °C was approximately 35 vol.%. At 350 °C, 100 vol % was recovered.

A pronounced reduction in OSI was observed immediately after preparing the Cu-doped samples of 2 ppm and 4 ppm. The B20-1 sample, containing Cu at a concentration of less than 1 ppm, had an OSI result of 22 h. The samples B20-2 ($\leq$2 ppm of Cu) and B20-3 ($\leq$4 ppm of Cu) both showed results of 1 h, well below EN 590's minimum OSI value of 20 h. Furthermore, although the increment in metal content from 1 to 2 ppm was enough to weaken oxidation stability from 22 h to 1 h, the increase from 2 to 4 ppm did not cause any further deterioration in OSI.

4. Discussion

Fazal et al. [18] stated that neat biodiesel is corrosive when it is in contact with metals. They also stated that the corrosion rate is higher in contact with Cu than with other metals.

Oni et al. [19] studied corrosion behaviors of mild steel (MS), Al and Cu. According to them, neat biodiesel was more reactive with metals than its blends. The same statement was also made as in the study [18]: Cu was the most prone to corrode when it was immersed in biodiesel.

The study of Kugelmeier et al. [15] showed that biodiesel–diesel blends of B7, B15 and B30 did not cause corrosion in carbon and stainless steels, neither did aluminum corrode during the tests.

The corrosive effect of added Cu on steels was not clearly seen in our study. Nevertheless, the Cu content of all the blends increased in the case of the carbon steel, and the color of the B20 fuel changed.

These results also showed that even low concentrations of Cu—from 2 to 4 ppm—can notably cause a significant deterioration in the fuel quality. Doping with Cu did not have any effect on the kinematic viscosity, nor the density of the fresh samples. However, the addition of Cu had a major effect on fuel quality in terms of oxidation stability of the fresh B20 samples, with the OSI result decreasing from 22 h to 1 h. This is outside EN 590's OSI limit.

Acidity may restrict the uptake in the market of alternative fuels, especially biodiesels. It seems that the increased content of Cu in the B20-2 and B20-3 fuel samples slightly raised their acid numbers, but these were still well within the requirement for 100% biodiesel set in EN 14214 with <0.50 mg KOH/g.

The obtained results show that Cu may have a deleterious catalytic effect on the oxidation stability, and even minor metallic contamination may cause serious damage to the B20 fuel quality. Metals at a certain concentration initiate or accelerate the oxidation reaction but are not consumed in the reaction. The same observation was also made by Jain and Sharma [12] and Yang [14]. Jain and Sharma in particular showed that the higher the metal contents, the lower the oxidation stability. However, they concluded that metal concentrations higher than 2 ppm did not lead to further reductions of OSI. The present study supports that finding, but more detailed chemical analysis would be required to approve this statement. Kugelmeier et al. [15] studied corrosion behaviors of metallic materials in contact with biodiesel–diesel fuel blends. The authors found that the corrosion rate of Cu was higher than that of other studied materials. This was also shown by Fazal et al. [18]. At the same time, Kugelmeier et al. [15] found relatively low OSI values, notably less than 30 min for biodiesel blends in contact with Cu. In the present study, it is presumed that Cu had an effect on the marked reduction of oxidation stability, while it affected only one out of the three tested steel qualities.

The knowledge of biodiesel's corrosive effect on metals provides an opportunity to study antioxidants' effect on biodiesel stability in more detail. Serqueira et al. [32] studied the efficiency of corrosion inhibition of antioxidants for carbon steel and Cu. The main findings were that treating biodiesel with antioxidants forms protective layers of carbonaceous solids over metals, and propyl gallate increases the oxidation stability of residual cooking oil biodiesel exposed to carbon steel and Cu. On the other hand, copper oxide CuO, at a concentration of 100 mg/L, has also been used as an additive in biodiesel blends to improve diesel engine performance and cause a marginal emissions reduction [33]. Most probably, part of the enhancement of this additive in the combustion process is due to its oxygen content. Still, adding Cu to fuel may also have beneficial effects on its combustion properties.

5. Conclusions

The primary task of this study was to investigate the corrosion of carbon steel, stainless steel and a special alloy immersed in metal-doped biodiesel–diesel blends. UCOME, treated with antioxidant BHT, and LFO were mixed based on volume to make B20 fuel. Three B20 fuel batches were prepared with three different concentrations of Cu (1 ppm, $\leq$2 ppm and $\leq$4 ppm). Steel samples were then put into the fuel blends and stored at 50 °C for 692 h. The Cu analysis of fuels was conducted twice during the storing period to see if Cu will dissolute from steels to fuels. Visual evaluations of samples were carried out five times during the immersion. The kinematic viscosity, density, distillation properties, oxidation stability and acid number were measured for the three batches of B20 fuel blends to see if the addition of Cu content from 1 to 2 and 4 ppm will have an effect on the mentioned properties.

Based on the performed work, the following conclusions could be drawn:

- When the fuel is contaminated by Cu, it will not amplify the corrosion of the selected steels. The concentration of Cu increased slightly in all samples containing carbon steel, not only in those which were doped with metals.
- The Cu concentration rose in all three fuel blends when CS steel was present.
- The dissolution of Cu seems not to depend only on its concentration in the steel.
- Based on the visual evaluations, only minor signs of corrosion were seen in the studied steel qualities. However, the color of the fuel samples changed during the immersion.
- Even a minor metal contamination may cause serious damage to fuel quality in terms of the oxidative stability. An incremental increase in metal content from 1 to 2 ppm was sufficient to weaken oxidation stability from 22 h to 1 h. A further increment from 2 to 4 ppm did not additionally affect the results.
- The addition of Cu to the blend did not affect fuel density.
- The increased content of Cu slightly raised the acid numbers, but the results were still safely below the limits set for biodiesels, ranging from 0.12 to 0.15 mg KOH/g.

Author Contributions: Conceptualization, K.S. and C.N.; methodology, K.S. and C.N.; investigation, J.K. and H.W.-A.; resources, S.N.; writing—original draft preparation, K.S., J.K. and T.O.; writing—review and editing, C.N. and S.N.; visualization, J.K.; supervision, S.N.; project administration, S.N.; funding acquisition, S.N. All authors have read and agreed to the published version of the manuscript.

Funding: This research received no external funding.

Data Availability Statement: No new data were created, except for what is written in the article.

Conflicts of Interest: The authors declare no conflict of interest.

References

1. Mizik, T.; Gyarmati, G. Economic and Sustainability of Biodiesel Production—A Systematic Literature Review. *Clean Technol.* **2021**, *3*, 2. [CrossRef]
2. Zhu, L.; Cheung, C.S.; Zhang, W.G.; Huang, Z. Compatibility of different biodiesel composition with acrylonitrile butadiene rubber (NBR). *Fuel* **2015**, *158*, 288–292. [CrossRef]
3. Ovaska, T.; Niemi, S.; Sirviö, K.; Nilsson, O.; Portin, K.; Asplund, T. Effects of alternative marine diesel fuels on the exhaust particle size distributions of an off-road diesel engine. *Appl. Therm. Eng.* **2019**, *150*, 1168–1176. [CrossRef]
4. Zamboni, G.; Capobianco, M. Definition of the optimal content of used cooking oil methyl ester in blends fuelling a turbocharged diesel engine. *Biomass Bioenergy* **2021**, *150*, 106098. [CrossRef]
5. Chaves, E.S.; dos Santos, E.J.; Araujo, R.G.O.; Oliveira, J.V.; Frescura, V.L.A.; Curtius, A.J. Metals and phosphorus determination in vegetable seed used in the production of biodiesel by ICP OES and ICP-MS. *Microchem. J.* **2010**, *96*, 71–76. [CrossRef]
6. Romih, N.; Grabner, B.; Lakota, M.; Ribarič-Lasnik, C. Distribution of Cd, Pb, Zn, Mo and S in juvenile and mature *Brassica napus* L. var. *napus*. *Int. J. Phytoremediation* **2012**, *14*, 282–301. [CrossRef]
7. Lepri, F.G.; Chaves, E.S.; Vieira, M.A.; Ribeiro, A.S.; Curtius, A.J.; DeOliveira, L.C.C.; DeCampos, R.C. Determination of trace elements in vegetable oils and biodiesel by atomic spectrometric techniques—A review. *Appl. Spectrosc. Rev.* **2011**, *46*, 175–206. [CrossRef]
8. Peralta-Videa, J.R.; Lopez, M.L.; Narayan, M.; Saupe, G.; Gardea-Torresdey, J. The biochemistry of environmental heavy metal uptake by plants: Implications for the food chain. *Int. J. Biochem. Cell Biol.* **2009**, *41*, 1665–1677. [CrossRef]

9. Merker, G.; Teichmann, R. *Grundlagen Verbrennungsmotoren*; Springer: Wiesbaden, Germany, 2014; p. 1132.
10. Cárdenas, J.; Orjuela, A.; Sánchez, D.L.; Narváez, P.C.; Katryniok, B.; Clark, J. Pre-treatment of used cooking oils for the production of green chemicals: A review. *J. Clean. Prod.* **2021**, *289*, 125129. [CrossRef]
11. Sirviö, K.; Nuortila, C.; Niemi, S.; Heikkilä, S.; Ovaska, T. The corrosion behavior of steels in contact with metal doped biodiesel-diesel blends. In Proceedings of the CIMAC Congress Paper 011, Busan, Republic of Korea, 12–16 June 2023.
12. Jain, S.; Sharma, M.P. Effect of metal contents on oxidation stability of biodiesel/diesel blends. *Fuel* **2014**, *116*, 14–18. [CrossRef]
13. Knothe, G.; Steidley, K. The effect of metals and metal oxides on biodiesel oxidative stability from promotion to inhibition. *Fuel Process. Technol.* **2018**, *177*, 75–80. [CrossRef]
14. Yang, Z.; Hollebone, B.P.; Wang, Z.; Yang, C.; Landriault, M. Factors affecting oxidation stability of commercially available biodiesel products. *Fuel Process. Technol.* **2013**, *106*, 366–375. [CrossRef]
15. Kugelmeier, C.L.; Marcos Monteiro, M.R.; da Silva, R.; Kuri, S.E.; Sordi, V.L.; Rovere, C.A.D. Corrosion behavior of carbon steel, stainless steel, aluminum and copper upon exposure to biodiesel blended with petrodiesel. *Energy* **2021**, *226*, 120344. [CrossRef]
16. Hoang, A.T.; Tabatabaei, M.; Aghbashlo, M. A review of the effect of biodiesel on the corrosion behavior of metals/alloys in diesel engines. *Energy Sour. Part A Recover. Util. Environ. Eff.* **2020**, *42*, 2923–2943. [CrossRef]
17. Yamagata, H. *The Science and Technology of Materials in Automotive Engines*; Elsevier: Amsterdam, The Netherlands; Woodhead Publishing: Cambridge, UK, 2005; ISBN 9781855737426.
18. Fazal, M.A.; Suhaila, N.R.; Haseeb, A.S.M.A.; Rubaiee, S. Sustainability of additive-doped biodiesel: Analysis of its aggressiveness toward metal corrosion. *J. Clean. Prod.* **2018**, *181*, 508–516. [CrossRef]
19. Oni, B.A.; Sanni, S.E.; Ezurike, B.O.; Okoro, E.E. Effect of corrosion rates of preheated Schinzochytrium sp. microalgae biodiesel on metallic components of a diesel engine. *Alex. Eng. J.* **2022**, *61*, 7509–7528. [CrossRef]
20. SAE Mobilus. Perkins Tackles Decarbonization of Off-Highway Machines. Available online: https://www.sae.org/news/2023/02/perkins-engines-q-and-a (accessed on 1 March 2023).
21. Metallinjalostajat, ry. *Teräskirja*, 9th ed.; Bookwell Oy: Helsinki, Finland, 2014. (In Finnish)
22. *EN 14538*; Fat and Oil Derivatives. Fatty Acid Methyl Ester (FAME). Determination of Ca, K, Mg and Na Content by Optical Emission Spectral Analysis with Inductively Coupled Plasma (ICP OES). Finnish Standards Association SFS: Helsinki, Finland, 2007.
23. *EN 14107*; Fat and Oil Derivatives-Fatty Acid Methyl Esters (FAME)-Determination of Phosphorus Content by Inductively Coupled Plasma (ICP) Emission Spectrometry. Finnish Standards Association SFS: Helsinki, Finland, 2003.
24. Novotny-Farkas, F.; Böhme, W.; Stabinger, H.; Belitsch, W. Customer Portrait. In *The Stabinger Viscometer: A New and Unique Instrument for Oil Service Laboratories*; Anton Paar: Vienna, Austria, 2010; p. 4.
25. *ASTM D704*; Standard Test Method for Dynamic Viscosity and Density of Liquids by Stabinger Viscometer (and the Calculation of Kinematic Viscosity). ASTM International (ASTM): West Conshohocken, PA, USA, 2021.
26. *EN 15751*; Automotive Fuels. Fatty Acid Methyl Ester (FAME) Fuel and Blends with Diesel Fuel. Determination of Oxidation Stability Accelerated Oxidation Method. Finnish Standards Association SFS: Helsinki, Finland, 2014.
27. *EN 3405*; Petroleum Products. Determination of Distillation Characteristics at Atmospheric Pressure. Finnish Standards Association SFS: Helsinki, Finland, 2019.
28. *EN 14104*; Fat and Oil Derivatives. Fatty Acid Methyl Esters (FAME). Determination of Acid Value. Finnish Standards Association SFS: Helsinki, Finland, 2003.
29. *EN 14103:2003*; Fat and Oil Derivatives. Fatty Acid Methyl Esters (FAME). Determination of Ester and Linolenic Acid Methyl Ester Contents. Finnish Standards Association SFS: Helsinki, Finland, 2003.
30. *EN 14214:2012+A1:2014*; Finnish Petroleum Federation. Liquid Petroleum Products—Fatty Acid Methyl Esters (FAME) for Use in Diesel Engines and Heating Applications—Requirements and Test Method. Finnish Standards Association SFS: Helsinki, Finland, 2014.
31. *EN 590*; Automotive Fuels. Diesel. Requirements and Test Methods. Finnish Standards Association SFS: Helsinki, Finland, 2013.
32. Serqueira, D.S.; Pereira, J.F.S.; Squissato, A.L.; Rodrigues, M.A.; Lima, R.C.; Faria, A.M.; Richter, E.M.; Munoz, R.A.A. Oxidative stability and corrosivity of biodiesel produced from residual cooking oil exposed to copper and carbon steel under simulated storage conditions: Dual effect of antioxidants. *Renew. Energy* **2021**, *164*, 1485–1495. [CrossRef]
33. Khidr, M.E.; Megahed, T.F.; Ookawara, S.; Elwardany, A.E. Effects of aluminum and copper oxides nanoparticles as fuel additives on diesel engine operating characteristics. *Atmos. Pollut. Res.* **2023**, *14*, 101721. [CrossRef]

energies

MDPI

Review

Investigation of Underground Coal Gasification in Laboratory Conditions: A Review of Recent Research

Ján Kačur *, Marek Laciak, Milan Durdán and Patrik Flegner

Institute of Control and Informatization of Production Processes, Faculty BERG, Technical University of Košice, Němcovej 3, 042 00 Košice, Slovakia; marek.laciak@tuke.sk (M.L.); milan.durdan@tuke.sk (M.D.); patrik.flegner@tuke.sk (P.F.)
* Correspondence: jan.kacur@tuke.sk; Tel.: +421-55-602-5176

Abstract: The underground coal gasification (UCG) technology converts coal into product gas and provides the option of environmentally and economically attractive coal mining. Obtained syngas can be used for heating, electricity, or chemical production. Numerous laboratory coal gasification trials have been performed in the academic and industrial fields. Lab-scale tests can provide insight into the processes involved with UCG. Many tests with UCG have been performed on ex situ reactors, where different UCG techniques, the effect of gasification agents, their flow rates, pressures, and various control mechanisms to improve gasification efficiency and syngas production have been investigated. This paper provides an overview of recent research on UCG performed on a lab scale. The study focuses on UCG control variables and their optimization, the effect of gasification agents and operating pressure, and it discusses results from the gasification of various lignites and hard coals, the possibilities of steam gasification, hydrogen, and methane-oriented coal gasification, approaches in temperature modeling, changes in coal properties during gasification, and environmental risks of UCG. The review focuses on laboratory tests of UCG on ex situ reactors, results, and the possibility of knowledge transfer to in situ operation.

Keywords: underground coal gasification; UCG; laboratory experiments; syngas; UCG control; coal properties; environmental impacts; lignite gasification; high-pressure gasification; steam gasification

1. Introduction

Underground coal gasification (UCG) is a technology that converts coal into a synthetic gas (i.e., syngas) by heating. Currently, more than 909 trillion tons of coal are in stock worldwide, and only 15% of the available coal is for traditional mining methods. UCG technology is still evolving and provides an alternative to conventional underground coal mining. This technology can be attractive from an environmental and economic point of view and may have extensive use in the future. The technology is also less expensive than conventional coal mining. UCG allows using coal resources that would otherwise be economically or technically unfeasible to extract through conventional mining methods. UCG can lead to lower emissions compared to traditional coal mining and combustion because the process takes place underground and can include carbon capture and storage (CCS) technologies. UCG can potentially reduce the carbon footprint associated with traditional coal utilization. By converting coal in situ, UCG can facilitate the capture and storage of carbon dioxide (CO_2) produced during the process, reducing greenhouse gas emissions.

Additionally, syngas production through UCG can be combined with the technology of carbon capture and storage (CCS), which reduces the emission of CO_2 from industrial processes. Although countries worldwide are placing increased emphasis on reducing greenhouse gas emissions, mitigating climate change, and diversifying their energy mix away from fossil fuels, UCG is discussed as a potential technology that could provide a bridge between traditional coal-based energy systems and cleaner energy alternatives. UCG can be integrated with renewable energy sources to enhance overall energy efficiency. For instance, the syngas produced from UCG can be a backup fuel for renewable energy systems, ensuring a continuous and reliable energy supply. Additionally, surplus renewable energy can be utilized in UCG operations during periods of low demand, optimizing energy utilization. The main benefits of UCG compared to conventional coal utilization methods are listed in [1].

On the other hand, it must be said that there is also criticism of UCG, which is based on the fear of groundwater contamination, subsidence, and the release of harmful gases during the UCG process. The environmental risks associated with UCG should have regulatory frameworks which need to be stringent to ensure safety and environmental protection. Moreover, the global energy situation is shifting towards renewable energy sources due to increasing concerns about climate change. Solar and wind energy have grown substantially and have become more economically viable. Governments, industries, and investors are focusing more on sustainable energy solutions and reducing their reliance on fossil fuels. In this context, the prospects for UCG can be uncertain, as it remains an intermediate solution with potential environmental risks, competing against the rapid advancements in cleaner and more sustainable energy technologies.

However, it is likely that, in times of energy crisis, there may be a lack of energy, and countries may return to using fossil fuels. In addition, some countries have stopped using nuclear power, and renewable sources are insufficient to meet the population's energy needs. Moreover, renewable energy technologies depend on the weather (e.g., sun and wind) to generate energy. They are still significantly new to the market and lack the much-needed efficiency. Setting up renewable energy generation facilities requires a substantial financial outlay.

In a chemical view, gasification is the conversion of bigger organic macromolecules of solid fuel to smaller volatile or gas molecules consisting of syngas fuel. This conversion is obtained by heating the solid fuel to temperatures above 750 °C. Such temperatures are achieved by partial combustion of the solid fuel or indirectly by heating it with an overheated mean. UCG is performed as an auto-thermal process, in which, with the help of injected gasification agent from an injection well, heat is generated in the coal deposit through combustion reactions with coal. When coal is heated, it releases volatile substances, leading to the production of combustible gases. Raw-unprocessed fuel (i.e., solid coal) is converted into combustible syngas, mainly containing CO, CH_4, and H_2. The gasification process also generates heat, CO_2, and H_2O. The primary chemical reactions in coal gasification include drying, pyrolysis, combustion, and gasification of solid hydrocarbons. UCG essentially represents the acquisition of a spatially and thermally distributed reaction zone in a coal seam, in which regions of coal oxidation, coal reduction, and coal pyrolysis occur (see Figure 1).

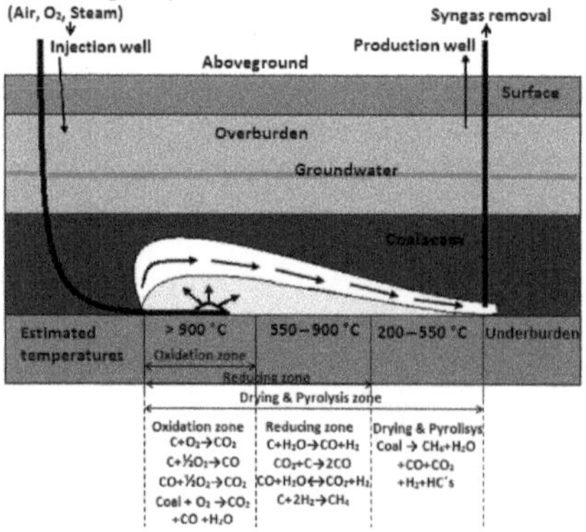

Figure 1. Reaction zones of UCG [2]. Reproduced under CC BY 4.0, accessed on 24 June 2023.

The principle of UCG technology was well illustrated in [3], and the description of the leading chemical processes of UCG has been well described in [4–6].

In an in situ UCG test, the coal is converted into syngas underground in its natural location within a georeactor. A typical in situ UCG test involves drilling injection and production wells into the coal seam. The coal is ignited, and the injection well supplies oxidants (such as air, oxygen, or steam) to the coal seam to initiate the gasification process. Next, the syngas formed is extracted through the production well and can be further processed for various applications. In situ UCG tests provide insights into the behavior of coal, gasification reactions, and the potential for underground resource utilization.

Ex situ UCG involves performing UCG tests aboveground, typically in laboratory gasifiers or reactors. In this method, coal samples are taken from underground coal seams and brought to the surface for gasification experiments. The coal bedding in the ex situ reactor usually corresponds to the underground seam under conditions of geometric similarity. Ex situ UCG tests allow researchers to study and analyze the gasification process in a more controlled and manageable setting. The results obtained from ex situ tests help us to understand the fundamental aspects of coal gasification and optimize the process parameters. In addition, valuable data from laboratory tests can be used to design mathematical models of UCG and perform simulations [7].

Both approaches are valuable in the development and understanding of UCG technology. In situ UCG tests are crucial for evaluating the viability of underground coal gasification in specific geological formations and providing site-specific data. Ex situ UCG tests, on the other hand, help in fundamental research, process optimization, and developing efficient gasification techniques.

This review study focuses on UCG trials performed in laboratory conditions. The motivation for writing this review was to provide a comprehensive view of recent research and achievements in the field. The study aims to overview current trends in UCG laboratory research and the achieved results. Experimental research using ex situ reactors plays an essential role in advancing UCG technology. In addition, the study wants to increase interest in this technology. The scope of the study is wide-ranging. The study aims to provide the reader with automated UCG control and optimization knowledge. Researchers are developing various supporting UCG automatic control algorithms that stabilize or optimize process variables. Such control algorithms make it possible to eliminate the human factor when deciding on control interventions. Moreover, laboratory-scale experiments allow researchers to investigate various parameters and conditions to optimize the UCG process.

Furthermore, the study provides an overview of the field of modeling temperatures in UCG. The UCG operation can be optimized by understanding temperature variations to maximize gas production and minimize undesirable byproducts. This study reviews the effects of different gasification agents (such as air, oxygen, or steam) on the gasification efficiency and syngas composition. The influence of gasification agents, and their flows and pressures, on syngas composition, tar concentration, and cavity growth is also investigated in this study. High-pressure gasification and multiphase gasification with oxygen and steam have great potential. Much attention in UCG research is devoted to steam gasification, methane-oriented gasification, and hydrogen-oriented gasification. In many countries, lignite is mined, while deep, inaccessible deposits or deposits with tectonic faults remain unmined by conventional mining techniques. However, these deposits can be mined with UCG technology.

Therefore, the study provides an overview of recent research in the gasification of various types of lignite with high humidity (i.e., ortho-lignites and meta-lignites). The study compares the calorific values of the produced syngas using different techniques and operating conditions. Laboratory research also investigates the environmental and safety aspects of UCG. These aspects need to be monitored during the operation of the UCG. This study, therefore, includes an overview of research in the field of analysis of pollutants and potential ecological impacts. The research focuses on potential groundwater pollution, analysis of gasification residues, and elimination of risks associated with gas leakage and poisoning in the vulnerable UCG area.

Overall, laboratory-scale research on UCG in ex situ reactors plays a crucial role in understanding the gasification process, which is improved by this technology, and addressing environmental and safety concerns. The findings from such studies contribute to the development of UCG technology and provide valuable insights for potential industrial-scale implementation.

2. Automated Control of UCG

Improving UCG technology also requires implementing advanced automatic control tools. Recent research has focused on developing adaptive control (AC), extreme seeking control (ESC), optimal control, or model predictive control (MPC). Such algorithms can calculate the optimal flow of gasification agents to the UCG reactor. The objective of the control system is the maximization of syngas calorific value or tracking desired calorific values (i.e., setpoints). The control system can be model-free or model-based.

Kostúr et al. [8,9] proposed two experimental reactors (i.e., marked as G1 and G2) to test UCG and develop automated control based on the programmable logic controller (PLC), and supervisory control and acquisition (SCADA). The ex situ reactor G1 had the shape of a cuboid and G2 of a truncated cylinder. The proposed ex situ reactors (i.e., the syngas generator) allow the simulation of UCG in laboratory conditions with controlled air and oxidizer flow. Also designed was the UCG control by an exhaust ventilator on the outlet from the ex situ reactor. These reactors were created as steel vessels to simulate the underground coal seam (see Figure 2).

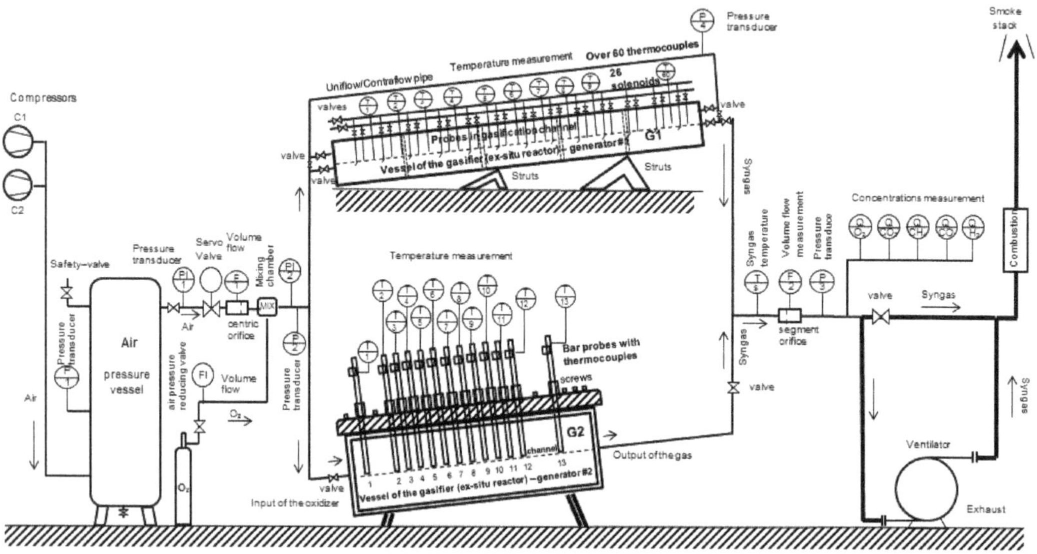

Figure 2. Ex situ reactors for testing UCG control and gasification agents optimization [2]. Reproduced under CC BY 4.0.

Ex situ reactors were used for testing overpressure control by the regulated flow of injected gasification agents and under pressure control by the regulated sucking pressure. In in situ UCG, the second-mentioned UCG control method can eliminate syngas losses to the surrounding layers, as under negative pressure, air enters the georeactor from the surface through various cracks and fissures in the overlying rock layers and through the injection well. This control method, also called burnout control, was successfully applied in Calamity Hollow at the site of an abandoned shallow drift mine in Pittsburgh [10]. The idea of combined under-pressure and over-pressure UCG control applied in practice is shown in Figure 3.

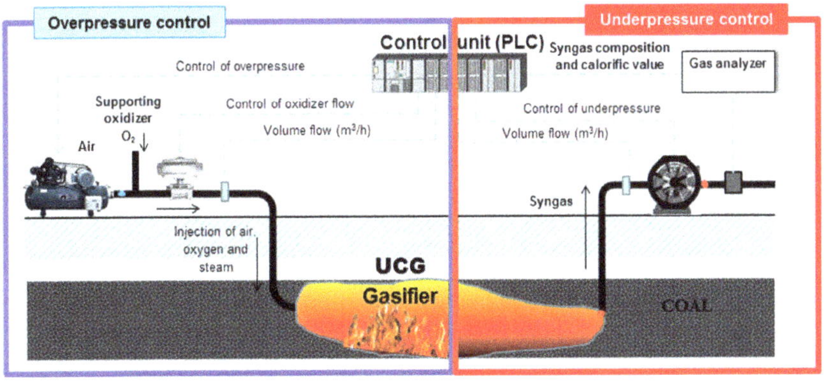

Figure 3. Principle of under-pressure and over-pressure UCG control [2]. Reproduced under CC BY 4.0.

2.1. Stabilization of Temperature and Oxygen in Syngas

The gasification temperature is critical in determining the reaction kinetics, gas composition, and tar formation during UCG. Optimization involves identifying the temperature range that promotes desired gasification reactions while minimizing undesired byproducts and ensuring reactor integrity. Temperature stabilization on the optimal setpoint with an adaptive controller can provide the optimal flow of gasification agents to keep the intended temperature within the oxidation zone. The most severe problem in the control of UCG is the long-term maintenance of the syngas calorific value at the desired values. Hence, it becomes essential to stabilize either the underground temperature or the concentration of oxygen in the produced syngas. Adding oxygen to the oxidation zone can improve the energy efficiency of the UCG, but it can also increase the amount of CO_2. In addition, too much oxidizer can lower the temperature in the oxidation zone due to the cooling of the coal. Therefore, it is necessary to search for the optimal flow rate of the oxidizer. At elevated temperatures, endothermic reactions (such as the Boudouard reaction: $CO_2 + C = 2CO$) become dominant over exothermic reactions, resulting in a decrease in CH_4 concentration and an increase in CO and H_2. The maximum effect of CO_2 consumption is at a temperature of 1000 °C. Increasing the temperature to levels above 1000 °C is necessary to achieve a higher proportion of combustible components in the syngas. As the temperature rises, there is an increase in CO production (see Figure 4), and this component continues to be the prevailing one. The $CO/(CO+CO_2)$ ratio, which is an essential indicator of the progress of gasification, also rises. However, a higher temperature must be maintained for a long time to produce as much heating gas as possible. The above shows that an algorithm is needed to stabilize the temperature to the selected setpoint.

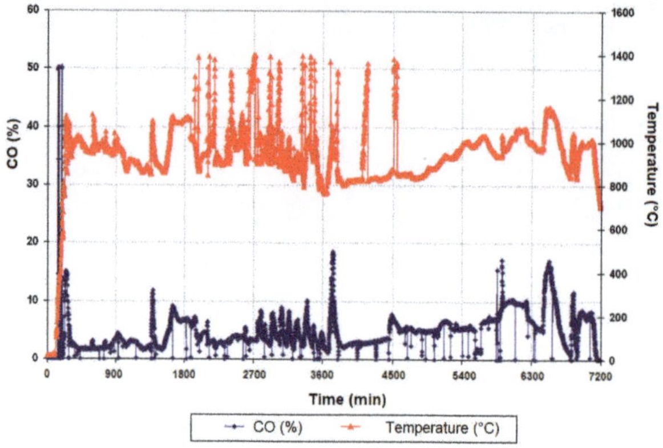

Figure 4. Behavior of temperature and concentration of CO in produced syngas during the UCG experiment [11].

The stabilization level of UCG control based on the adaptive discrete controller that continuously calculated the optimal airflow or exhaust ventilator power to stabilize the underground temperature or oxygen concentration in syngas at the desired value was proposed in [2,12]. The equation of the used proportional-integration controller has the following form:

$$\begin{aligned} \Delta u(k) = u(k) - u(k-1) = K_P\left[e(k) + \left(\frac{T_0}{T_I} - 1\right)e(k-1)\right] = \\ K_P e(k) - K_P\left(1 - \frac{T_0}{T_I}\right)e(k-1) = q_0 e(k) + q_1 e(k-1), \end{aligned} \tag{1}$$

where $\Delta u(k)$ represents an increase in the control variable $u(k)$ in the step k (i.e., gasification agent flow rate or exhaust fan power), $e(k)$ is the control error, $y(k)$ is the controlled variable to be stabilized (i.e., measured temperature, oxygen content in syngas), K_P and T_I are the proportional gain and integration constant, the parameter T_0 represents the sampling period (s), and q_0 and q_1 are parameters of the discrete controller. The adaptation of the controller was based on repeated discrete system model identification and the recalculation of controller parameters K_P, T_I [2].

Figure 5a shows the oxidation zone's regulated airflow and stabilized temperature. Figure 5b shows the stabilization of oxygen in syngas. In both cases, the discrete proportional-integral (PI) controller algorithm was adapted when stabilization quality decreased and because the coal bed changes its properties when it passes through different phases of UCG. The stabilized temperature or oxygen in syngas resulted in an increased heating value of syngas.

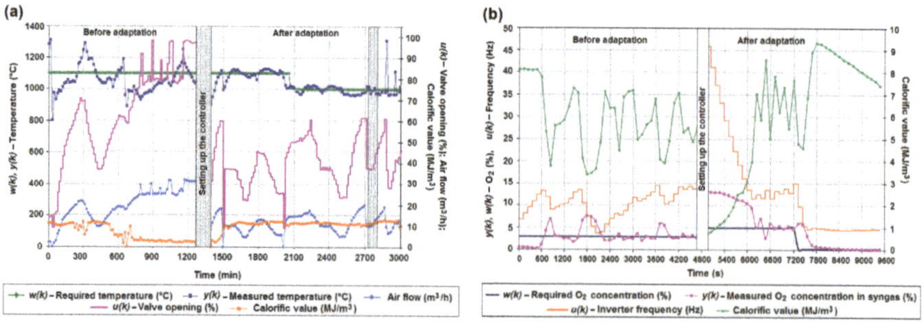

Figure 5. Adaptive stabilization of: (**a**) temperature in oxidation zone, (**b**) oxygen in syngas [2,12].

2.2. Model-Free Maximization of the Syngas Calorific Value

Uncertainty is a common feature of many systems, and model-free extreme search control has been proven to be a relevant approach to avoid typical problems related to optimization with a model, e.g., time- and resource-consuming derivation and identification of dynamic models and lack of robustness of optimal control [13–15]. Using a perturbation signal at the system input and observing its effect at the output to estimate the slope of a nonlinear static map can be traced back to the work of the French engineer Leblanc in 1922 [16]. Ariyur and Krstic popularized the extremum-seeking control based on this principle [17] and successfully applied it in various fields (e.g., [18–20]).

Kostúr and Kačur [21] proposed a complex UCG control system based on two levels, i.e., stabilization and optimization. While the stabilization level was based on discrete proportional-integrating (PI) controllers, the optimization level was based on a continuous search for the extremum of the objective function. Optimal control based on the principle of a simple extreme controller and control based on the principle of a simple gradient method with constraints were proposed. Figure 6 illustrates a schematic diagram of the UCG complex control system.

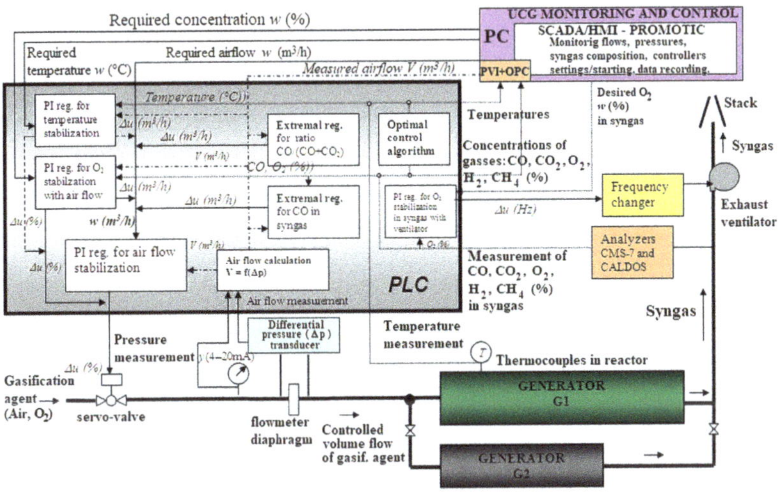

Figure 6. Control system scheme for experimental UCG [22].

The extremum-seeking control (ESC) based on a simplified extremum-seeking controller calculated the optimal airflow $u(k+1)$ (i.e., for next control step $k+1$) injected to the ex situ reactor to maximize the average concentration of CO in syngas (i.e., the objective function $y(k)$). The equation of the simple extremum-seeking controller applied in ex situ UCG has the following form [11]:

$$u(k+1) = u(k) + \mathrm{sgn}(\Delta u(k+1)) \cdot \Delta V/m, \quad (2)$$

where $u(k+1)$ represents the calculated optimized airflow that enters the PI controller as a new setpoint (m^3/h), ΔV is empirically determined by the flow stabilization quality and $m = 1$, $\mathrm{sgn}(\Delta u(k+1))$ is positive if $\Delta u(k) = u(k) - u(k-1)$ and $\Delta y(k) = y(k) - y(k-1)$ are the same (i.e., positive or negative), otherwise $\mathrm{sgn}(\Delta u(k+1))$ is negative. If $\mathrm{sgn}(\Delta u(k+1))$ changes from (+) to (–) or from (–) to (+) then it is need to modify the parameter m (i.e., $m = m+1$) and the algorithm continues from the beginning.

The monitoring system was designed in the SCADA system Promotic (see Figure 7), which provides measured data, the controllers' setup, and turning on various control algorithms.

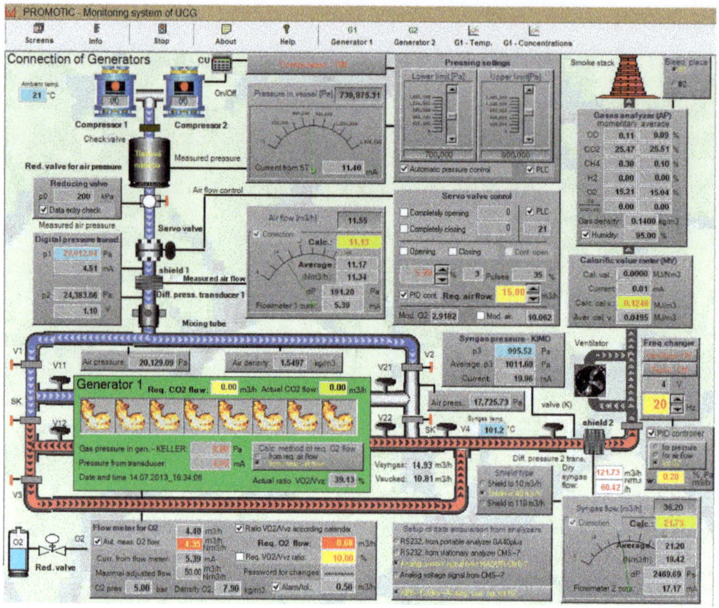

Figure 7. Monitoring system of ex situ reactors [11,22].

Figure 8 shows the continual maximization of CO concentration in syngas during ex situ coal gasification. As seen, the controller successfully increased the CO concentration from 14% to 30%, resulting in the syngas calorific value increasing [11,21–23].

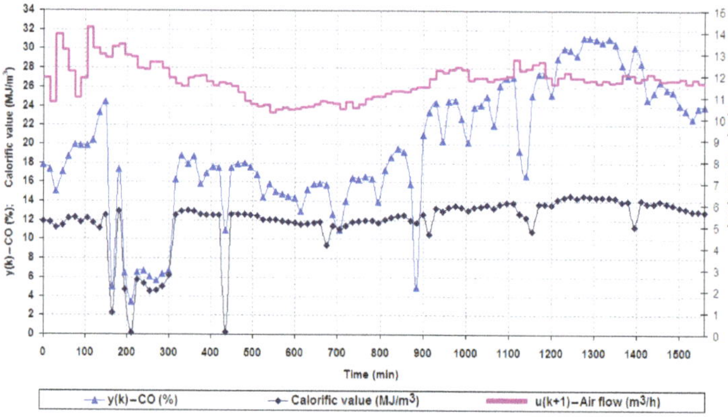

Figure 8. Model-free ESC based on extremal controller [22].

In the ESC of UCG with three manipulation variables (i.e., u_1 is the desired airflow (m^3/h) or servo valve opening, u_2 is the desired oxygen flow to the oxidation mixture (m^3/h), and u_3 represents the controlled under-pressure (Pa) or sucking ventilator power frequency (Hz)), the gradient method with constraints was used to optimize the vector $\bar{u} = (u_1 u_2 u_3)^{\mathrm{T}}$. The optimized manipulation variables for each new control step i were calculated according to the following equation:

$$\bar{u}^{i+1} = \bar{u}^i + h \cdot \nabla J(\bar{u}^i) = \bar{u}^i + h \cdot \left(\frac{\partial J}{\partial u_1^i}, \frac{\partial J}{\partial u_2^i}, \frac{\partial J}{\partial u_3^i} \right)^T, \tag{3}$$

where the cost function $J(\bar{u})$ was expressed as the average syngas calorific values recorded in the history buffer. The vector of gradients $\nabla J(\bar{u}^i)$ is continually calculated in the steady-state by loading perturbations on manipulation variables during gasification. Afterward, a new action intervention $\bar{u}^{i+1}$ was computed using the gradient method so that the cost function converges to extreme (i.e., maximum). The value of the iterative constant h is chosen to ensure that the values in $\bar{u}^{i+1}$ will lead to the existence of the objective function and its continual maximization (i.e., $J(\bar{u}^{i+1}) > J(\bar{u}^i)$). Constraints represent limitations for manipulation variables (i.e., u_1, u_2, and u_3) and selected output variables (e.g., allowed concentration of oxygen in syngas) The principle of model-free ESC based on the gradient method applied on UCG is shown in Figure 9 [12,24].

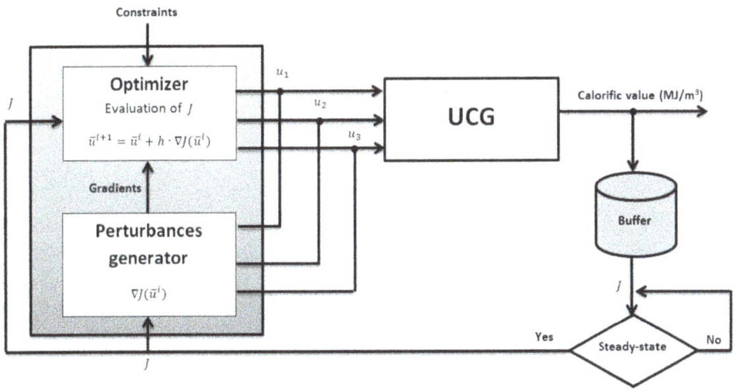

Figure 9. Principle of model-free ESC based on the gradient method [12]. Reproduced under CC BY 4.0.

The presented optimal control was tested during gasification on an ex situ reactor where the syngas's heating value was elevated from 4.4 MJ/m^3 to 8 MJ/m^3 (see Figure 10). Figure 11a shows the ex situ reactors (i.e., larger green and smaller gray), coal model preparation in the sizeable reactor G1 (see Figure 11b), and the result from gasification with optimal control (see Figure 11c,d). The coal model was bedded only on the right side of the reactor.

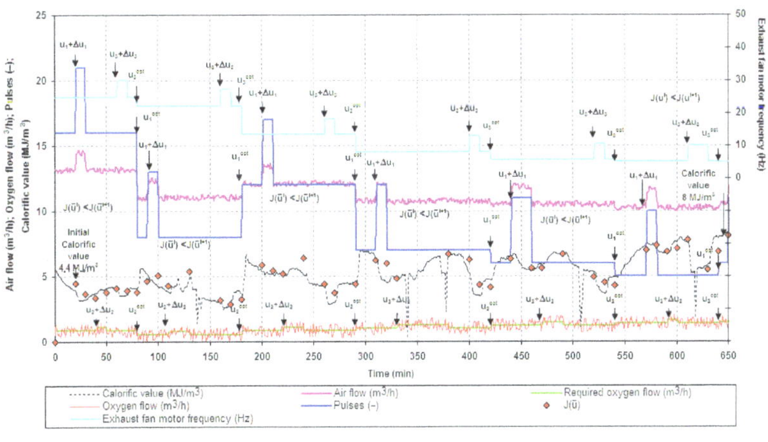

Figure 10. The course of model-free maximization of the syngas calorific value [24]. Reproduced under CC BY 4.0.

Figure 11. Preparation and results of lab-scale coal gasification: (**a**) ex situ reactor before experiment, (**b**) artificial coal seam bedding, (**c**) residues after gasification, and (**d**) gasified coal sample [24]. Reproduced under CC BY 4.0.

2.3. *Model-Based Robust Control*

Model-based robust control involves using a mathematical representation of the UCG process to create a controller that can effectively handle uncertainties and variations within the system. This advanced control is more complex than the mentioned PI control or SMC. Uppal and colleagues introduced a one-dimensional packed bed model for UCG, which they combined with a robust sliding mode controller (SMC) in a closed-loop configuration. This SMC approach aimed to maintain the desired syngas calorific value by adjusting the flow rates of the gasification agents, even in the presence of disturbances and model-related uncertainties. The input of the UCG process was the airflow rate during injection, while the output was the heating value of the produced gas. Figure 12 shows the principle of SMC application in UCG.

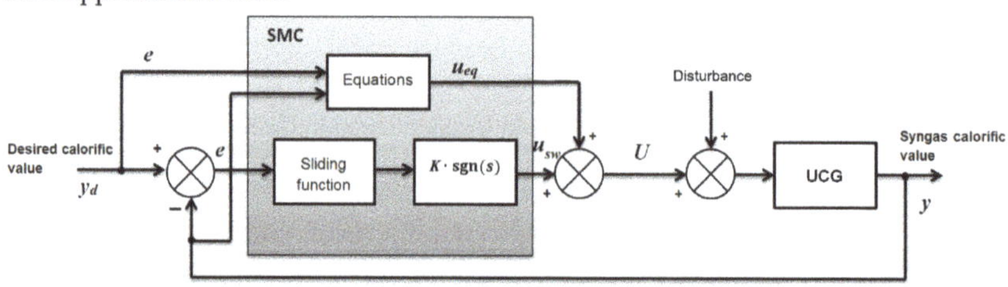

Figure 12. Block diagram of sliding mode control (SMC) [12]. Reproduced under CC BY 4.0.

The control law of SMC is represented by the following equation [25]:

$$U = u_{eq} + u_{sw} = L[C_T\gamma_1 - \gamma_2] + K \cdot \mathrm{sgn}(s),\ \ s = s(x,t) = e, \tag{4}$$

where U represents the vector of control variables (i.e., injected flow rate of H_2O, O_2, and N_2 (moles/cm^2.s)), u_{eq} and u_{sw}, which represent the equivalent and switching parts of the control effort. Parameters t, x, s, e, and K represent the time (s), state vector, sliding function, control error, and positive constant, which determine the speed of trajectory converging to the sliding surface. Parameter L represents the length of the UCG reactor (cm). Parameter C_T represents the sum of the syngas component concentration and γ_1, γ_2 depends on the rates of the chemical reactions.

To address uncertainties stemming from coal and char ultimate analysis and the steam-to-oxygen ratio, an optimized version of the one-dimensional packed bed UCG model was proposed. Another advancement represents the robust dynamic integral sliding mode control (DISMC), which was recently developed to ensure the desired syngas calorific value is accurately tracked. Unknown states were estimated through a gain-scheduled modified Utkin observer (GSMUO) for effective model-based control. Compared with the integral sliding mode control (ISMC), described in a previous work by Uppal, the newly proposed controller exhibited enhanced performance. Notably, compared to the conventional SMC algorithm, the euro-adaptive sliding mode control (NASMC), introduced by Khattak et al. [26], demonstrated superior syngas calorific value tracking capabilities.

2.4. Tracking the Syngas Calorific Value by Model Predictive Control

Another control principle proposed in [27] was based on model predictive control (MPC) (see Figure 13). A model predictive control (MPC) represents a robust control technique that can be applied to multivariable linear and nonlinear processes. UCG is also among such processes. This principle requires an internal prediction model in state space. In general, some applications of MPC for the gasification industry can be found in the literature (e.g., [28–32]). Unfortunately, only insufficient evidence of this technique can be found for UCG [27,33]. In each control step, the MPC finds the optimized manipulation variables (i.e., oxygen flow, airflow, and regulated under pressure) while observing technological constraints and minimizing the cost function J (i.e., measured the syngas's heating value). The optimization in the MPC algorithm was based on quadratic programming. To imitate UCG, a data-driven machine learning model was used in simulations. The new optimal values will ensure tracking of the desired calorific value of the syngas. Such a control method is also called adaptive MPC (APMC) [12,27].

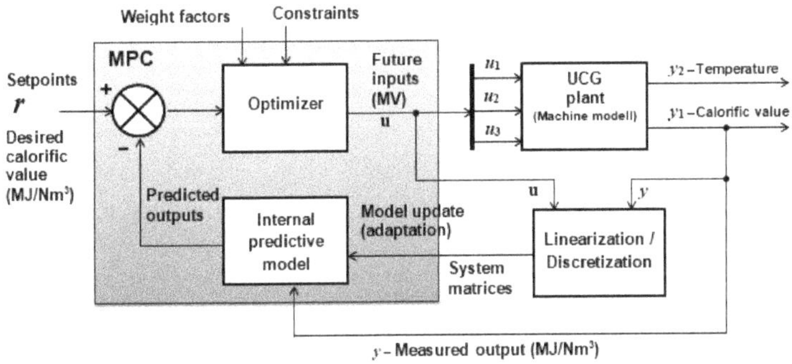

Figure 13. Principle of an adaptive MPC applied on experimental UCG [12,27]. Reproduced under CC BY 4.0.

The simulation model designed in Matlab Simulink in each new control step calculated a new predictive autoregressive–moving-average model (ARX) from the linearized machine learning model. The ARX model is transformed into the state space for MPC. A model based on multivariate adaptive regression splines (MARSs) was used for UCG process imitation in simulation (see Figure 14).

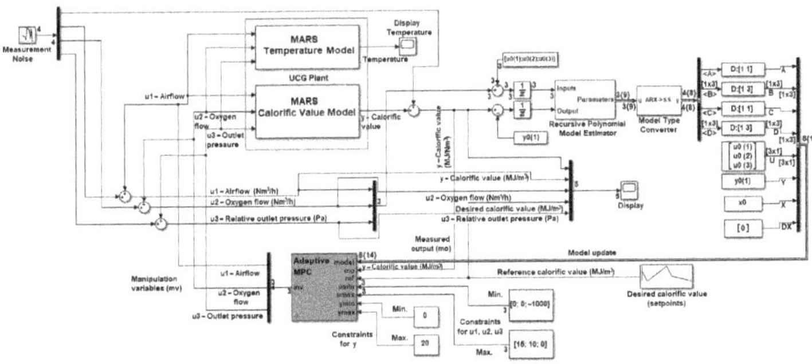

Figure 14. Simulation model of adaptive MPC [12]. Reproduced under CC BY 4.0.

Figure 15 presents the course of the syngas's heating value tracking and the behavior of optimized airflow by MPC and the discrete PI controller. Figure 16 shows the behavior of the syngas's heating value tracking by three optimized manipulation variables (i.e., air and oxygen volume flow and sucking relative pressure; see Figure 17). Results show that a higher quality of calorific value tracking was obtained using three optimized manipulation variables instead of one. The MPC performed better when compared with the discrete PI controller [27].

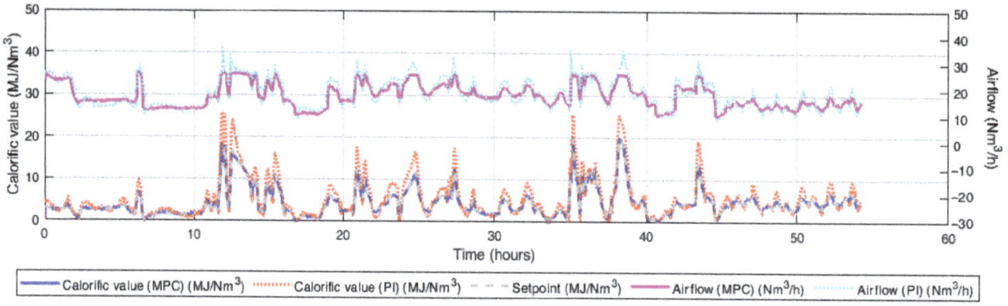

Figure 15. Syngas's heating value tracking by MPC and PI controller with one manipulation variable [27]. Reproduced under CC BY 4.0.

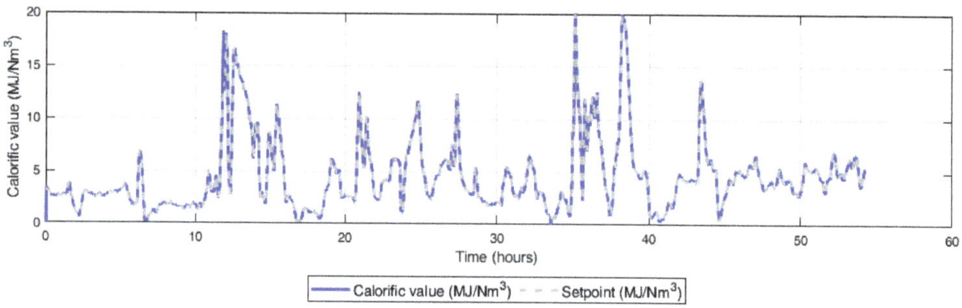

Figure 16. Syngas's heating value tracking by MPC with three manipulation variables [27]. Reproduced under CC BY 4.0.

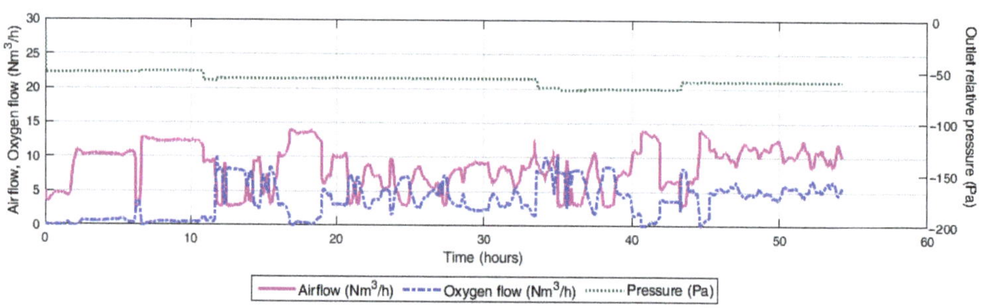

Figure 17. Optimized manipulation variables by MPC in simulation [27]. Reproduced under CC BY 4.0.

Recently, Chaudry et al. [33] proposed a constrained linear model for model predictive control (MPC) of the UCG process. The unknown states of the MPC internal predictive model were reconstructed using a linear adaptive Kalman filter (AKF) and an unscented Kalman filter (UKF). The proposed control was compared with MPC based on an unscented Kalman filter predictor, with MPC based on a gain-scheduled modified Utkin observer (GSMUO) [34], and with the control based on dynamic integral sliding mode control (DISMCGSMUO). The simulation was performed in an open-loop and closed-loop (see Figure 18) [35]. Research has shown that MPC with a linear adaptive Kalman filter (AKF) improves MPC in the absolute relative root-mean-squared error. In addition, the proposed MPC is more robust to changes in initial measurement values and process covariances [33].

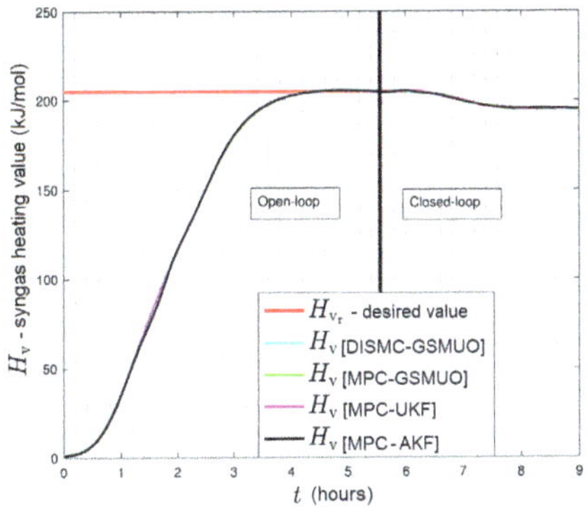

Figure 18. Syngas tracking by MPC with various state estimators and with sliding mode control [33]. Reproduced under CC BY 4.0.

2.5. Model-Based Maximization of the Syngas Calorific Value

The ESC was also experimentally verified based on regression models in the linear and nonlinear form [23,36]. The regression models, whose parameters were continuously adapted based on measured data, calculated the optimal air (V_{air}) and oxygen flow (V_{O_2}) to keep the calorific value within the required limits (i.e., 1–3 MJ/m^3, 3–6 MJ/m^3, or >6 MJ/m^3). The regression analysis was based on least squares. The first type of the model has the following form:

$$\begin{aligned} u_1 \equiv & V_{\text{air}}(k) = a_0 + a_1 \cdot V_{\text{air}}(k-1) + a_2 \cdot \varphi_{\text{CO}}(k-1) \\ & + a_3 \cdot \varphi_{\text{CO}_2}(k-1) + a_4 \cdot \varphi_{\text{CH}_4}(k-1) + a_5 \cdot T(k-1), \end{aligned} \tag{5}$$

$$\begin{aligned} u_2 \equiv & V_{\text{O}_2}(k) = a_0 + a_1 \cdot V_{\text{O}_2}(k-1) + a_2 \cdot \varphi_{\text{CO}}(k-1) \\ & + a_3 \cdot \varphi_{\text{CO}_2}(k-1) + a_4 \cdot \varphi_{\text{CH}_4}(k-1) + a_5 \cdot T(k-1), \end{aligned} \tag{6}$$

where φ_{i} is the measured concentration of CO, CO_2, and CH_4 in the syngas (%) and T represents coal temperature in the gasification channel (°C).

For example, Figure 19 shows online control using the proposed model that continuously calculates the optimal injected airflow based on five measured parameters. As a result, the control system maximized the syngas calorific value up to 10 MJ/m^3 [23,36].

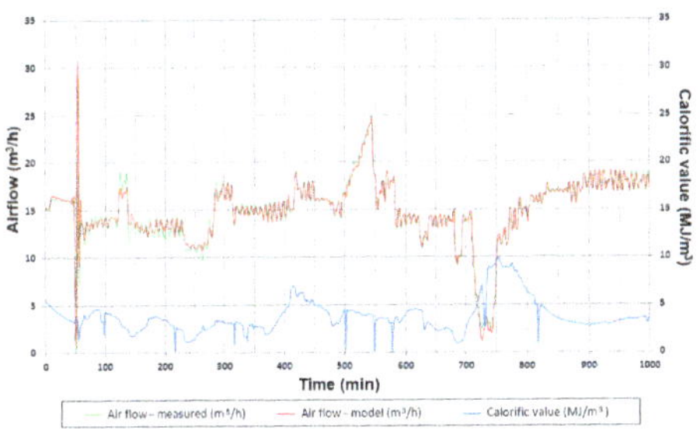

Figure 19. Controlled air flow and maximized syngas calorific value by the adapted regression model [12]. Reproduced under CC BY 4.0.

2.6. Syngas Composition Prediction

Recently, a thermodynamic UCG model was developed to find the syngas composition at a known temperature and pressure [37,38]. The optimization task in the model was based on the method of Langrage multipliers. This model can predict the amount of injected gasification agents to the geo-reactor [39,40].

The regarded system consists of chemical equations of evaporation, pyrolysis, and heterogeneous and homogeneous reactions. The task is to determine the number of kilomoles of the system with n equations that are stabilized in a state of equilibrium near a given temperature during a relatively short time. The model is based on minimizing the system's total Gibson energy G at the known temperature and pressure. Minimized is the following equation:

$$G = \sum_{j=1}^{N} n_j \mu_j, \tag{7}$$

where μ_j is the standard chemical potential, calculated as

$$\mu_j = \mu_j^0 + RT \ln\left(\frac{n_j}{\sum_{j=1}^{N} n_j} \cdot \frac{p}{p^o} \right), \tag{8}$$

where R is the universal gas constant, T is the temperature, p is the pressure of the system (Pa), and p^0 is the standard pressure. The model (i.e., its unknown parameters, b_k and a_{kj}) can be found analytically by minimization of the Langrage function:

$$L(\bar{n}, \bar{\lambda}) = \sum_{j=1}^{N} n_j \mu_j + \sum_{k=1}^{M} \lambda_k \left(b_k - \sum_{j=1}^{N} a_{kj}, n_j \right). \tag{9}$$

The model for known pressure and temperature can determine the composition of the products of gasification reactions at equilibrium (see Figure 20) [8,9]. An alternative method for forecasting syngas composition relies on the use of partial differential equations [41].

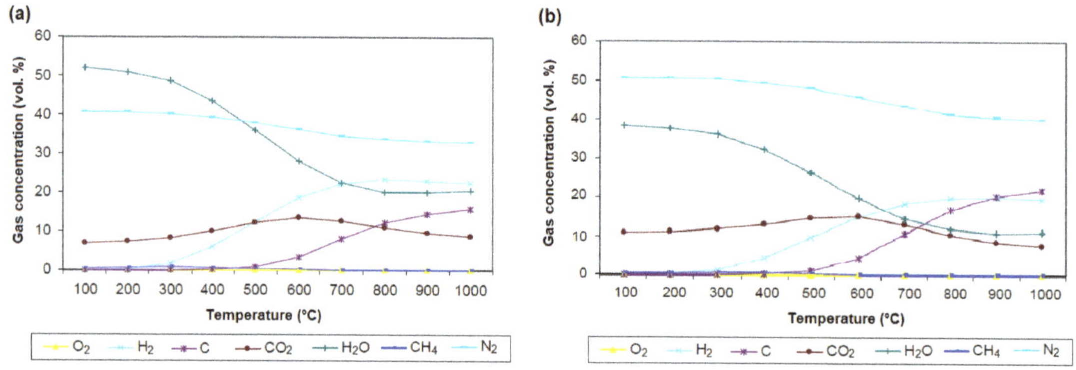

Figure 20. Simulated syngas composition by thermodynamic model: (**a**) lignite with 25% moisture; (**b**) lignite with 35% moisture [9].

3. Effect of Gasification Agent on UCG

The choice and composition of the gasification agent, which can be air, oxygen, or steam, can significantly impact the gasification process. Optimizing the gasification agent involves determining the optimal oxygen ratio to coal, or steam to coal, to achieve the desired gasification reactions and syngas composition. Controlling the oxidant ratio (i.e., air, oxygen, or steam) to coal is crucial for optimizing gasification. The stoichiometric balance of the oxidant-to-fuel ratio influences the gasification efficiency, syngas composition, and heat release. Optimizing this ratio allows for efficient utilization of the available energy in the coal.

3.1. Effect of Additional Oxygen on the Calorific Value of Syngas

Air is usually used as the primary oxidizer during gasification experiments on ex situ reactors. Because, in actual gasification operations, oxygen-enriched air is used to improve syngas production, Kačur et al. [11,42] performed three ex situ trials where the impact of additional oxygen injected on the syngas calorific value was investigated. In the experimental reactor (see Figure 11), the lignite blocks from the Cigel' mine (Slovakia) were gasified. The received coal had a total moisture of 22.25% and a calorific value of 13.74 MJ/kg. Figure 21 presents the time behavior of the syngas calorific value of the syngas calculated from the composition and the percentage ratio of the oxygen volume flow to the airflow (i.e., $V_{O_2}/(V_{air} + V_{O_2}) \cdot 100$). The figure shows that the syngas's heating value also elevates with the increasing ratio. Table 1 indicates the average syngas composition and heating value at different average temperatures and the considered ratio. The results show increased gasification efficiency with higher added oxygen to the oxidation mixture. Still, the economic return of UCG must be considered, considering that oxygen production can be expensive.

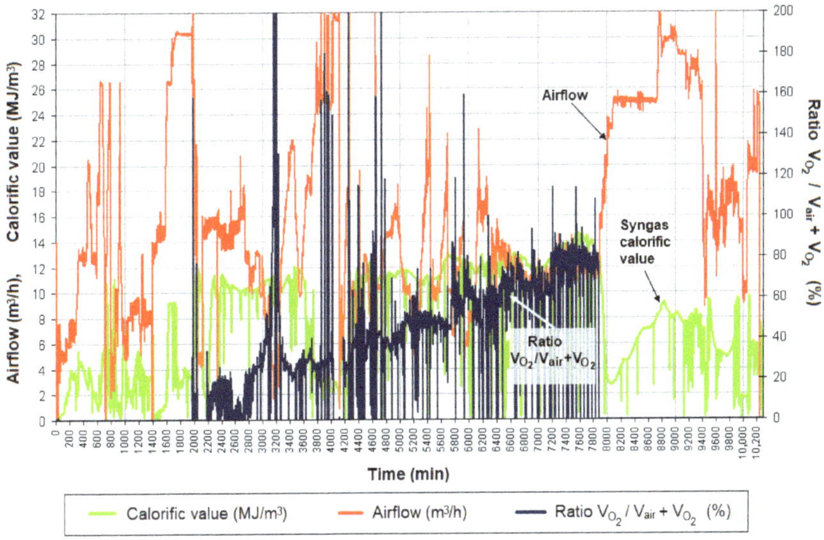

Figure 21. Behavior of syngas's heating value and ratio of $V_{O_2}/(V_{air}+V_{O_2})$ during the experiment [11].

Table 1. Results of ex situ experiments with adding oxygen to the oxidation mixture [11].

Exp.	Coal (kg)	Air (m^3)	O_2 (m^3)	Gasification Agent	Ratio $V_{O_2}/(V_{air}+V_{O_2})$ (%)	Temp. (°C)	O_2 (%)	CO (%)	CH_4 (%)	Cal. Value (MJ/m^3)
#1	766	1878	0	Air	0	1221	3.66	3.3	4.08	2.29
					0	765	0.02	0.82	0.02	0.11
#2	351	1497	147	Air + O_2	5	1022	0.19	14.07	3.67	4.29
					17	1146	0.01	26.87	9.54	8.18
					31.89	1089	0.62	16.34	14.16	8.21
				Air	0	1128	0.29	23.09	8.58	7.39
					0	804	1.15	3.05	2.27	1.71
#3	684	2667	486	Air + O_2	11.05	1118	1.58	10.8	20.11	10.94
					39.29	1257	1.02	11.98	20.54	10.96
					42.76	1038	0.15	25.8	20.59	13.79
				Air	0	890	3	4.85	19.5	8.64
					0	704	0.09	3.95	5.42	2.88

Feng et al. [43] confirmed that gasification with oxygen, unlike gasification with air, brings higher temperatures in the oxidation zone and a higher quality of syngas, and the reaction zone is closer upstream. In addition, they found that, although an increased gasifier flow can improve syngas quality, it can also cause coal cooling. Moreover, they experimented with the initial gasification channel length (L_0). Their findings revealed that igniting the gasification at $3/4L_0$ results in higher temperatures than ignition at $1/2L_0$, leading to a shorter effective syngas production time (t_e). However, this time increases as the gasification channel increases. The reaction zone is closer to the upstream with a wider gasification channel. In addition, they found that, although an increased gasifier flow can improve syngas quality, it can also cause coal cooling. The increased oxygen concentration in the oxidation mixture was evaluated as an effective tool for increasing the calorific value while increasing the flow rate of the input oxidizer will extend the effective time of syngas production. By increasing the oxygen flow, it was possible to increase the temperature of the oxidation zone up to 1300 °C and the syngas's heating value to 12.1 MJ/m^3. When air was injected, the oxidation zone's temperature was lower, resulting in a lower calorific value of syngas. Other researchers, e.g., Zagorscak et al. [44] found that when the flow rate increased from 6 Nm^3/h to 10 Nm^3/h during air gasification, the proportion of CO and CO_2 in syngas increased. And, the average heating value and maximum heating value of gas produced under air at 10 m^3/h were lower than those under air 6 m^3/h. Stanczyk et al. [45] found that compared with air gasification, in oxygen gasification, the proportion of CO, CO_2, H_2, and CH_4 syngas and heating value is higher, which is consistent with the results obtained in [43].

3.2. *Effect of Gasification Agent on Tar Concentration*

The tars from UCG are black, viscous liquids with visible inclusions of dust and high-molecular-weight agglomerates. Smaller-scale experiments are useful for studying the tar evolution mechanisms in coal gasification. Researchers usually sample from the syngas or the neighboring water. Wiatowski et al. [46,47] carried out a series of measurements on the yields, composition, heating value, density, and viscosity of tar samples in the UCG trial at Mine "Wieczorek" and "Barbara" (Poland). Xu et al. [48] investigated the relationship between tar behaviors, including its yields, viscosity, and composition at low pyrolysis temperature, and tar formation in a fixed bed reactor. Xu et al. [49] found the tar yield decreased with the increase in pyrolysis temperature on a high-temperature tube furnace in UCG conditions.

In their study, Dong et al. [50] investigated the spatial and temporal changes of tar during ex situ coal gasification. They carried out multiple experiments, varying the flow rates of gasification agents (i.e., oxygen and air), and analyzed the tar composition at different locations and time intervals. The artificial coal bed consisted of bituminous coal cut into blocks with gasification channels and channels for tar removal. The coal moisture was 14.39%, and the ash content was 5.67%. They found that the concentration of tar in the reaction zone decreased during gasification with oxygen. Also, the percentage of PAHs (i.e., polycyclic aromatic hydrocarbons) fell when the oxygen flow rate increased from 10 to 15 L/min.

Moreover, with this increase in oxygen flow, a decrease in carbon emissions in the gases and an increase in the percentage of tar were observed. The tar concentration was found to be much lower in gasification with air than in gasification with oxygen. In addition, when the airflow rate increased (i.e., from 10 to 50 L/min), a more even distribution of concentration and tar composition occurred. Carbon emissions also decreased, but the percentage of tar-polluting substances increased. The highest temperature of more than 1300 °C was reached during gasification with pure oxygen. With air, this temperature was only reached at a flow rate of 50 L/min.

Pankiewicz-Sperka et al. [51] have found that the higher values of PAHs were in the case of wastewater from semi-anthracite while from bituminous coal gasification PAHs values are in lower ranges. Studies have shown that concentrations of phenols, BTEX, and PAHs decrease with increasing pressure. Other results showed that the yield and viscosity of tar increased with the increase in heating rate and pyrolysis temperature [48]. In addition, the tar yield under a hydrogen atmosphere was observed to be higher than that under a nitrogen, carbon dioxide, methane, and carbon monoxide atmosphere. The tar yield increased gradually with the increasing of H_2 flow rate [49].

3.3. *Effect of Gasification Agent on Cavity Growth and Syngas Production*

The experiments reported in [52,53] showed the cavity growth under certain operating conditions in a horizontal channel of a coal block, through which the flow of gas takes place. It was assumed that the cavity size and shape are likely to substantially impact the gasification extent. The effect of the gasification agent on cavity enlargement during coal gasification was investigated by Daggupati et al. [54]. Moreover, they studied various UCG operating parameters required to convert coal to syngas (i.e., initial burn time, steam-to-oxygen ratio, feed water temperature). They gasified an artificial coal seam with a moisture of 40%. The mixture of steam and oxygen was injected to support ex situ coal gasification. The experiments have shown that the optimum oxygen-to-steam ratio depends on the type of coal being gasified. Several experiments showed that this optimal ratio is 2.5. During the test, they produced syngas with a heating value of 178 kJ/mol and content of H_2 in the syngas of up to 38%. Moreover, the gasification cavity's growth rate was observed to be relatively higher than that of the cavity of coal combustion. It was mainly due to the higher speed of the reaction gases in the case of gasification. The speed of the cavity enlargement also significantly affects the coal spalling, increasing the reaction surface. It was also found that the cyclic steam injection into the cavity could cause thermal shocks due to structural failures of the coal. Shu-qin et al. [55] have also reported that the optimum value of the ratio of injected steam and oxygen for lignite is circa 2.5. Hettema et al. [56] have experimentally demonstrated that the surface material cracks and breaks due to local temperature variations and they have also shown the effect of steam pressure on thermal spalling using laboratory-scale experiments.

4. High-Pressure Coal Gasification

Among the essential operational parameters of UCG, the process pressure stands out as particularly significant. The pressure during UCG operations impacts the gasification rate, syngas composition, and the thermodynamic equilibrium of the reactions. Pressure optimization involves determining the optimal range that provides high gasification efficiency, syngas quality, and reactor stability. A higher operating pressure can positively affect the concentration of methane and hydrogen, ultimately affecting the syngas's calorific value [57]. Existing numerical models considered UCG up to a pressure of 0.5 MPa [58–61]. Recently, a UCG test was conducted in Alberta (Canada) with an operating pressure of up to 12 MPa. Increased operating pressure mainly affects the kinetics and balance of gasification. For example, recent research has shown that reaction rates increase by 1.5–2.0 MPa [62]. Experiments by Roberts and Harris [63] with coal reactivity in the presence of CO_2 and H_2O at pressures ranging from 0.1 to 3.0 MPa demonstrated that reaction rates elevate as the pressure increases. Still, the rate of increase decreases or may be constant at higher pressures. Studies of the effect of pressure on pyrolysis have shown that pressure reduces the extent of pyrolysis [64]. The study [65] showed that increased pressure has a negligible effect on gaseous products and coal pyrolysis but a more significant impact on chemical reactions. Research conducted in [66–68] indicates that, as the coal seam depth increases, the hydrostatic pressure is elevated, and there is a corresponding need to utilize higher pressures in the gasification process. The deeper coal allows for significantly elevated pressures in the georeactor, leading to higher methane content and, consequently, a higher calorific value of the gas, as observed in Bhutto's study [1].

4.1. Methane-Oriented Gasification of Semi-Anthracite and Hard Coal

The research on the gasification of hard coal in an underground environment at an elevated pressure of 0.5 MPa was conducted by Wiatowsky et al. [69]. The coal used for gasification was sourced from Poland's Upper Silesia coal basin. The gasification was supported using pure oxygen, air, and oxygen-enriched air. Their experiment showed that the increased pressure most affected the thermodynamic conditions during gasification with only oxygen. Higher methane and carbon dioxide concentrations were achieved at a higher pressure than gasification under atmospheric pressure. It was found that the concentration of combustible components in syngas and the syngas calorific value increased in proportion to the oxygen content in the gasification agent. During gasification with oxygen, the average calorific value of syngas was 8.1 MJ/Nm^3, and the average energy efficiency of gasification was 60.5%. Changing the gasification agent in the air resulted in a decrease in temperatures, the concentration of combustible components in the syngas, and the average calorific value of the syngas (2.1 MJ/Nm^3) (see Table 2).

An increased concentration of methane at higher operating pressure was also noted by Kapusta et al. [70]. They have performed gasification with two types of coal (i.e., "Six Feet" semi-anthracite (Wales, United Kingdom) and "Wesoła" hard coal (Poland)) at different operating pressures (i.e., 20 and 40 bar) with pure oxygen and a mixture of oxygen and steam. The experiment also showed that 40 bars gradually increased the syngas production rate (see average CH_4 at 20 and 40 bar in Table 2). Furthermore, the efficiency increased with higher gasification pressure in both cases. It was found that a mixture of oxygen and steam supported heating better than oxygen alone (see Figure 22). During the experiments, no significant influence of the pressure on the temperatures in the reactor was recorded.

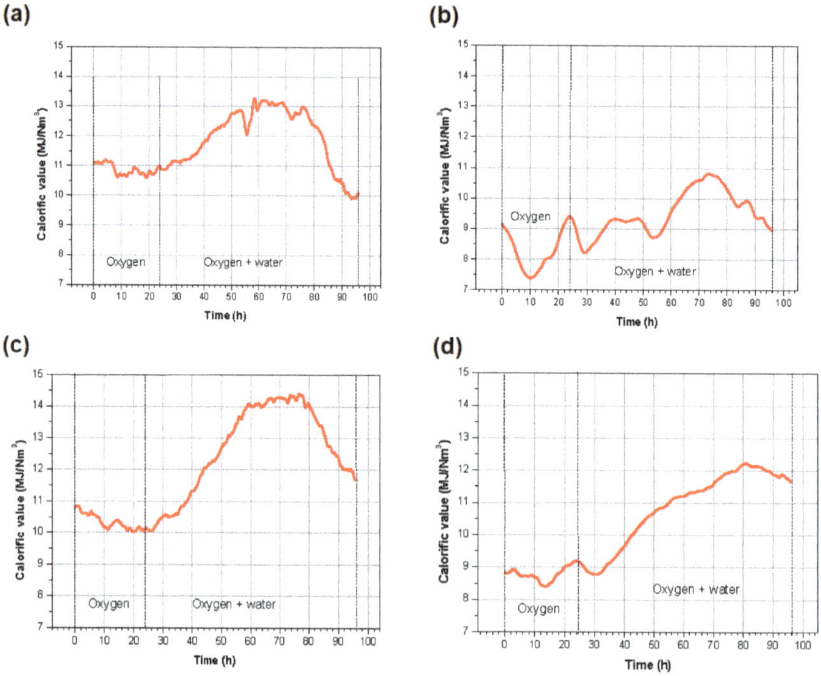

Figure 22. Changes in gas calorific value throughout gasification experiments: (**a**) "Six Feet" semi-anthracite at 20 bar; (**b**) "Wesoła" hard coal at 20 bar; (**c**) "Six Feet" semi-anthracite at 40 bar; (**d**) "Wesoła" hard coal at 40 bar [71]. Reproduced under CC BY 4.0.

Table 2. Results of laboratory UCG tests with different types of coal and operating pressures.

Coal	Origin	Coal Characteristic (As Received)					Gasification Agent	Pressure (Bar)	Average Syngas Composition (*vol.*%)							Average Calorific Value (MJ/Nm^3)	Reference
		Moisture W_t^r (%)	Ash A_t^r (%)	Volatiles V^r (%)	Total Sulphur S_t^r (%)	Calorific Value Q_i^r (kJ/kg)			CO_2	N_2	H_2	CH_4	CO	C_2H_6	H_2S		
"Six Feet" Semi-anthracite	South Wales, UK	1.15	4.61	9.92	1.55	33,416	$O_2 + H_2O$	20	36.3	0.4	19.2	15.8	27.2	0.7	0.4	11.7	[70]
							$O_2 + H_2O$	40	41.6	0.6	14.1	19.1	23.2	1.1	0.3	12.1	[70]
							O_2, $O_2 + H_2O$, Air, Air + O_2, Air + $O_2 + H_2O$	Atmos.	28.9	18.8	18.1	2.3	31.6	0.1	0.3	6.9	[44]
							O_2, $O_2 + H_2O$, Air, Air + O_2, Air + $O_2 + H_2O$	30	38.4	16.9	12.0	11.8	20.1	0.6	0.3	8.5	[44]
		0.88	2.99	13.42	0.7	-	$H_2O + O_2$	36	52.0	-	20.9	15.3	6.3	-	-	8.6	[72]
"Wesoła" Hard Coal	Upper Silesia, Poland	3.60	8.74	27.67	0.31	28,798	$O_2 + H_2O$	20	46.3	0.7	21.6	10.9	19.5	0.6	0.4	9.2	[70]
							$O_2 + H_2O$	40	46.1	0.7	17.7	14.8	19.3	0.9	0.5	10.4	[70]
							$H_2O + O_2$	36	56.3	-	15.7	14.6	6.3	-	-	7.7	[72]
		3.88	14.79	-	0.72	24,638	O_2	0.5	36.6	0.5	32.1	3.7	27.0	0.1	-	8.1	[69]
							Air	0.5	17.5	68.9	8.2	2.0	3.3	0.0	-	2.1	[69]
							Air + O_2	0.5	21.3	53.7	9.3	2.6	12.9	0.2	-	3.3	[69]
Ortho-lignite	Oltenia, Romania	45.64	8.86	25.78	1.49	10,642	O_2	10	75.5	1.1	13.9	3.6	4.1	0.2	1.5	3.9	[73]
Meta-lignite	Velenje, Slovenia	31.62	4.29	43.67	0.51	13,615	O_2	35	75.3	1.3	10.5	8.7	2.4	0.6	1.2	5.0	[73]

Likewise, a positive effect of increased operating pressure (i.e., from 20 to 36 bar) and doubling the amount of injected steam concerning oxygen on methane production was reported by Sadasivam et al. [72]. The research also revealed that raising the temperature from 650 °C to 850 °C enhanced methane production. Under the conditions of this optimal temperature, an operating pressure of 36 bar, and a H_2O:O_2 ratio of 2:1, the average concentrations of CH_4 in syngas were measured at 15.34 *vol.*% for semi-anthracite coal and 14.64 *vol.*% for bituminous coal. CH_4 concentration increased by 94% and 129% for "Six Feet" and "Wesoła" coal. From a practical point of view, maintaining the optimum temperature for methane production depends on the amount of water and hydrogen in the georeactor. Therefore, it is necessary to assess the hydrogeology of the coal deposit.

Further gasification trials published [44] with "Six Feet" semi-anthracite coal confirmed that, under atmospheric pressure, syngas with a lower calorific value (6.92 MJ/Nm^3) was produced compared to under increased pressure 30 bar (8.49 MJ/Nm^3) (see Table 2). The linked vertical wells (LVWs) technique combined with the controlled retracting injection point (CRIP) was used for gasification. Likewise, the energy efficiency of gasification was lower in atmospheric gasification (51.72%) than in high-pressure gasification (57.67%), which was caused by higher methane production. The syngas contained a slightly lower calorific value when using oxygen as a gasification agent at 30 bar. Syngas contained more methane but less H_2 and CO than in atmospheric gasification with oxygen.

The positive effect of increased gasification pressure was also recorded in ex situ hydrogasification. Recently, Kapusta et al. [74] have performed experimental trials of methane-oriented UCG using hydrogen. Hydrogen was used as the gasification medium. They investigated the effect of coal quality and gasification pressure on hydrogasification. In underground coal hydrogasification (UCHG), coal is gasified under increased pressure, which is related to the depth of the seam. This operation should not exceed the hydrostatic pressure of the water to avoid contamination of the surrounding layers, as reported in [57]. The experiments were performed on different types of coal, semi-anthracite "Six Feet" and hard coal "Wesoła" (Poland). It was found that the operating pressure and the physicochemical properties of coal affect hydro-gasification. Syngas obtained from "Six Feet" semi-anthracite demonstrated higher methane content at the specified pressures. Specifically, during the H_2 phase of the experiment, the average methane concentration for "Six Feet" semi-anthracite was recorded at 24.12% at 20 bar and 27.03% at 40 bar.

On the other hand, in the gasification of "Wesoła" black coal, CH_4 concentrations were higher by 19.28% and 21.71% at 20 and 40 bar. Experiments have shown that the composition of syngas from UCHG depends not only on the coal grade but also on the gasification pressure. A negative correlation was found between gasification pressure (i.e., O_2 and H_2) and syngas production rate and yield.

4.2. Ortho-Lignite and Meta-Lignite Gasification

Comparison of gasification low-quality Ortho-lignite Oltenia (Romania) and Meta-lignite Velenje (Slovenia) was recently performed by Wiatowski et al. [73]. Both types of coal had a high moisture content. Oxygen was used as a gasification agent with a pressure of 10 bar (i.e., for Oltenia lignite) and 35 bar (i.e., for Velenje lignite). The results showed that operating pressure and physicochemical properties significantly affect gasification. The higher content of moisture in Oltenia lignite enabled it to efficiently produce H_2 and CH_4 over an extended duration. Although the gasification of Oltenia brown coal brought a more stable production of syngas, and the methane and carbon dioxide concentrations were increased at a pressure of 10 bar, this type of coal contained more moisture, which led to significant heat loss due to excessive water evaporation. Overall, Velenje lignite's gasification produced better results than the Oltenia lignite process (see Table 2).

Table 2 shows summary results of previously discussed laboratory tests with different types of coal and operating pressures. The summary shows that the highest average calorific value of syngas (12.1 MJ/Nm^3) was recorded in the gasification of semi-anthracite coal under a pressure of 40 bar with the help of oxygen and steam, and the lowest (2.1 MJ/Nm^3)

during the gasification of hard coal with air. In general, at lower operating pressures, the average calorific value of the syngas was lower.

5. Gasification of Lignite with Different Gasification Agents

Lignite coal has a relatively high moisture content compared to other coal types. The presence of moisture can affect the efficiency of the gasification process, as it requires additional energy to evaporate the water present in the coal. It can reduce the overall energy efficiency and make the process less economical. In addition, lignite is low-rank coal with a lower calorific value than higher-rank coal such as bituminous or anthracite coal. The lower energy content of lignite can limit the amount of useful gas produced during the gasification process. It means that more lignite may need to be gasified to obtain the same amount of energy, resulting in increased operational costs and potentially lower gas quality. Several ex situ experiments with lignite gasification have been performed. Optimum gasification conditions, various gasification reagents, and gasification techniques were investigated. The following subsections discuss the experiments performed and the results obtained.

5.1. Gasification of Lignite with High Moisture

The combination of low-heating coal with higher moisture can be critical in the decision to start UCG operation. Kostúr et al. [75] experimentally gasified low-calorific lignite (i.e., from mine Cigel, Slovakia) with moisture of 38.2%. They used only air as a gasification agent. An experiment was also performed with technical oxygen added to the air. The trials showed that higher coal moisture had a beneficial impact on water vapor formation. In turn, it led to the cooling of the oxidation zone on the one hand, while on the other hand, it contributed to the higher enthalpy of gases in the subsequent gasification areas.

Additionally, the increased water vapor facilitated increased methane concentrations due to chemical reactions involving oxygen, water vapor, and carbon. However, the downside was the augmented consumption of thermal energy generated in the oxidation zone, primarily used for water evaporation. The calorific value of the syngas was 2.5–5 MJ/Nm^3 when only air was injected. When a mixture of air and oxygen was used, the calorific value reached 10 MJ/Nm^3 (see Table 3).

The results from gasification of Slovak lignite were also reported by Laciak et al. [76]. The experiments used the same gasification technique and the same ex situ reactors as in [75]. The air was used as the primary gasification agent, and technical oxygen was added to injected air in a short period of the experiment. The average calorific value of produced syngas was 3.27 MJ/Nm^3 when injected with air and 4.13 MJ/Nm^3 when injected with oxygen (see Table 3). During the experiment, approximately 30% of the chemical energy of the coal was transformed into the energy of the produced syngas.

A comparison of lignite and hard coal gasification was performed by Stańczyk et al. [45]. The gasified lignite had a moisture content of 53.0% and hard coal had a moisture content of 1.6%. The coal was gasified with air and a mixture of oxygen-enriched air. The experiments showed that gasification only with the air caused a drop in temperatures and a decrease in the calorific value of the syngas, and eventually, the gasification reactions stopped.

On the other hand, gasification by the mixture of oxygen and air with a ratio of 4:2 produced syngas with a higher calorific value. In the gasification of hard coal, the quality of the produced syngas was higher than in the case of lignite for all gasification agents (see Table 3 and 4).

Large bulk lignite samples were experimentally gasified by Kapusta [71]. The research focused on the possibilities of gasification of two different types of lignites, i.e., "Velenje" meta-lignite (Slovenia) and "Oltenia" ortho-lignite (Romania), with an average moisture content of 31.6 wt% and 45.6 wt%. Temperatures in the reactor were measured using 14 sensors (see Figure 23). Air, oxygen, and water vapor, injected individually or supplied as a gas mixture, were used as gasification agents (see Figure 24a,b).

Table 3. Overview of results from lignite gasification.

Coal	Origin	Coal Characteristic (As Received)					Gasification Agent	Average Syngas Composition (*vol.*%)								Average Calorific Value (MJ/Nm^3)	Reference
		Moisture W_t^r (%)	Ash A_t^r (%)	Volatiles V^r (%)	Total Sulphur S_t^r (%)	Calorific Value Q_i^r (kJ/kg)		CO_2	H_2	CH_4	CO	O_2	N_2	H_2S	C_2H_6		
Lignite	Cigel, Slovakia	38.20	9.40	50.00	-	13,400	Air	-	-	6.52	-	-	-	-	-	2.86	[75]
							Air + O_2	-	11.66	19.99	5.81	-	-	-	-	9.69	[75]
		22.25	20.47	34.59	1.93	13,740	Air	14.95	2.78	6.44	5.25	4.78	-	-	-	3.15	[76]
							O_2	14.95	2.78	6.44	5.25	4.78	-	-	-	4.13	[76]
	Bełchatów, Poland	53.00	4.7	-	1.1	9316	O_2 (1st stage)	64.9	16.1	1.3	7.00	3.6	7	0.1	0.02	3.10	[77]
							H_2O	39.5	46.3	4.4	4.8	0.58	2.4	1.8	0.26	7.80	[77]
							O_2 (2nd stage)	54	26.4	2.7	9.9	1.8	4.7	0.53	0.08	5.20	[77]
							O_2	63.6	19.2	1.7	6.2	2	7	0.34	0.05	3.49	[45]
							Air	12.1	2.5	0.17	1.3	5.7	78.2	0.02	0.01	0.49	[45]
							Air + O_2	49.4	23.1	2.3	6.3	1.47	16.4	0.42	0.07	4.18	[45]
	Malkara Pirinççesme, Turkey	25.17	17.95	28.74	-	15,100	O_2	42.9	20.9	5.1	18.3	1.6	11.1	-	-	5.14	[78]
							Air + O_2	30.6	15.7	3.3	14.8	3.6	32	-	-	3.87	[78]
		22.83	15.67	28.9	3.68	15,680	O_2	44.21	19.26	3.68	16.97	-	9.03	-	-	4.28	[79]
							H_2O	52.12	27.17	5.38	4.38	3.73	7.22	-	-	4.56	[79]
	Shenbei, China	11.87	22.89	32.88	32.36	-	Air	-	11.57	1.36	6.05	-	-	-	-	2.49	[80]
							Air + O_2	-	22.39	2.08	22.1	-	-	-	-	5.98	[80]
"Velenje" Meta-lignite	Slovenia	31.62	4.29	43.67	0.51	13,615	Air, O_2, H_2O, O_2 + H_2O	52.5	21	4.3	18.6	1.0	2.0	0.5	0.2	6.40	[71]
"Oltenia" Ortho-lignite	Romania	45.64	8.86	25.78	1.49	10,642	Air, O_2, H_2O, O_2 + H_2O	63.3	21.3	2.7	10.2	0.1	1.5	0.6	0.2	4.80	[71]
Ortho-lignite	Turów, Poland	46.52	3.18	-	0.15	12,656	O_2	45.3	29.8	5.2	15.5	0.2	3.7	0.08	0.16	7.20	[81]

Table 4. Overview of results from hard and black coal gasification.

Coal	Origin	Coal Characteristic (As Received)					Gasification Agent	Average Syngas Composition (*vol.*%)								Average Calorific Value (MJ/Nm^3)	Reference
		Moisture W_t^r (%)	Ash A_t^r (%)	Volatiles V^r (%)	Total Sulphur S_t^r (%)	Calorific Value Q_i^r (kJ/kg)		CO_2	H_2	CH_4	CO	O_2	N_2	H_2S	C_2H_6		
"Piast" Black Coal	Piast, Poland	4.70	16.30	30.10	0.83	22,719	O_2	27.27	30.93	3.53	37.11	-	0.8	0.27	0.09	9.41	[82]
Hard coal	Bielszowice, Poland	1.60	2.20	-	0.28	33,370	O_2	27.9	31.5	3.2	33.2	0.71	3.2	0.2	0.13	8.58	[45]
							Air	14.7	11.8	2.8	10.3	0.7	59.6	0.08	0.09	3.63	[45]
							Air + O_2	23	18.7	4.2	17.7	0.6	35.9	0.18	0.09	5.74	[45]

Figure 23. Placement of sensors on the side of the ex situ reactor [71]. Reproduced under CC BY 4.0.

UCG tests were performed under near-atmospheric pressure conditions. Velenje lignite's gasification resulted in an average heating value of 6.4 MJ/Nm^3 and a gasification efficiency of 44.6%. During the gasification of Oltenia, lignite was 4.8 MJ/Nm^3, and the gasification efficiency was 33.4% (see Figure 24c,d). The primary reason for this difference is the higher moisture and ash content in the Oltenia lignite sample utilized in the test.

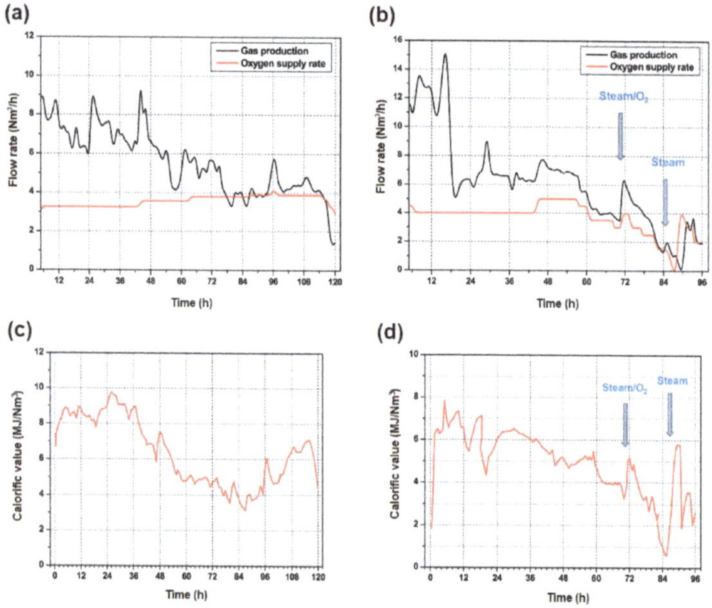

Figure 24. Results of lignite gasification: (**a**) controlled flows of oxidizers in Velenje lignite gasification; (**b**) controlled flows of oxidizers in Oltenia lignite gasification; (**c**) syngas calorific value in Velenje lignite gasifictaion; (**d**) syngas calorific value in Oltenia lignite gasification [71]. Reproduced under CC BY 4.0.

Experiments showed that the lignite's physicochemical properties significantly influence the in situ gasification process. Research also recommends gasifying lignite with lower moisture content and higher energy density [71].

The results from the gasification of ortho-lignite with higher humidity were also presented in the work [81]. The average moisture content of the coal was 46.5%. The artificial coal seam was gasified under atmospheric pressure with a higher oxygen flow rate of up to 5 m^3/h. The average heating value of syngas was 7.2 MJ/Nm^3 (see Table 3), and the gasification efficiency was 59%, which is close to the gasification of black coal [46,83]. The results prove that the given type of wet brown coal (i.e., Miocene ortho-lignite) can be gasified with oxygen with good syngas quality. It was found that the high moisture content of lignite eliminates the need to supply additional water and the gasification agent in the initial stages of the gasification process. Excess water can reduce the efficiency of gasification due to significant heat losses. The experiment shows that the poor mechanical strength of the lignite in the deposit could influence the mining of UCG lignite deposits, as it may cause the collapse of the seam roof and uncontrolled gas flow. Therefore, for such coal seams, a geomechanical analysis of the stability of the overlying layers is necessary.

Previous studies (e.g., [84]), which presented results from the gasification of similar wet Miocene lignites, achieved a syngas calorific value of only 5 MJ/Nm^3 and a low UCG efficiency. However, these field tests were performed at high pressures (approximately hydrostatic) to increase syngas quality and UCG efficiency [1]. In addition, passing studies on lignites were performed on shorter coal seams, resulting in heat loss to evaporate water. It is assumed that the geo-rector dimensions influence the gasification efficiency of high-moisture lignites.

5.2. Three-Stage Lignite Gasification

Three-stage lignite gasification was tested by Stańczyk et al. [77]. In this technique, the coal was ignited in the first stage, then heated and gasified with oxygen, and in the last stage, steam gasification was carried out. They gasified coal with a moisture content of 53%. The results showed that the given coal was not suitable for steam gasification. During gasification, significant heat losses occur as a result of heat consumption and evaporation of water from coal. During gasification with oxygen in the stable operation of the reactor, the average calorific value was 5.2 MJ/m^3. The analysis results showed the regular development of individual zones of the cavity during the experiment. Therefore, it is necessary to control the flow of water in the lignite layer and the surrounding layers with a suitable technique. It is also recommended that the lignite layer be thicker to prevent heat dissipation because it increases with the contact surface between the reaction space and the adjacent layer. The comprehensive review of cavity growth measurement techniques was well-discussed in [85]. Laboratory studies on cavity growth and product gas composition in the context of UCG was also studied in [54].

5.3. Gasification of Turkish Lignite

Gur et al. [78] gasified Turkish lignite with moisture 25.17%. Oxygen, air, steam, and their mixtures in different proportions were used as gasification agents. The test found that most calorific syngas was produced in gasification with oxygen (6–9 MJ/Nm^3) (see Table 3). In the pure oxygen phase, the syngas had a heating value of 5.14 MJ/m^3. When injecting a mixture of air with technical oxygen, a lower-quality syngas was produced (i.e., 3.87 MJ/m^3). The experiment revealed that the optimal air/oxygen ratio by volume was 2:1.

Turkish brown coal has a relatively low moisture content, predisposing it to low hydrogen production. In further research, Gur et al. [79] performed an ex situ UCG hydrogen-oriented experiment with Turkish lignite with a moderate moisture of 22.83. Oxygen and steam were injected into the ex situ reactor. When only oxygen was injected, the average heating value was 4.28 MJ/Nm^3. When using steam as a gasification agent, the average heating value was 4.56 MJ/Nm^3. The highest recorded syngas's heating value was 8.30 MJ/Nm^3. However, it should be noted that steam was added only at the stage when the lignite block was preheated (i.e., by the two-stage gasification method) to extend the hydrogen production time. In the first phase, the combustion reactions were supported, and the coal was heated by injecting air or oxygen-enriched air. In the second phase, only water vapor was injected, which supports hydrogen production. This technique was also investigated in [77,86].

5.4. Gasification of Chinese Lignite

Liu et al. [80] have researched the effect of different gasification agents on the gasification efficiency of Chinese coal in a cylindrical ex situ reactor (see Figure 25). The coal seam model consisted of coal lumps connected by clay and powdered coal (see Figure 26). The reverse gasification technique was employed, comparing the use of air solely and oxygen-enriched air as gasifying agents.

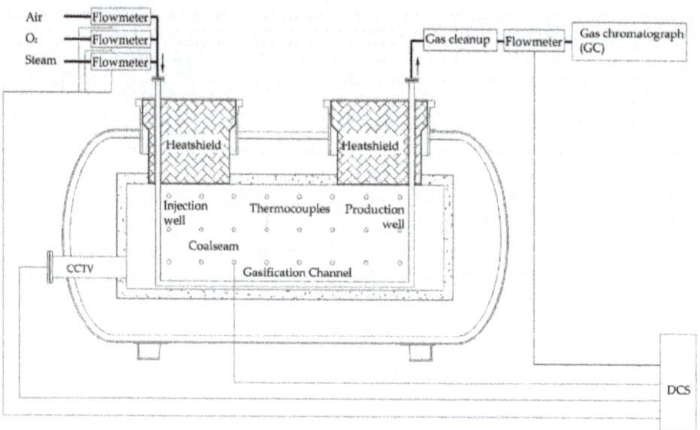

Figure 25. Ex situ reactor proposed for reverse coal gasification [80]. Reproduced under CC BY 4.0.

Figure 26. Coal seam in the test [80]. Reproduced under CC BY 4.0.

The research results showed that a proportion of combustible components in the syngas that is higher than the total calorific value can be achieved using oxygen with a concentration higher than 21% in the oxidizing mixture.

The course of gasification and composition of syngas is presented in Figure 27. Figure 27a shows the increased concentration of the combustible syngas component and average calorific value when enriched oxygen was employed as the gasifying agent. The average calorific value, in this case, was 1430.19 kcal/Nm^3 (i.e., 5983.12 kJ/Nm^3) (see Table 3). It turned out that gradually increasing the gasifying agent's flow rate also increases the combustion front's speed (see Figure 27b). Moreover, it was shown that the gasification rate was higher when using oxygen-enriched air than when using air as a single oxidizer. Moreover, the enriched mixture induces a faster flame front movement towards the gas-injection well in reverse gasification, where the oxidant is injected from one well. In reverse gasification, the entire coal seam is positioned in the reaction zone, resulting in significantly longer reduction and dry distillation zones than when air is utilized (refer to Figure 28).

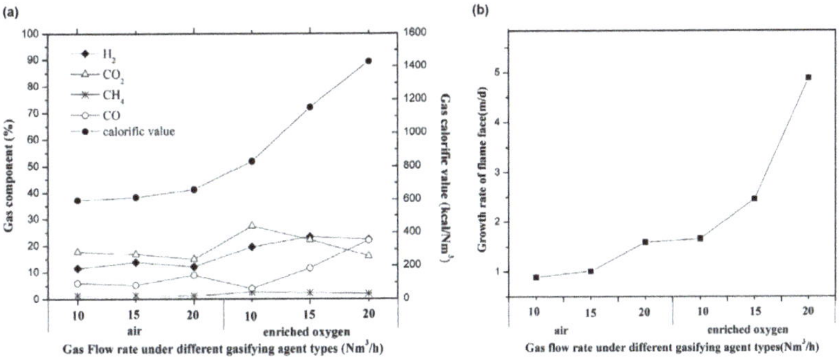

Figure 27. Results of the reverse UCG experiment: (**a**) the changes of syngas composition and calorific value when enriched oxygen and air were used as gasification agents; (**b**) changes in the growth rate of the combustion front [80]. Reproduced under CC BY 4.0.

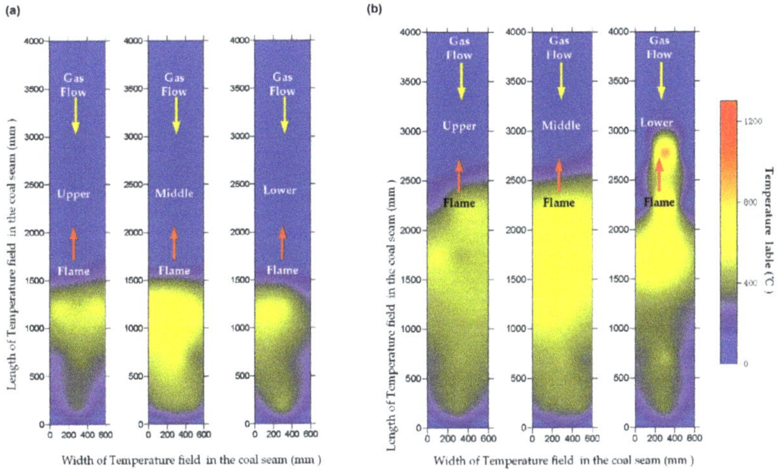

Figure 28. Temperature fields in reverse gasification at flow rates of gasification agent of 10 Nm^3/h: (**a**) gasification agent was only air; (**b**) gasification agent was enriched oxygen [80]. Reproduced under CC BY 4.0.

The results from previously discussed experiments are summarized in Table 3. The summary shows coal analysis, used gasification agent, final product analysis, and syngas calorific value. The great potential to be used in practice has three-state or two-state steam gasification reported in [77,79]. Alternating the oxygen of steam as a gasification agent contributed in several experiments to the stable production of syngas with a calorific value of 5–7 MJ/Nm^3.

6. Gasification of Hard Coal

Hard coal is often considered suitable for underground gasification. This type of coal has high carbon content (i.e., typically above 80%) and low volatile matter content, which refers to the components of coal that vaporize or transform into gases at relatively low temperatures. Low volatile matter content reduces the risk of excessive gas pressure or uncontrolled combustion during the gasification process, making it safer and more manageable. Hard coal exhibits relatively high porosity, with many tiny pores and crevices. This porous nature facilitates the flow of gases and liquids within the coal bed, allowing for better distribution of heat and reactants during gasification.

Moreover, hard coal has a high energy density, meaning it contains a large amount of energy per unit of volume or mass. This characteristic makes it an efficient fuel source, as it can produce a substantial amount of syngas when subjected to gasification. Hard coal is known for its stability and resistance to deformation under high temperature and pressure conditions.

Gasification of Hard Coal Seam with a Siderite Layer

Siderite is an iron carbonate mineral occasionally found in coal seams. Its presence in coal seams can be attributed to several factors (e.g., paleoenvironmental conditions, organic matter decay, or diagenetic processes). Wiatowski et al. [82] have performed a gasification test in an ex situ reactor (Figure 29) with low moisture hard coal and a 2 cm siderite layer (see Figure 30), using technical oxygen with a constant flow rate of 4.5 Nm^3/h as an oxidizer. The surroundings of the coal were simulated with sand with 11% moisture.

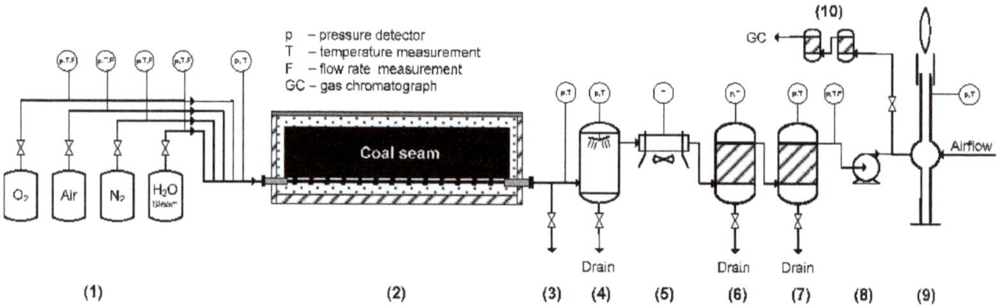

Figure 29. Scheme of ex situ reactor: (1) reagent injection system, (2) ex situ reactor, (3) tar sampling, (4) water scrubber, (5) air cooling of syngas, (6,7) gas separators, (8) centrifugal suction fan, (9) thermal combustor, (10) gas purification chromathography analysis (GC) [82,87,88]. Reproduced under CC BY 4.0.

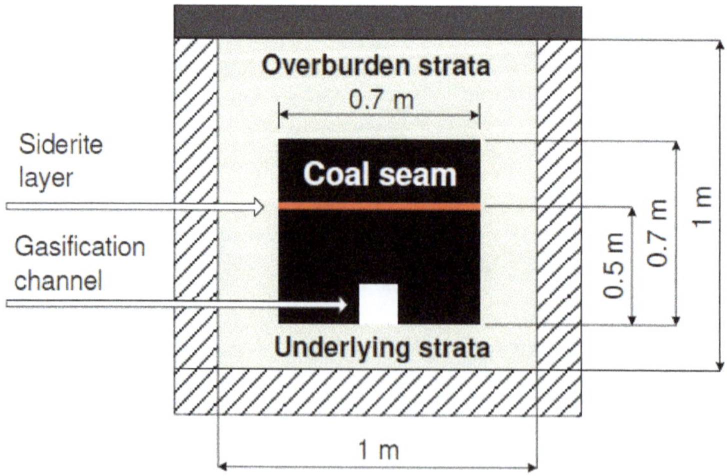

Figure 30. Cross-section of a coal seam model with a siderite layer [82]. Reproduced under CC BY 4.0.

The results showed that siderite has no direct effect on the course of gasification. On the other hand, the water contained in the sand had the most significant influence on the gasification efficiency. The average calorific value of the produced syngas was 9.71 MJ/m^3 (see Figure 31).

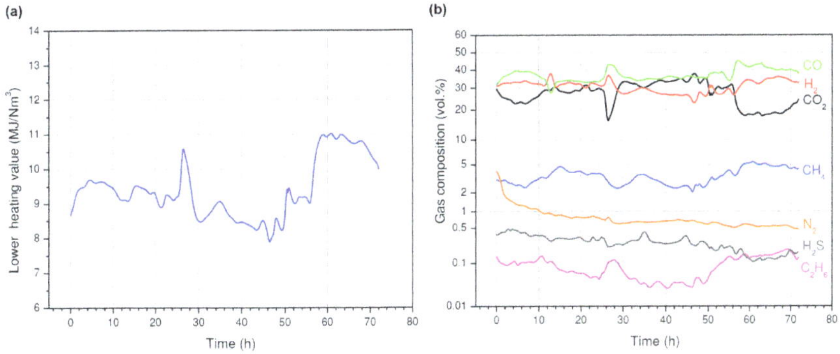

Figure 31. Measured process variables: (**a**) syngas calorific value; (**b**) syngas composition [82]. Reproduced under CC BY 4.0.

Another hard coal gasification result with a 2 cm siderite layer was reported in [87]. The configuration of the artificial coal seam seam was similar to [82]. As listed in Table 4, the produced syngas had an average heating value of 9.41 MJ/m^3. The calculations showed that only 34.1% of the bedded siderite was decomposed into CO_2, which has little effect on the CO_2 balance during gasification. The depicted isolines with the concentration of CO_2 in the siderite layer after gasification provide a view of the location of the high-temperature zones in the reactor (see Figure 32).

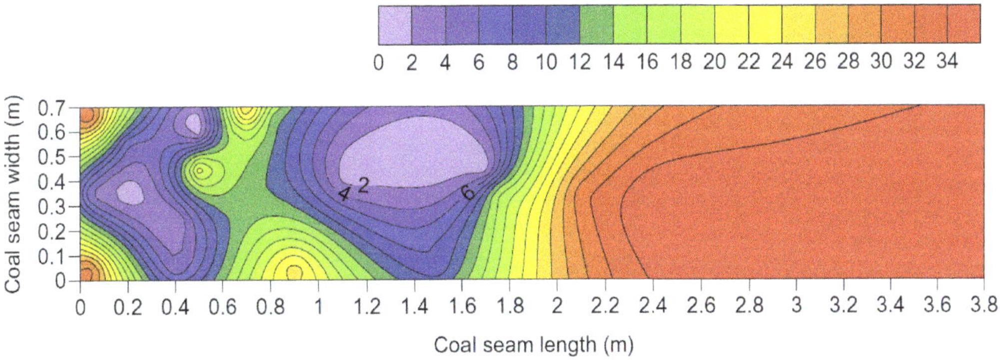

Figure 32. Isolines of carbon dioxide content in siderite layer after gasification process (wt%) [87]. Reproduced under CC BY 4.0.

7. Modeling Temperatures in UCG

Temperature modeling helps to optimize the UCG process by providing insights into temperature distribution and profiles within the coal seam. It helps to identify regions of high and low temperatures, enabling the adjustment of process parameters to achieve more efficient and controlled gasification. The UCG operation can be optimized by understanding temperature variations to maximize gas production and minimize undesirable byproducts. By understanding temperature dynamics, appropriate measures can be taken to ensure the stability and integrity of the UCG process, reducing the risk of accidents or operational failures. Temperature modeling helps to understand the temperature profiles and gradients in the surrounding rock formations and aquifers, thus enabling the evaluation of potential heat transfer and groundwater contamination risks. In addition, temperature modeling allows for real-time process monitoring and control during UCG operations. By predicting and analyzing the thermal effects, steps can be taken to minimize or mitigate the impact on the surrounding environment.

Although thermocouples in ex situ reactors can measure temperatures, their utilization in in situ coal gasification can be problematic, given that the reactor is underground. Therefore, various proxies (e.g., radon emanation [89,90], carbon isotopes [91], or acoustic emission measurement [92]) and model-based soft sensors are used to determine the underground temperature. The model-based temperature prediction is discussed in the following sections.

7.1. Modeling Based on the Theory of Non-Stationary Heat Conduction

A temperature model based on the theory of non-stationary heat conduction and measured ex situ data was developed by Kostúr et al. [93]. The temperature prediction model relied on Fourier's partial equations for the solid phase, encompassing coal and rock. The model is based on the following equation:

$$\frac{\partial(\rho_j c_j t_j)}{\partial \tau} = \frac{\partial\left(\lambda_j \frac{\partial t_j}{\partial x}\right)}{\partial x} + \frac{\partial\left(\lambda_j \frac{\partial t_j}{\partial y}\right)}{\partial y} \quad \text{for } j = 1, \cdots, L, \tag{10}$$

where c represents the specific heat capacity (J.kg^{-1}), ρ is the density (kg.m^{-3}), x, y are the coordinates (m), t is the temperature (°C), λ is the thermal conductivity (W.m.K^{-1}), τ is the time (s), j is the index of j-th layer.

The system of equations (10) can solve each layer in overburden, including coal. It means that the mathematical model for coal and overburden consists of L layers (i.e., coal, sandstones, claystone, ...).

The mathematical model was solved using the first boundary condition, defined by the measured temperature on the interface of the gasification channel and the coal. The model considers only very low heat transfer by convection in the y axis direction due to the limited permeability of underground rocks for gas streaming. Heat transfer by radiation and convection in the x axis direction was mainly relevant within the gasification channel. The proposed 2-D model to simulate gasification front movement was regarded for one gasification channel. The input to the model was the ambient temperature T_0 and temperatures measured in the gasification channel (see Figure 1). The model's output was the coal's predicted temperature at the selected position (e.g., T_3, see Figure 33). The results showed excellent accordance between the simulated and measured temperature (see Figure 34).

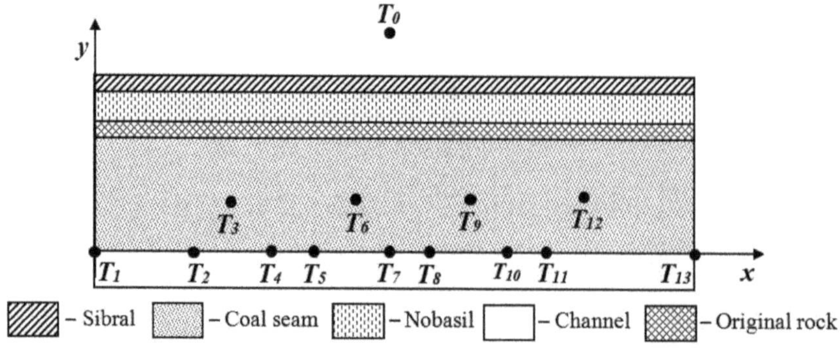

Figure 33. Thermocouples placement in ex situ reactor. Reproduced under CC BY 4.0.

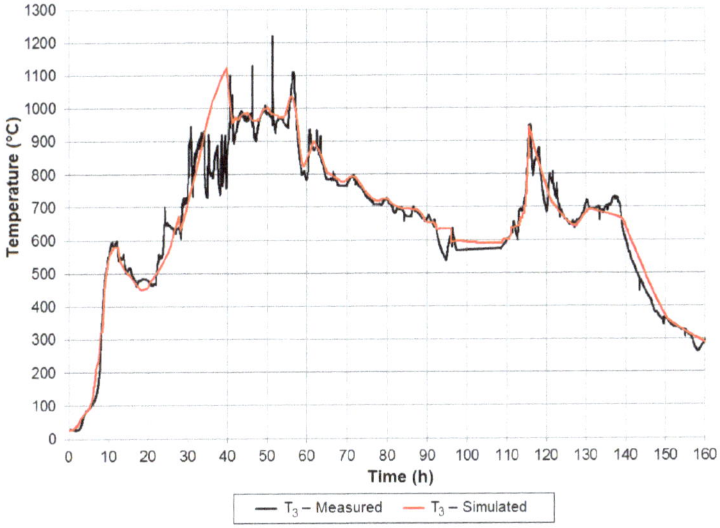

Figure 34. Measured and simulated temperature T_3. Reproduced under CC BY 4.0.

7.2. Data-Driven Modeling Based on Regression Models

For temperature prediction, regression models calculated from measured ex situ data were also proposed. Durdán et al. [94] modeled the temperature in a small ex situ reactor using multiple linear regression models. Compared to the previous approach [93], applying thermo-physical parameters to the calculation is unnecessary. They performed two UCG experiments with Slovak Cigel's coal glued with tar, coal dust, and water (see Figure 35). The experiments differed in the weight of bedded coal (i.e., 214 kg vs. 472 kg), the diameter of the gasification channel (i.e., 20 mm vs. 40 mm), the maximum temperature reached (1200 °C vs. 1400 °C), and the calorific value of the syngas (i.e., 2.5 MJ/m^3 vs. 5 MJ/m^3). The model preparation and results after gasification are shown in Figure 36. Recorded process data were used in temperature modeling by regression analysis.

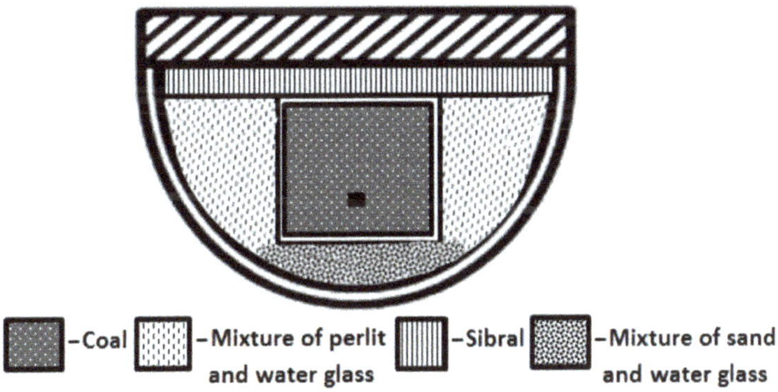

Figure 35. Cross-section view on model preparation in ex situ reactor [94]. Reproduced under CC BY 4.0.

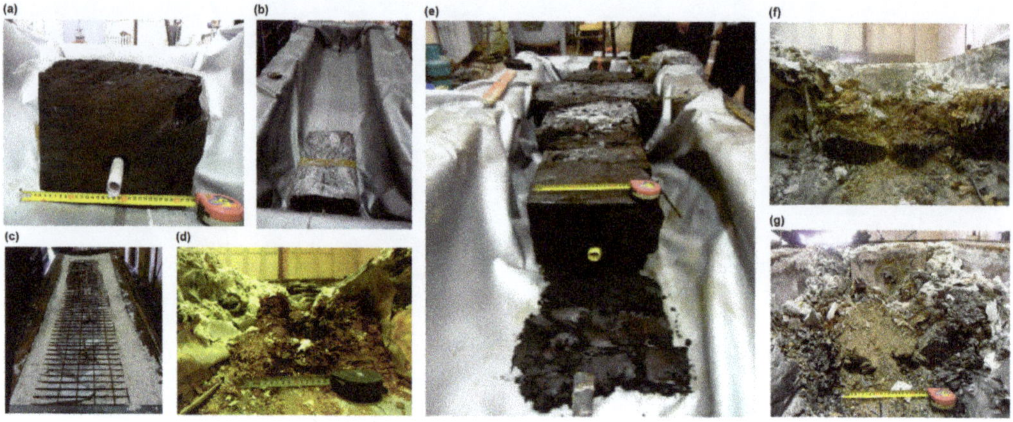

Figure 36. Model preparation in ex situ reactor for experiment #1: (**a**) gasification channel preparation; (**b**) coal block bedding; (**c**) metal grid installation; (**d**) result of gasification. Model preparation for experiment #2: (**e**) coal block bedding and gasification channel preparation; (**f**) a layer of coal in the second meter that remained unburned; (**g**) a layer of coal in the third meter that remained unburned [94]. Reproduced under CC BY 4.0.

Various measured variables from experimental gasification (i.e., oxygen and airflow, exhaust fan speed, the calorific value calculated from syngas composition, and measured coal temperatures) were used as independent variables in regression analysis. The regression parameters were calculated using the least squares method for different modeled temperatures in the gasification channel. The simulation results achieved were outstanding. Figure 37 shows the excellent agreement of the modeled and measured temperature in the gasification channel. From the comparison, we can see that the regression model achieved better performance in the case of the first experiment.

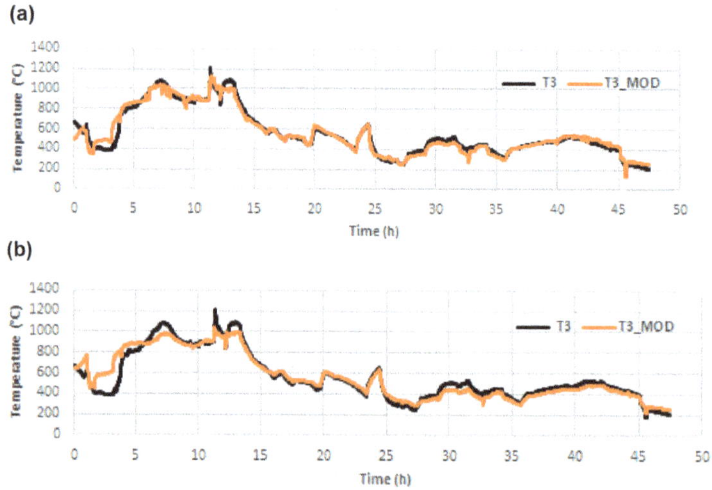

Figure 37. Behavior of measured and modeled channel temperature: (**a**) data-drive regression model based on experiment #1; (**b**) data-drive regression model based on experiment #2 [94]. Reproduced under CC BY 4.0.

7.3. Machine Learning Approach

Machine learning is a type of artificial intelligence (AI) that allows software applications to become more accurate at predicting outcomes without being explicitly programmed to do so. Machine learning algorithms use historical data as input to predict new output values. Kačur et al. [95] proposed and verified several data-driven machine learning models for forecasting the syngas's heating value and temperature in the oxidizing zone. The models investigated were back propagation neural networks (BPNNs) [95], multivariate adaptive regression splines (MARSs) [27,95,96], and support vector regression (SVR) as a part of support vector machines (SVM) [97]. The models were trained on data from experimental trials on ex situ UCG reactors. The experimental data were divided into training and testing sets, and statistical indicators evaluated the performance of the models (Figure 38). For example, Figure 39a shows the SVR prediction of underground temperature in the oxidizing zone based on measured syngas composition. These models achieved higher prediction performance than classical multivariate linear regression models.

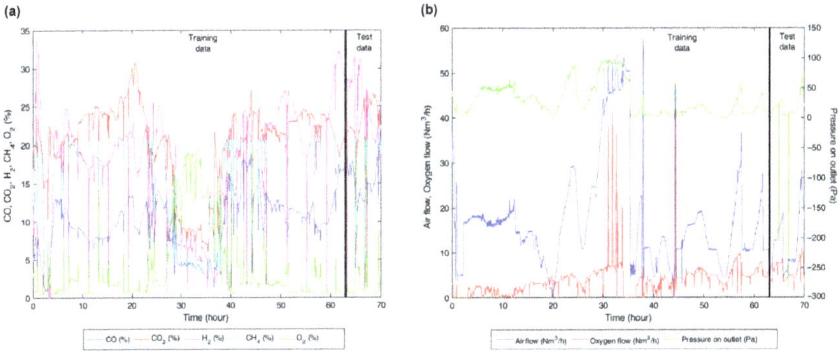

Figure 38. Training and testing observation: (**a**) syngas composition; (**b**) manipulation variables [95]. Reproduced under CC BY 4.0.

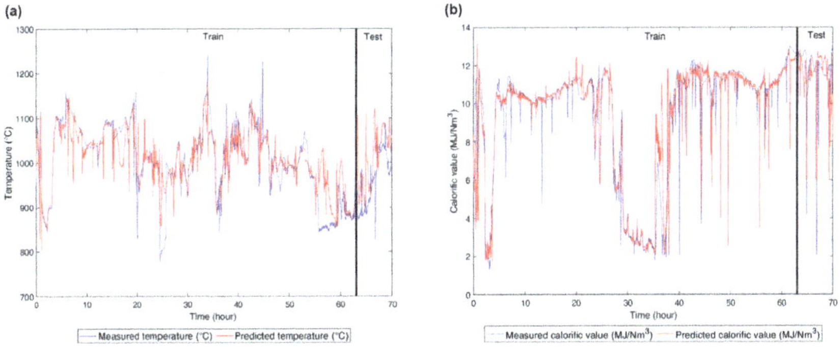

Figure 39. Results of prediction by SVM on known and unknown input observations (i.e., training and testing): (**a**) underground temperature; (**b**) syngas calorific value [95]. Reproduced under CC BY 4.0.

On the second side, the syngas's heating value can be predicted from the volume flows of gasification agents measured (i.e., steam, oxygen, and airflow) (see Figure 38b). The size of the training set and the number of input observations, as well as the setting of the learning parameters have an impact on the prediction performance both in the phase of training on known data and in the stage of testing the model on unknown input data (e.g., adjustable can be the number of neurons in the hidden layers of the NN, the types of the kernel functions in SVM, or the types of the MARS model). Learned or trained models, however, can be used as an effective tool for temperature soft sensing or calorific value prediction in advanced control algorithms. Figure 40 shows principle of adapted soft sensor based on a data-driven machine model [27,95].

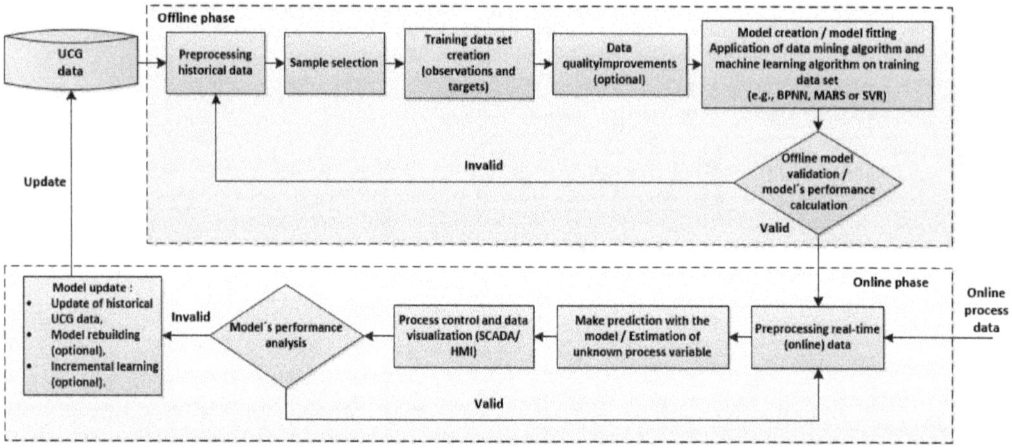

Figure 40. The principle of soft-sensing in UCG [95]. Reproduced under CC BY 4.0.

For example, in Support Vector Regression (SVR) the data $\mathbf{x}$ are mapped into a high-dimensional feature space, to do linear regression in this space. Non-linear SVR finds the optimal function $f(\mathbf{x})$ in the transformed predictor space. The prediction function has the following form [98,99]:

$$y = f(\mathbf{x}) = \sum_{i=1}^{l} (\alpha_i - \alpha_i^*)\mathbf{K}(\mathbf{x}_i, \mathbf{x}) + b, \tag{11}$$

where y is target variable, $\mathbf{K}$ is the kernel matrix of inner products between all pairs of points $\{\mathbf{x}_i\}_{i=1}^{l}$, α and α^* are non-negative Lagrange multipliers for each observation $\mathbf{x}$, $\mathbf{x} = (\mathbf{x}_1, \mathbf{x}_2, \ldots, \mathbf{x}_l)$ is the vector of the observations where l denotes the size of sample. A threshold b can be determined from the Lagrange multipliers.

Observations with nonzero Lagrange multipliers are called support vectors. The prediction of new values is based on a function that depends only on the support vectors. Lagrange coefficients can be found by minimization of the following function [95,98,99]:

$$L(\alpha) = \frac{1}{2}\sum_{i=1}^{l}\sum_{j=1}^{l} (\alpha_i - \alpha_i^*)\left(\alpha_j - \alpha_j^*\right)\mathbf{K}(\mathbf{x}_i, \mathbf{x}_j) + \varepsilon \sum_{i=1}^{l} (\alpha_i + \alpha_i^*) - \sum_{i=1}^{l} y_i(\alpha_i^* - \alpha_i), \tag{12}$$

subject to the constraint and Karush–Kuhn–Tucker complementary conditions required to obtain the optimum.

The minimization problem can be solved by common quadratic programming techniques, e.g., the chunking and working set method, sequential minimal optimization (SMO), or iterative single data algorithm (ISDA) [98,99].

For comparison, the MARS model is based on the so-called basis functions, which express the mutual interaction of input observations. The MARS model can be expressed by the following equation [100]:

$$y = f(\mathbf{x}) + \varepsilon = c_0 + \sum_{m=1}^{M} c_m B_m(\mathbf{x}) + \varepsilon, \tag{13}$$

where y represents the output variable, $\mathbf{x}$ is the vector of input variables, M is the number of basis functions in the model (i.e., number of spline functions, c_0 is the coefficient of the constant basis function B_0, and sum is over the basis functions B_m).

A good comparison of SVR and MARSs was also made in [101]. The research showed that both methods perform similarly depending on the training set. Although in [95], the SVR-based machine model performed slightly better in temperature prediction, MARS-type models are more portable and can be quickly implemented into control algorithms. For example, the MARS model has been successfully applied to test AMPC [27] and ESC [96] for UCG imitation and prediction model adaptation. The disadvantage of MARS-type models is the high computational complexity in the machine-modeling phase. The disadvantage of the neural network is frequent overfitting and the need for the experimental setup of the network [95].

7.4. Application of Geostatistical Methods

Laciak et al. [102] and Kostur et al. [9] have applied geostatistical methods in spatio-temporal modeling of temperature changes of UCG experimental trials. They used the Kriging method for modeling temperature fields in a georeactor. Kriging is a geostatistical estimation procedure to interpolate or predict values at unobserved locations within a spatial domain. The Kriging process involves constructing a mathematical model that characterizes the spatial correlation of the variable across the study area. Utilizing the principles of geostatistics and Isatis's geostatistical environment, a spatiotemporal model was developed to analyze temperature changes within the experimental generator during laboratory trials of UCG (see Figure 41).

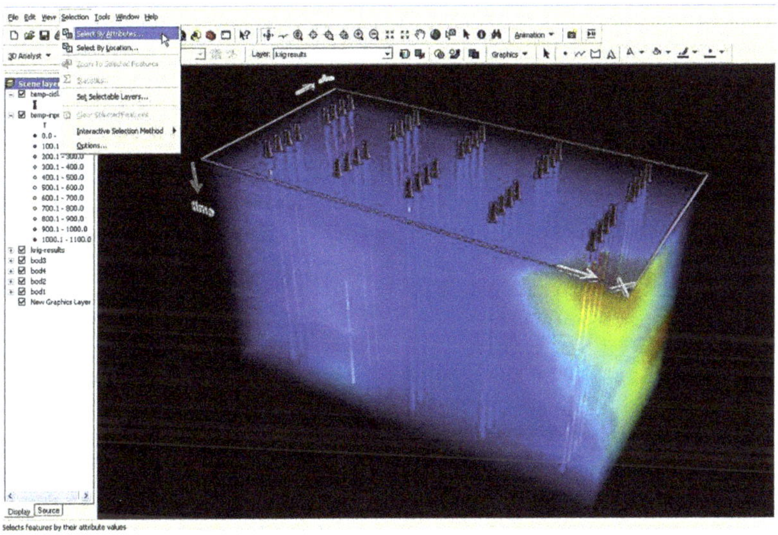

Figure 41. Spatio-time model of temperature distribution in the ex situ reactor [9].

This model was created from measured temperatures from an ex situ UCG experiment. The vertical dimension in a spatial-time model of temperature represents time. A time-space model of temperature changes or temperature sections was derived from this model, which was used to analyze the movement of the combustion front in the UCG reactor. Individual time sections of temperatures in the reactor were analyzed using ArcGIS-ArcMap software, in which an animation of the movement of temperature zones in time and space was created. Such an animation provides insight into the dynamics of the process. The model can calculate the movement vectors of temperature zones. From the space-time visualization of the gasified layer, it is possible to transform even the in situ UCG. It is possible to identify places with high temperatures or places with tectonic faults. It is also possible to locate drying, oxidation, and reduction zones.

8. Research of Gasification Techniques

Various gasification techniques, such as linked vertical wells (LVWs), retreating injection points (CRIPs), or reverse gasification, are tested in the UCG experimental tests. These techniques are used alone or in combination to enhance the permeability of a coal bed and to increase the efficiency of UCG. Some methods for connecting injection and production wells have been well documented in [11,103]. In the following sections, applications of these techniques in ex situ tests will be reviewed.

8.1. Reverse Gasification

Dobbs and Krantz [104] investigated the combustion front propagation in UCG and designed one of the first experimental gasification reactors. They used a technique of reverse gasification. The research also aimed to determine the amount and composition of syngas produced under different pressures in the reactor. This UCG research used a pressure vessel as a gasification container. The container was loaded with blocks of sub-bituminous coal and a mixture of oxygen and nitrogen-supported reverse combustion under a pressure of 200 kPa. It was found that, in reverse gasification, the combustion front moves upstream of the oxidizer flow. The fuel for combustion is gases that are gradually released from the coal into the generator space while the porosity of the coal changes. A critical insight in reverse combustion is that the combustion reaction is confined to only a very thin reaction zone. The reverse gasification was also tested by Liu et al. [80] in the gasification of Chinese lignite.

8.2. Gasification with Axial Injection and Moving Reaction Point

Su et al. [92] designed a coaxial UCG system with a horizontal well, which was tested on an ex situ reactor (see Figure 42). They used horizontal borehole for gasification agent injection and syngas extraction. Figure 23 shows a schematic of the experimental reactor with the coaxial-hole model and the equipment for measurement and regulation. The received coal had fixed moisture, 2.10%.

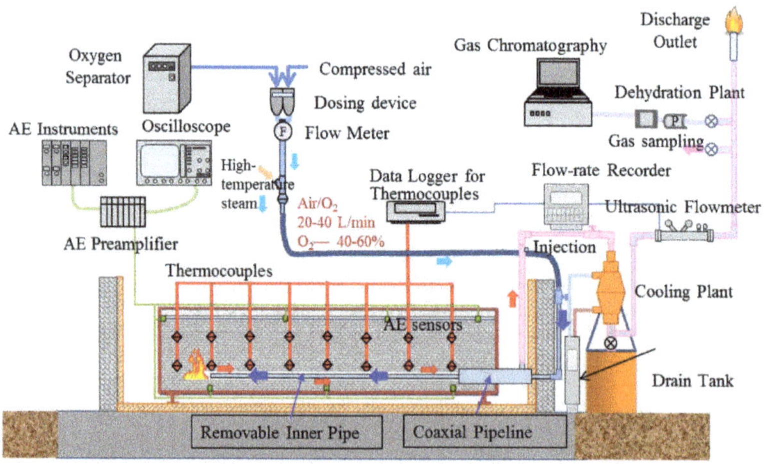

Figure 42. Ex situ UCG reactor with the coaxial-hole model [92]. Reproduced under CC BY 4.0.

During the gasification experiment, they achieved an average heating value of 6.85 MJ/Nm^3 and a gasification efficiency of 65.43%.

It was found that the coaxial UCG method achieved the same results as the traditional UCG so that it can be applied to abandoned underground coal seams. The experiment shows that the locations of acoustic emission (AE) sources aligned with those derived from temperature profiles, indicating that numerous acoustic emissions were generated in regions experiencing thermal stress (refer to Figure 43). Furthermore, it was shown that changing the position (shifting) of the injection reaction point during gasification affected the calorific value of the syngas positively (see Figure 44).

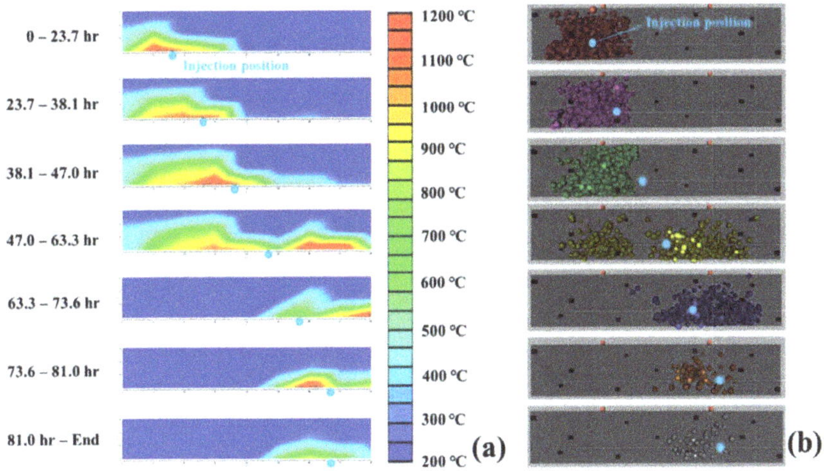

Figure 43. Temperature fields monitoring: (**a**) temperature distribution; (**b**) location of acoustic emissions sources [92]. Reproduced under CC BY 4.0.

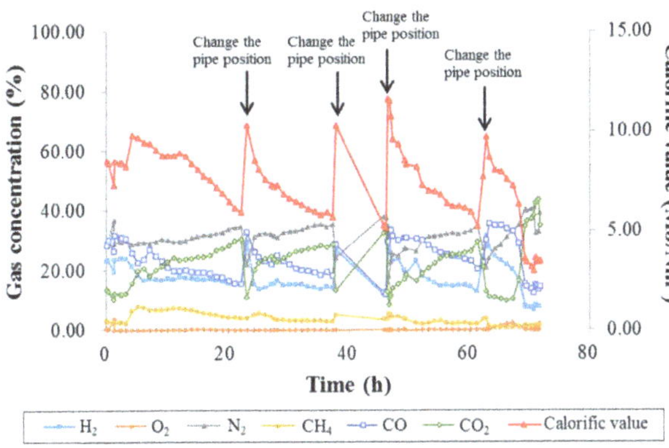

Figure 44. Composition and calorific value of syngas [92]. Reproduced under CC BY 4.0.

The coaxial UCG technique with a horizontal inlet well was also studied by Hamanala et al. [105]. In tests with UCG, the model was composed of coal with two types of coal. Coal with a heating value of 30.18 MJ/kg and moisture content of 2.9%, and coal with a calorific value of 22.66 MJ/kg and a moisture content of 2.2% (i.e., type 2) were used. These coals were gasified using air and technical oxygen and obtained an average calorific value of 8.05 MJ/m^3 and 6.91 MJ/m^3. It was found that the reaction zone gradually increased along the wall of the coaxial opening. Moreover, Su et al. [106] investigated three gasification techniques, i.e., a coaxial model using a coaxial pipeline as a gasification channel, a coaxial model using a coaxial pipeline combined with a bottom cross-hole, and a linking-hole model using a horizontal V-shaped cross-hole. The proximate analysis showed that the moisture content of the coal was 2.10%, and the gross calorific value of the gasified coal was 32.12 MJ/kg. They performed three UCG trials where oxygen-enriched air of about 60%, 50%, and 60% was used as a gasification agent. The largest AEs were recorded when temperature fluctuated between 400–900 °C. In these three experiments, calorific values of 5.11 MJ/Nm^3 in test 1, 5.51 MJ/Nm^3 in test 2, and the highest 8.43 MJ/Nm^3 in test 3, where a horizontal V-shaped linking-hole model was used, were achieved.

8.3. Technique CRIP and LWM

The combination of the controlled retreating injection point (CRIP) along with the linked vertical wells (LVWs) technique was simulated by Zagorcak et al. [44] where semi-anthracite coal was gasified under atmospheric pressure, and an increased pressure of 30 bar. The results indicate that, by correct dosing of reactants, stable quality of syngas both with the LVM technique and with the CRIP technique can be ensured. In both methods, as the cavity grows, the calorific value of the syngas decreases, which must then be compensated by the correct dosing of the oxidizer and not exceed the hydrostatic pressure, to avoid gas losses and leaks into the environment. The most important factors that influence the composition of syngas and its stable production are the gasification agent used. It has been found that the most suitable gasifying agent is a mixture of oxygen and water, unless natural water is available in excess. However, it should be noted that the gradual reduction in calorific value cannot be avoided, and, therefore, the oxygen injection point (CRIP) must be shifted. The LVWs technique was applied in practice at the Hanna and Hoe Creek deposits during in situ UCG, while, when gasifying highly volatile bituminous coal with air, the average calorific value of syngas was in the range of 3.4–5.3 MJ/m^3 and when gasifying with a mixture of oxygen and vapors 6.9–9.0 MJ/m^3 [107]. The CRIP technique was applied to the gasification of sub-bituminous coal in the Rocky Mountain deposit, where it was gasified with a mixture of oxygen and steam, while the calorific value of the produced syngas was 9.5 MJ/m^3 [108]. Similarly, this technique was also used at the El Tremedal site [109], where, with a mixture of oxygen and steam, the calorific value of syngas was up to 10.9 MJ/m^3.

9. Change in the Coal Properties during Gasification

During the gasification process, several changes occur in the properties of coal. These changes result from the chemical and physical transformations that occur when coal reacts with gasifying agents like oxygen, steam, or air. The key changes in coal properties during gasification include a decrease in carbon content, release of gaseous products (e.g., CO, CO_2, CH_4, H_2, and other hydrocarbons), increased porosity, loss of volatiles, change in calorific value, and formation of tar and ash. Gasification substantially changes coal properties, converting solid coal into a mixture of gases and byproducts. The specific changes depend on various parameters, including the gasification technique used, the composition of the gasifying agent, and the characteristics of the coal being gasified. The following subsection discusses recent research on coal's optical and petrographic changes.

9.1. Change in the Optical Properties of Coal during Gasification

Previous studies [110–112] showed that light reflectance and vitrinite birefringence begin to change when coal is heated. Variations in optical properties are associated with thermal alterations occurring within the internal structure of Vitrinite and its resultant transformation products, such as mesophases and matrix. The type and magnitude of these alterations are contingent upon the coal rank and the coking susceptibility of the initial material. Vitrinite's reflectivity is a highly reliable indicator of the magnitude of alterations occurring within the georeactor and its immediate vicinity.

Nowak et al. [113] have investigated changes in the optical properties of coal during gasification in an ex situ reactor (see Figure 45) under pressures of 10 and 40 bars. Technical oxygen was used as a gasification agent. The optical characteristics of both raw coal and the residues resulting from gasification were examined, encompassing the determination of the average random reflectance of Vitrinite in the raw coal and its transformation products after gasification. Microscopic examinations were performed using an optical light microscope.

Figure 45. Reactor prepared for gasification [113]. Reproduced under CC BY 4.0.

As the distance from the reactor chamber inlet increased, it was observed that the true maximum reflectance and bireflectance decreased. Furthermore, the extent to which temperature affected the coal depended on the gasification pressure, with a distance of 60 m for 10 bar and 35 m for 40 bar.

Following coal gasification at 10 bar pressure, transformed coal components were detected within the reactor. These included vitrinite grains exhibiting visible pores, coke grains with varying pore sizes and pronounced relief, and mesophase, leading to significant morphological changes in the liptinite macerals. The rate of morphological changes in coal components decreased with increasing distance from the reactor inlet.

Under a gasification pressure of 40 bar, degassed vitrinite grains exhibited visible pores, while coke grains displayed various pore sizes and high relief. The influence of temperature on the morphology of coal components decreased as the distance from the reactor chamber entrance increased (see Figure 46).

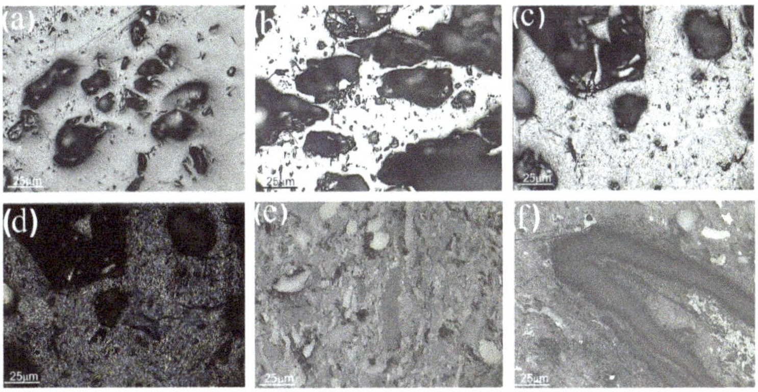

Figure 46. The residue after gasification of coal at a pressure of 40 bar: (**a**) Vitrinite with visible pores after degassing; (**b**) a discernible coal–coke structure altered by heat; (**c**) heat-altered coal-mesophase (matrix), is observable (parallel Nicols); (**d**) the same field as photo c (crossed nicols); (**e**) trimacerite displaying unchanged color and morphology compared to the raw sample; (**f**) macrosporinite exhibiting unchanged color and morphology of liptinite when compared to the raw sample [113]. Reproduced under CC BY 4.0.

Similar changes in optical properties and the structure of residues after coal gasification were also observed in previous studies [114–117].

The investigation revealed that as the distance (d) from the reactor inlet increased, the maximum temperature exerted on the coal and the true maximum reflectance declined. Figure 47 depicts the correlation between the I_R ratio and the distance from the reactor entrance. Here, I_R represents the ratio of the true maximum reflectance values between the raw sample and the gasification residue. The ratio exhibited an upward trend as the distance increased. An I_R value of 1 signifies that the reflectance value of the gasification residue matches that of the raw sample.

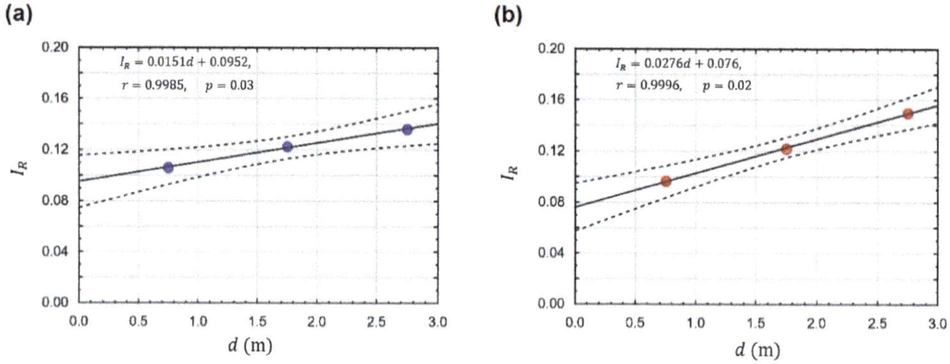

Figure 47. The relation between the I_R ratio and the distance from the inlet of the reactor chamber at (**a**) a pressure of 10 bar and (**b**) a pressure of 40 bar [113]. Reproduced under CC BY 4.0.

9.2. Change in the Petrographic Composition of Coal during Gasification

The impact of the petrographic composition of coal on the production of char was studied in [118–120], and the petrographic characteristics of char were studied in [121,122]. However, the mentioned papers focus on the products of the surface gasification of coal, while petrographic changes resulting from the UCG have been studied only in one research study. Under the influence of thermal processes, the coal in the reactor is transformed. At the same time, after gasification, it is possible to distinguish the area of ash, coke, dried coal, cracked transversely to stratification, and the dried lignite area in the cavity. Petrographic analysis can help to predict the suitability of a coal seam for UCG. Bielowicz [123] investigated the change in the petrographic composition of lignite with high moisture content (46.52%) during the ex situ lignite gasification using oxygen. The results showed that the most significant changes in the coal were observed in the cavity and along the gasification channel that was most exposed to the gasification agent. It was found that the sapropelic brown layers were mostly degassed and dried. In addition, the gelation and fusain layers were not observed there. Outside of the cavity, the bituminous lignite was altered to a lesser extent. Its partial drying and degassing were visible. In the case of lignite transformations, strong gelifications and the formation of fusains were most often observed.

9.3. Changes in Heat-Affected Zone from the UCG

Prabu and Jayanti [124] analyzed heat-affected zones with high ash coals during ex situ gasification. These zones of dry and volatile depleted porous zones create coal to a certain depth, which causes its spalling. Coal samples from different locations of heat-affected zones from experimental coal gasification were analyzed. The extent of change depended on coal type and gasification conditions. These changes in the coal affect its reactivity and the syngas's composition. The research results showed that in the heat-affected zone, typically, there is a significant reduction of volatile substances over the entire surface of the cavity in the coal, one of the reasons being the loss of reactivity. The reactivity of the coal is mainly lost during the initial heating of the coal and the loss of the surface area of the coal through pyrolysis. The loss of volatile substances and moisture within the heat-affected zone directly impacts the composition of the syngas, while the Boudoard reaction influences the gasification rate. The loss of volatile substances can lead to incomplete utilization of the coal seam as it becomes less reactive. Similar studies simulate the UCG cavity geometries using laboratory-scale bore-hole combustion and gasification experiments on wood and coal [125,126]. Moreover, measurements of the Brunauer–Emmett–Teller (BET) surface area of the char prepared at different temperatures [127] showed a pronounced decrease in the available surface area which was attributed to the loss of reactivity in CO_2 gasification.

10. Research of Environmental and Health Impacts of Underground Coal Gasification

Environmental control measures, such as gas cleaning and emissions reduction techniques, are crucial to minimize the release of pollutants and ensure compliance with environmental regulations. It is essential to analyze coal residue tar to understand its composition and potential contaminants, to prevent the impact of underground coal gasification (UCG) on the environment and human health. Environmental analysis should regard the relationship between the physicochemical composition of wastewater, coal properties, and gasification pressure. Additionally, predicting the release of pollutants from the residues after UCG is essential, as it helps identify potential contamination risks. The UCG process can be optimized by supplying additional CO_2, which helps to reduce greenhouse gas emissions and promotes more controlled and efficient gasification to mitigate the environmental impact. In addition, proper control of operating parameters can eliminate the risks associated with gas leakage into the critical area of the UCG. Leakage of produced gas can result in an explosion or the poisoning of people. These topics have also been studied in the research papers that will be discussed next.

10.1. Analysis of Wastewater, Coal Residues, and Tar from UCG

One of the most critical issues during the UCG process is wastewater production and treatment. Many hazardous compounds can contaminate condensed gasification wastewater. In the literature, an evaluation of the biological treatability of wastewater from the UCG pilot installation in the Hanna coal deposit (Wyoming) can be found [128]. Zhang et al. propose the pretreatment of wastewater generated during coal gasification by acidification demulsion [129]. Many toxic compounds in UCG wastewater are challenging to decompose if only biological methods are used [130]. Thomas et al. present the possibility of phenol removal from UCG effluents by using coagulation–flocculation and the H_2O_2/UV process [131]. Treatment of coal gasification wastewater by catalytic oxidation with trace ozone is another promising technique [132]. Recently, Pankiewicz-Sperka et al. [51] performed a qualitative and quantitative wastewater characterization on various UCG experiments in an ex situ reactor (see Figures 29 and 48) with "Six Feet" semi-anthracite and "Wesoła" bituminous at 20 and 40 bars. Oxygen and water vapor were used as gasification agents.

Figure 48. Ex situ reactor prepared for atmospheric pressure coal gasification [88]. Reproduced under CC BY 4.0.

The results showed the relationship between the physicochemical composition of wastewater and the characteristics of coal, including gasification pressure. It was found that coal properties and gasification pressure significantly influence organic pollutants, mainly phenols, BTEX (i.e., benzene, toluene, ethylbenzene, and xylene), and PAHs (i.e., polycyclic aromatic hydrocarbons). Phenols had the highest concentrations in bituminous coal (29.25–49.5 mg/L), while these concentrations were lower in semi-anthracite coal gasification (2.1–29.7 mg/L). The opposite result was recorded in the case of BTEX, where the concentrations were lower (2514.3–1354.4 g/L). There was found a similar relationship in the case of the PAHs' concentration, with higher concentrations observed during the gasification of semi-anthracite coal (362–1658 g/L), while in the case of gasification of bituminous coal, PAH concentrations in wastewater were lower (407–1090 g/L) (see Figure 49). Furthermore, the experimental investigation revealed that the concentrations of phenols, BTEX, and PAHs decreased as the pressure increased (see Figure 50).

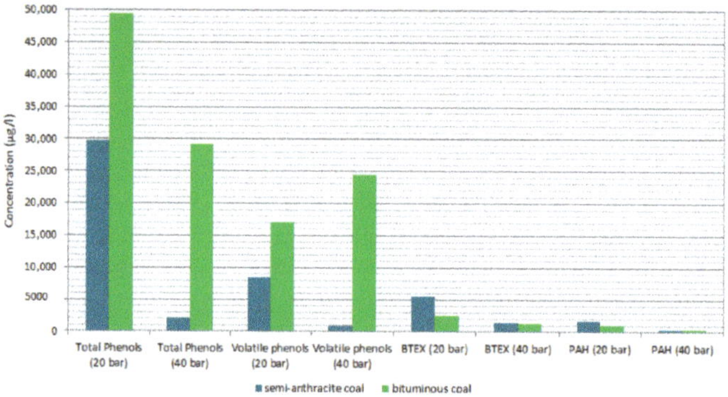

Figure 49. Impact of coal type on concentrations of selected organic contaminants in gasification wastewater [51]. Reproduced under CC BY 4.0.

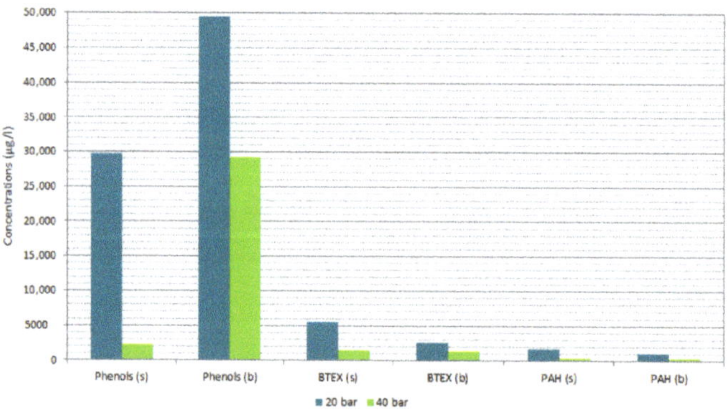

Figure 50. Impact of pressure on selected organic contaminants in wastewater from the gasification of semi-anthracite (s) and bituminous coal (b) [51]. Reproduced under CC BY 4.0.

Another chemometric investigation of wastewater can be found in [133]. This research revealed that, apart from phenolics, PAHs were identified as the most widespread organic pollutants which could be formed in the process of UCG. The amount and types of PAHs and their homologs depend on the type of the gasification process. The BTEX group seriously threatens the underground water system for its strong and long-term toxic effects.

Analysis of tar from ex situ coal gasification was recently performed by Wiatowski [88]. They gasified "Piast" hard coal with a moisture of 4.7%. It was found that the content of BTEX fractions in the tar was 82.6%, while the dominant component was benzene. PAHs had a concentration of 14.7% and phenols 2.7%. The tar yield was determined to be 12.3 kg per ton of coal, corresponding to 1.23% of the total mass of the gasified coal. Figure 51 showcases the findings concerning the total sum of BTEX, PAHs, phenol concentration, mass flow rate, and proportion of tar groups. The high benzene content stemmed from secondary processes within the tar, such as hydrocracking and steam reforming. The tar yield per ton of coal was 1.8%, approximately three times less than that observed during coking. Moreover, the results indicated that the residence time of tar decreased from 1 s at the beginning of the experiment to 0.35 s after 70 h.

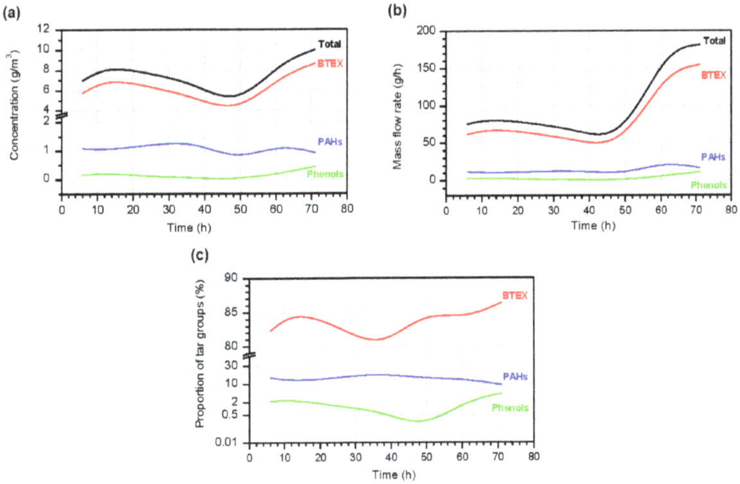

Figure 51. Total sum of contaminants from UCG: (**a**) concentration, (**b**) mass flow rate, and (**c**) proportion of tar groups [88]. Reproduced under CC BY 4.0.

10.2. Pollutants Release from the Residues Remaining after UCG

Groundwater contamination from UCG operations is one of the most critical risks that can occur. This topic is crucial for this process to be carried out on an industrial scale with appropriate environmental safeguards. Strugała-Wilczek et al. [134] performed laboratory tests of solid residues from ex situ underground gasification of ortho- and meta-lignites. The experiments aimed to determine the effect of brown coal on the composition of residues after gasification and contaminants that can leach into the water environment from the UCG cavity.

The gasification results revealed that the residues from lignite gasification exhibit non-uniform characteristics and show varying levels of transformation, confirming the discrepancies in ash content. The ortho-lignite from Oltenia exhibited higher concentrations of the analyzed elements than the meta-lignite Velenje, especially notable in the case of As, B, Co, Cu, Ni, Al, and Fe. The contents of tested elements in Oltenia lignite were higher compared to the world average trace elements content in coals (coal Clarke values) [135]. It turned out that, after lignite gasification, no dangerous trace elements such as Cr, Cd, Hg, and Pb were present in the water extracts, which indicates that lignite gasification is safe for the aquatic environment. Despite the elevated metal and metalloid content in Oltenia ortho-lignite, the aqueous extracts derived from Velenje meta-lignite exhibited higher concentrations of dissolved substances than those obtained from Oltenia.

10.3. Prediction of Poisoning and Explosion in the UCG

Laciak et al. [136] have performed two gasification experiments to investigate gas leaks based on material balance. One on a sizeable ex situ reactor (G1) in the shape of a steel cuboid (see Figure 11a) and the other on a small reactor (G2) in the form of a truncated cylinder (see Figure 52). The bed model was prepared in a different way in each reactor. Figure 53a illustrates a cross-section of the bed model in the first reactor (G1), and Figure 53b demonstrates the placement of the model in the small reactor (G2). There was gasified lignite with total moisture of 13.8%. The air was used as the primary gasification agent, and oxygen was added as an auxiliary oxidizer in case of decreased calorific value or temperatures.

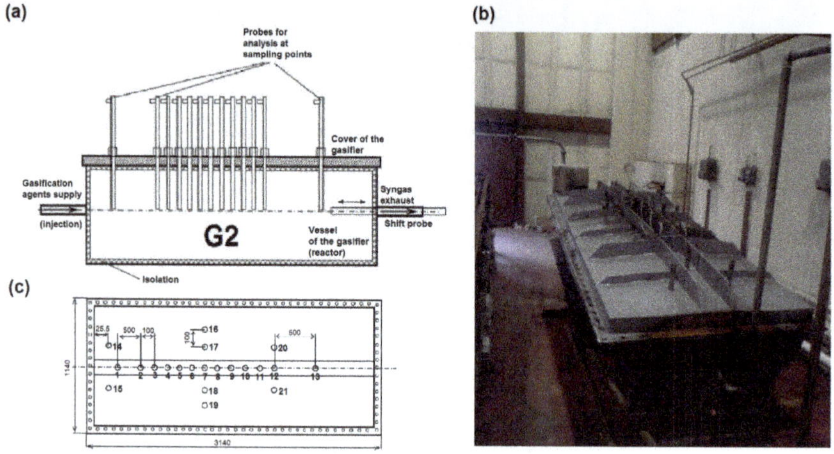

Figure 52. Scheme of small ex situ reactor: (a) side view; (b) top view; (c) photo [136,137]. Reproduced under CC BY 4.0.

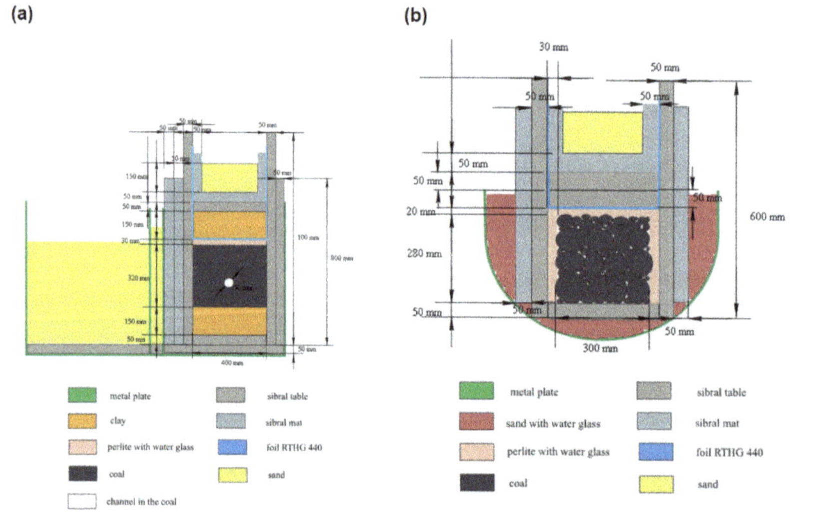

Figure 53. Cross-section of artificial coal seam in the: (a) first experiment and (b) second experiment [136]. Reproduced under CC BY 4.0.

For each experiment, the material balance was established by applying the principle of mass conservation in ex situ reactors during the experiments and determining the effective gasification time. This balance was subsequently adjusted to consider the conservation of element atoms in chemical reactions. The measured syngas composition was used to determine the mass of individual chemical elements in the syngas.

The findings revealed that the most substantial losses occurred in the pipeline at the entrance to the ex situ reactor. Higher losses were in the first experiment. In both experiments, the most significant losses at the atomic level were found for nitrogen (see Figures 54 and 55).

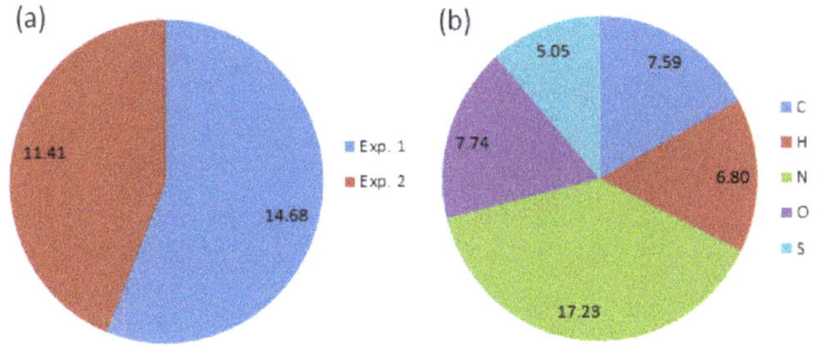

Figure 54. Calculation results: (**a**) Percentage of losses in the experiments; (**b**) average percentage of atom losses [136]. Reproduced under CC BY 4.0.

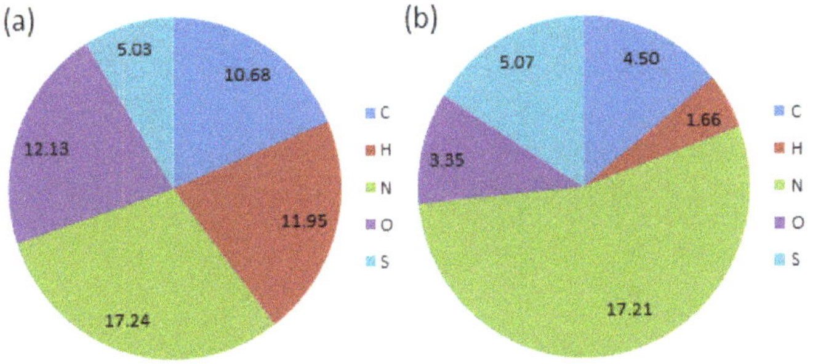

Figure 55. Atom losses in percentages: (**a**) the first experiment; (**b**) the second experiment [136]. Reproduced under CC BY 4.0.

Risk factors must be taken into account when handling syngas and generating syngas. For example, a mixture of air and leaked syngas can be explosive. A high concentration of CO in the operator's area can cause poisoning. Syngas leakage can occur through various leaks and cracks in the surrounding layers of the UCG zone. In previous research, Laciak et al. [137] analyzed the impact of the leakage of produced gas from UCG on the vulnerable area from the point of view of CO poisoning and explosion hazard. Four UCG experiments were performed in a small ex situ reactor, the schematic of which is shown in Figure 52. The bedding of coal blocks in individual experiments is illustrated in Figure 56. The total moisture of the received coal was 20.4%. The results from the material balance showed that the most considerable syngas losses were in the case of experiment #1 (15.4%) and the smallest in experiment #2 (7.9%).

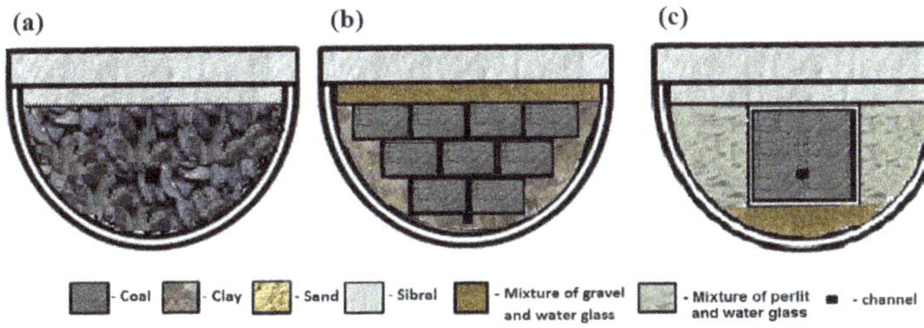

Figure 56. Scheme of coal bedding for the (**a**) first experiment, (**b**) second experiment, (**c**) third and fourth experiment [137]. Reproduced under CC BY 4.0.

Using data measurements from ex situ coal gasification, a mathematical model of gas mixing used to predict dangerous concentrations of CO was proposed. This model can calculate the syngas's composition based on the volume flows of input gasification agents, the volume flow, and the produced syngas. The proposed mixing model has the following form:

$$\begin{aligned} \frac{dX(1)}{d\tau} &= (Q_{IN} \cdot X_{IN}(1) - Q_{\mathrm{OUT}} \cdot X(1)) \cdot \frac{1}{V}, \\ \frac{dX(2)}{d\tau} &= (Q_{IN} \cdot X_{IN}(2) - Q_{\mathrm{OUT}} \cdot X(2)) \cdot \frac{1}{V}, \\ &\vdots \\ \frac{dX(n)}{d\tau} &= (Q_{IN} \cdot X_{IN}(n) - Q_{\mathrm{OUT}} \cdot X(n)) \cdot \frac{1}{V}, \end{aligned} \tag{14}$$

where $X(1)$ represents the concentration of the first component of the internal gas element, $X_{IN}(1)$ represents the concentration of the first component of gasification agent, $X(n)$ represents the concentration of the n-th component of the internal gas element, $X_{IN}(n)$ represents the concentration of the n-th component of inlet gas, Q_{IN} represents the volume flow of the gasification agent, Q_{OUT} is the volume flow of the syngas, and V is the volume of element, $X_{IN}(CO, CH_4, H_2, CO_2, N_2, O_2)$.

The specific concentration of CO can be critical in terms of poisoning or explosion. The crucial concentrations of gaseous components that can lead to an explosion were determined as H_2 at 4%, CH_4 at 5%, and CO at 12.5%.

The simulation results with the model showed that the critical value of CO in syngas poisoning was reached at a specific syngas composition. Therefore, it is necessary to eliminate the risk of gas leakage into vulnerable areas of the UCG. The simulations indicated neither experiment reached a concentration level that could trigger an explosion. The third experiment achieved the highest concentration of hazardous gases (refer to Figure 57a). Even when the volume flow of fresh air was reduced to 1 m^3/h, the limits of explosive components did not surpass the critical values (see Figure 57b). Figure 58 shows the behavior of the predicted concentration of CO and its critical level for individual experiments.

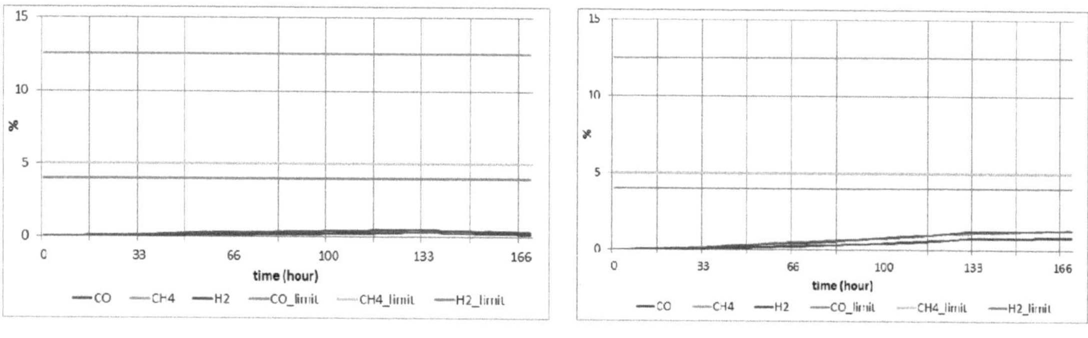

(a) (b)

Figure 57. Results of simulation of the potential explosion due to a gas leak during the third experiment: (**a**) airflow, 25 m^3/h; (**b**) airflow 1 m^3/h [137]. Reproduced under CC BY 4.0.

Moreover, a predictive model was introduced to estimate when CO gas poisoning reaches a critical level. This critical time represents the duration required for the CO gas concentration to reach the critical threshold in a specific area where a syngas leak is expected. The regression model used to compute the critical time of CO poisoning was formulated as follows:

$$t_{\text{critical}} = a_0 + a_1 V_{\text{space}} + a_2 V\_flow_{\text{air}} + a_3 \%V_{\text{leak syng}} + a_4 \frac{\%V_{\text{leak syng}}}{V\_flow_{\text{air}}} \tag{15}$$

where a_i ($i = 1, \ldots, 4$) are the regression parameters; V_{space} is the volume space of regarded vulnerable area; V_flow_{air} represents airflow supplied to the space; $V_{leak\ syng}$ represents the amount of leaking syngas as a percentage.

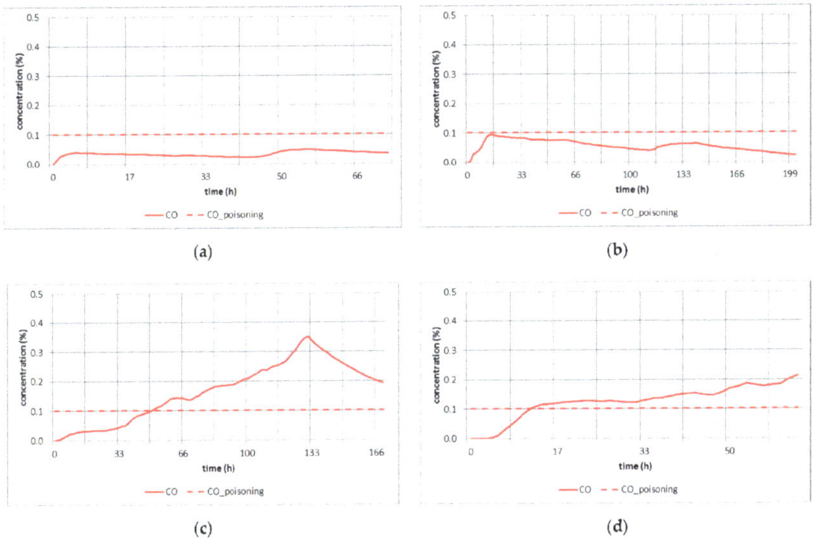

Figure 58. The potential risk of CO poisoning in the (**a**) first experiment, (**b**) second experiment, (**c**) third experiment, and (**d**) fourth experiment [137]. Reproduced under CC BY 4.0.

Figure 59 presents the results of the third model's application. The graph displays the critical time computed using the gas mixing model (GMM) and the critical time obtained from the regression model (StM).

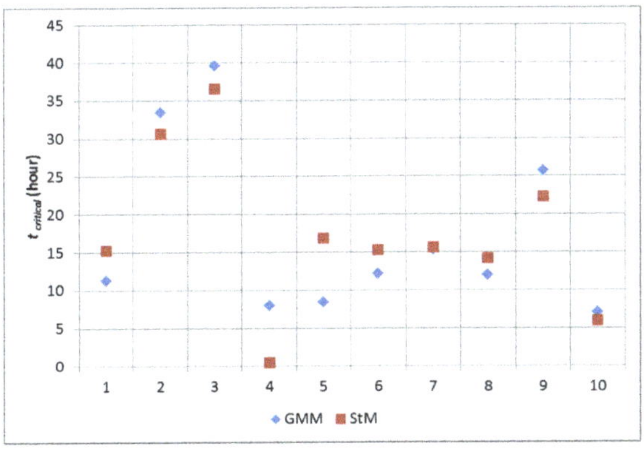

Figure 59. The calculated critical time for CO poisoning as determined by the regression model [137]. Reproduced under CC BY 4.0.

Based on this regression model, supplying fresh air to the critical region of the UCG can be dynamically controlled. The control algorithm can regulate fresh air flow to the area, ensuring that the critical concentration of CO in that space is not surpassed. The algorithm cooperates with the gas mixing model, which calculates the critical time of CO poisoning. Figure 60 shows the control system simulation according to the proposed scheme for the two considered UCG experiments. The rendering presents the airflow Q_{air} regulated by the control algorithm to prevent CO gas poisoning in a vulnerable area. In the figure, the critical or limit value of CO is represented by a dashed line.

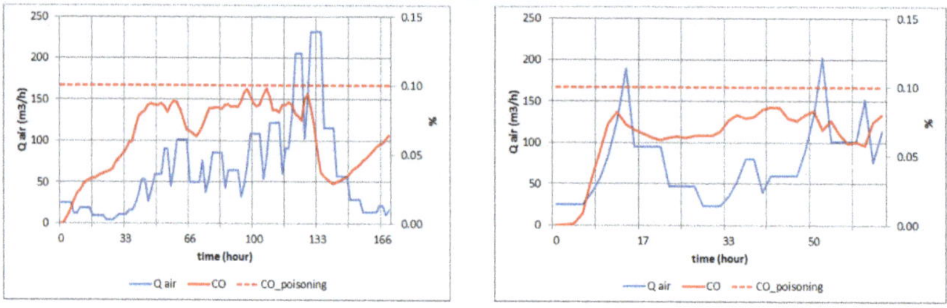

(a) (b)

Figure 60. Controlled concentration of CO in the (**a**) third experiment, and (**b**) fourth experiment [137]. Reproduced from MDPI under CC BY 4.0

An assessment of the harmful effects of syngas from UCG was also carried out in [138]. The analysis and experiments showed that the coal seam's permeability affects the georeactor's thermal performance. The cause is gas leakage through the coal layers towards the top surface (i.e., an experiment with crushed coal). To assess the harmful effects of syngas from UCG, a limit value of 10% was considered as the maximum syngas leakage during the experiment, and in the simulations with the gas mixing model, a space of approximately 1600 m^3 was considered, in which the ex situ reactors were placed. It was found the risk of syngas leakage and CO poisoning can be reduced when an additional ventilation system is installed in a UCG vulnerable zone and the control system performs a regular check of the environment. The research revealed that it is necessary to monitor the gas tightness of the gasification reactor, the concentration of CO in the vulnerable area, and ensure the availability of additional ventilation to prevent poisoning. Predicting syngas composition during various operational interventions will allow us to avoid dangerous situations in UCG.

Unfortunately, there is only low evidence of gas leak monitoring in practice. A comprehensive geochemical survey conducted at the North Knobs Wyoming UCG site [139] has provided conspicuous evidence regarding the escape of gases produced during in situ coal combustion. These resulting gases migrated predominantly within loosely compacted and fractured sandstone layers above the combusted coal seam. Within 3 to 5 days after ignition of the burns, these leaked gases were detected at the surface. The evidence based on localized high-value "hot spots" indicates that some leaked products have reached the surface via pathways associated with imperfect casing cementing in various wells on the facility. Economically, the loss incurred through such leakage likely falls within acceptable thresholds. Extensive analyses throughout the survey consistently revealed no indications of these hazardous gases approaching noteworthy or dangerous levels.

Consequently, it is believed that minimal quantities of toxic gases generated within a well-structured and correctly managed in situ coal gasification facility would not yield detrimental outcomes from safety or environmental perspectives [139].

10.4. Gasification with Supplying an Additional CO_2

UCG generates CO_2 emissions, contributing to climate change and environmental concerns. However, the CO_2 captured can be used as the supporting gasification agent in UCG, preventing it from being released into the atmosphere.

UCG technology holds significant promise in producing environmentally friendly energy by incorporating carbon capture and storage (CCS) techniques [1,140]. Multiple investigations have explored the coal gasification process using CO_2 as a gasification agent. Irfan et al. [141] performed a concise survey of coal gasification studies conducted within a CO_2 environment. Duan et al. [142] introduced a CO_2-recycling strategy within UCG, presenting a two-stage approach for optimizing UCG process variables under CO_2 recycling conditions. Their findings highlighted that introducing CO_2 as a feed gas reduced reaction temperatures considerably compared to steam. In a pilot-scale UCG experiment spanning 60 days at the Wieczorek mine in Poland, Mocek et al. [143] employed oxygen, air, and CO_2 as gasifying agents. Their study indicated the substantial influence of operational pressure on efficient coal gasification in in situ conditions. Investigating bituminous coals across operating pressures from 1 to 3 MPa using a steam and CO_2 mixture, Konstantinou et al. [144] observed that elevated UCG operating pressures resulted in increased gasification rates and efficiency. gasification was performed with a H_2O/CO_2 mass ratio of 2:1, resulting in an average syngas calorific value of 7.8 MJ/Nm^3.

Gasification with a CO_2/O_2 mixture has been minimally investigated, but recently Kumari and Vairakannu [145] performed laboratory experiments based on CO_2 oxy-fuel combustion with a focus on UCG. The experiment revealed that injection of CO_2/O_2 gas can eliminate problems with superheated steam fed into deep coal seams. However, for the needs of UCG control, it is necessary to set the optimal value of CO_2/O_2 because a high concentration of CO_2 can cause the flame extinguishing of the oxidation zone of the georeactor.

The results showed this gasification technique's potential for fresh and dry coal. Increasing the CO_2/O_2 molar ratio (up to 0.6) resulted in a decrease in the CO/H_2 ratio and the content of CO and H_2 in the syngas. Conversely, decreasing the CO_2/O_2 molar ratio to 0.4 led to an increase in the volume of CO and H_2, as well as the CO/H_2 ratio (i.e., reaching a value of 1). Therefore, the optimal CO_2/O_2 molar ratio for efficient CO_2 dry reforming was estimated to be 0.4. The experiment showed that, during gasification with pure oxygen, the highest syngas calorific value was recorded at 298 kJ/mol, which is a similar value to the gasification of fresh coal with the ratio of CO_2/O_2 agents. Gasification with pure oxygen resulted in a higher methane concentration in syngas (averaging 22.12%), primarily due to the conversion of methane into syngas through the dry reforming reaction.

11. Conclusions

The impact of experimental coal gasification in ex situ reactors on in situ underground gasification (UCG) is significant, as it allows researchers and engineers to gain valuable insights and data that can be applied to improve the technology of UCG. Conducting experimental research in ex situ reactors is generally safer and more cost-effective compared to full-scale in situ UCG operations. It allows researchers to test and optimize gasification conditions without the risks associated with underground operations. The controlled ex situ setting enables researchers to isolate specific variables and systematically explore their effects on the gasification process. Furthermore, the data obtained from these experiments can be used to validate computational models and simulations, enhancing the accuracy of predictive tools for assessing UCG performance and potential environmental impacts. This study reviewed the current trends of UCG laboratory research and the achieved results. In summary, experimental research using ex situ reactors plays an essential role in advancing UCG technology. It provides a platform for understanding the gasification process, optimizing operating conditions, improving safety, and designing environmentally sustainable UCG processes. Considering the achieved results and the potential for further research, it is necessary to continue further ex situ tests. It is important that the knowledge gained from the experimental laboratory research of UCG can be transferred to the practice of in situ gasification. The knowledge gained from such research contributes to the successful implementation and widespread adoption of UCG as a cleaner and more efficient energy production method.

Author Contributions: Conceptualization, J.K.; Supervision, M.L.; Writing—original draft preparation, J.K.; Investigation, J.K.; Investigation, M.D.; Resources, P.F.; Writing—Review and Editing, J.K., M.L. and P.F. All authors have read and agreed to the published version of the manuscript.

Funding: This work was supported by the Slovak Research and Development Agency under contract No. APVV-18-0526 and APVV-22-0508.

Institutional Review Board Statement: Not applicable.

Informed Consent Statement: Not applicable.

Data Availability Statement: Not applicable.

Acknowledgments: We appreciate the support from the Slovak Research and Development Agency under contract No. APVV-18-0526 and APVV-22-0508.

Conflicts of Interest: The authors declare no conflict of interest.

References

1. Bhutto, A.W.; Bazmi, A.A.; Zahedi, G. Underground coal gasification: From fundamentals to applications. *Prog. Energy Combust. Sci.* **2013**, *39*, 189–214. [CrossRef]
2. Kačur, J.; Kostúr, K. Approaches to the Gas Control in UCG. *Acta Polytech.* **2017**, *57*, 182–200. [CrossRef]
3. Perkins, G.M.P. Mathematical Modelling of Underground Coal Gasification, Academic Dissertation. Ph.D. Thesis, University of New South Wales, Faculty of Science and Engineering, Sydney, Australia, 2005.
4. Littlewood, K. Gasification: Theory and application. *Prog. Energy Combust. Sci.* **1977**, *3*, 35–71. [CrossRef]
5. Hobbs, M.L.; Radulovic, P.T.; Smoot, L.D. Combustion and gasification of coals in fixed-beds. *Prog. Energy Combust. Sci.* **1993**, *19*, 505–586. [CrossRef]
6. Thimsen, D.P.; Maurer, R.E. *Coal Devolatilization in a Moving-Bed Gasifier*; Technical Report GS-6797, Research Project 2525-14; Technical Report; EPRI—Electric Power Research Institute: Palo Alto, CA, USA, 1990.
7. Ranade, V.; Mahajani, S.; Samdani, G. Laboratory Studies on Underground Coal Gasification. In *Computational Modeling of Underground Coal Gasification*; CRC Press: Boca Raton, FL, USA, 2019; pp. 121–149. [CrossRef]
8. Kostúr, K.; Šašváry, T.; Ďurove, J.; Vavrek, P.; Kondela, J.; Kačur, J.; Laciak, M.; Durdán, M.; Blišťanová, M. *Research Report of the Project "Underground Gasification by Thermal Decomposition No. APVV-0582-06 for Year 2007"*; Technical Report; Technical University of Košice, Faculty BERG: Košice, Slovakia, 2007.
9. Kostúr, K.; Šašváry, T.; Durdán, M.; Kačur, J.; Koštial, I.; Kuffa, S.; Laciak, M.; Mikula, J.; Blišťanová, M.; Ďurove, J.; et al. *Research Report of the Project "Underground Gasification by Thermal Decomposition no. APVV-0582-06 for Year 2008"*; Technical Report; Technical University of Košice, Faculty BERG and HBP a.s. Prievidza: Košice, Slovakia, 2008.
10. Chaiken, R.F.; Martin, J.W. In Situ Gasification and Combustion of Coal. *SME Mining Engineering Handbook*; Academia Press: Cambridge, MA, USA, 1998; pp. 1954–1970.
11. Kačur, J. Optimálne Riadenie Procesov Splyňovania Uhlia v Podzemí (en: Optimal Control of Coal Gasification Processes in Underground). Ph.D. Thesis, Technical University of Košice, Faculty BERG, Košice, Slovakia, 2009.
12. Kačur, J.; Laciak, M.; Durdán, M.; Flegner, P.; Frančáková, R. A Review of Research on Advanced Control Methods for Underground Coal Gasification Processes. *Energies* **2023**, *16*, 3458. [CrossRef]
13. Dewasme, L.; Wouwer, A.V. Model-Free Extremum Seeking Control of Bioprocesses: A Review with a Worked Example. *Processes* **2020**, *8*, 1209. [CrossRef]
14. Ye, M.; Hu, G. A robust extremum seeking scheme for dynamic systems with uncertainties and disturbances. *Automatica* **2016**, *66*, 172–178. . [CrossRef]
15. Guay, M.; Beerens, R.; Nijmeijer, H. A time-varying extremum-seeking control approach for discrete-time systems with application to model predictive control. *IFAC Proc. Vol.* **2014**, *47*, 1023–1028. [CrossRef]
16. Leblanc, M. Sur l'électrification des chemins de fer au moyen de courants alternatifs de fréquence élevée. *Rev. Générale L'Électricité* **1922**, *12*, 275–277.
17. Ariyur, K.B.; Krstić, M. *Real-Time Optimization by Extremum-Seeking Control*; John Wiley & Sons, Inc.: Hoboken, NJ, USA, 2003. [CrossRef]
18. Ghaffari, A.; Krstic, M.; Seshagiri, S. Power Optimization for Photovoltaic Microconverters Using Multivariable Newton-Based Extremum Seeking. *IEEE Trans. Control Syst. Technol.* **2014**, *22*, 2141–2149. [CrossRef]
19. Hellström, E.; Lee, D.; Jiang, L.; Stefanopoulou, A.G.; Yilmaz, H. On-Board Calibration of Spark Timing by Extremum Seeking for Flex-Fuel Engines. *IEEE Trans. Control Syst. Technol.* **2013**, *21*, 2273–2279. [CrossRef]
20. Koeln, J.P.; Alleyne, A.G. Optimal subcooling in vapor compression systems via extremum seeking control: Theory and experiments. *Int. J. Refrig.* **2014**, *43*, 14–25. [CrossRef]
21. Kostúr, K.; Kačur, J. Extremum Seeking Control of Carbon Monoxide Concentration in Underground Coal Gasification. *IFAC-PapersOnLine* **2017**, *50*, 13772–13777. [CrossRef]
22. Kačur, J. *Optimal Control of Underground Coal Gasification Processes*; VSB—Technical University of Ostrava: Ostrava, Czech Republic, 2012; pp. 1–95. ISBN 978-80-248-3218-0.
23. Kačur, J. Riadenie Procesov Podzemného Splyňovania Uhlia (en: Control of Underground Coal Gasification Processes). Habilitation Thesis, Technical University of Košice, Faculty BERG, Košice, Slovakia, 2016.

24. Kačur, J.; Laciak, M.; Durdán, M.; Flegner, P. Model-Free Control of UCG Based on Continual Optimization of Operating Variables: An Experimental Study. *Energies* **2021**, *14*, 4323. [CrossRef]
25. Arshad, A.; Bhatti, A.I.; Samar, R.; Ahmed, Q.; Aamir, E. Model development of UCG and calorific value maintenance via sliding mode control. In Proceedings of the 2012 International Conference on Emerging Technologies, Islamabad, Pakistan, 8–9 October 2012; IEEE: Piscataway, NJ, USA, 2012; pp. 1–6. [CrossRef]
26. Khattak, M.; Uppal, A.A.; Khan, Q.; Bhatti, A.I.; Alsmadi, Y.M.; Utkin, V.I.; Chairez, I. Neuro-adaptive sliding mode control for underground coal gasification energy conversion process. *Int. J. Control* **2021**, *95*, 2337–2348. [CrossRef]
27. Kačur, J.; Flegner, P.; Durdán, M.; Laciak, M. Model Predictive Control of UCG: An Experiment and Simulation Study. *Inf. Technol. Control* **2019**, *48*, 557–578. [CrossRef]
28. Al Seyab, R.; Cao, Y. Nonlinear model predictive control for the ALSTOM gasifier. *J. Process. Control* **2006**, *16*, 795–808. [CrossRef]
29. Bequette, B.W.; Mahapatra, P. *Model Predictive Control of Integrated Gasification Combined Cycle Power Plants*; Technical Report; Rensselaer Polytechnic Institute: Troy, NY, USA, 2010. [CrossRef]
30. Xu, Q.; Li, D.; Tan, W. Model predictive control for an IGCC gasifier. In Proceedings of the Proceedings of the 33rd Chinese Control Conference, Nanjing, China, 28–30 July 2014; IEEE: Piscataway, NJ, USA, 2014; pp. 7742–7746. [CrossRef]
31. Zhang, S.; Bentsman, J.; Lou, X.; Neuschaefer, C. Wavelet multiresolution model based generalized predictive control for Hybrid Combustion-Gasification Chemical Looping process. In Proceedings of the 2012 IEEE 51st IEEE Conference on Decision and Control (CDC), Maui, HI, USA, 10–13 December 2012; IEEE: Piscataway, NJ, USA, 2012; pp. 2409–2414. [CrossRef]
32. Hou, Z.; Liu, S.; Yin, C. Local learning-based model-free adaptive predictive control for adjustment of oxygen concentration in syngas manufacturing industry. *IET Control Theory Appl.* **2016**, *10*, 1384–1394. [CrossRef]
33. Chaudhry, A.M.; Uppal, A.A.; Bram, S. Model Predictive Control and Adaptive Kalman Filter Design for an Underground Coal Gasification Process. *IEEE Access* **2021**, *9*, 130737–130750. [CrossRef]
34. Uppal, A.A.; Butt, S.S.; Bhatti, A.I.; Aschemann, H. Integral Sliding Mode Control and Gain-Scheduled Modified Utkin Observer for an Underground Coal Gasification Energy Conversion Process. In Proceedings of the 2018 23rd International Conference on Methods & Models in Automation & Robotics (MMAR), Miedzyzdroje, Poland, 27–30 August 2018; IEEE: Piscataway, NJ, USA, 2018; pp. 357–362. [CrossRef]
35. Uppal, A.A.; Butt, S.S.; Khan, Q.; Aschemann, H. Robust tracking of the heating value in an underground coal gasification process using dynamic integral sliding mode control and a gain-scheduled modified Utkin observer. *J. Process. Control* **2019**, *73*, 113–122. . [CrossRef]
36. Laciak, M.; Kačur, J. Automatizovaný systém riadenia podzemného splyňovania uhlia v laboratórnych podmienkach (en: Automated control system of underground coal gasification in laboratory conditions). *AT&P J.* **2009**, *8*, 47–52.
37. Laciak, M.; Kačur, J.; Kostúr, K. Simulation Analysis for UCG with Thermodynamical Model. In Proceedings of the 9th International Carpathian Control Conference ICCC'2008, Sinaia, Romania, 25–28 May 2008; University of Craiova, Faculty of Automation, Computers and Electronics: Craiova, Romania, 2008; pp. 358–361.
38. Laciak, M.; Kačur, J.; Kostúr, K. The verification of thermodynamic model for UCG process. In Proceedings of the ICCC 2016: 17th International Carpathian Control Conference, High Tatras, Slovakia, 29 May–1 June 2016; pp. 424–428. [CrossRef]
39. Laciak, M.; Ráškayová, D. The using of thermodynamic model for the optimal setting of input parameters in the UCG process. In Proceedings of the ICCC 2016: 17th International Carpathian Control Conference, High Tatras, Slovakia, 29 May–1 June 2016; IEEE: Piscataway, NJ, USA, 2016; pp. 418–423. [CrossRef]
40. Ráškayová, D.; Laciak, M.; Mudarri, T. The System of Optimization Quantity of Oxidizers in UCG Process with Thermodynamic Model. In Proceedings of the 2017 18th International Carpathian Control Conference (ICCC), Sinaia, Romania, 28–31 May 2017; IEEE: Piscataway, NJ, USA, 2017; pp. 76–80. [CrossRef]
41. Ji, P.; Gao, X.; Huang, D.; Yang, Y. Prediction of Syngas Compositions in Shell Coal Gasification Process via Dynamic Soft-sensing Method. In Proceedings of the Proceeding of 10th IEEE International Conference on Control and Automation (ICCA), Hangzhou, China, 13–14 June 2013; pp. 244–249. [CrossRef]
42. Kačur, J.; Durdán, M.; Laciak, M.; Flegner, P. Impact analysis of the oxidant in the process of underground coal gasification. *Measurement* **2014**, *51*, 147–155. [CrossRef]
43. Feng, L.; Zhou, S.; Xu, X.; Qin, B. Importance evaluation for influencing factors of underground coal gasification through ex situ experiment and analytic hierarchy process. *Energy* **2022**, *261*, 125116. [CrossRef]
44. Zagorščak, R.; Sadasivam, S.; Thomas, H.R.; Stańczyk, K.; Kapusta, K. Experimental study of underground coal gasification (UCG) of a high-rank coal using atmospheric and high-pressure conditions in an ex situ reactor. *Fuel* **2020**, *270*, 117490. [CrossRef]
45. Stańczyk, K.; Howaniec, N.; Smoliński, A.; Świądrowski, J.; Kapusta, K.; Wiatowski, M.; Grabowski, J.; Rogut, J. Gasification of lignite and hard coal with air and oxygen enriched air in a pilot scale ex situ reactor for underground gasification. *Fuel* **2011**, *90*, 1953–1962. [CrossRef]
46. Wiatowski, M.; Kapusta, K.; Świądrowski, J.; Cybulski, K.; Ludwik-Pardała, M.; Grabowski, J.; Stańczyk, K. Technological aspects of underground coal gasification in the Experimental "Barbara" Mine. *Fuel* **2015**, *159*, 454–462. [CrossRef]
47. Wiatowski, M.; Kapusta, K.; Muzyka, R. Study of properties of tar obtained from underground coal gasification trials. *Fuel* **2018**, *228*, 206–214. [CrossRef]
48. Xu, Y.; Wang, J.; Zhang, G.; Zhang, X.; Zhang, Y. The characteristics and mechanism for the formation of tars from low temperature pyrolysis of lignite. *J. Energy Inst.* **2021**, *99*, 248–255. [CrossRef]

49. Xu, M.; Xin, L.; Liu, W.; Hu, X.; Cheng, W.; Li, C.; Wang, Z. Study on the physical properties of coal pyrolysis in underground coal gasification channel. *Powder Technol.* **2020**, *376*, 573–592. [CrossRef]
50. Dong, M.; Feng, L.; Zhou, Q.; Zhou, S.; Xu, X.; Qin, B. Spatial and temporal evolution of tar during ex situ underground coal gasification. *Fuel* **2022**, *317*, 123423. [CrossRef]
51. Pankiewicz-Sperka, M.; Kapusta, K.; Basa, W.; Stolecka, K. Characteristics of Water Contaminants from Underground Coal Gasification (UCG) Process—Effect of Coal Properties and Gasification Pressure. *Energies* **2021**, *14*, 6533. [CrossRef]
52. Shannon, M.; Thorsness, C.; Hill, R. *Early Cavity Growth during Forward Burn*; 5:28302, Report no. UCRL-84584. U.S. DOE, Livermore, CA; Technical Report; Lawrence Livermore National Laboratory: Livermore, CA, USA, 1980.
53. Thorsness, C.; Hill, R. *Coal Block Gasification: Laboratory Results and Field Plans*; 81:12021, Report no. UCRL-85906. U.S. DOE, Livermore, CA; Technical Report; Lawrence Livermore National Laboratory: Livermore, CA, USA, 1981.
54. Daggupati, S.; Mandapati, R.N.; Mahajani, S.M.; Ganesh, A.; Sapru, R.K.; Sharma, R.K.; Aghalayam, P. Laboratory studies on cavity growth and product gas composition in the context of underground coal gasification. *Energy* **2011**, *36*, 1776–1784. [CrossRef]
55. Liu, S.Q.; Wang, Y.Y.; Zhao, K.; Yang, N. Enhanced-hydrogen gas production through underground gasification of lignite. *Min. Sci. Technol.* **2009**, *19*, 389–394. [CrossRef]
56. Hettema, M.; Wolf, K.H.; DePater, C. The influence of steam pressure on thermal spalling of sedimentary rock: Theory and experiments. *Int. J. Rock Mech. Min. Sci.* **1998**, *35*, 3–15. [CrossRef]
57. Burton, E.; Friedmann, J.; Upadhye, R. *Best Practices in Underground Coal Gasification*; Report No. W-7405-Eng-48; Technical Report; Lawrence Livermore National Laboratory: Livermore, CA, USA, 2013.
58. Massaquoi, J.G.M.; Riggs, J.B. Mathematical modeling of combustion and gasification of a wet coal slab–I: Model development and verification. *Chem. Eng. Sci.* **1983**, *38*, 1747–1756. [CrossRef]
59. Massaquoi, J.G.M.; Riggs, J.B. Mathematical modeling of combustion and gasification of a wet coal slab–II: Mode of combustion, steady state multiplicities and extinction. *Chem. Eng. Sci.* **1983**, *38*, 1757–1766. [CrossRef]
60. Park, K.Y.; Edgar, T.F. Modeling of early cavity growth for underground coal gasification. *Ind. Eng. Chem. Res.* **1987**, *26*, 237–246. [CrossRef]
61. Perkins, G.; Sahajwalla, V. A Mathematical Model for the Chemical Reaction of a Semi-infinite Block of Coal in Underground Coal Gasification. *Energy Fuels* **2005**, *19*, 1679–1692. [CrossRef]
62. Mühlen, H.J.; Heinrich van Heek, K.; Jüntgen, K. Kinetic studies of steam gasification of char in the presence of H_2, CO_2 and CO. *Fuel* **1985**, *64*, 944–949. [CrossRef]
63. Roberts, D.G.; Harris, D.J. A Kinetic Analysis of Coal Char Gasification Reactions at High Pressures. *Energy Fuels* **2006**, *20*, 2314–2320. [CrossRef]
64. Anthony, D.B.; Howard, J.B. Coal devolatilization and hydrogastification. *AIChE J.* **1976**, *22*, 625–656. [CrossRef]
65. Makino, M.; Toda, Y. Factors affecting methane evolution on pyrolysis of coal under pressure. *Fuel* **1979**, *58*, 231–234. [CrossRef]
66. Wall, T.F.; Liu, G.S.; Wu, H.W.; Roberts, D.G.; Benfell, K.E.; Gupta, S.; Lucas, J.A.; Harris, D.J. The effects of pressure on coal reactions during pulverised coal combustion and gasification. *Prog. Energy Combust. Sci.* **2002**, *28*, 405–433. [CrossRef]
67. Bell, D.A.; Towler, B.F.; Fan, M. *Coal Gasification and Its Applications*; Elsevier BV: Amsterdam, The Netherlands, 2010; p. 416.
68. Perkins, G.; Sahajwalla, V. Modelling of Heat and Mass Transport Phenomena and Chemical Reaction in Underground Coal Gasification. *Chem. Eng. Res. Des.* **2007**, *85*, 329–343. [CrossRef]
69. Wiatowski, M.; Kapusta, K.; Ludwik-Pardała, M.; Stańczyk, K. Ex situ experimental simulation of hard coal underground gasification at elevated pressure. *Fuel* **2016**, *184*, 401–408. [CrossRef]
70. Kapusta, K.; Wiatowski, M.; Stańczyk, K.; Zagorščak, R.; Thomas, H.R. Large-scale Experimental Investigations to Evaluate the Feasibility of Producing Methane-Rich Gas (SNG) through Underground Coal Gasification Process. Effect of Coal Rank and Gasification Pressure. *Energies* **2020**, *13*, 1334. [CrossRef]
71. Kapusta, K. Effect of Lignite Properties on Its Suitability for the Implementation of Underground Coal Gasification (UCG) in Selected Deposits. *Energies* **2021**, *14*, 5816. [CrossRef]
72. Sadasivam, S.; Zagorščak, R.; Thomas, H.R.; Kapusta, K.; Stańczyk, K. Experimental study of methane-oriented gasification of semi-anthracite and bituminous coals using oxygen and steam in the context of underground coal gasification (UCG): Effects of pressure, temperature, gasification reactant supply rates and coal rank. *Fuel* **2020**, *268*, 117330. [CrossRef]
73. Wiatowski, M.; Kapusta, K.; Stańczyk, K.; Stańczyk, K. Efficiency assessment of underground gasification of ortho- and meta-lignite: High-pressure ex situ experimental simulations. *Fuel* **2019**, *236*, 221–227. [CrossRef]
74. Kapusta, K.; Wiatowski, M.; Thomas, H.R.; Zagorščak, R.; Sadasivam, S.; Masum, S.; Kempka, T.; Otto, C.; Basa, W.; Szyja, M.; et al. Experimental simulations of methane-oriented underground coal gasification using hydrogen—The effect of coal rank and gasification pressure on the hydrogasification process. *Int. J. Hydrogen Energy* **2023**, *48*, 921–932. [CrossRef]
75. Kostúr, K.; Laciak, M.; Durdán, M.; Kačur, J.; Flegner, P. Low-calorific gasification of underground coal with a higher humidity. *Measurement* **2015**, *63*, 69–80. [CrossRef]
76. Laciak, M.; Kostúr, K.; Durdán, M.; Kačur, J.; Flegner, P. The analysis of the underground coal gasification in experimental equipment. *Energy* **2016**, *114*, 332–343. [CrossRef]
77. Stańczyk, K.; Smoliński, A.; Kapusta, K.; Wiatowski, M.; Świądrowski, J.; Kotyrba, A.; Rogut, J. Dynamic experimental simulation of hydrogen oriented underground gasification of lignite. *Fuel* **2010**, *89*, 3307–3314. [CrossRef]

78. Gur, M.; Eskin, N.; Okutan, H.; Arısoy, A.; Böke, E.; Altıntaş, U.; Büyükşirin, A.Y.O.; Canbaz, E.D.; Yıldırım, O. Experimental results of underground coal gasification of Turkish lignite in an ex situ reactor. *Fuel* **2017**, *203*, 997–1006. [CrossRef]
79. Gür, M.; Canbaz, E.D. Analysis of syngas production and reaction zones in hydrogen oriented underground coal gasification. *Fuel* **2020**, *269*, 117331. [CrossRef]
80. Liu, H.; Chen, F.; Wang, Y.; Liu, G.; Yao, H.; Liu, S. Experimental Study of Reverse Underground Coal Gasification. *Energies* **2018**, *11*, 2949. [CrossRef]
81. Kapusta, K.; Wiatowski, M.; Stańczyk, K. An experimental ex situ study of the suitability of a high moisture ortho-lignite for underground coal gasification (UCG) process. *Fuel* **2016**, *179*, 150–155. [CrossRef]
82. Wiatowski, M.; Kapusta, K.; Nowak, J.; Szyja, M.; Basa, W. An Ex Situ Underground Coal Gasification Experiment with a Siderite Interlayer. Course of the Process, Production Gas, Temperatures and Energy Efficiency. *Int. J. Coal Sci. Technol.* 2020, *preprint*. [CrossRef]
83. Wiatowski, M.; Stańczyk, K.; Świądrowski, J.; Kapusta, K.; Cybulski, K.; Krause, E.; Grabowski, J.; Rogut, J.; Howaniec, N.; Smoliński, A. Semi-technical underground coal gasification (UCG) using the shaft method in Experimental Mine "Barbara". *Fuel* **2012**, *99*, 170–179. [CrossRef]
84. Bielowicz, B.; Kasiński, J.R. The possibility of underground gasification of lignite from Polish deposits. *Int. J. Coal Geol.* **2014**, *131*, 304–318. [CrossRef]
85. Fang, H.; Liu, Y.; Ge, T.; Zheng, T.; Yu, Y.; Liu, D.; Ding, J.; Li, L. A Review of Research on Cavity Growth in the Context of Underground Coal Gasification. *Energies* **2022**, *15*, 9252. [CrossRef]
86. Hongtao, L.; Feng, C.; Xia, P.; Kai, Y.; Shuqin, L. Method of oxygen-enriched two-stage underground coal gasification. *Min. Sci. Technol.* **2011**, *21*, 191–196. [CrossRef]
87. Wiatowski, M.; Kapusta, K.; Nowak, J.; Szyja, M.; Basa, W. An exsitu underground coal gasification experiment with a siderite interlayer: Course of the process, production gas, temperatures and energy efficiency. *Int. J. Coal Sci. Technol.* **2021**, *8*, 1447–1460. [CrossRef]
88. Wiatowski, M. An Experimental Study on the Quantitative and Qualitative Characteristics of Tar Formed during Ex Situ Coal Gasification. *Energies* **2023**, *16*, 2777. [CrossRef]
89. Wu, J.M. Radon Distribution under the Mine and the Application of Radon Measuring in the Monitoring of the Natural Fire Zone. Ph.D. Thesis, Shanxi Institute of Mining and Technology, Yangquan, China, 1994.
90. Wu, H.S. The transferring effects of Radon moving. *Geophys. J.* **1995**, *14*, 136.
91. Brasseur, A.; Antenucci, D.; Bouquegneau, J.M.; Coëme, A.; Dauby, P.; Létolle, R.; Mostade, M.; Pirlot, P.; Pirard, J.P. Carbon stable isotope analysis as a tool for tracing temperature during the El Tremedal underground coal gasification at great depth. *Fuel* **2002**, *81*, 109–117. [CrossRef]
92. Su, F.Q.; Hamanaka, A.; Itakura, K.I.; Deguchi, G.; Zhang, W.; Nan, H. Evaluation of a Compact Coaxial Underground Coal Gasification System Inside an Artificial Coal Seam. *Energies* **2018**, *11*, 898. [CrossRef]
93. Kostúr, K.; Laciak, M.; Durdan, M. Some Influences of Underground Coal Gasification on the Environment. *Sustainability* **2018**, *10*, 1512. [CrossRef]
94. Durdán, M.; Benková, M.; Laciak, M.; Kačur, J.; Flegner, P. Regression Models Utilization to the Underground Temperature Determination at Coal Energy Conversion. *Energies* **2021**, *14*, 5444. [CrossRef]
95. Kačur, J.; Durdán, M.; Laciak, M.; Flegner, P. A Comparative Study of Data-Driven Modeling Methods for Soft-Sensing in Underground Coal Gasification. *Acta Polytech.* **2019**, *59*, 322–351. [CrossRef]
96. Kačur, J.; Laciak, M.; Durdán, M.; Flegner, P. Application of multivariate adaptive regression in soft-sensing and control of UCG. *Int. J. Model. Identif. Control* **2019**, *33*, 246. [CrossRef]
97. Kačur, J.; Laciak, M.; Durdán, M.; Flegner, P. Utilization of Machine Learning method in prediction of UCG data. In Proceedings of the 2017 18th International Carpathian Control Conference (ICCC), Sinaia, Romania, 28–31 May 2017; IEEE: Piscataway, NJ, USA, 2017; pp. 278–283. [CrossRef]
98. Vapnik, V.N. Constructing Learning Algorithms. In *The Nature of Statistical Learning Theory*; Springer: New York, NY, USA, 1995; pp. 119–166. [CrossRef]
99. Müller, K.R.; Smola, A.J.; Rätsch, G.; Schölkopf, B.; Kohlmorgen, J.; Vapnik, V. Predicting time series with support vector machines. In *Lecture Notes in Computer Science*; Springer: Berlin/Heidelberg, Germany, 1997; pp. 999–1004. [CrossRef]
100. Hastie, T.; Tibshirani, R.; Friedman, J. *The Elements of Statistical Learning—Data Mining, Inference, and Prediction*, 2nd ed.; Springer: New York, NY, USA, 2009. [CrossRef]
101. Chugh, M.; Thumsi, S.S.; Keshri, V. A Comparative Study Between Least Square Support Vector Machine(Lssvm) and Multivariate Adaptive Regression Spline(Mars) Methods for the Measurement of Load Storing Capacity of Driven Piles in Cohesion Less Soil. *Int. J. Struct. Civ. Eng. Res.* **2015**, *4*, 189–194. [CrossRef]
102. Laciak, M.; Vízi, L.; Kačur, J.; Durdán, M.; Flegner, P. Application of geostatistical methods in spatio-temporal modelling of temperature changes of UCG experimental trial. *Measurement* **2021**, *171*, 108826. [CrossRef]
103. Gregg, D.W.; Edgar, T.F. Underground coal gasification. *AIChE J.* **1978**, *24*, 753–781. [CrossRef]
104. Dobbs, R.L.; Krantz, W.B. *Combustion Front Propagation in Underground Coal Gasification*; Final Report, Work Performed under Grant No. DE-FG22-86PC90512; Technical Report; University of Colorado, Boulder Department of Chemical Engineering: Boulder, CO, USA, 1990. [CrossRef]

105. Hamanaka, A.; Su, F.Q.; Itakura, K.I.; Takahashi, K.; Kodama, J.I.; Deguchi, G. Experimental study on evaluation of underground coal gasification with a horizontal hole using two different coals. *Fuel* **2021**, *305*, 1556. [CrossRef]
106. Su, F.Q.; Hamanaka, A.; Itakura, K.I.; Zhang, W.; Deguchi, G.; Sato, K.; Takahashi, K.; Kodama, J.I. Monitoring and evaluation of simulated underground coal gasification in an ex situ experimental artificial coal seam system. *Appl. Energy* **2018**, *223*, 82–92. [CrossRef]
107. Cena, R.J.; Thorsness, C.B. *Underground Coal Gasification Data Base*; Technical Report UCID-19169; Technical Report; Lawrence Livermore National Laboratory, University of California: Berkeley, CA, USA, 1981. [CrossRef]
108. Boysen, J.E.; Canfield, M.T.; Covell, J.R.; Schmit, C.R. *Detailed Evaluation of Process and Environmental Data from the Rocky Mountain I Underground Coal Gasification Field Test*; Technical Report No. GRI-97/0331; Technical Report; Gas Research Institute: Chicago, IL, USA, 1998.
109. Chappell, R.; Mostade, M. The El Tremedal underground coal gasification field test in Spain. First trial at great depth and high pressure. In Proceedings of the Fifteenth Annual International Pittsburgh Coal Conference, Pittsburgh, PA, USA, 14–18 September 1998; Curran Associates, Inc.: Red Hook, NY, USA, 1998; pp. 1–18.
110. Goodarzi, F.; Murchison, D. Optical properties of carbonized vitrinites. *Fuel* **1972**, *51*, 322–328. [CrossRef]
111. Jiménez, A.; Iglesias, M.J.; Laggoun-Defarge, F.; Suárez-Ruiz, I. Effect of the increase in temperature on the evolution of the physical and chemical structure of vitrinite. *J. Anal. Appl. Pyrolysis* **1999**, *50*, 117–148. [CrossRef]
112. Murchison, D.G. Petrographic aspects of coal structure: Reactivity of macerals in laboratory and natural environments. *Fuel* **1991**, *70*, 296–315. [CrossRef]
113. Nowak, J.; Kokowska-Pawłowska, M.; Komorek, J.; Wiatowski, M.; Kapusta, K.; Adamczyk, Z. Optical Properties of Coal after Ex Situ Experimental Simulation of Underground Gasification at Pressures of 10 and 40 bar. *Energies* **2022**, *15*, 8824. [CrossRef]
114. Róg, L. Vitrinite reflectance as a measure of the range of influence of the temperature of a georeactor on rock mass during underground coal gasification. *Fuel* **2018**, *224*, 94–100. [CrossRef]
115. Malumbazo, N.; Wagner, N.; Bunt, J. The petrographic determination of reactivity differences of two South African inertinite-rich lump coals. *J. Anal. Appl. Pyrolysis* **2012**, *93*, 139–146. [CrossRef]
116. Malumbazo, N.; Wagner, N.; Bunt, J. The impact of particle size and maceral segregation on char formation in a packed bed combustion unit. *Fuel* **2013**, *111*, 350–356. [CrossRef]
117. Malumbazo, N.; Wagner, N.; Bunt, J.; Niekerk, D.V.; Assumption, H. Structural analysis of chars generated from South African inertinite coals in a pipe-reactor combustion unit. *Fuel Process. Technol.* **2011**, *92*, 743–749. [CrossRef]
118. Furimsky, E.; Palmer, A.; Kalkreuth, W.; Cameron, A.; Kovacik, G. Prediction of coal reactivity during combustion and gasification by using petrographic data. *Fuel Process. Technol.* **1990**, *25*, 135–151. [CrossRef]
119. Wagner, N.; Matjie, R.; Slaghuis, J.; Heerden, J. Characterization of unburned carbon present in coarse gasification ash. *Fuel* **2008**, *87*, 683–691. [CrossRef]
120. Oboirien, B.; Engelbrecht, A.; North, B.; Cann, V.M.; Verryn, S.; Falcon, R. Study on the structure and gasification characteristics of selected South African bituminous coals in fluidised bed gasification. *Fuel Process. Technol.* **2011**, *92*, 735–742. [CrossRef]
121. Gray, R. To Defend Your Empire and the Faith. Edited by P. E. H. Hair. Liverpool: Liverpool University Press, 1990. Pp. iii+298. £12.50, paperback. *J. Afr. Hist.* **1991**, *32*, 535–536. [CrossRef]
122. Alonso, M.; Borrego, A.; Álvarez, D.; Menéndez, R. A reactivity study of chars obtained at different temperatures in relation to their petrographic characteristics. *Fuel Process. Technol.* **2001**, *69*, 257–272. [CrossRef]
123. Bielowicz, B. Change of the petrographic composition of lignite during the ex situ lignite gasification. *Fuel* **2017**, *206*, 219–229. [CrossRef]
124. Prabu, V.; Jayanti, S. Heat-affected zone analysis of high ash coals during ex situ experimental simulation of underground coal gasification. *Fuel* **2014**, *123*, 167–174. [CrossRef]
125. Prabu, V.; Jayanti, S. Simulation of cavity formation in underground coal gasification using bore hole combustion experiments. *Energy* **2011**, *36*, 5854–5864. [CrossRef]
126. Prabu, V.; Jayanti, S. Laboratory scale studies on simulated underground coal gasification of high ash coals for carbon-neutral power generation. *Energy* **2012**, *46*, 351–358. [CrossRef]
127. Adschiri, T.; Furusawa, T. Relation between CO_2-reactivity of coal char and BET surface area. *Fuel* **1986**, *65*, 927–931. [CrossRef]
128. Bryant, C.W.; Cawein, C.C.; King, P.H. Biological Treatability of In Situ Coal Gasification Wastewater. *J. Environ. Eng.* **1988**, *114*, 400–414. [CrossRef]
129. Zhang, W.; Ma, J.; Yang, S.; Zhang, T.; Li, Y. Pretreatment of Coal Gasification Wastewater by Acidification Demulsion. *Chin. J. Chem. Eng.* **2006**, *14*, 398–401. [CrossRef]
130. Shi, J.; Han, Y.; Xu, C.; Han, H. Biological coupling process for treatment of toxic and refractory compounds in coal gasification wastewater. *Rev. Environ. Sci. Bio/Technol.* **2018**, *17*, 765–790. [CrossRef]
131. Thomas, M.; Zdebik, D.; Niewiara, E. Removing Phenols from Post-Processing Wastewater Originating from Underground Coal Gasification Using Coagulation-Flocculation and the H2O2/UV Process. *Pol. J. Environ. Stud.* **2018**, *27*, 2757–2763. [CrossRef] [PubMed]
132. Zhang, G.; Wan, X.; Li, W.; Yang, Y.; Wang, L. Advanced treatment of coal gasification wastewater by catalytic oxidation with trace ozone. *Chin. J. Environ. Eng.* **2013**, *7*, 263–267.

133. Smoliński, A.; Stańczyk, K.; Kapusta, K.; Howaniec, N. Chemometric Study of the Ex Situ Underground Coal Gasification Wastewater Experimental Data. *Water Air Soil Pollut.* **2012**, *223*, 5745–5758. [CrossRef]
134. Strugała-Wilczek, A.; Kapusta, K.; Pankiewicz-Sperka, M. Pollutants release from the residues remaining after underground gasification (UCG) of ortho- and meta-lignites. *Fuel* **2022**, *327*, 125126. [CrossRef]
135. Ketris, M.; Yudovich, Y. Estimations of Clarkes for Carbonaceous biolithes: World averages for trace element contents in black shales and coals. *Int. J. Coal Geol.* **2009**, *78*, 135–148. [CrossRef]
136. Laciak, M.; Durdán, M.; Kačur, J.; Flegner, P. The Underground Coal Gasification Process in Laboratory Conditions: An Experimental Study. *Energies* **2023**, *16*, 3266. [CrossRef]
137. Laciak, M.; Durdán, M.; Kačur, J.; Flegner, P.; Benková, M. The Prediction of Possibilities of CO Poisoning and Explosion during Syngas Leakage in the UCG Process. *Processes* **2021**, *9*, 1912. [CrossRef]
138. Durdán, M.; Laciak, M.; Kačur, J.; Flegner, P.; Kostúr, K. Evaluation of synthetic gas harmful effects created at the underground coal gasification process realized in laboratory conditions. *Measurement* **2019**, *147*, 106866. [CrossRef]
139. Jones, V.T.; Thune, H.W. Surface Detection of Retort Gases From an Underground Coal Gasification Reactor in Steeply Dipping Beds Near Rawlins, Wyoming. In Proceedings of the 57th Annual Fall Technical Conference and Exhibition of the Society of Petroleum Engineers of AIME, New Orleans, LA, USA, 26–29 September 1981; (SPE) Society of Petroleum Engineers of AIME: London, UK, 1982; pp. 1–24. [CrossRef]
140. Self, S.J.; Reddy, B.V.; Rosen, M.A. Review of underground coal gasification technologies and carbon capture. *Int. J. Energy Environ. Eng.* **2012**, *3*, 16. [CrossRef]
141. Irfan, M.F.; Usman, M.R.; Kusakabe, K. Coal gasification in CO2 atmosphere and its kinetics since 1948: A brief review. *Energy* **2011**, *36*, 12–40. [CrossRef]
142. Duan, T.H.; Lu, C.P.; Xiong, S.; Fu, Z.B.; Chen, Y.Z. Pyrolysis and gasification modelling of underground coal gasification and the optimisation of CO_2 as a gasification agent. *Fuel* **2016**, *183*, 557–567. [CrossRef]
143. Mocek, P.; Pieszczek, M.; Świądrowski, J.; Kapusta, K.; Wiatowski, M.; Stańczyk, K. Pilot-scale underground coal gasification (UCG) experiment in an operating Mine "Wieczorek" in Poland. *Energy* **2016**, *111*, 313–321. [CrossRef]
144. Konstantinou, E.; Marsh, R. Experimental study on the impact of reactant gas pressure in the conversion of coal char to combustible gas products in the context of Underground Coal Gasification. *Fuel* **2015**, *159*, 508–518. [CrossRef]
145. Kumari, G.; Vairakannu, P. Laboratory scale studies on CO_2 oxy-fuel combustion in the context of underground coal gasification. *J. CO_2 Util.* **2017**, *21*, 177–190. [CrossRef]

energies

Article

Polyolefin Pyrolysis in Multilayer Fluidized Beds: An Innovative Approach to Obtain Valuable Alternative Fuels

Witold Żukowski [1,*], Krystian Leski [2], Gabriela Berkowicz-Płatek [1] and Jan Wrona [3]

1 Faculty of Chemical Engineering and Technology, Cracow University of Technology, Warszawska 24, 31-155 Kraków, Poland; gabriela.berkowicz@pk.edu.pl
2 Doctoral School, Cracow University of Technology, Warszawska 24, 31-155 Kraków, Poland; krystian.leski@doktorant.pk.edu.pl
3 Faculty of Environmental Engineering and Energy, Cracow University of Technology, Warszawska 24, 31-155 Kraków, Poland; jan.wrona@pk.edu.pl
* Correspondence: witold.zukowski@pk.edu.pl

Abstract: This paper presents the application of two versions of the multilayer fluidized bed made out of two materials with significantly different densities. The first type of fluidized bed was composed of raw cenospheres and quartz sand. The second type of fluidized bed was composed of cenospheres coated with iron oxides and quartz sand. A variable vertical density profiles in the prepared fluidized beds were confirmed, making them suitable for processing polymeric materials, specifically, polyolefins with a density below 1 g/cm^3. The polyolefin pyrolysis process was investigated in both versions of the fluidized bed at temperatures of 520, 540, 560, and 590 °C. The products of the pyrolysis were monitored using high-resolution infrared spectroscopy (with a resolution of 1 cm^{-1}). While the process is organized in these fluidized beds, the absence of the accumulation of solid residues is notable. The results show that the pyrolytic gaseous mixture is composed of numerous compounds, namely, unsaturated and saturated aliphatic hydrocarbons and benzene. The possibility of producing a gas rich in ethylene, propylene, and 1-butene during the pyrolysis was demonstrated. Additionally, during the pyrolysis of both polymers, the production of benzene was shown with yields, ranging from 5%$_{wt.}$ in the fluidized bed made out of raw cenospheres to 11 %$_{wt.}$ in the fluidized bed made out of cenospheres modified by iron oxides. Due to the complex nature of the resulting pyrolytic gas, it is suggested that we process the created gaseous mixtures entirely in a steam conversion process, making them a potential source of hydrogen.

Keywords: fluidized bed; pyrolysis; polyolefins; hydrocarbons

1. Introduction

Creating a modern, resource-efficient, competitive, and net-zero greenhouse gas emission economy is the present focus of the European Green Deal [1,2]. The circular economy aims to minimize waste generation and maximize resource efficiency by promoting the reuse, recycling, and repurposing of waste materials and by-products [3,4]. Pyrolysis of plastic wastes involves breaking down high-molecular-weight polymers into smaller molecules, resulting in the production of valuable alternative fuels and chemical raw materials [5,6]. The process holds promise in advanced recycling and has captured the attention of numerous scientists, emphasizing its potential impact on sustainable development [7,8].

The pyrolysis of plastic wastes on a small laboratory scale is already well investigated. However, the commercialization of the process on a larger scale requires overcoming certain challenges. Impurities and variability of the feedstock, the formation of harmful by-products, energy requirements, the quality of end products, and the scaling of the process are the main issues that need to be dealt with [9–11].

Waste plastics often contain contaminants, such as paper labels, food residues, and other materials that are not made out of plastics and require different processing conditions.

These contaminants can affect the efficiency of the pyrolysis process and the quality of the end products [12]. Furthermore, different types of plastics are characterized by varying compositions, melting points, and thermal degradation behaviors. The feedstock variability can make it challenging to optimize the pyrolysis conditions for consistent and high-quality output [13]. Plastic waste pyrolysis is often the source of hazardous gases and volatile organic compounds (VOCs) emission into the environment [14]. Proper emission control systems and treatment methods are necessary to minimize the release of harmful substances and ensure compliance with environmental regulations [15].

One of the main challenges in scaling the pyrolysis up to industrial dimensions is ensuring a uniform temperature and effective mixing of the input materials. Maintaining a constant temperature in the reactor during the pyrolysis of polyolefins is particularly challenging due to their low thermal conductivity [16]. Uneven heating of the pyrolyzed material results in areas that are not reaching the required temperature for complete processing, while others may be overheated. Consequently, the pyrolysis process leads to the formation of undesirable products, such as soot [17], low-calorific value gases [18], or toxic chemical compounds [19]. Another difficulty is ensuring the appropriate quality of the end product, which must meet specific standards for later utilization [20]. The quality of pyrolysis products, such as oil or gas, may vary depending on the process organization [21]. Obtaining consistent and high-quality degradation products that can be effectively utilized in various industries is an additional issue [22]. Achieving a high degree of fuel conversion into desired alternative fuels is particularly important during the waste pyrolysis process [23,24]. Factors such as heat transport, reaction kinetics, residence time, and reactor design play a significant role in maximizing waste conversion into useful products while simultaneously minimizing energy consumption and optimizing resource utilization [25]. Furthermore, scaling up small laboratory or pilot projects to large-scale industrial operations poses challenges related to logistics, raw material supply chains, and integration with existing waste management infrastructure [26].

To achieve proper mixing during the pyrolysis process, and thereby better control process conditions to transform waste plastics into alternative fuels and the desired chemical raw materials in a repeatable manner, the organization of the process in a fluidized bed (FB) is proposed. This is a crucial element for the commercialization of pyrolysis, ensuring uniform process conditions. In the present study, multilayer fluidized beds were created, characterized by a variable vertical density profile. Three layers were distinguished in the prepared fluidized beds: the upper layer with the lowest density, the lower layer with the highest density, and the middle layer characterized by variable density. The innovative approach of the multilayer fluidized beds provided the ability to feed polymers from the top of the reactor, allowing them to freely immerse in the FB volume by passing through the low-density layer. Degradation of fuel particles and the release of gaseous products occurred in the layer with variable density, adapted to the density of polyolefins. The high-density layer protected the bottom sieve from the pyrolysis of polyolefins at the gas distributor and thus from clogged gas inlets.

2. Materials and Methods

2.1. Materials

Multilayer fluidized beds were prepared using quartz sand and cenospheres. In the study, raw cenospheres and cenospheres coated with an iron oxides layer were utilized. The natural quartz sand with a grain size of 120–200 μm was selected.

Cenospheres (i.e., spheres of aluminosilicate filled with gas) were obtained from fly ash from the combustion of hard coal in a pulverized coal boiler at the Połaniec Power Plant (Zawada, Poland). The cenospheres were boiled in distilled water to isolate perforated cenospheres from undamaged ones. Subsequently, undamaged cenospheres were collected from the top layer above the solution, dried, and sieved to a fraction of 140–160 μm. The hydrothermal method of separating undamaged cenospheres from perforated ones is shown in Figure 1. The method is divided into I–IV steps:

- Step I: Perforated (A) and undamaged (B) cenospheres float on the water surface. Water is mixed continuously.
- Step II: In the boiling process, water vapor fills the interior of the perforated cenospheres due to gas exchange. Oxygen and nitrogen diffuse out of the damaged particles.
- Step III: Cenospheres are completely filled with water vapor.
- Step IV: During the cooling process, the water vapor condenses inside the perforated grains, generating a vacuum and filling damaged cenospheres with liquid water, which leads these cenospheres to sink to the bottom of the vessel.

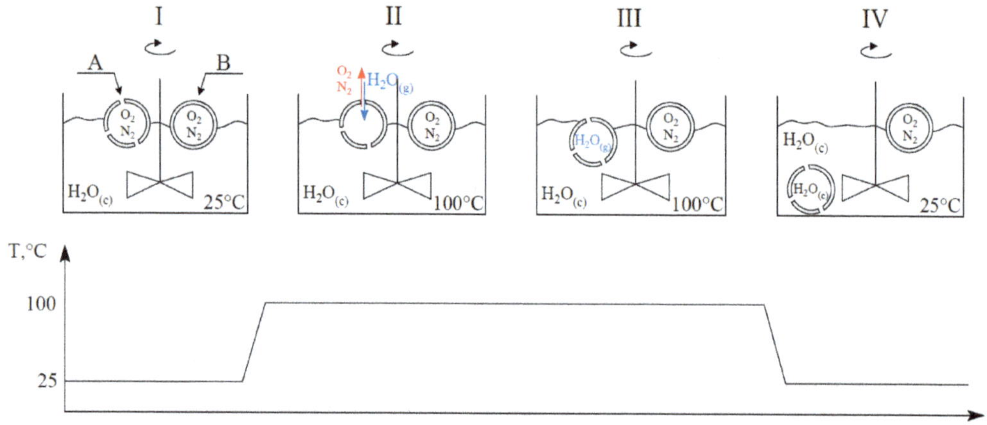

Figure 1. Scheme of isolating perforated cenospheres (**A**) from undamaged ones (**B**). Cenospheres not to scale.

The raw cenospheres were coated with iron through the gas decomposition process of $Fe(CO)_5$, as described in [27]. The iron content deposited on the cenospheres was $8\%_{wt}$. The whole layer of iron was oxidized to a mixture of Fe_2O_3 and Fe_3O_4 oxides. The oxidation was conducted in the fluidization state at 500 °C for 30 min. The process was completed when the measured oxygen concentration at the reactor outlet was equal to the value in the atmospheric air.

Commercial sheets of polyethylene (PE) and polypropylene (PP) made out of pure polymers with a thickness of 1 mm were used in the pyrolysis tests. Discs weighing approximately 20 mg were cut from the sheets. The density of PE was 0.94 g/cm^3, and the density of PP was 0.90 g/cm^3. The PerkinElmer 2400 Series II C/H/N analyzer was chosen to investigate the elemental composition of polyethylene and polypropylene. The CHN analysis of polyolefin samples showed that the C/H content in the polymers equals $85.5\%_{wt.}/14.1\%_{wt.}$ and $85.6\%_{wt.}/14.2\%_{wt.}$ for PE and PP, respectively. In this study, pure polyolefins were intentionally used. Experiments with pure samples are important in terms of reference points (benchmarks) and evaluating the effectiveness of various thermal recycling methods. Nitrogen Premium (with a purity of 99.995%) was purchased from Air Products (Allentown, PA, USA) and used as a fluidizing agent.

2.2. Methods

Thermogravimetric analysis of PP and PE samples was conducted on a TGA 8000 analyzer (by Perkin Elmer, Waltham, MA, USA) in a nitrogen atmosphere. The analysis of the chemical composition of gases produced in the polyolefin pyrolysis process was performed using the atmosFIR spectrometer (by Protea^{TM}, Middlewich, UK). Matlab® R2020b, Update 8 software was used for the deconvolution of pyrolysis gas FTIR spectra based on reference spectra provided by the FTIR analyzer manufacturer.

2.3. Experimental Procedure

The laboratory stand for the pyrolysis of polyolefins in multilayer fluidized beds is presented in Figure 2. A quartz reactor with an external diameter of 80.0 mm and a height of 500.0 mm was utilized in the research. The quartz tube was placed on a gas distributor, which was a perforated plate made out of stainless steel. The plate had evenly distributed holes with a diameter of 0.5 mm. Between the quartz tube and the nickel–chromium plate, serving as a gas distributor, a ring-shaped seal made of heat-resistant soft fibrous material was placed. The entire reactor was mounted on a frame, which allowed for the precise pressing of the quartz tube to the metal plate using pressure exerted by four springs pressing the upper, horizontal mounting plate, in which the upper part of the quartz tube was centrally mounted. To prevent micro gas leaks, the junction between the quartz tube and the sieve bottom was additionally sealed with a high-temperature silicone sealant. The temperature in the reactor was controlled by a thermocouple located 8 cm above the sieve bottom. The open-top design of the reactor allowed for the free dosing of polyolefin samples into the fluidized bed volume and the collection of gases for analysis. In the studies, samples of polyolefins were manually dosed into the reactor. Thanks to the open-top design of the reactor, with the use of special feeders, the risk of premature fuel degradation and nozzle clogging was avoided. The gaseous products were directed to a heated cell of the FTIR spectrometer with an optical path length of 4.2 m and a volume of 300 cm^3. Gas samples from pyrolysis and infrared radiation were transmitted through the cell, allowing for the acquisition of interferograms using a Michelson interferometer. The interferograms were then transformed into IR absorption spectra of the gas mixture using Fourier transformation. During reactor operation, the spectra of the gases exiting the fluidized bed reactor were continuously monitored. The absence of an increase in absorbance in the range from 900 to 5000 cm^{-1} indicated the end of the gas emission and readiness to dose another sample (periodically dosing).

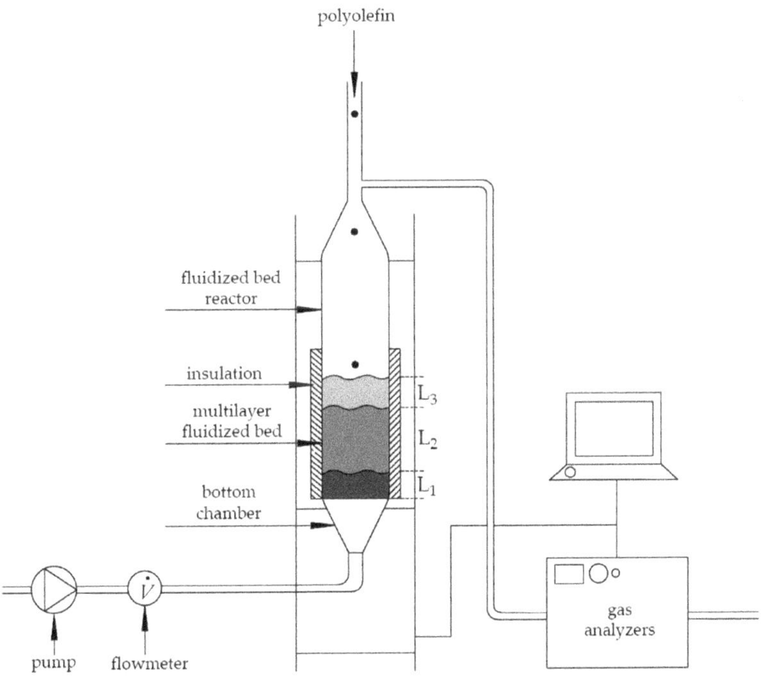

Figure 2. Laboratory stand for polyolefins pyrolysis in the multilayer fluidized beds: L_1—high-density layer; L_2—decomposition layer; L_3—low-density layer.

In this research, two types of multilayer fluidized beds, created from powders characterized by different densities (Table 1), were used.

Table 1. Characterization of the multilayer fluidized beds investigated in this research.

Fluidized Bed Version	Mass, g	Particle Diameter, µm	Static Height in the FB, cm	Bulk Density, g/cm^3
Fluidized bed I		quartz sand		
	360	120–200	6	1.36
		raw cenospheres		
	160	140–160	8	0.46
Fluidized bed II		quartz sand		
	360	120–200	6	1.36
		modified cenospheres		
	160	140–160	8	0.47

3. Results

3.1. The Minimum Flduization Velocity of the Multialyer Fluidized Beds

The minimum fluidization velocities in the multilayer fluidized beds were determined according to the following steps [28]:

(a) Initiating the data acquisition, starting the flow of nitrogen, and setting the flow regulator to the lowest possible value (to prevent powder grains from overflowing through the holes in the bottom sieve);
(b) Placing the lower layer of the multilayer fluidized bed (quartz sand) into the reactor;
(c) Placing the upper layer of the fluidized bed (cenospheres) into the reactor;
(d) Gradually increasing the gas flow while recording the pressure drop in the fluidized bed at a given flow rate through the pressure drop sensor located in the bottom chamber;
(e) Slowly reducing the gas flow to the minimum value (after achieving the fluidized state).

These steps allowed for the determination of the fluidization curves of the multilayer fluidized beds. Two inflection points were observed in the curves: the first point was at the gas surficial velocities of 0.81 cm/s (for raw cenospheres) and 0.83 cm/s (for modified cenospheres), which corresponds to the cenospheres' minimum fluidization velocity obtained in the multilayer fluidized bed; the second point, at the gas surficial velocity of 2.60 cm/s, corresponds to the minimum fluidization velocity of quartz sand and determines the beginning of the fluidization of the lower layer of the multilayer bed, which is equivalent to reaching of the fluidized state of the entire multilayer fluidized bed. On this basis, it was assumed that the minimum fluidization velocity of the multilayer fluidized beds is determined based on the minimum fluidization velocity of the material with the highest density in the multilayer system.

3.2. The Density Profile of the Multilayer Fluidized Beds

Figure 3 illustrates the density profile of a multilayer fluidized bed made out of raw cenospheres and quartz sand. The density profile was obtained by fluidizing the two-component FB with nitrogen at a flow rate 60% higher than the minimum fluidization velocity of quartz sand (i.e., the lowest layer); thus, $u/u_{mf,\,sand} = 1.6$. Due to the similarity of the fluidized beds made out of raw cenospheres and chemically modified cenospheres, the density profiles of both fluidized beds are equivalent.

Vertical density profiles in the multilayer fluidized beds were examined based on the pressure drop within the fluidized beds. To accomplish this, a measurement probe was constructed. The probe consisted of 6 differential pressure sensors connected to brass tubes forming the probe's frame. The ends of the tubes were placed at various heights from each other, allowing for pressure drop measurements at different heights from the bottom sieve inside the fluidized beds. The experimental procedure for determining vertical density profiles in the multilayer fluidized beds was divided into several parts [28]:

(a) Initiating the data acquisition and setting the nitrogen flow to the lowest possible value;
(b) Placing of lower and upper layers of the multilayer fluidized bed into the reactor;
(c) Adjusting the gas flow to achieve $u/u_{mf,\ sand}$ = 1.6 and fluidizing the multilayer fluidized beds for 10 min to achieve the desired mixing;
(d) Recording pressure drop values at different heights inside the fluidized bed, for 10 min, at a frequency of 10 Hz;
(e) Slowly reducing the gas flow to the minimum value (after achieving the fluidized state).

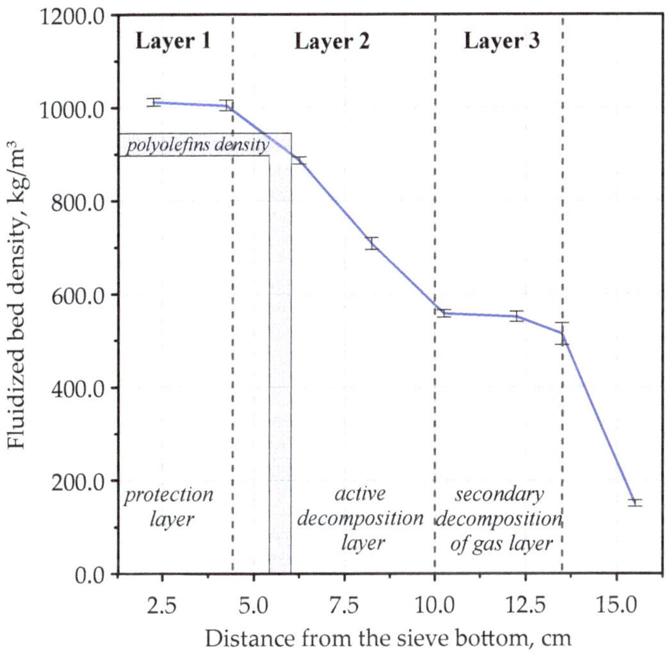

Figure 3. Vertical density profile of multilayer fluidized bed ($u/u_{mf,\ sand}$ = 1.6). Distance from the sieve bottom, where the buoyancy force balances the gravitational force acting on the pyrolyzed material particle, is additionally marked.

The methodology for calculating the vertical density profiles in the multilayer fluidized beds involved determining the pressure drop values at different heights within the fluidized beds and then converting the pressure drop values into density values. The pressure in the pseudo-liquid layer of the fluidized bed is described by Equation (1):

$$\Delta p = g \cdot h \cdot \rho. \quad (1)$$

After introducing the differential form and making minor algebraic transformations, the formula for density calculation is obtained in the form of Equation (2):

$$\rho(h) = [g^{-1} \cdot dp(h)] \cdot dh^{-1}, \quad (2)$$

where:

$\rho(h)$—fluidized bed density at height h [kg/m^3];

dp(h)—change of differential pressure in the fluidized bed at height h [Pa];
dh—height difference between two heights in the fluidized bed [m].

The density profile consists of three characteristic regions representing individual layers in the multilayer fluidized bed. The first layer (L_1) is located at the bottom of the FB and extends up to 4.5 cm above the gas distributor. The density of the layer remains constant at 1008.2 ± 2.6 kg/m^3, indicating that L_1 is primarily composed of quartz sand grains (sand bulk density: approx. 1300 kg/m^3) with a small contribution from the cenospheres. Layer L_1 serves as a protection against the contact of pyrolyzed polymer (density: approx. 920 kg/m^3) with the bottom sieve, continually cooled by the cold fluidizing gas. Layer L_2 (above L_1) is characterized by a linear decrease in density from 1004.5 kg/m^3 to 557.4 kg/m^3 as raised from the bottom sieve. Layer L_2 is located between 4.5 and 10 cm above the gas distributor and represents the region where pyrolysis of polyolefins occurs. In layer L_2, the mixing of cenospheres and sand also takes place, ensuring that it is composed of grains of both powders. Layer L_3 (above L_2) exhibits a relatively constant density of 540.7 ± 19.1 kg/m^3, indicating that it mainly consists of cenospheres grains (cenospheres bulk density: approx. 450 kg/m^3) with a small contribution from quartz sand. Layer L_3 is situated between 10 and 13.5 cm above the gas distributor and is referred to as the zone of additional decomposition processes as the decomposition of gas bubbles formed from polyolefin pyrolysis in layer L_2 occurs in layer L_3. The rare zone, which is located above the L_3 layer, is characterized by a low presence of solid particles. In the rare zone, free spaces between grains predominate, resulting in low-density values reaching below 200 kg/m^3.

The selected masses of the two materials in the multilayer fluidized bed and the applied ratio of $u/u_{mf,\,sand}$ allowed for the creation of a variable vertical density profile in the FB. The variable density provides space for the pyrolysis of lightweight polyolefins inside the fluidized bed. Polyethylene and polypropylene samples were characterized by densities higher than the density of layer L_3. This allows for polyolefin particles to immerse into the FB to a height corresponding to their density. It also protects against fuel processing in the freeboard, which means that the advantages of fluidization are fully utilized. A space in layer L_2, located approximately 5 to 6 cm above the gas distributor, was characterized by density matching the density of polyolefins. According to Archimedes' principle, this space is where the buoyancy force balances the gravitational force acting on the pyrolyzed material particles, keeping particles in that zone. It is the preferred zone where the pyrolysis processes of PE and PP occur.

3.3. Polyolefin Thermogravimetric Analysis

Thermogravimetric analysis of PP and PE was conducted in a nitrogen atmosphere with a gas flow rate of 60 mL/min and a heating rate of 100 °C/min. Samples weighing 5.0 mg were used in the experiments. The purpose of the analysis was to determine the lower limit of the temperature range for conducting the pyrolysis in the fluidized bed reactor. During the fluidized pyrolysis, where samples experience rapid heating upon contact with the hot fluidized bed material, a relatively fast heating rate was chosen for the TG analysis. The thermogram (TG) and differential thermogram (DTG) of PP and PE samples are presented in Figure 4. It was observed that the initial decomposition temperature corresponding to 1% decomposition of PP and PE was 400 °C. It was also noted that the final decomposition temperature corresponding to 99% decomposition was 495 °C for PP and 515 °C for PE. Based on the obtained curves, four fluidized bed process temperatures for PE and PP were selected, all above the maximum decomposition temperature of the samples, to ensure an efficient degradation. Temperatures ranging from 530 °C to 590 °C were chosen at 20 °C intervals and are marked on the graph with vertical dashed lines.

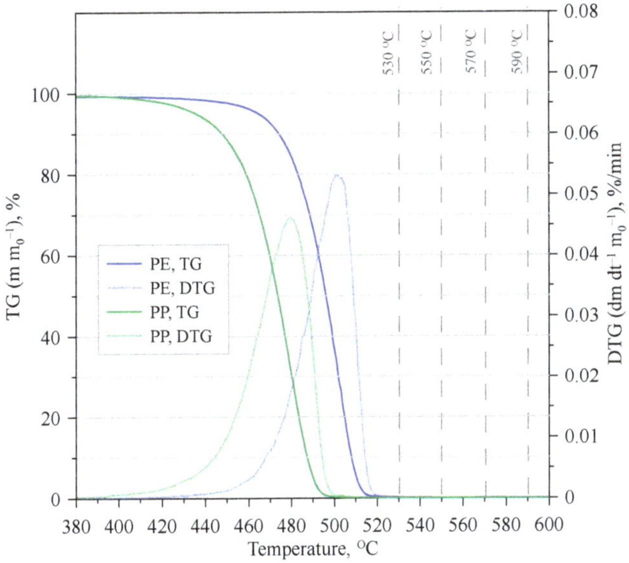

Figure 4. Thermogram and differential thermogram of PE and PP under nitrogen atmosphere.

3.4. Polyolefin Pyrolysis in the Fluidized Beds

As the temperature of the process changes, the chemical parameters of fluids, such as density and viscosity, change. The minimum fluidization velocities of the multilayer fluidized beds were additionally calculated in the process temperatures. In Table 2, the u_{mf} values of the multilayer fluidizeds bed at temperatures of 530, 550, 570, and 590 °C based on the Grace correlation are presented with the nitrogen velocities used in the experiments. Due to our assumption that the minimum fluidization velocity of the multilayer fluidized bed is equal to the heaviest component, u_{mf} values were calculated for quartz sand.

Table 2. Minimum fluidization velocities of the multilayer fluidized beds and the operation velocities of nitrogen used in the research.

Temperature, °C	$u_{mf,\ multilayer\ fluidized\ bed}$ Based on Grace Equation, cm/s	Quartz Sand Fluidization Number $u/u_{mf,\ sand}$	Nitrogen Velocity under Process Conditions, cm/s
530	1.17		1.87
550	1.14	1.6	1.82
570	1.08		1.73
590	1.09		1.68

During the pyrolysis of polyolefins, FTIR spectra of gaseous products were recorded every 20 s. The registered spectra represented the sum of the absorbance of optically active compounds in the gaseous mixture. Based on a set of spectra provided by the FTIR analyzer manufacturer, a deconvolution of the data was performed. The procedure of minimizing the sum of squares of differences between the sample spectrum and the sum of reference spectra with known concentrations was used. The original program written in MATLAB® was used in the deconvolution using the method described in [29]. An example of a recorded spectrum of the gas mixture produced during the pyrolysis of PE in the multilayer fluidized bed made out of raw cenospheres and quartz sand at 550 °C is shown in Figure 5 as a gray line. Two specific regions suitable for deconvolution and concentration calculations were selected from the spectrum. In the analysis, wavelength ranges where no optical activity of components was observed were excluded, as well as regions where intense signals from water vapor were detected (water vapor was analyzed in a narrower wavelength range from 1250 to 1400 cm^{-1}). The calculated spectrum (blue line) was overlaid on the sample spectrum, achieving very good agreement between the two spectra.

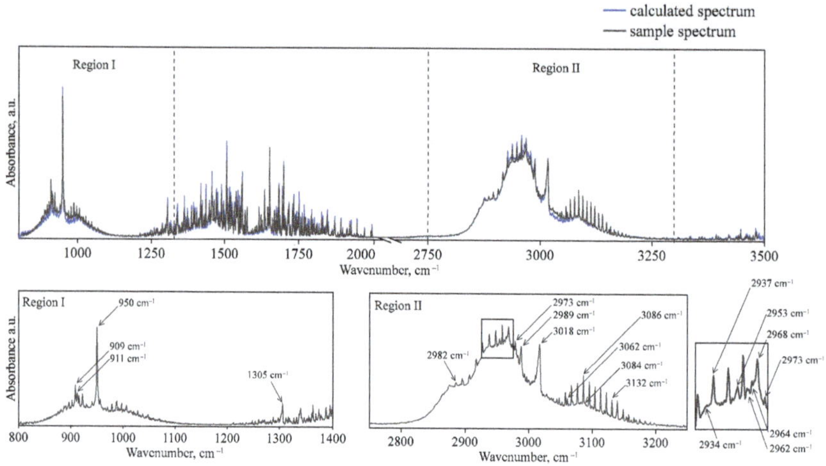

Figure 5. FTIR spectra of the flue gases registered during polyethylene (PE) pyrolysis (gray line) and calculated spectrum (blue line).

Figure 6 presents the deconvolution of the PE spectrum divided into regions I and region II. In region I, the strongest signal at a wavenumber of 950 cm^{-1} is attributed to the presence of ethylene in the product's mixture [30]. The peak at 909 cm^{-1} is characteristic of 1,3-butadiene and arises from the rotational fine structure in the molecule of the diene [31]. Another absorption peak in region I, at a wavenumber of 911 cm^{-1}, is characteristic of 1-butene and propylene [32]. The absorption bands of these compounds are associated with vibrations in the methylene group, typical for hydrocarbon molecules. Methane was also identified in region I through its characteristic spectrum structure, with a peak at a wavenumber of 1305 cm^{-1} [33]. In the sample spectrum, characteristic signals stretching in the wavenumber range of 1400–2000 cm^{-1} were attributed to the presence of water vapor [34].

In region II, selected for analysis (2750–3250 cm^{-1}), many organic compounds exhibit optical activity. One of the most characteristic absorption bands in this range is attributed to methane. The CH_4 band, in the form of densely spaced signals (comb structure), extends in the wavenumber range of 2800–3200 cm^{-1}, with the strongest signal at a wavenumber of 3018 cm^{-1}, corresponding to the stretching vibrations of C-H bonds [35]. Other components of the comb structure appeared on both sides of the maximum, including at 3086 cm^{-1}. Additionally, other alkanes and alkenes were optically active in the wavenumber range of 2700–3200 cm^{-1}. Due to the stoichiometry of polyolefins, the products of the process were dominated by unsaturated aliphatic hydrocarbons. Ethylene, previously identified at 950 cm^{-1}, also exhibited a distinct peak at 2989 cm^{-1} and a series of narrow peaks at 3062 and 3132 cm^{-1}, attributed to the bending vibrations of carbon–hydrogen bonds [30]. In the studied range, propene's spectrum appeared as a broad band with a sharp signal at 2953 cm^{-1}. The infrared spectra of aliphatic hydrocarbons with five or more carbon atoms are characterized by a broad, triple band between 2850 and 3100 cm^{-1}. Aromatic compounds in the wavenumber range of 3000–3150 cm^{-1} show optical activity due to the stretching vibrations of =C–H bonds. The possibility of benzene formation seems likely due to the high temperature and hydrogen deficiency in the process [36].

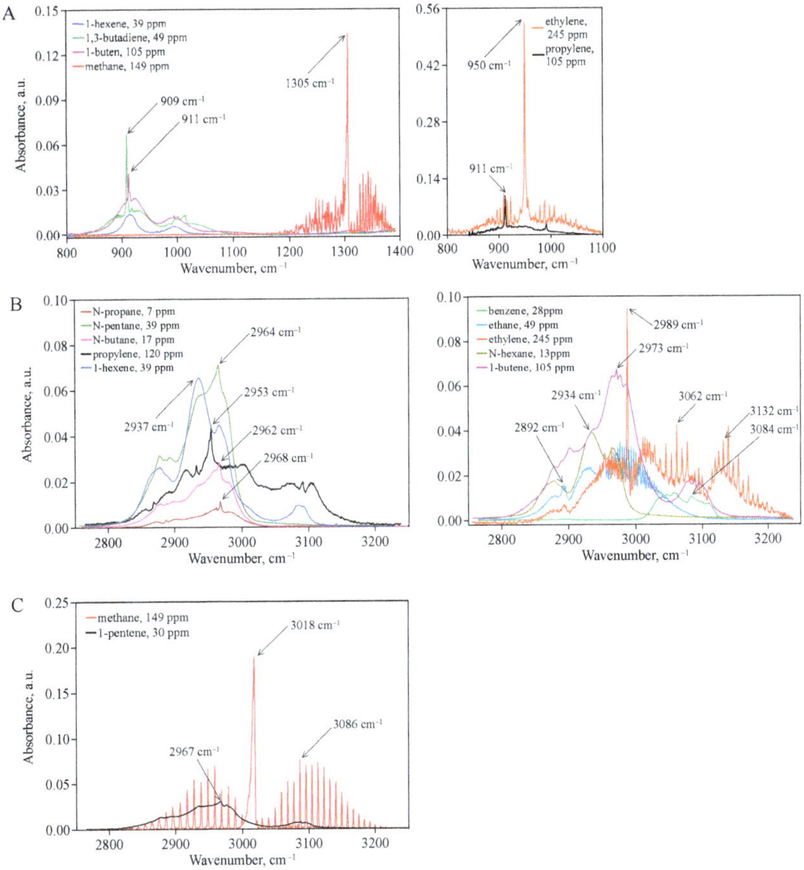

Figure 6. Individual components of the calculated spectrum correspond to the sample spectrum recorded during polyethylene (PE) pyrolysis in region I (**A**) and region II (**B**,**C**) at 550 °C.

The preliminary spectrum analysis showed various chemical compounds in the product mixtures, such as benzene, ethylene propylene, ethane, 1-butene, 1-hexene, 1-pentene, 1,3-butadiene, methane, N-butane, N-hexane, N-pentane, propane, and also water. Software developed in MATLAB® was used for the spectrum deconvolution of the samples. Minimization of the objective function, defined based on the least squares method, allows for the determination of a calculated spectrum that closely approximates the sample spectrum using reference spectra and known concentrations. Through iterative calculations, a series of coefficients were obtained. The compound concentrations were calculated based on the obtained coefficients and concentrations of reference standards.

In Figure 7, the conversions of polyolefins into different product groups (aliphatic unsaturated hydrocarbons, aliphatic saturated hydrocarbons, and aromatic hydrocarbons) during pyrolysis in the multilayer fluidized bed made out of raw cenospheres and quartz sand are presented.

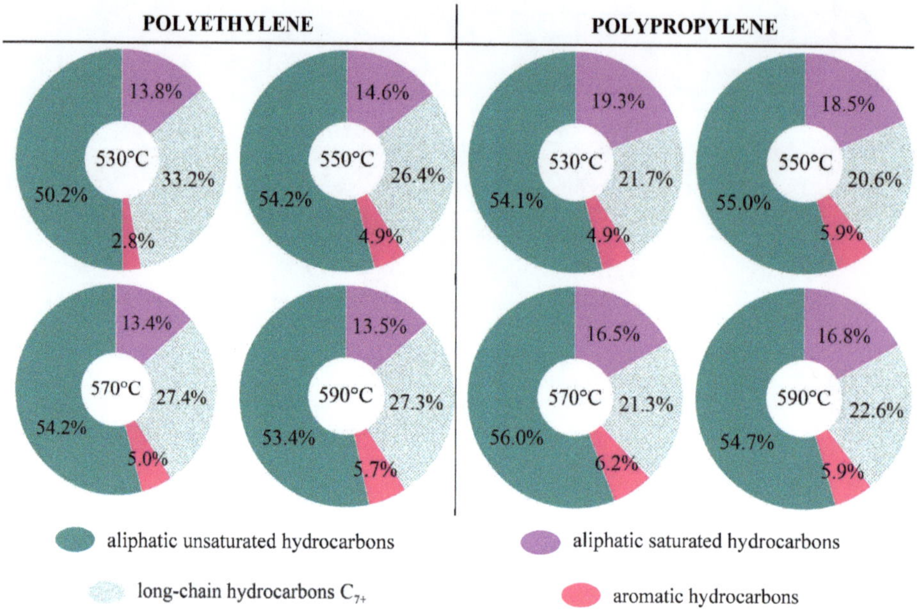

Figure 7. Polyethylene and polypropylene conversion ($\%_{wt.}$) to product groups during polyolefin pyrolysis in the multilayer fluidized bed made out of raw cenospheres and quartz sand.

Within the aromatic hydrocarbons, only benzene was detected, with an average content of $4.6 \pm 1.1\%_{wt.}$ in the case of PE pyrolysis and $5.7 \pm 0.5\%_{wt.}$ in the case of PP pyrolysis. These values are similar to those reported by some literature sources [37,38]. Aliphatic hydrocarbons analyzed by the FTIR spectrometer were characterized by up to six carbon atoms per molecule. Long-chain products were not analyzed using FTIR spectroscopy because this fraction was condensed before flue gases were introduced into the cuvette of the FTIR spectrometer to avoid contamination of its mirrors. However, their presence was calculated by the carbon balance (see Figure 7). The carbon conversion to C_{7+} products averaged 27.3–33.2$\%_{wt.}$ in the case of PE and 20.6–22.6$\%_{wt.}$ in the case of PP. The literature also suggests the presence of liquid (and solid under non-FB conditions) long-chain polyolefins pyrolysis products [39,40]. Additionally, the visual observation of tubes directing the gases to the analyzer proved that liquid process products were condensed.

As can be seen in Figure 7, the average conversion of carbon from polyethylene and polypropylene to aliphatic unsaturated hydrocarbons was $53.0 \pm 1.6\%_{wt.}$ and $55.0 \pm 0.7\%_{wt.}$ respectively. Alongside the aliphatic unsaturated hydrocarbons, saturated aliphatic hydrocarbons were identified in the gas mixtures. The carbon conversion from polyethylene to saturated compounds was approximately $13.8 \pm 0.8\%_{wt.}$, and it was $17.8 \pm 1.2\%_{wt.}$ in the case of polypropylene.

As can be seen in Figure 8, in the case of polyethylene pyrolysis, ethylene was the main component of the aliphatic unsaturated hydrocarbons fraction, decreasing from 52.1% at 530 °C to 32.4% at 590 °C. Additionally, 1-butene constituted an average of 22.9 ± 6.9% in the abovementioned fraction. At temperatures of 550 °C and 570 °C, a significant part of the unsaturated fraction was 1-hexene, with a value of approximately 16%. The content of other aliphatic unsaturated hydrocarbons in the mixture, namely, propene, 1,3-butadiene, and 1-pentene, was significantly lower, totaling approximately 25% of the entire fraction.

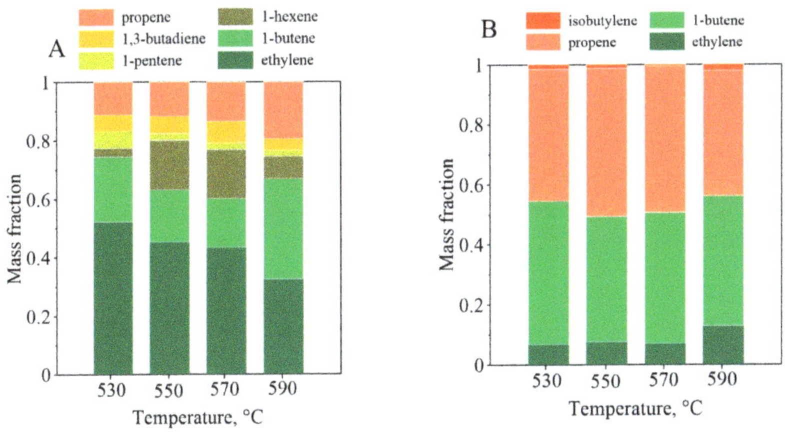

Figure 8. Composition of aliphatic unsaturated hydrocarbon fraction obtained during polyethylene (**A**) and polypropylene (**B**) pyrolysis in the fluidized bed made out of raw cenospheres and quartz sand.

Pyrolysis of polypropylene in the multilayer fluidized bed made out of raw cenospheres and quartz sand led to the recovery of a monomer with an average content of 46.0 ± 3.2% in the aliphatic unsaturated hydrocarbon fraction. Similar to polyethylene, a relatively high conversion of polymeric carbon to 1-butene was observed during PP pyrolysis, reaching an average value of 44.1 ± 2.2% of the abovementioned fraction. The content of other aliphatic unsaturated hydrocarbons, such as ethylene and isobutylene, was low and amounted to about 10%.

Within saturated aliphatic hydrocarbons, methane, ethane, and other aliphatic hydrocarbons such as propane, N-butane, N-pentane, and N-hexane were detected (see Figure 9). In the degradation products, N-hexane and N-pentane were identified in the highest concentration, and in the case of PE pyrolysis, constituted 52.6 ± 5.8% of the aliphatic saturated fraction and 64.8 ± 5.7% in the case of PP degradation.

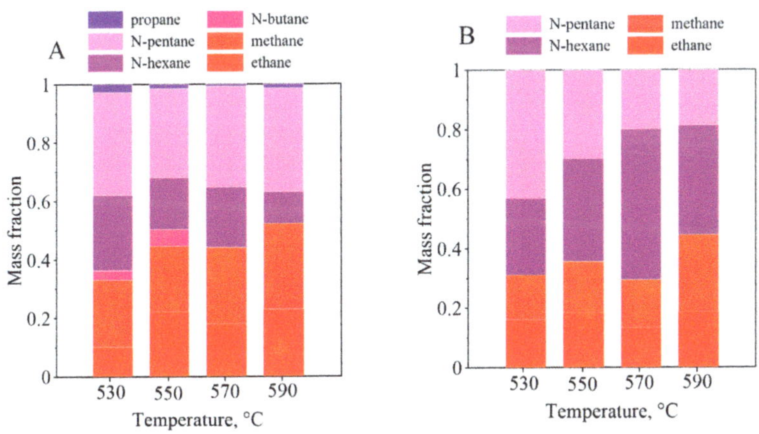

Figure 9. Composition of aliphatic saturated hydrocarbon fraction obtained during polyethylene (**A**) and polypropylene (**B**) pyrolysis in the fluidized bed made out of raw cenospheres and quartz sand.

Figures 10–12 present the conversions of polyolefins into product groups (aliphatic saturated hydrocarbons, aliphatic unsaturated hydrocarbons, aromatic hydrocarbons) and individual compounds during pyrolysis conducted in the fluidized bed made out of modified cenospheres and quartz sand. The pyrolysis process of polyolefins differed when using cenospheres coated with the iron oxide layer to form the fluidized bed. This variation was explored because d-block metal oxides can catalytically influence degradation processes, and iron oxides are one of the most low-cost materials easily applied to raw cenospheres [41,42]. Additionally, iron oxides can be the source of oxygen, which can lead to a change in the composition of the degradation gaseous products.

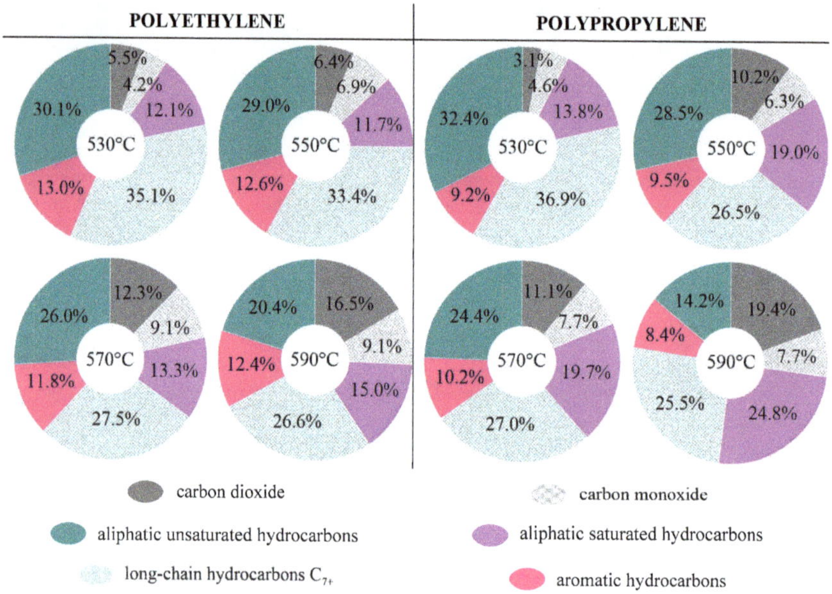

Figure 10. Polyethylene and polypropylene conversion ($\%_{wt.}$) to product groups and carbon oxides during pyrolysis in the multilayer fluidized bed made out of modified cenospheres and quartz sand.

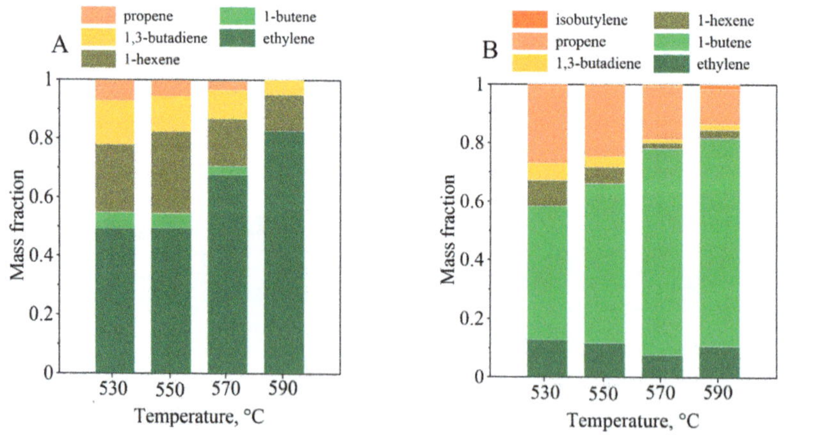

Figure 11. Composition of aliphatic unsaturated hydrocarbon fraction obtained during polyethylene (**A**) and polypropylene (**B**) pyrolysis in the fluidized bed made out of modified cenospheres and quartz sand.

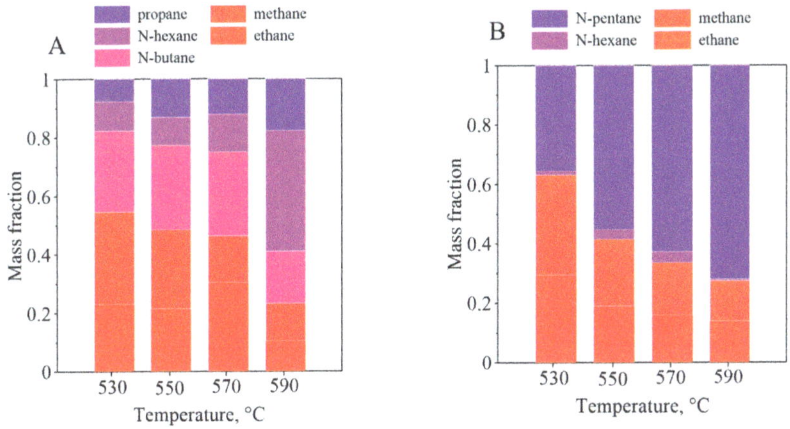

Figure 12. Composition of aliphatic saturated hydrocarbon fraction obtained during polyethylene (**A**) and polypropylene (**B**) pyrolysis in the fluidized bed made out of modified cenospheres and quartz sand.

Both effects were observed during experiments; catalytic effects were observed due to the presence of a micrometric layer of iron oxides applied to the cenospheres particles in the multilayer fluidized bed, and CO/CO_2 were identified among the pyrolysis products. The conversion of polymeric carbon to carbon dioxide and carbon monoxide increased with temperature, with CO_2 exhibiting a faster growth rate. In the pyrolysis products of polyethylene, the conversion of carbon to CO_2 increased from 5.5%$_{wt.}$ at 530 °C to 16.5%$_{wt.}$ at 590 °C, and for polypropylene, it increased from 3.1%$_{wt.}$ at 530 °C to 19.4%$_{wt.}$ at 590 °C. The average conversion to CO during PE pyrolysis ranged from 4.2%$_{wt.}$ to 9.1%$_{wt.}$, and for PP, it ranged from 4.6%$_{wt.}$ to 7.7%$_{wt}$. Furthermore, the conversion of polymeric carbon to benzene (also the only aromatic product) increased in comparison to the multilayer fluidized bed made out of raw cenospheres and quartz sand. The average carbon conversion to benzene changed in the case of PE and PP from 4.6%$_{wt.}$ and 5.7%$_{wt.}$, respectively, to 12.4%$_{wt.}$ and 9.3%$_{wt.}$ in the case of the fluidized bed made out of modified cenospheres and quartz sand. The increase in carbon conversion to aromatic products during pyrolysis in a catalytic environment is already known in the literature. Chen et al. [43] demonstrated that the impregnation of Fe on the Pt/Al_2O_3 catalyst during polyethylene pyrolysis improves oil production efficiency and promotes the formation of aromatic compounds and alkenes. Shen et al. [44] proved that polyethylene pyrolysis using a Ga/HZSM-5 catalyst allows for achieving a process selectivity towards BTEX of 77.0%.

In the fluidized bed made out of modified cenospheres and quartz sand, primarily long-chain hydrocarbons were obtained. The PE conversation to C_{7+} fraction varied from 35.1%$_{wt.}$ to 26.6%$_{wt.}$ and from 36%$_{wt.}$ to 25.5%$_{wt.}$ in the case of PP. The fraction of aliphatic unsaturated hydrocarbons with up to six carbon atoms per molecule was also obtained. The conversion of polymer carbon to the abovementioned fraction decreased with increasing temperature. For PE, the fraction decreased from 31.0%$_{wt.}$ to 20.4%$_{wt.}$, and for PP, from 32.4%$_{wt.}$ to 14.2%$_{wt}$. Alongside unsaturated hydrocarbons, aliphatic saturated hydrocarbons were also identified in the gas mixtures, with their quantity increasing with temperature from 12.1%$_{wt.}$ to 15.0%$_{wt.}$ in the case of polyethylene and from 13.8%$_{wt.}$ to 24.8%$_{wt.}$ in the case of polypropylene.

Among the aliphatic unsaturated hydrocarbons obtained during PE pyrolysis, the main component of the fraction was gas rich in ethylene. Its content increased from 49.3% at 530 °C to 82.6% at 590 °C in the whole fraction. The majority of the unsaturated fraction obtained during PP pyrolysis constituted 1-butene. The content of the compound increased with the increasing temperature from 45.6% at 530 °C to 70.9% at 590 °C. This suggested that selectivity towards ethylene and 1-butene increased rapidly in comparison to the fluidized bed made out of raw cenospheres and quartz sand.

The aliphatic saturated hydrocarbon fraction obtained during PP pyrolysis constituted mainly N-pentane, increasing from 35.6% to 72.1% with increasing temperature. In a similar concentration, menthane and ethane were detected at each applied temperature. In the case of the saturated fraction obtained via PE pyrolysis, all components were detected in similar concentrations in temperatures ranging from 520 °C to 570 °C, and only N-hexane showed the highest content in the fraction at the highest temperature of 590 °C, with a value of 41.3%.

4. Conclusions

1. The paper introduces an alternative method for conducting the polyolefin pyrolysis process in a fluidized bed reactor.

In this study, the possibility of creating multilayer fluidized beds was demonstrated. Density measurements of the multilayer fluidized beds revealed a variable vertical density profile. The chosen materials and used ratio of $u/u_{mf,\ sand}$ (equal to 1.6) were suitable for the pyrolysis of polyolefins. According to Archimedes' principle, in the investigated multilayer FB, a balance between gravitational and buoyancy forces acting on polymer particles would occur in the central part of the fluidized bed. Based on density measurements, three characteristic FB layers were distinguished. Layer L_1 was located at the bottom of the multilayer fluidized bed and had a constant density of about 1000 kg/m^3. The first layer served as protection for the bottom sieve against the falling of processed polymers and the clogging of the gas distributor. In layer L_2, where the density varied from 1000 kg/m^3 to 550 kg/m^3, the distribution of the particles of polymers took place. Layer L_3, with a density of about 500 kg/m^3, was situated at the top of the fluidized bed. Polyolefins were immersed freely into the FB volume, passing through the third, low-density layer, where the decomposition of gases produced during pyrolysis in layer L_2 also occurred.

2. The new approach used in the investigation avoids the formation of a solid phase, a common challenge in pyrolysis processes.

The specially designed fluidized bed reactor was constructed using a transparent quartz tube in this research. This allowed for the observation of the fluidization state during the pyrolysis process. It was possible to monitor changes in the coloration of quartz sand and cenospheres, which could serve as early indicators of the formation of solid pyrolysis products (soot). The outlet of the reactor (where finer soot particles could accumulate) and FB materials were additionally investigated after the completion of the pyrolysis to identify any particles that were distinct from the initial materials. No changes in the coloration of the material particles and foreign soot particles in the FB and at the reactor outlet were observed. Additionaly, the color of the multilayer fluidized bed consisting of raw cenospheres grains and quartz sand after pyrolysis process was varified. Upon detailed color analysis, it was not observed that sand and cenospheres grains were covered with soot. The color of the sand remained beige, while the cenospheres were light gray.

3. Hydrocarbon mixtures containing CO and CO_2 were obtained in the case of a multilayer fluidized bed made out of modified cenospheres.

Quantitative and qualitative FTIR analysis of products from pyrolysis revealed the formation of complex gas mixtures. Mainly unsaturated aliphatic hydrocarbons were obtained during pyrolysis in the multilayer fluidized bed made out of raw cenospheres and quartz sand. If the goal of pyrolysis is to produce a gas rich in ethylene, the reasonable solution would be to pyrolyze polyethylene in the fluidized bed made out of raw cenospheres. However, if the aim of the process is to obtain 1-butene and propene, polypropylene should be processed in the same multilayer fluidized bed consisting of raw cenospheres; the average content of product rich in the desired gases was a 90% aliphatic unsaturated hydrocarbons fraction. In the multilayer fluidized bed made out of modified cenospheres coated with iron oxides and quartz sand, alongside aliphatic and aromatic hydrocarbons, the presence of carbon oxides was also detected. The mixture containing CO, formed after the pyrolysis process, could be directed to various industrial applications, including the production of syngas and chemicals (e.g., formic acid and methanol) and iron ore reduction in metallurgical industries. The hydrocarbons fraction, in this case, was lower than in the FB consisting of raw materials. However, the selectivity towards ethylene and 1-butene significantly increased. At 590 °C, during PE pyrolysis, approximately 80% of the aliphatic unsaturated hydrocarbons fraction was ethylene, and during PP, approximately 70% of the same fraction was 1-butene. Benzene was detected as the only aromatic hydrocarbon identified in the gaseous products. The average carbon conversion to benzene in the multilayer fluidized bed made out of raw cenospheres was 5%, increasing to a value of approximately 11% in the multilayer fluidized bed consisting of modified materials. Thus, if pyrolysis aims to produce benzene, the reasonable method would be to use cenospheres coated with iron oxides.

4. The gas mixtures obtained during pyrolysis are suitable for various catalytic processes. Depending on the chosen catalytic process for gas mixtures, saturated hydrocarbons and hydrogen can be produced.

The complex nature of the mixture of pyrolysis products poses potential challenges in physically separating the resulting chemical gas products. The simplest solution would be to process the mixture as a whole. Utilizing the generated hydrocarbon mixture in a steam conversion process to produce hydrogen could be an application for the obtained mixture.

5. The general benefits of using multilayer fluidized beds.
    (I) The low-density (upper) layer of the multilayer fluidized beds allows for the free dosing of fuel samples into the fluidized bed volume without the use of special feeders. The premature fuel degradation and nozzle clogging in the feeders are eliminated.
    (II) The variable-density (middle) layer acts as a space for material pyrolysis, characterized by even temperature, good mass and heat exchange, and intensive mixing, and imposing a specific residence time for reagents inside the fluidized beds. According to Archimedes' principle, polyolefins freely sink into the multilayer fluidized beds' depth, where the density is like that of PE and PP. The decomposition process then takes place inside the fluidized bed layer.
    (III) The high-density (bottom) layer protects the reactor from the falling of the polyolefins to the bottom sieve during the operation and from the clogging of the gas inlets.

Author Contributions: Conceptualization, K.L. and G.B.-P.; methodology, W.Ż., K.L. and G.B.-P.; software, W.Ż.; validation, G.B.-P.; formal analysis, K.L. and J.W.; investigation, W.Ż., K.L., G.B.-P. and J.W.; resources, W.Ż. and J.W.; data curation, K.L., G.B.-P. and J.W.; writing—original draft preparation, K.L.; writing—review and editing, W.Ż., K.L., G.B.-P. and J.W.; visualization, K.L. and G.B.-P.; supervision, W.Ż. and G.B.-P.; project administration, W.Ż. and J.W.; funding acquisition, W.Ż. and K.L. All authors have read and agreed to the published version of the manuscript.

Funding: The research was supported by a subsidy of the Ministry of Science and Higher Education of Poland for Cracow University of Technology supporting the development of young scientists and PhD students. Interial number: C09.31.734.2023.

Data Availability Statement: The data presented in this study are available on reasonable request.

Conflicts of Interest: The authors declare no conflict of interest. The funders had no role in the design of the study; in the collection, analyses, or interpretation of data; in the writing of the manuscript, or in the decision to publish the results.

References

1. Golub, G.; Tsyvenkova, N.; Kukharets, S.; Holubenko, A.; Omarov, I.; Klymenko, O.; Mudryk, K.; Hutsol, T. European Green Deal: An Experimental Study of the Biomass Filtration Combustion in a Downdraft Gasifier. *Energies* **2023**, *16*, 7490. [CrossRef]
2. Nitkiewicz, T.; Ociepa-Kubicka, A. Eco-Investments-Life Cycle Assessment of Different Scenarios of Biomass Combustion. *Ecol. Chem. Eng. S* **2018**, *25*, 307–322. [CrossRef]
3. Garcia-cortes, S. Repurposing End-of-Life Coal Mines with Business Models. *Energies* **2023**, *16*, 7617.
4. Barmparitsas, N.; Karellas, S.; Pallis, P.; Thanopoulos, S.; Kobelt, D. An Innovative Heating, Ventilation, Air Conditioning and Refrigeration Circular Economy System for Reducing Carbon Dioxide Emissions in Europe via Extensive Reuse of Existing Fluorinated Gases. *Energies* **2023**, *16*, 7705. [CrossRef]
5. Leski, K.; Berkowicz-Płatek, G. Pyrolysis of Plastic Wastes as a Way of Obtaining Valuable Chemical Raw Materials. *Chem. Rev. Lett.* **2021**, *4*, 92–97. [CrossRef]
6. Holubčík, M.; Klačková, I.; Ďurčanský, P. Pyrolysis Conversion of Polymer Wastes to Noble Fuels in Conditions of the Slovak Republic. *Energies* **2020**, *13*, 4849. [CrossRef]
7. Anuar Sharuddin, S.D.; Abnisa, F.; Wan Daud, W.M.A.; Aroua, M.K. A Review on Pyrolysis of Plastic Wastes. *Energy Convers. Manag.* **2016**, *115*, 308–326. [CrossRef]
8. Harussani, M.M.; Sapuan, S.M.; Rashid, U.; Khalina, A.; Ilyas, R.A. Pyrolysis of Polypropylene Plastic Waste into Carbonaceous Char: Priority of Plastic Waste Management amidst COVID-19 Pandemic. *Sci. Total Environ.* **2022**, *803*, 149911. [CrossRef]
9. López, A.; de Marco, I.; Caballero, B.M.; Laresgoiti, M.F.; Adrados, A. Pyrolysis of Municipal Plastic Wastes: Influence of Raw Material Composition. *Waste Manag.* **2010**, *30*, 620–627. [CrossRef]
10. Schwartz, N.R.; Paulsen, A.D.; Blaise, M.J.; Wagner, A.L.; Yelvington, P.E. Analysis of Emissions from Combusting Pyrolysis Products. *Fuel* **2020**, *274*, 117863. [CrossRef]
11. Sciazko, M.; Mertas, B.; Kosyrczyk, L.; Sobolewski, A. A Predictive Model for Coal Coking Based on Product Yield and Energy Balance. *Energies* **2020**, *13*, 4953. [CrossRef]
12. Brown, R.M.; Hoover, A.N.; Klinger, J.L.; Wahlen, B.D.; Hartley, D.; Lee, H.; Thompson, V.S. Decontamination of Mixed Paper and Plastic Municipal Solid Waste Increases Low and High Temperature Conversion Yields. *Front. Energy Res.* **2022**, *10*, 834832. [CrossRef]
13. Ganguly, A.; Brown, R.C.; Wright, M.M. Investigating the Impacts of Feedstock Variability on a Carbon-Negative Autothermal Pyrolysis System Using Machine Learning. *Front. Clim.* **2022**, *4*, 842650. [CrossRef]
14. Sanito, R.C.; Chen, C.H.; You, S.J.; Yang, H.H.; Wang, Y.F. Volatile Organic Compounds (VOCs) Analysis from Plasma Pyrolysis of Printed Circuit Boards (PCB) with the Addition of $CaCO_3$ from Natural Flux Agents. *Environ. Technol. Innov.* **2023**, *29*, 103011. [CrossRef]
15. European Commission A Green Deal Industrial Plan for the Net-Zero Age. 2023. Available online: https://commission.europa.eu/system/files/2023-02/COM_2023_62_2_EN_ACT_A%20Green%20Deal%20Industrial%20Plan%20for%20the%20Net-Zero%20Age.pdf (accessed on 21 February 2024).
16. Jia, Y.; Mao, Z.; Huang, W.; Zhang, J. Effect of Temperature and Crystallinity on the Thermal Conductivity of Semi-Crystalline Polymers: A Case Study of Polyethylene. *Mater. Chem. Phys.* **2022**, *287*, 126325. [CrossRef]
17. Wang, X.; Jin, Q.; Zhang, J.; Li, Y.; Li, S.; Mikulčić, H.; Vujanović, M.; Tan, H.; Duić, N. Soot Formation during Polyurethane (PU) Plastic Pyrolysis: The Effects of Temperature and Volatile Residence Time. *Energy Convers. Manag.* **2018**, *164*, 353–362. [CrossRef]
18. Onwudili, J.A.; Insura, N.; Williams, P.T. Composition of Products from the Pyrolysis of Polyethylene and Polystyrene in a Closed Batch Reactor: Effects of Temperature and Residence Time. *J. Anal. Appl. Pyrolysis* **2009**, *86*, 293–303. [CrossRef]
19. Font, R.; Aracil, I.; Fullana, A.; Martín-Gullón, I.; Conesa, J.A. Semivolatile Compounds in Pyrolysis of Polyethylene. *J. Anal. Appl. Pyrolysis* **2003**, *68–69*, 599–611. [CrossRef]
20. Antoniou, N.A.; Zorpas, A.A. Quality Protocol and Procedure Development to Define End-of-Waste Criteria for Tire Pyrolysis Oil in the Framework of Circular Economy Strategy. *Waste Manag.* **2019**, *95*, 161–170. [CrossRef]
21. Islam, M.N.; Joardder, M.U.H.; Hoque, S.M.N.; Uddin, M.S. A Comparative Study on Pyrolysis for Liquid Oil from Different Biomass Solid Wastes. *Procedia Eng.* **2013**, *56*, 643–649. [CrossRef]
22. Supriyanto; Usino, D.O.; Ylitervo, P.; Dou, J.; Sipponen, M.H.; Richards, T. Identifying the Primary Reactions and Products of Fast Pyrolysis of Alkali Lignin. *J. Anal. Appl. Pyrolysis* **2020**, *151*, 104917. [CrossRef]
23. Oliveri, L.M.; Chiacchio, F.; D'Urso, D.; Matarazzo, A.; Cutaia, L.; Luciano, A. Circular Economy and Industrial Symbiosis in Sicily. *IFIP Adv. Inf. Commun. Technol.* **2022**, *663 IFIP*, 432–439. [CrossRef]
24. Baral, B.; Raine, R. *Performance and Emissions of a Spark Ignition Engine Running on Gasoline Adulterated with Kerosene*; SAE Technical Paper; SAE International: Warrendale, PA, USA, 2009; Volume 2009-Decem. [CrossRef]
25. Igliński, B.; Kujawski, W.; Kiełkowska, U. Pyrolysis of Waste Biomass: Technical and Process Achievements, and Future Development—A Review. *Energies* **2023**, *16*, 1829. [CrossRef]
26. Peters, J.; May, J.; Ströhle, J.; Epple, B. Flexibility of CFB Combustion: An Investigation of Co-Combustion with Biomass and RDF at Part Load in Pilot Scale. *Energies* **2020**, *13*, 4665. [CrossRef]
27. Żukowski, W.; Berkowicz, G. The Combustion of Liquids and Low-Density Solids in a Cenospheric Fluidised Bed. *Combust. Flame* **2019**, *206*, 476–489. [CrossRef]

28. Berkowicz-Płatek, G.; Żukowski, W.; Leski, K. Combustion of Polyethylene and Polypropylene in the Fluidized Bed with a Variable Vertical Density Profile. *Energy* **2024**, *286*, 129611. [CrossRef]
29. Berkowicz, G.; Majka, T.M.; Żukowski, W. The Pyrolysis and Combustion of Polyoxymethylene in a Fluidised Bed with the Possibility of Incorporating CO_2. *Energy Convers. Manag.* **2020**, *214*, 112888. [CrossRef]
30. Merrill, P.B.; Madix, R.J. Hydrogenation of Weakly Rehybridized Ethylene on Fe(100)−H: Ethyl Group Formation. *J. Am. Chem. Soc.* **1996**, *118*, 5062–5067. [CrossRef]
31. Vouvoudi, E.C.; Rousi, A.T.; Achilias, D.S. Effect of the Catalyst Type on Pyrolysis Products Distribution of Polymer Blends Simulating Plastics Contained in Waste Electric and Electronic Equipment. *Sustain. Chem. Pharm.* **2023**, *34*, 101145. [CrossRef]
32. Abdel-Rahman, M.K.; Trenary, M. Propyne Hydrogenation over a Pd/Cu(111) Single-Atom Alloy Studied Using Ambient Pressure Infrared Spectroscopy. *ACS Catal.* **2020**, *10*, 9716–9724. [CrossRef]
33. Azzolina-Jury, F.; Thibault-Starzyk, F. Mechanism of Low Pressure Plasma-Assisted CO_2 Hydrogenation over Ni-USY by Microsecond Time-Resolved FTIR Spectroscopy. *Top. Catal.* **2017**, *60*, 1709–1721. [CrossRef]
34. Hakkarainen, T.; Mikkola, E.; Laperre, J.; Gensous, F.; Fardell, P.; Le Tallec, Y.; Baiocchi, C.; Paul, K.; Simonson, M.; Deleu, C.; et al. Smoke Gas Analysis by Fourier Transform Infrared Spectroscopy—Summary of the SAFIR Project Results. *Fire Mater.* **2000**, *24*, 101–112. [CrossRef]
35. Plyler, E.K.; Tidwell, E.D.; Blaine, L.R. Infrared Absorption Spectrum of Methane from 2470 to 3200 cm^{-1}. *J. Res. Natl. Bur. Stand. Sect. A Phys. Chem.* **1960**, *64A*, 201. [CrossRef] [PubMed]
36. Kınaytürk, N.K.; Kalaycı, T.; Tunalı, B.; Altuğ, D.T. A Spectroscopic Approach to Compare the Quantum Chemical Calculations and Experimental Characteristics of Some Organic Molecules; Benzene, Toluene, P-Xylene, P-Toluidine. *Chem. Phys.* **2023**, *570*, 111905. [CrossRef]
37. Cho, M.H.; Jung, S.H.; Kim, J.S. Pyrolysis of Mixed Plastic Wastes for the Recovery of Benzene, Toluene, and Xylene (BTX) Aromatics in a Fluidized Bed and Chlorine Removal by Applying Various Additives. *Energy Fuels* **2010**, *24*, 1389–1395. [CrossRef]
38. Namioka, T.; Saito, A.; Inoue, Y.; Park, Y.; Min, T.J.; Roh, S.A.; Yoshikawa, K. Hydrogen-Rich Gas Production from Waste Plastics by Pyrolysis and Low-Temperature Steam Reforming over a Ruthenium Catalyst. *Appl. Energy* **2011**, *88*, 2019–2026. [CrossRef]
39. Maqsood, T.; Dai, J.; Zhang, Y.; Guang, M.; Li, B. Pyrolysis of Plastic Species: A Review of Resources and Products. *J. Anal. Appl. Pyrolysis* **2021**, *159*, 105295. [CrossRef]
40. Calero, M.; Solís, R.R.; Muñoz-Batista, M.J.; Pérez, A.; Blázquez, G.; Ángeles Martín-Lara, M. Oil and Gas Production from the Pyrolytic Transformation of Recycled Plastic Waste: An Integral Study by Polymer Families. *Chem. Eng. Sci.* **2023**, *271*, 118569. [CrossRef]
41. Pal, P.; Ting, J.M.; Agarwal, S.; Ichikawa, T.; Jain, A. The Catalytic Role of D-Block Elements and Their Compounds for Improving Sorption Kinetics of Hydride Materials: A Review. *Reactions* **2021**, *2*, 22. [CrossRef]
42. Ranjbar, N.; Kuenzel, C. Cenospheres: A Review. *Fuel* **2017**, *207*, 1–12. [CrossRef]
43. Chen, Z.; Xu, L.; Zhang, X. Upgrading of Polyethylene to Hydrocarbon Fuels over the Fe-Modified Pt/Al_2O_3 Catalysts at a Mild Condition without External H_2. *Chem. Eng. J.* **2022**, *446*, 136213. [CrossRef]
44. Shen, X.; Zhao, Z.; Li, H.; Gao, X.; Fan, X. Microwave-Assisted Pyrolysis of Plastics with Iron-Based Catalysts for Hydrogen and Carbon Nanotubes Production. *Mater. Today Chem.* **2022**, *26*, 101166. [CrossRef]

energies

MDPI

Article

Investigating the Applicability of the Tafel Equation in Polymer Electrolyte Membrane Electrolyzers via Statistical Analysis

Fabian Scheepers [1,*] and Werner Lehnert [2]

1 Forschungszentrum Juelich GmbH, Institute of Energy Technologies, IET-4, Electrochemical Process Engineering, 52425 Juelich, Germany
2 RWTH Aachen University, 52062 Aachen, Germany
* Correspondence: f.scheepers@fz-juelich.de; Tel.: +49-2461-61-2177

Abstract: Analyzing PEM electrolyzer polarization curves via voltage breakdown analysis involves decomposing contributions from underlying processes, typically based on the assumption of reaction kinetics that may be expressed by means of the Tafel equation. When extrapolating the corresponding straight line to high current densities, there is a discrepancy between the measurement and model, which is often attributed to mass transport resistance. In addition to the qualitative description of this mass transport resistance, a consistent quantification is difficult to obtain from the measurement results. Accordingly, the approach to the breakdown analysis of the polarization curves is strongly based on assumptions that evade experimental verification. In this study, an alternative statistical method is introduced that permits the falsifiability of the standard approach. By means of experiments at different hydrogen partial pressures and a subsequent data fit, it is possible to extract the kinetic behavior without prior specification. The results indicate that behavior corresponding to the Tafel equation cannot be proven wrong. In addition, transport coefficients can be evaluated that fall between those of membranes and porous transport layers, indicating that the catalyst layer predominantly contributes to the mass transport resistance.

Keywords: voltage breakdown analysis; PEM-water electrolysis; Tafel equation; statistical analysis

1. Introduction

The polarization curve of a polymer electrolyte membrane (PEM) water electrolyzer links the applied cell voltage to the initiated current density of the cell [1]. As the cell voltage comprises the predominant part of the overall plant efficiency, it is important to understand its origin [2]. Therefore, investigating which processes contribute to the curve is of elemental interest. The separation and assignment of voltage drops is called voltage breakdown analysis [3].

Based on the usual description, the cell voltage, U_{cell}, can be broken down into contributions from the reversible cell voltage, U_{rev}, the overpotentials from the reaction kinetics, η_{kin}, electric resistance, η_{res}, and mass transport, η_{mt} [4]:

$$U_{cell} = U_{rev} + \eta_{kin} + \eta_{res} + \eta_{mt} \tag{1}$$

When the overpotential of the electric resistance, which can be obtained from electrochemical impedance spectroscopy, is subtracted, the iR-corrected cell voltage results from it:

$$U_{iR-free} = U_{rev} + \eta_{kin} + \eta_{mt} \tag{2}$$

The reaction mechanisms are frequently said to follow Butler–Volmer kinetics [5]. The hydrogen evolution reaction is supposed to be negligibly small and the oxygen evolution reaction is said to conform to a high overpotential approximation called the Tafel equation [6]. This approach requires negligible mass transport resistance, which is assumed to vanish at low current densities of up to 100 $mAcm^{-2}$ [7]. In this case, the overvoltage rises linearly with the logarithm of the current density. The slope of this straight line is called the Tafel slope. If the fit obtained is extrapolated to high current densities, a difference from the actual measurement data can be observed [8]. This difference is often referred to as overpotential due to the mass transport [9].

It can be concluded that this approach brings the Tafel equation into the spotlight of the voltage breakdown analysis. For this reason, it is important to draw attention to certain aspects, in particular the membrane electrode assembly. The catalyst layers are expanded materials with stochastic properties [10]. This means that the structure, morphology, and chemical composition vary between localities [11]. Within these layers, fluids, heat, and electric charges must be transferred, which are obviously subject to transport resistances [12]. Due to the expansion of the layers, the values of the resistances also depend on the locality, as the transport path lengths vary due to the stochastic nature of the materials. In a recent publication, the crucial role of ionic and electric conductivity of the catalyst layers have been mentioned once again and it was shown how this leads to deviations from the Tafel equation's -straight line at high current density [13]. It must be concluded that the electric field, fluid concentrations, temperature, and local current densities are distributed in order to minimize the energetic state [14]. As a consequence of the local properties, a polarization curve is unable to resolve them, but it only indicates integral information about the entire system. The loss of local information ultimately makes it improbable that a precise description, as defined by Equation (1), is possible. In particular, the Tafel equation, which itself only approximates the Butler–Volmer equation, can never describe the reaction kinetics in the way it has been developed for [15]. Additionally, it locks the understanding of the cell polarization into a predetermined reaction mechanism, whereas physical boundary conditions such as the cell temperature or concentrations are assumed to be fixed. In conclusion, it is questionable to apply the Tafel equation to a polarization curve in order to achieve charge transfer coefficients or exchange current densities at the three-phase boundary. This does not exclude the possibility that the entire catalyst layer shows an exponential increase in the current density with cell voltage; however, it must not be viewed as a given. Experiments employing rotating disc electrodes have shown how difficult it is to achieve this behavior, even under optimal conditions [16]. Consequently, the poor verifiability is critical, especially since the experimental assessment of the mass transfer resistance is often based on the prior assumption of the Tafel equation. Although the increase in potential due to mass transport phenomena has been described by an additional half-cell potential, the consistency of the increase with mass transport laws has not been verified based on experiments [17].

To conclude this discussion, voltage breakdown analysis is crucial but eludes falsifiability, which is epistemologically significant. The original reason why this challenge is difficult to tackle is due to the under-determination inherent in Equation (2). In general, this would be solved through variation in the experiments, which are linearly independent. However, the high complexity of the electrochemical setup limits the possibility of being able to do so. Accordingly, models from close experiments and qualitative considerations plays an undesirably significant role.

The aim of this methodological contribution is to enable partial access to falsifiability. For this purpose, all of the assumptions of the standard approach are taken, including that the cathodic kinetic overpotential is negligible. This leads to the possibility that measurements at different cathode pressures can be used to reduce the degree of under-determination in Equation (2). The description of the influence of the hydrogen partial pressure in accordance with the laws of mass transport and its consideration in the equation of the reversible cell voltage enables an evaluation procedure that does not require predetermination of the kinetic behavior. Consequently, the presented approach is suitable for falsifying both the linear trend of the logarithmic current density with the cell voltage and a consistency of the shape of the overpotential due to mass transport resistances. It is important to reiterate that this approach is incapable of verification, but only falsification.

2. Model

Gas transport in the cell—During its operation, an electrolysis cell continuously produces hydrogen at the cathode and oxygen at the anode through the splitting of water. Each gas is produced in the corresponding catalyst layer and is transported through the porous transport layer into the flow field, where it is released from the cell. This mass transport requires a pressure or concentration gradient between the location where it originates and the one to which it is transported. Although the situation in the flow field is known because it can be defined in the experiment, it is impossible to directly measure the concentration in the nanoporous catalyst layer [17]. However, it can be said with certainty that the system is far from a steady state and that various mechanisms overlap locally or change in the short term. Assuming primary pore sizes on the order of 10^{-8} m [18], molecular flow (mf), i.e., (Knudsen-)diffusion [19], could be the dominant transport mechanism in the pores. Additionally, gas phase development and transport according to the continuum flow (cf) could appear in secondary pores when reaching pore sizes on the order of 10^{-7} m and inside the porous transport layers [20]. Furthermore, various transport mechanisms likely overlap; however, identifying the precise mechanism is not critical for the approach presented. More importantly, molecular and continuum flows have distinct relationships between molar flow and pressure gradient or concentration gradient, respectively, and encompass other effects. Although an overlap between the mechanisms can be assumed for the actual flow, this work considers both cases individually. This allows for investigating the predominant effect and prevents the incorporation of too many unknown parameters. The two basic relationships between the molar flux, $\dot{n}$, and concentration, c, or pressure gradient, p, are

$$\dot{n}_i \sim \nabla c_i \sim \nabla p_i \text{ (molecular flow)} \tag{3}$$

$$\dot{V}_i \sim \nabla p \rightarrow \dot{n}_i \sim \overline{p_i} \cdot \nabla p \text{ (continuum flow)} \tag{4}$$

It has been noted that concentrations in liquid and gas pressure can be converted into each other by means of Henry's law [21]. In continuum flow, the absolute pressure of the system decreases in the direction of the flow, which is accounted for by using an average pressure, $\overline{p_i}$, between the inlet and outlet [22].

Reversible cell voltage—The reversible cell voltage, U_{rev}, results from the difference in the electrode potentials according to the Nernst equation. It describes the standard cell voltage, U°, and demonstates how it is affected by changes in the species' activities, a_i. It contains the gas constant, R, the Faraday constant, F, as well as the temperature, T. When assuming an excess energy-free system, pressure-independent solubility coefficients, and an ideal gas law, the activities can be substituted by either the concentration or pressure referred to by the standard values. If the water activity is one, the reversible cell voltage is usually expressed as follows [23]:

$$U_{rev} = U^\circ + \frac{RT}{2F}\ln\left(\frac{a_{H_2}^{cat}\sqrt{a_{O_2}^{an}}}{a_{H_2O}^{an}}\right) \cong U^\circ + \frac{RT}{2F}\ln\left(\frac{c_{H_2}^{cat}}{c_{H_2}^{cat\,\circ}}\sqrt{\frac{c_{O_2}^{an}}{c_{O_2}^{an\,\circ}}}\right) \cong U^\circ + \frac{RT}{2F}\ln\left(\frac{p_{H_2}^{cat}}{p_0}\sqrt{\frac{p_{O_2}^{an}}{p_0}}\right) \tag{5}$$

Typically, the partial pressures at the set operating conditions are used for the calculation. However, this does not correspond to reality because they refer to the conditions in the flow fields and not to those in the catalyst layers. The actual concentration that must be inserted into Equation (5) is the bulk concentration of products close to the surface of the catalyst just outside the diffuse double layer [24]. This is different from the flow field concentration according to the explanation in the previous section. Therefore, the typical approach neglects the mass transport at this point and reintroduces it elsewhere as overpotential from the mass transport resistance in Equation (1). This standard approach must be performed because the additional concentration drop required for sufficient mass transport in the electrodes cannot be determined experimentally and can therefore only be determined indirectly by assuming Tafel kinetics or pure modeling. However, the reversible cell voltage can generally be transformed into Equation (6):

$$U_{rev} \cong U^{\circ} + \frac{RT}{2F}\ln\left(\frac{c_{H_2}^{cat,ff} + \Delta c_{H_2}^{cat}}{c_{H_2}^{cat\ \circ}}\sqrt{\frac{c_{O_2}^{an,ff} + \Delta c_{O_2}^{an}}{c_{O_2}^{an\ \circ}}}\right) \cong U^{\circ} + \frac{RT}{2F}\ln\left(\frac{p_{H_2}^{cat,ff} + \Delta p_{H_2}^{cat}}{p_0}\sqrt{\frac{p_{O_2}^{an,ff} + \Delta p_{O_2}^{an}}{p_0}}\right) \quad (6)$$

In this equation, the gas concentrations in the electrodes are split into the sum of the flow field concentrations and additional concentrations increase towards the corresponding catalyst layers, Δc_i^{el}, due to mass transport. It is noted that the concentration drop is zero if the current density is zero and is expected to increase linearly with it according to Equations (1) and (2). In the standard voltage breakdown analysis, this value is always zero as no concentration drop is considered. The concentration drop can also be expressed as pressure drop, Δp_i^{el}, in the corresponding equation where the concentration is replaced by the pressure.

Access to the concentration drop—The fundamental idea of the new approach is to perform cell measurements which only differ in the cathode pressures and to use the assumptions made in the standard approach to explain the difference between the measurements. These are explained in the following:

As assumed in the standard evaluation procedure and supported by hydrogen pump experiments, the hydrogen evolution reaction on platinum is very rapid and causes only a negligible overpotential of a few mV per A cm^2 [25], which, according to Butler–Volmer kinetics, should evolve linearly with the current density [26]. A change in the cathodic gas pressure is not going to lead to a significantly different kinetic overpotential so that the according overpotential is negligible at any cathode pressure [27]. The oxygen evolution reaction, on the other hand, should be almost unchanged, since no adjustment is made between the experiments at the anode. For this reason, the anodic reaction kinetics should be very similar and so should the kinetic overpotential. Apart from kinetics, an increase in the cathode pressure should not lead to a significant change in the equilibrium potential at the anodic half-cell. Contrary to the previous points, the cathode pressure affects the reaction equilibrium of the corresponding half-cell. When summarizing the assumptions made, the iR-free cell voltage as a function of the cathode pressure can be rearranged to form Equation (7):

$$U_{iR-free}\left(p_{H_2}^{cat}\right) \cong \frac{RT}{2F}\ln\left(\frac{p_{H_2}^{cat,ff} + \Delta p_{H_2}^{cat}}{p_0}\right) + \underbrace{U_0 + \frac{RT}{4F}\ln\left(\frac{p_{O_2}^{an,ff} + \Delta p_{O_2}^{an}}{p_0}\right) + \eta_{kin}^{an} + \eta_{kin}^{cat}}_{\approx constant\ value:\ A} \quad (7)$$

Accordingly, a change in the cathode pressure should cause an increase in the iR-free curves corresponding to the change in the cathodic half-cell potential. If the standard approach, which means $\Delta p_{H_2}^{cat} = 0$, was a good approximation, the offset between the iR-free curves should vanish at very low current densities. Here, we postulate that even at a low current density, an offset is measurable due to mass transport resistance, which is taken into account as $\Delta p_{H_2}^{cat} > 0$. In order to distinguish normal data scattering between the measurements and reduce the risk of misinterpretation of the deviations, the implementation of $\Delta p_{H_2}^{cat}$ must be consistent with mass transport laws and its determination must be statistically significant. If this cannot be guaranteed, it will be likely that a change in the cathode pressure either affects other operating parameters or at least one of the parameters assumed to have fixed values is actually pressure-dependent.

Consistency with mass transport laws is based on the following: according to Faraday's law, keeping the current density constant between the measurement means that the gas production is the same in all cases, as is the molar flux towards the flow fields. As the cell configuration is same, the gas must be released through the same catalyst layers, porous transport layers, and flow fields. When assuming a molecular flow as the dominant mass transport mechanism, the pressure drop should be proportional to the amount of hydrogen produced (see Equation (3)), but independent of the total pressure. In different words: (a) $\Delta p_{H_2}^{cat}$ must increase linearly with the current density and (b) $\Delta p_{H_2}^{cat}$ must be the same for all measurements if the same current density is applied. Accordingly, Equation (7) can be fitted by Equation (8), where $\hat{A}$ is the constant value from Equation (7) and $\hat{B}_{mf} = \Delta p_{H_2}^{cat}/p_0$ applies:

$$\hat{U}_{mf}\left(p_{H_2}^{cat,ff}\right) = \hat{A}_{mf} + \frac{\mathrm{RT}}{2\mathrm{F}}\ln\left(\frac{p_{H_2}^{cat,ff}}{p_0} + \hat{B}_{mf}\right) \tag{8}$$

In contrast, if continuum flow is assumed as dominant transport mechanism, the situation is more difficult because the molar flux is proportional to the product of the average partial pressure and the total pressure gradient (see Equation (4)). By means of some conversion steps (see Appendix A, Equations (A1)–(A5)), a fit that is consistent with continuum flow laws is

$$\hat{U}_{cf}\left(p_{H_2}^{cat,ff}\right) = \hat{A}_{cf} + \frac{\mathrm{RT}}{2\mathrm{F}}\ln\left(\frac{\sqrt{p_{H_2}^{cat,ff^2} + 2\cdot\hat{B}_{cf}}}{p_0}\right) \tag{9}$$

The pressure drop can be calculated accordingly:

$$\Delta p_{H_2}^{cat} = \sqrt{p_{H_2}^{cat,ff^2} + 2\cdot\hat{B}_{cf}} - p_{H_2}^{cat,ff} \tag{10}$$

Both Equations (8) and (9) become the original Equation (5), if $\hat{B} = 0$.

Falsifiability—The presented approach allows for falsification of the standard voltage breakdown analysis. First, the coefficients of determination for $\hat{A}$ and $\hat{B}$ are checked if they differ significantly from one. In this case, the differences between the iR-free cell voltages measured at different cathode pressures cannot be described solely as a consequence of mass transport. This would mean that unconsidered parameters are dominant; in particular, this could constitute a change in temperature, effects that are induced by the absolute pressure, or a general misinterpretation of the reversible cell voltage.

The second check is based on the evolution of $\hat{A}$ as a function of the current density. This parameter contains the anodic kinetic overpotentials and the anodic half-cell potential. The latter must be eliminated, as it can be assumed that the anode is affected by mass transport in a similar way compared to the cathode. Unfortunately, this cannot be evaluated experimentally by the use of a different measurement setup in which the cathode pressure is kept constant while the anode pressure is changed. This is due to the assumption that the reaction kinetics at the anode are significantly influenced by a change in the anode gas and liquid concentrations [28]. However, when considering the eightfold molar mass of oxygen, half of the production rate, and similar transport lengths in comparison with hydrogen, it would result in the following pressure drops for the anode compared with the cathode (see Appendix A, Equations (A6)–(A9)):

$$\Delta p_{mf,O_2}^{an} = \sqrt{2}\cdot\Delta p_{mf,H_2}^{cat} \tag{11}$$

$$\Delta p_{cf,O_2}^{an} = \frac{\Delta p_{cf,H_2}^{cat}}{2} \tag{12}$$

The corresponding pressure drop can be used to correct the originally evaluated fit-parameter, $\hat{A}$, in order to obtain a corrected value, $\hat{A}_{corr}$, where the activity term of the anodic half-cell potential is eliminated (see Appendix A Equation (A10)). This remaining part should only contain the fixed standard cell voltage, U°, and the kinetic overpotentials. The latter is what is assumed to follow the Tafel equation in the standard approach. Accordingly, $\hat{A}_{corr}(\log(i))$ should indicate a linear course over the entire current density range in order to support the applicability of the standard voltage breakdown analysis. Here, some deviation must be tolerated because of the uncertainty in determining the mass transport resistance of the anode catalyst layer. However, a linear fit of $\hat{A}_{corr}(\log(i))$ must not result in low coefficients of determination but the Tafel character must be obvious in principle. If this were not the result, constant Butler–Volmer kinetics over the entire current density range would be highly questionable. Accordingly, unconsidered effects that develop with the current density would become relevant, such as temperature development or changes in the electro-osmotic drag [29] or those resulting from the expanded structure of the catalyst layer [13]. In addition, it could provide an indication of the effects of the unconsidered kinetics of the hydrogen evolution reaction or indicate a change in the kinetics of the oxygen evolution reaction.

A third checking option results from the course of the current density dependency of parameter $\hat{B} = \hat{B}(i)$. Regarding Equations (8) and (9), $\hat{B}$ should contain the law of mass transport and link it to the polarization curve differences when changing the cathode pressure. Consequently, when the current density changes, the function $\Delta p_{H_2}^{cat}(\hat{B}(i))$ should also be in accordance with mass transport laws (see Equations (3) and (4)). Otherwise, the results would contradict each other and the model would be falsified. On the other hand, if a precise determination of $\hat{B}$ is possible and $\Delta p_{H_2}^{cat}(i, \hat{B})$ behaves as expected, the model withstands its falsification.

It should be repeated that the absence of falsification does not constitute a verification of the model, but merely a validation of its use within descriptive rather than predictive analyses.

3. Demonstration

Measurement data—Measurement data published previously by Martin et al. were used for demonstrating the approach [30]. A 4 cm^2 cell was used for those experiments, with a catalyst loading of 1.05 ± 0.05 mg_{Ir} cm^{-2} and 0.20 ± 0.02 mg_{Pt} cm^{-2}. Bekaert titanium 350 μm and Freudenberg H23I2 were used as the anode and cathode porous transport layers. The compression of the active area was varied between 10 and 85 μm. The polarization curves were measured between 0 and 3.6 A cm^{-2} at 80 °C and different cathode pressure levels of 1, 5, 10, and 15 bar while the anode pressure was kept atmospheric. That study investigated the effect of cell compression on the hydrogen crossover; therefore, the cell preparation was assembled in a very controlled manner, which is why the data seem viable to use in this study. In the hope of suppressing the effects of mechanically introduced stochastics in the functional layers as much as possible, the recorded measurement data were used for a cell compression of 10 μm. More details about the experimental part can be viewed in the original publication [30], while the results are summarized in the next paragraph.

The results of the original study showed qualitatively that the iR-free cell voltage increased with the applied cathode pressure as expected, due to Equation (5) (see Figure 1a). These curves were corrected by the logarithmic term of the reversible cell voltage and illustrated a roughly semi-logarithmic behavior, while the curves came close to each other in the low current density range (see Figure 1b).

In contrast to the uncorrected data, where the overpotential increases with higher cathode pressure, the corrected data showed the opposite behavior, indicating the highest overpotential for the curves evaluated at the lowest cathode pressure. Furthermore, linear fits between 0.01 and 0.1 A cm^{-2} of the semi-logarithmic plotted curves indicated good agreement with the expected Tafel kinetic behavior ($R^2 > 99.7\%$). When extrapolating the trend to high current densities, significant shifts from the straight line occurred at several hundred mA cm^{-2}. These deviations were attributed to the onset of mass transport resistance, which increased up to about 30 mV at 3.6 A cm^{-2} and 0.1 MPa (see Equation (2)).

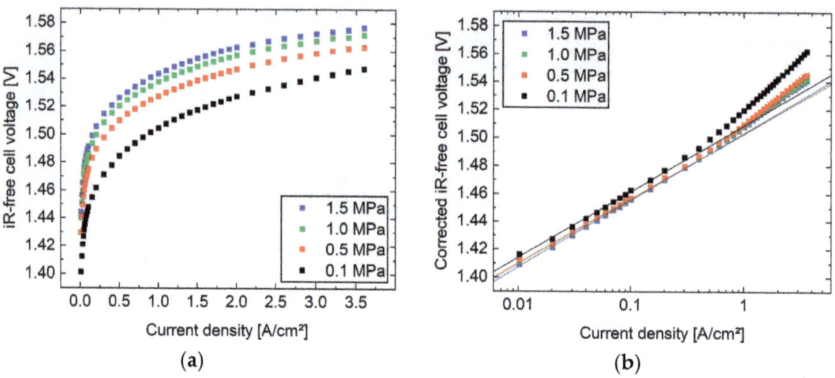

Figure 1. (**a**) Experimental data of iR-free cell voltage and their Tafel-fits before and (**b**) after subtraction of the logarithmic term of the reversible cell voltage according to Equation (5) (figure adapted with permission from Ref. [30]. CC BY 4.0, 2022, The Authors).

However, the assumptions made and the results obtained contradict this, as follows. According to Equation (2), subtracting the reversible cell voltage and assuming a negligible mass transport overpotential below 100 mA cm^{-2} leaves only the activation overpotential. As the cathodic overvoltage is assumed to be negligible but the anode is operated under the same conditions, there should be no offset between the straight lines, as seen in Figure 1b. From this, we can derive that the overpotential due to mass transport resistance differs from zero, and consequently, the reversible cell voltage according to Equation (5) should be replaced with Equation (6), even at a low current density. Alternatively, the model is incomplete, or the data are scattered randomly. All these possibilities can be checked by the presented statistical approach.

Statistical approach—Depending on whether the molecular flow (see Equation (8)) or continuum flow (see Equation (9)) is assumed, the iR-free cell voltage can be fitted accordingly, which results in decent levels of accuracy for all current densities (see Figure 2a,b). Under the assumption of mass transport by molecular flow, the coefficient of determination is above 99.8% and accompanied by a mean residual of less than 0.7 mV between the fit and measurement data (see Figure 2c). More specifically, over half of the fits achieved a coefficient of determination greater than 99.97% and a mean residual of less than 0.3 mV, indicating a near perfect fit of the data. Similar but slightly worse results were received assuming a continuum flow with $R^2 > 98.7\%$ and mean residuals of between 0.3 and 1.3 mV (see Figure 2d). In both cases, values below 0.03 A cm^{-2} and above 2 A cm^{-2} exhibit slightly worse accuracies, potentially indicating boundary effects on the measurement. According to the literature, the gas transport becomes ever more hindered by gas bubbles [31], which relates to the pore structure of the porous transport layers [32]. However, most likely, the deviations can be attributed to the expanded structure of the catalyst layer and the resulting shifts in the reaction zone, which become particularly dominant with increasing cell voltage [13].

From these results, it can be concluded that the difference between the iR-free cell voltages can be adequately described by the introduced approach across a wide range of current densities. This means that the standard approach of voltage breakdown analysis withstands the first check for falsification. Therefore, it is suggested that the unconsidered effects that result from hydrogen partial pressure changes are insignificant compared to the considered ones.

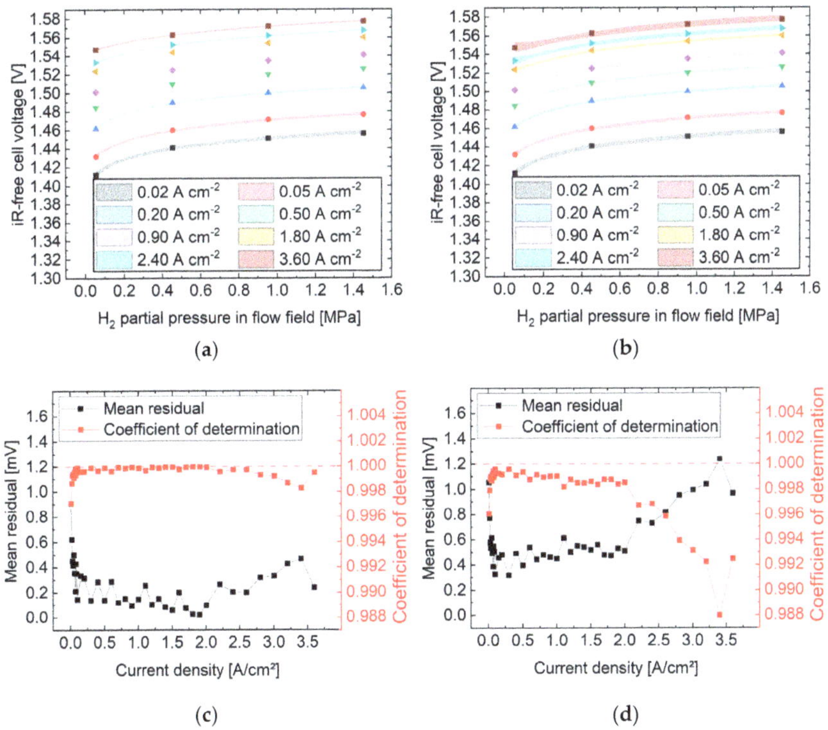

Figure 2. Fits of the iR-free cell voltage as a function of the hydrogen partial pressure in the flow field according to (**a**) Equation (8) (molecular flow) and (**b**) Equation (9) (continuum flow). The corresponding coefficients of determination and the mean residuals are illustrated as a function of the current density for both (**c**) the molecular flow and (**d**) the continuum flow.

Despite the near perfect fits, good agreement between the measurement data and fitting is insufficient for a comprehensive evaluation (see falsifiability). Accordingly, fit-parameter $\hat{B} = \hat{B}(i)$ must lead to a pressure increase that follows the considered mass transport laws. When evaluating the assumption of molecular flow as the dominant mass transport mechanism, the parameter develops linearly ($R^2 = 99.2\%$) with the current density (see Figure 3a). Due to its definition (see Equation (8)), it must be proportional to the partial pressure increase, thus fulfilling the required linear behavior between the pressure drop and molar flow (see Equation (3)). The slope of the fit is (38 $\pm$ 1) kPa cm^2 A^{-1}, which leads to a maximum pressure increase of (162 $\pm$ 3) kPa ($\hat{B}(i)$-fit) or (175 $\pm$ 16) kPa ($\hat{B}\left(p_{H_2}^{cat,ff}\right)$-fit), respectively. This pressure increase means permeability due to diffusion in the range of 10^{-12} mol (m s Pa)$^{-1}$, which is two orders of magnitude higher than the Nafion permeability to hydrogen (see Appendix A Equation (A8)) [33]. Aside from this, a (22 $\pm$ 1) kPa offset is evaluated, which contradicts the proportionality expected. However, three reasons for this can be suggested. First, it could indicate a capillary pressure that must be exceeded in order to enable gas transport through open voids in the catalyst layer [17]. Second, the MEA could consume more heat than is supplied, meaning that the temperature would be lower than actually measured and controlled as the cell temperature [34]. This would indicate that the MEA is operated below its thermo-balanced voltage [35]. However, this pressure increase would correspond to a temperature drop of about 13 °C and should result in a decrease in the ionic conductivity of the membrane of about 5 mOhm cm^2 [36], which in turn is difficult to resolve experimentally. A third reason, and probably the predominant effect, is the test station specific pressure drop, which exists between the cell outlet and the test station outlet. In addition to the offset, the evaluation also reveals that the determination of $\hat{B}_{mf}$ becomes less accurate above 2.6 A cm^{-2} (see also Figure 2c), as already mentioned for the overall fit quality in the previous paragraph.

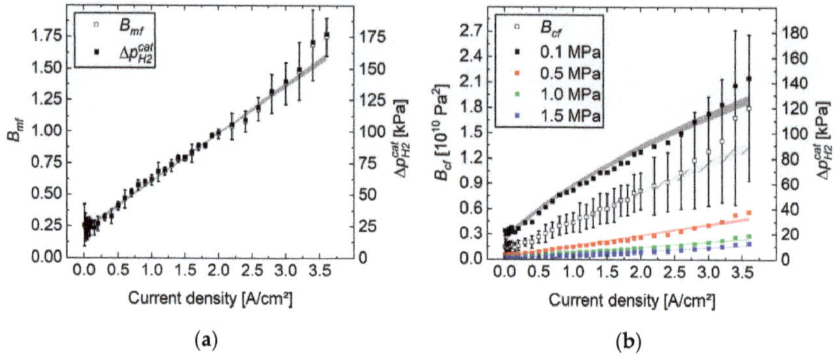

Figure 3. The fit-parameter $\hat{B}$ and the hydrogen partial pressure increase as functions of the current density for (**a**) the molecular flow and (**b**) the continuum flow. In the case of continuum flow, the pressure increase depends on the absolute pressure at the cathode. The confidence bands are then fitted according to mass transport laws.

In comparison, if a continuum flow is presumed as the predominant mass transfer mechanism, the situation above 2 A cm^{-2} is even worse at any current density (see Figure 3b). Even if the coefficient of determination is still high at R^2 = 97.9% for the fit, the confidence level of individual data points is fairly low. However, in calculating the pressure drop, the expected non-linear behavior exhibits 0.1 MPa. In the cases of higher flow field pressures, $\Delta p^{cat}_{H_2} \ll p^{cat,ff}$ is found; therefore, the current density dependence approximates a linear pressure increase. Using these slopes, the intrinsic permeability according to Darcy's law can be estimated, which results in values on the order of 10^{-18} m^2 (see Appendix A Equation (A9)). This is much lower than the permeability of porous transport layers [37], and at a similar order of magnitude that is reported for Nafion membranes [38]. Therefore, it must be assumed that the gas permeates through a void-free catalyst layer [17].

In summary, not only can $\hat{B} = \hat{B}\left(p^{cat,ff}_{H_2}\right)$ be evaluated with sufficient accuracy but $\hat{B} = \hat{B}(i)$ also indicates the expected shape according to mass transport laws. Although it is not fully clear from these results how mass transport works within the electrode, the values of the resistance-determining constants are within the expected ranges. Accordingly, the standard approach cannot be falsified.

Based on the results obtained so far, the course of the fit parameter $\hat{A}$ is discussed next. This parameter indicates the kinetic overpotential, and, according to the standard approach, it would be expected to follow the Tafel equation. In contrast, it is assumed that the expanded structure of the electrode will lead to a bending of the curve. Surprisingly, the semi-logarithmic illustrations indicate good agreements for both mechanisms with the Tafel equation. Moreover, after correcting these values, $\hat{A}_{mf,corr}$ and $\hat{A}_{cf,corr}$ follow near perfect linear trends, whereas evaluating under the assumption of molecular flow (R^2 = 99.9%) is slightly better than assuming continuum flow (R^2 = 99.8%) (see Figure 4a,b). Therefore, the assumption that the kinetics of PEM electrolyzers follow Butler–Volmer kinetics across the entire current density range cannot be refuted but is verified. The detailed analysis reveals an average deviation of 0.9 mV (molecular flow) or 1.2 mV (continuum flow) and the maximum deviation is 2.2 mV, excluding the outliers at 10 mA cm^{-2}; the origin may be related to the experiments themselves. However, it can be seen that the residuals are distributed non-randomly, which indicates additional unrevealed processes. These should be more than one, as two inflection points are seen in the residual curve. It is suggested that these factors may be related to temperature management, the agglomerate structure of the electrode, the neglected cathodic activation overpotential, electro-osmotic drag, as well as gas bubble formation and water oversaturation inside the cell. In any case, these contributions surprisingly seem to not play a significant role in the typical evaluation. Furthermore, the evaluated slopes that result from fitting $\hat{A}$ are (50.1 ± 0.4) mV dec^{-1} (molecular flow) and (51.4 ± 0.8) mV dec^{-1} (continuum flow). These values are about 5–10% higher than the values that were obtained from the standard approach [30].

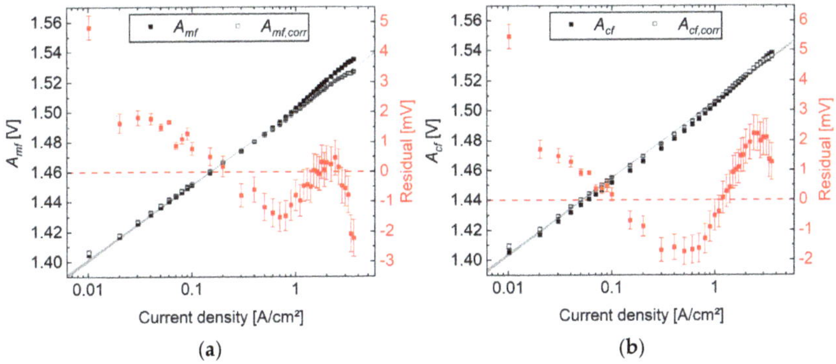

Figure 4. Original and corrected fit-parameter $\hat{A}$ as function of the current density when assuming (**a**) molecular flow and (**b**) continuum flow. The residuals illustrate the difference between the corrected values and its linear fit in a semi-logarithmic scale.

Further remarks—As demonstrated, no obvious contradiction can be produced, and so the standard approach is partially validated. However, it must be taken into account that the approach is only a purely mathematical fitting procedure and is therefore neither a prediction tool nor a model in the strict sense.

In principle, the approach can also be viewed as an option for describing polarization curves. As can be seen here, the method leads to a significant improvement regarding the evaluation accuracy. This gain results from the additional known variable that can be measured without inducing a dominant change in the rest of the system. Accordingly, the degree of under-determination in Equation (2) is reduced by one and additional information can be extracted from the measurement data, such as the permeability coefficient. Arguing in the opposite direction, incorporating additional variables or mechanisms into the model would not enhance its predictive power, as this would increase the model's degrees of freedom, automatically resulting in higher coefficients of determination. However, since the model already explains nearly 100% of the data, the rationale for adding extra mechanisms cannot be justified without having very precise descriptions of these mechanisms, which are not accessible. Aside from all of its limitations, the approach checks the kinetic behavior of an electrolysis cell, as well as demonstrates that mass transport resistances are not negligible at any current density. In addition to this, more insights into the gas transport mechanism are provided. As the demonstration is limited to four pressure variations in this study, it can be suggested that deviations occur if an extended pressure scan is performed. This allows even more insights into the processes and can help to illuminate the residual curve (see Figure 4).

In this study, measurement data were used where the cell compression is low. With increasing mechanical compression, the determination of $\hat{A}_{corr}$ is of a similar quality, whereas the deviation in $\hat{B}$ becomes slightly higher but is still acceptable. This indicates that the evaluation procedure can be used for different kinds of measurement setups. In spite of this, the results show the expected behavior, such as a steeper slope of the additional pressure in the electrode above 2 A cm^{-2} (see Figure A1). It is hypothesized that this is due to the mechanical deformation of the layers, which broadens the local property distribution of the reaction, as well as the gas, fluid, and heat transport. Accordingly, a change in the current density should be correlated to non-linear changes in these processes that cannot be resolved by a model that is based on integral information. Because of this integral character, the model only describes effective properties under which water splitting and hydrogen production take place. In particular, the reaction penetration length tends to shift as a function of the current density, consequently leading to pressure curves that contradict gas transport laws [39]. Due to the lack of spatial information, it is also not possible to use the data to explain the hydrogen permeation through the membrane based on the pressure drop determined by this approach (see Figure A2). In order to do this, the actual pressure at the interface between the catalyst layer and membrane must be known and not the calculated effective pressure. In summary, models that take local properties into account are necessary in order to explain polarization curves.

When concluding these further remarks, it is evident that the approach can help describe polarization curves with improved precision. It can also help to generate polarization curves that are closer to measured curves by incorporating a linear pressure function into the equation of the reversible cell voltage (see Equation (8)). However, this should not be confused with using the approach as a prediction tool. Since only integral, but no spatial, information of the cell potential is available, physical or electrochemical processes cannot be extracted from these curves. All obtained data and parameters describe global information. In particular, at high current densities, slight deviations occur, most likely due to the expanded structure of the catalyst layers. Finally, it is essential that other research groups check the validity of the presented model by applying it to their experimental data. It is suggested that dedicated pressure scans be measured, varying the pressure at a constant current density, to evaluate whether a fit according to Equation (8) is reliable. It can be assumed that the presented model can be implemented into other voltage breakdown analyses, with its strength particularly evident in the low- to mid-current density range.

4. Conclusions

When using the standard method for voltage breakdown analysis of PEM electrolyzers, two basic assumptions are typically made. First, the electrochemical reaction follows Butler–Volmer kinetics over the entire current density range; second, the low current density regime is unaffected by mass transport. In this study, an alternative approach to voltage breakdown analysis was introduced, which grants partial falsifiability of the standard approach.

The new approach is based on implementing the pressure increase due to gas transport laws into the equation of the reversible cell voltage. Here, both the molecular and continuum flow can be assumed to be considerable mechanisms. When applying this model to measurements at the same current density but different cathode pressures, the pressure drop between the catalyst layer and flow field can be fitted. As a result, three criteria can be formulated that potentially allow for checking the applicability of the standard approach. However, the demonstration does not indicate a contradiction: the corresponding coefficient of determination approximates unity for all current densities, the change in the pressure drop with different current densities follows the assumed gas transport laws, and the course of the iR-free overpotential cannot prove that assuming Butler–Volmer kinetics is false. Accordingly, the standard approach is validated but not verified.

This novel evaluation method clarifies the residuals for more than 90% of the residuals observed in the standard approach, ultimately reducing the residuals to less than 2.2 mV. It indicates a voltage increase with logarithmic current density, which is 5–10% higher than typically evaluated, due to a pressure increase of about 38 kPa cm^2 A^{-1} inside the functional layer system. This value corresponds to a permeability on the order of 10^{-18} m^2 or 10^{-12} mol (m s Pa)$^{-1}$, which falls between the values characteristic of PTLs and membranes. This suggests a predominant transport resistance in the catalyst layer, although this conclusion remains speculative. However, more ideas must be developed regarding how to reduce the number of unknown variables in the base equation. Additionally, other research groups are encouraged to test this new method on their data.

Author Contributions: Conceptualization, F.S. and W.L.; Methodology, W.L.; Validation, F.S.; Formal analysis, F.S.; Investigation, F.S.; Writing—original draft, F.S.; Writing—review & editing, F.S. and W.L.; Visualization, F.S.; Supervision, W.L. All authors have read and agreed to the published version of the manuscript.

Funding: Funded by the Deutsche Forschungsgemeinschaft (DFG, German Research Foundation)—491111487.

Data Availability Statement: The original contributions presented in the study are included in the article, further inquiries can be directed to the corresponding author.

Conflicts of Interest: The authors declare no conflict of interest.

Nomenclature

Parameter	Symbol	Unit
Latin		
Activity	a	
Concentration	c	mol l^{-1}
Diffusion coefficient	D	m^2 s^{-1}
Faraday constant	F	C mol^{-1}
Current density	i	A cm^{-2}
Intrinsic permeability	K	m^2
Length	l	m
Molecular weight	M	kg mol^{-1}
Amount of substance	n	mol
Permeability	P	mol $(m\ s\ Pa)^{-1}$
Pressure	p	Pa
Gas constant	R	J $(mol\ K)^{-1}$
Cell temperature	T	K
Cell voltage	U	V
Volume	V	m^3
Number of electrons	z	
Greek		
Overpotential	η	V
Dynamic viscosity	μ	Pa s
Indices		
Anode	an	
Cathode	cat	
Cell	cell	
Continuum flow	cf	
Corrected	corr	
Electrode	el	
Flow field	ff	
Substance	i	
iR-free	iR-free	
Kinetic	kin	
Molecular flow	mf	
Mass transport	mt	
Resistance	res	
Reversible	rev	

Appendix A

Implementation of mass transport into the reversible cell voltage—Starting with Equation (A4), it is

$$\dot{n}_i \sim \overline{p_i}\cdot\nabla p \text{with } \overline{p_i} = p_i^{ff} + \frac{\Delta p_i^{cl}}{2} \text{ and } \nabla p \sim \Delta p_i^{cl} \quad \text{(A1)}$$

When the current density is the same, it is

$$\dot{n}_i \sim \left(p_i^{ff} + \frac{\Delta p_i^{cl}}{2}\right)\Delta p_i^{cl} = const. = c \quad \text{(A2)}$$

Here, c is a constant that describe a specific mass transport resistance coefficient, which can be converted into

$$\Delta p_i^{cl} = \sqrt{p_i^{ff^2} + 2c} - p_i^{ff} \quad \text{(A3)}$$

When assuming continuum flow and half of the production rate of oxygen compared with hydrogen, Equation (A1) becomes

$$\dot{n}_i \sim 2\overline{p_i}\cdot\nabla p \quad \text{(A4)}$$

This leads to a pressure drop of

$$\Delta p_i^{cl} = \sqrt{p_i^{ff^2} + c} - p_i^{ff} \quad \text{(A5)}$$

In the case of molecular flow, the diffusion coefficient is proportional to the square root of the inverse molar mass:

$$\frac{D_{O_2}}{D_{H_2}} \sim \frac{\sqrt{\frac{1}{M_{O_2}}}}{\sqrt{\frac{1}{M_{H_2}}}} = \sqrt{\frac{M_{H_2}}{M_{O_2}}} = \frac{1}{2\sqrt{2}} \quad \text{(A6)}$$

As the oxygen production rate is only half the value of the hydrogen production rate, only half the mass must be transported, which means that the pressure gradient ratio is

$$D_{K,O_2}\cdot\nabla p_{O_2}^{el} = D_{H_2}\cdot\frac{\nabla p_{H_2}^{el}}{2} \Leftrightarrow \nabla p_{O_2}^{el} = \frac{D_{H_2}}{D_{O_2}}\cdot\frac{\nabla p_{H_2}^{el}}{2} = \sqrt{2}\cdot\nabla p_{H_2}^{el} \quad \text{(A7)}$$

Mass transfer coefficient—Molecular and continuum flow:

$$P = \frac{\dot{n}\Delta l}{\Delta p} = \frac{i}{\Delta p}\frac{\Delta l}{2\mathrm{F}} \approx \left(3.8\frac{\mathrm{Pa\,s\,m^2}}{\mathrm{C}}\right)^{-1}\cdot\frac{10^{-6}}{2\cdot 10^5}\frac{\mathrm{m\,mol}}{\mathrm{C}} \approx 1\cdot 10^{-12}\frac{\mathrm{mol}}{\mathrm{Pa\,m\,s}} \quad \text{(A8)}$$

$$K = \frac{\dot{V}}{\mathrm{A}}\frac{\mu\Delta l}{\Delta p} = \frac{i}{\Delta p}\frac{\mathrm{RT}}{2\mathrm{F}}\frac{\mu\Delta l}{\overline{p}} \approx \left(1\frac{\mathrm{Pa\,s\,m^2}}{\mathrm{C}}\right)^{-1}\cdot 0.0152\frac{\mathrm{J}}{\mathrm{C}}\cdot\frac{10^{-5}\cdot 10^{-6}}{10^5}\frac{\mathrm{Pa\,s\,m}}{\mathrm{Pa}} \approx 1\cdot 10^{-18}\mathrm{m}^2 \quad \text{(A9)}$$

Correction of $\hat{A}$—$\hat{A}_{corr}$ is the value of $\hat{A}$ corrected by the anodic activity term:

$$\hat{A}_{corr} = \hat{A} - \frac{\mathrm{RT}}{4\mathrm{F}}\ln\left(\frac{p_{O_2}^{ff,an} + \Delta p_{O_2}^{el}}{p_0}\right) \quad \text{(A10)}$$

High compressed cell—Results if the cell is compressed by 85 μm rather than 10 μm:

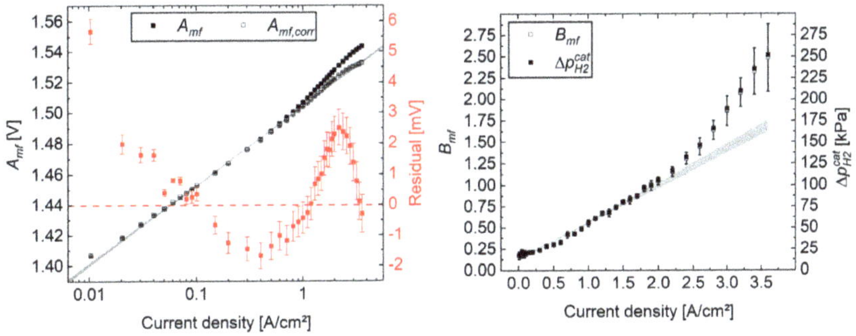

(a) (b)

Figure A1. (**a**) $\hat{A}$ (black dots) and the residuals (red dots) and (**b**) $\hat{B}$ (black dots) and the pressure increase (error band) as functions of the current density for an 8.5-fold cell compression of 85 μm.

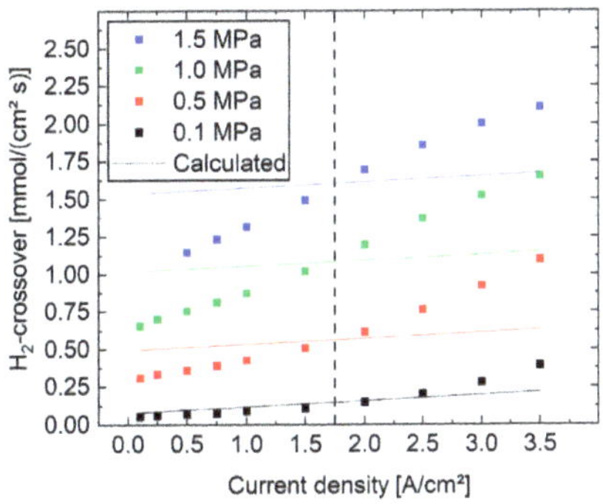

Figure A2. Hydrogen permeation across the membrane as a function of the current density. The values measured below 1.75 A cm^{-2} were higher than expected; the values above were lower.

References

1. Frensch, S.H.; Olesen, A.C.; Araya, S.S.; Kær, S.K. Model-supported characterization of a PEM water electrolysis cell for the effect of compression. *Electrochimica Acta* **2018**, *263*, 228–236. [CrossRef]
2. Scheepers, F.; Stähler, M.; Stähler, A.; Rauls, E.; Müller, M.; Carmo, M.; Lehnert, W. Improving the Efficiency of PEM Electrolyzers through Membrane-Specific Pressure Optimization. *Energies* **2020**, *13*, 612. [CrossRef]
3. Gerhardt, M.R.; Pant, L.M.; Bui, J.C.M.; Crothers, A.R.; Ehlinger, V.M.; Fornaciari, J.C.; Liu, J.; Weber, A.Z. Method—Practices and Pitfalls in Voltage Breakdown Analysis of Electrochemical Energy-Conversion Systems. *J. Electrochem. Soc.* **2021**, *168*, 074503. [CrossRef]
4. Bernt, M.; Siebel, A.; Gasteiger, H.A. Analysis of Voltage Losses in PEM Water Electrolyzers with Low Platinum Group Metal Loadings. *J. Electrochem. Soc.* **2018**, *165*, F305–F314. [CrossRef]
5. Choi, P.; Bessarabov, D.G.; Datta, R. A simple model for solid polymer electrolyte (SPE) water electrolysis. *Solid State Ionics* **2004**, *175*, 535–539. [CrossRef]
6. Järvinen, L.; Puranen, P.; Kosonen, A.; Ruuskanen, V.; Ahola, J.; Kauranen, P.; Hehemann, M. Automized parametrization of PEM and alkaline water electrolyzer polarisation curves. *Int. J. Hydrogen Energy* **2022**, *47*, 31985–32003. [CrossRef]
7. Suermann, M.; Gimpel, T.; Bühre, L.V.; Schade, W.; Bensmann, B.; Hanke-Rauschenbach, R. Femtosecond laser-induced surface structuring of the porous transport layers in proton exchange membrane water electrolysis. *J. Mater. Chem. A* **2020**, *8*, 4898–4910. [CrossRef]
8. Schuler, T.; Schmidt, T.J.; Büchi, F.N. Polymer Electrolyte Water Electrolysis: Correlating Performance and Porous Transport Layer Structure: Part II. Electrochemical Performance Analysis. *J. Electrochem. Soc.* **2019**, *166*, F555–F565. [CrossRef]
9. Ojong, E.T.; Kwan, J.T.H.; Nouri-Khorasani, A.; Bonakdarpour, A.; Wilkinson, D.P.; Smolinka, T. Development of an experimentally validated semi-empirical fully-coupled performance model of a PEM electrolysis cell with a 3-D structured porous transport layer. *Int. J. Hydrogen Energy* **2017**, *42*, 25831–25847. [CrossRef]
10. Verdin, B.; Fouda-Onana, F.; Germe, S.; Serre, G.; Jacques, P.; Millet, P. Operando current mapping on PEM water electrolysis cells. Influence of mechanical stress. *Int. J. Hydrogen Energy* **2017**, *42*, 25848–25859. [CrossRef]

11. Seip, T.; Lee, J.K.; Ge, M.; Lee, W.-K.; Shaigan, N.; Dinu, M.; Fatih, K.; Bazylak, A. Stochastic generation of electrolyzer anode catalyst layers. *Electrochimica Acta* **2023**, *462*, 142701. [CrossRef]
12. Lee, J.K.; Lee, C.; Fahy, K.F.; Zhao, B.; LaManna, J.M.; Baltic, E.; Jacobson, D.L.; Hussey, D.S.; Bazylak, A. Critical Current Density as a Performance Indicator for Gas-Evolving Electrochemical Devices. *Cell Rep. Phys. Sci.* **2020**, *1*, 100147. [CrossRef]
13. Padgett, E.; Bender, G.; Huag, A.; Lewinski, K.; Sun, F.; Yu, H.; Cullen, D.; Steinbach, A.; Alia, S.M. Catalyst Layer Resistance and Utilization in PEM Electrolysis. *J. Electrochem. Soc.* **2023**, *170*, acee25. [CrossRef]
14. Aubras, F.; Deseure, J.; Kadjo, J.-J.; Dedigama, I.; Majasan, J.; Grondin-Perez, B.; Chabriat, J.-P.; Brett, D. Two-dimensional model of low-pressure PEM electrolyser: Two-phase flow regime, electrochemical modelling and experimental validation. *Int. J. Hydrogen Energy* **2017**, *42*, 26203–26216. [CrossRef]
15. Dickinson, E.J.F.; Hinds, G. The Butler-Volmer Equation for Polymer Electrolyte Membrane Fuel Cell (PEMFC) Electrode Kinetics: A Critical Discussion. *J. Electrochem. Soc.* **2019**, *166*, F221–F231. [CrossRef]
16. Alia, S.M.; Anderson, G.C. Iridium Oxygen Evolution Activity and Durability Baselines in Rotating Disk Electrode Half-Cells. *J. Electrochem. Soc.* **2019**, *166*, F282–F294. [CrossRef]
17. Bernt, M.; Gasteiger, H.A. Influence of Ionomer Content in IrO_2/TiO_2 Electrodes on PEM Water Electrolyzer Performance. *J. Electrochem. Soc.* **2016**, *163*, F3179–F3189. [CrossRef]
18. García-Salaberri, P.A.; Sánchez-Ramos, A.; Das, P.K. On the optimal cathode catalyst layer for polymer electrolyte fuel cells: Bimodal pore size distributions with functionalized microstructures. *Front. Energy Res.* **2022**, *10*, 1058913. [CrossRef]
19. Abdin, Z.; Webb, C.; Gray, E. Modelling and simulation of a proton exchange membrane (PEM) electrolyser cell. *Int. J. Hydrogen Energy* **2015**, *40*, 13243–13257. [CrossRef]
20. Garcia-Navarro, J.; Schulze, M.; Friedrich, K.A. Understanding the Role of Water Flow and the Porous Transport Layer on the Performance of Proton Exchange Membrane Water Electrolyzers. *ACS Sustain. Chem. Eng.* **2019**, *7*, 1600–1610. [CrossRef]
21. Thorstenson, D.C.; Pollock, D.W. Gas transport in unsaturated porous media: The adequacy of Fick's law. *Rev. Geophys.* **1989**, *27*, 61–78. [CrossRef]
22. Islam, M.A.; Ulbricht, M. A New Interpretation of Gas Viscosity for Flow through Micro-Capillaries and Pores. *Small* **2023**, *19*, 2205827. [CrossRef] [PubMed]
23. Schalenbach, M.; Carmo, M.; Fritz, D.L.; Mergel, J.; Stolten, D. Pressurized PEM water electrolysis: Efficiency and gas crossover. *Int. J. Hydrogen Energy* **2013**, *38*, 14921–14933; Corrigendum in *Int. J. Hydrogen Energy* **2016**, *41*, 729–732. [CrossRef]
24. van Soestbergen, M. Frumkin-Butler-Volmer theory and mass transfer in electrochemical cells. *Russ. J. Electrochem.* **2012**, *48*, 570–579. [CrossRef]
25. Nordio, M.; Rizzi, F.; Manzolini, G.; Mulder, M.; Raymakers, L.; Annaland, M.V.S.; Gallucci, F. Experimental and modelling study of an electrochemical hydrogen compressor. *Chem. Eng. J.* **2019**, *369*, 432–442. [CrossRef]
26. Kucernak, A.R.J.; Zalitis, C.M. General Models for the Electrochemical Hydrogen Oxidation and Hydrogen Evolution Reactions: Theoretical Derivation and Experimental Results under Near Mass-Transport Free Conditions. *J. Phys. Chem. C* **2016**, *120*, 10721–10745. [CrossRef]
27. Wang, X.; Ahluwalia, R.K.; Steinbach, A.J. Kinetics of Hydrogen Oxidation and Hydrogen Evolution Reactions on Nanostructured Thin-Film Platinum Alloy Catalyst. *J. Electrochem. Soc.* **2013**, *160*, F251–F261. [CrossRef]
28. Lang, C.; Li, J.; Yang, K.R.; Wang, Y.; He, D.; Thorne, J.E.; Croslow, S.; Dong, Q.; Zhao, Y.; Prostko, G.; et al. Observation of a potential-dependent switch of water-oxidation mechanism on Co-oxide-based catalysts. *Chem* **2021**, *7*, 2101–2117. [CrossRef]
29. Medina, P.; Santarelli, M. Analysis of water transport in a high-pressure PEM electrolyzer. *Int. J. Hydrogen Energy* **2010**, *35*, 5173–5186. [CrossRef]
30. Martin, A.; Trinke, P.; Stähler, M.; Stähler, A.; Scheepers, F.; Bensmann, B.; Carmo, M.; Lehnert, W.; Hanke-Rauschenbach, R. The Effect of Cell Compression and Cathode Pressure on Hydrogen Crossover in PEM Water Electrolysis. *J. Electrochem. Soc.* **2022**, *169*, 014502. [CrossRef]
31. Diaz, D.F.R.; Valenzuela, E.; Wang, Y. A component-level model of polymer electrolyte membrane electrolysis cells for hydrogen production. *Appl. Energy* **2022**, *321*, 119398. [CrossRef]
32. Pushkarev, A.; Pushkareva, I.; Solovyev, M.; Prokop, M.; Bystron, T.; Rajagopalan, S.; Bouzek, K.; Grigoriev, S. On the influence of porous transport layers parameters on the performances of polymer electrolyte membrane water electrolysis cells. *Electrochimica Acta* **2021**, *399*, 139436. [CrossRef]
33. Schalenbach, M.; Hoefner, T.; Paciok, P.; Carmo, M.; Lueke, W.; Stolten, D. Gas Permeation through Nafion. Part 1: Measurements. *J. Phys. Chem. C* **2015**, *119*, 25145–25155. [CrossRef]
34. Möckl, M.; Bernt, M.; Schröter, J.; Jossen, A. Proton exchange membrane water electrolysis at high current densities: Investigation of thermal limitations. *Int. J. Hydrogen Energy* **2019**, *45*, 1417–1428. [CrossRef]
35. Schalenbach, M.; Tjarks, G.; Carmo, M.; Lueke, W.; Mueller, M.; Stolten, D. Acidic or Alkaline? Towards a New Perspective on the Efficiency of Water Electrolysis. *J. Electrochem. Soc.* **2016**, *163*, F3197–F3208. [CrossRef]
36. Yadav, R.; Fedkiw, P.S. Analysis of EIS Technique and Nafion 117 Conductivity as a Function of Temperature and Relative Humidity. *J. Electrochem. Soc.* **2012**, *159*, B340–B346. [CrossRef]
37. Lee, J.K.; Bazylak, A. Optimizing Porous Transport Layer Design Parameters via Stochastic Pore Network Modelling: Reactant Transport and Interfacial Contact Considerations. *J. Electrochem. Soc.* **2020**, *167*, 013541. [CrossRef]

38. Trinke, P.; Bensmann, B.; Reichstein, S.; Hanke-Rauschenbach, R.; Sundmacher, K. Hydrogen Permeation in PEM Electrolyzer Cells Operated at Asymmetric Pressure Conditions. *J. Electrochem. Soc.* **2016**, *163*, F3164–F3170. [CrossRef]
39. Kulikovsky, A.A. PEM Fuel Cell Impedance at Open Circuit. *J. Electrochem. Soc.* **2016**, *163*, F319–F326. [CrossRef]

energies

MDPI

Review

Microwave-Assisted Pyrolysis of Forest Biomass

I. Fernández [1,*], S. F. Pérez [1], J. Fernández-Ferreras [2] and T. Llano [2]

[1] Department of Electrical and Energy Engineering, ETSIIyT, Universidad de Cantabria, Avda. de los Castros 46, 39005 Santander, Spain; severiano.perez@unican.es

[2] Department of Chemistry and Process & Resources Engineering, ETSIIyT, Universidad de Cantabria, Avda. de los Castros 46, 39005 Santander, Spain; josefa.fernandez@unican.es (J.F.-F.); tamara.llano@unican.es (T.L.)

* Correspondence: fernandei@unican.es

Abstract: The global increase in energy consumption, driven by population growth and improved living standards, has led to a heavy reliance on fossil fuels, causing significant environmental concerns. This has prompted a shift toward sustainable energy sources, with biomass, especially lignocellulosic forest biomass, emerging as a key alternative due to its abundance and carbon-neutral potential. Microwave-assisted pyrolysis (MAP) is an efficient method for converting forest biomass into valuable bioproducts and bioenergy with reduced energy use. This review introduces biomass types, focusing on forest biomass and its role in global energy production. It compares MAP to conventional pyrolysis, highlighting the benefits of rapid, uniform heating and improved product yields. Key operational conditions, such as temperature, microwave power, biomass size, and catalyst ratios, are discussed in relation to their impact on product quality and yield. Despite its advantages, MAP faces challenges, particularly in temperature control, which can affect bio-oil yield and quality. High temperatures may cause unwanted secondary reactions, while low temperatures can lead to incomplete decomposition. Research into biomass dielectric properties and process modeling is essential in order to optimize MAP and scale it up for industrial use. Addressing bio-oil quality issues through catalytic upgrading is also critical for broader adoption.

Keywords: microwave-assisted pyrolysis; forest biomass; feedstocks; absorbers; temperature; power; catalyst

1. Introduction

The substantial population increase and improved quality of life have significantly increased energy consumption [1]. Historically, fossil fuels have met around 80% of the world's energy demand [2,3], generating their intensive usage, which has raised carbon emissions [4] and amplified global warming [5]. Consequently, the reliance on fossil fuels not only depletes natural resources [6], but also possesses severe environmental concerns. Major economies have pledged to achieve net-zero emissions by 2050 or 2060, facing the challenge of supporting advanced technologies like microwave-assisted pyrolysis, at minimal costs to produce economic biofuels and high value products. Although challenging, achieving net-zero emissions will bring economic and social benefits [7]. Additionally, the energy crisis has accelerated the shift towards a clean energy economy and increased investments in transforming the energy system [3]. For these reasons, ensuring energy security with a sustainable approach and reducing the dependency on fossil fuels is essential [8].

Interest in discovering new, affordable energy sources includes alternatives such as solar, wind, hydro, geothermal and biomass. One of the renewable resources which has the highest potential is biomass due to its energy recovery possibilities, and its valuable components [9]. Integrating biotechnological processes and biorefinery methods for its valorization are considered key aspects in the production of bioproducts (chemicals and biofuels) and bioenergy (electrical energy and heat) [10,11]. Lignocellulosic biomass, composed of cellulose, hemicellulose, and lignin [12], is the most sustainable, abundant, and cost-effective form of biomass [13] which has emerged as one of the most promising feedstocks. This biomass that refers to plant biomass [14] has an enormous potential to generate sustainable products with zero net carbon emissions. This kind of biomass derives from a wide range of sources such as municipal solid wastes, industrial wastes, agricultural residues, and forest biomass [15]. In recent years, interest in the use of lignocellulosic agricultural crops and forest residues as feedstocks for biofuels and bioproducts due to their abundant availability and non-food nature has grown [16,17]. Forest biomass (branches, trees, bark, needles, roots, trunks, leaves, etc.) is the primary biomass source in Europe that does not compete with the food supply, and its high demand as both a material and energy source has generated a competition between industries and the need of circularity improvement and resource efficiency to enhance sustainable development [18,19].

A clean and efficient use of lignocellulosic biomass to produce value-added products or energy requires the development of suitable conversion techniques [20]. These techniques have being investigated for years, with pyrolysis as one of them [21]. Pyrolysis is a thermochemical process that takes place in an inert atmosphere or with a low oxygen concentration, particularly appealing due to its low pollutant emissions and the variety of products it can produce [8,22]. Pyrolysis products include non-condensable gases, bio-oil, and biochar [23,24]. Biomass pyrolysis can be divided into different categories (flash, fast, and slow) depending on the heating rate and residence time; each category primarily aims to maximize either non-condensable gases, bio-oil, or biochar yields [25]. Slow heating rates (slow pyrolysis) generates bio-oil with a high water content. In contrast, high heating rates lead to a higher bio-oil yield and better biochar quality [26]. It has been detected that products' specificity requires more efficient or targeted heating methods and improvements [27]. For instance, microwave-assisted pyrolysis has appeared as a different method of heating which is easily operated through instant on/off control, improving product quality and yield [5,28]. This technique seems to be interesting due to its lower energy use and process time [29]. Although biomass is generally a poor microwave absorber, its microwave absorption capacity can be improved through inorganic substances or high humidity. The use of microwave absorbers (MWAs) enhances the pyrolysis temperature using low microwave power. MWAs can heat the surrounding biomass which modifies the quality and product yield. The simultaneous use of catalysts and MWAs can adapt the product distribution, increasing the concentration of specific components in non-condensable gases, bio-oil, and biochar, as well as the energy efficiency of MAP [30].

This paper reviews microwave-assisted pyrolysis (MAP), examining its characteristics and the forest biomass capacity to produce value-added products. This article also compares the products derived from different feedstocks and MAP operation conditions (microwave power, absorbers and catalysts use, temperature, residence time, and biomass size), as well as future directions in MAP use.

2. Biomass

Energy security, energy prices, health (emergence of diseases sensitive to global warming, and famine), and environmental concerns (climate change, global biodiversity loss, and soil degradation) [31–35] have turned towards bioenergy in a crucial part of different countries' strategy to reduce the dependence on fossil fuels and their products. Additionally, the low prices of forest products have favored the production of a wide range of high-value bioproducts such as bioenergy, biochemicals, biomaterials, and other marketable products [36] that can substitute similar products from fossil fuels which can help diminish GHG emissions and the dependency on energy imports, bring about efficient waste management, and contribute to bioeconomy development, while reducing the risk of fires [37–41]. On the other hand, biofuels emerge as the most economically viable alternative for replacing fossil fuels [42], especially in the shipping and aviation sectors [39,43,44], as well as a flexible option to electrify areas of the heat sector [31,39]. Biomass-based systems enhance energy accessibility in rural regions, which provides new opportunities for socio-economic development [45].

2.1. Biomass Types

Biomass, often referred to as organic waste, can be categorized into different groups: forest-woody residues, agri-food residues, animal residues, industrial biomass, and municipal solid residues. The management of biomass possesses a significant challenge in numerous countries. Inadequate collection and disposal practices contribute to social, economic, and environmental issues. Therefore, a suitable utilization of biomass is crucial. Since biomass comes from a diverse range of sources, their valorization strategy should be based on the composition, chemical characteristics, source, and quantity of these residues [46,47].

Within this framework, lignocellulosic biomass emerges as a leading resource, being the most common and renewable biomass on Earth [48]. Within this framework, lignocellulosic biomass, whose worldwide yield is around 200 billion tons, of which 8.2 billion tons/year are being used to fulfill society's needs (heat, energy, food, etc.) [49], emerges as a leading resource, being the most common and renewable biomass on Earth [48]. In 2017, approximately 1 billion tons of biomass were obtained in the EU; almost 95% of this biomass supply was lignocellulosic biomass (forestry, grazed biomass, agricultural crops, and their collected wastes) [50]. Moreover, lignocellulosic biomass is attracting increasing attention as a renewable feedstock for various applications (in the energy, food, and chemical industry). It can be converted into bioenergy through thermochemical and biochemical processes and is a key resource for biofuel production in transport. Additionally, it has the potential to replace petroleum-based plastics and petrochemicals, with applications in additive manufacturing, environmental remediation, and medical fields [51]. Softwood, hardwood, and grasses are the main kinds of lignocellulosic biomass. Although all these types of lignocellulosic biomass contain the same materials (cellulose, hemicellulose, and lignin) [52], as well as a variety of minor components (lipids, water, simple sugars, ash, proteins, starches, hydrocarbons, and extractives), depending on the plant species, growth stage, storage circumstances, cultivation conditions, soil minerals, fertilizers, biological type, or origin, the quantities may vary [15,25]. Subtle variations in the composition of softwood, hardwood, and grasses possess noticeable effects on the properties of their resulting high-value products when they are pyrolyzed. Cellulose comprises both high-ordered (crystalline) and low-ordered (amorphous) regions [53], with the ratio between these regions dependent on various factors, as previously explained. It undergoes decomposition within the range of 240–350 °C, yielding levoglucosan and anhydrocellulose. Hemicellulose, a shorter and amorphous polymer composed of various sugars, decomposes at 200–260 °C, resulting in a higher production of volatiles, less tar, and less char than cellulose. Lignin, an aromatic polymer, decomposes at 280–500 °C, yielding phenols. Lignin decomposition produces more char in comparison with cellulose [54].

Lignocellulosic biomass originates from a variety of sources such as paper, wood, and pulp industries, organic solid waste from recycling stations, or agricultural and forest residues [51].

2.2. Forest Biomass

Forests encompass approximately 30% of the Earth's land area, offering economic, social, ecological, and health benefits. They provide resources such as wood, energy, food, and medicine, while also acting as central pillars within their communities. Moreover, forests make possible climate change mitigation and adaptation, erosion prevention, air and water purification, and biodiversity conservation. Globally, around 1.6 billion people need forests for their sustenance. Thus, the sustainable management of forests is imperative. In view of the forest role, they must be considered comprehensively and integratively. They have to be protected, restored, and used in a sustainable way, promoting their governance and enhancing forest-based value chains, encouraging a sustainable trade of forest products which contribute to fulfillment of international commitments [55].

Forest residues and secondary forest products include portions obtained during wood processing activities, industrial manufacturing processes, and logging or sawmilling processes (sawdust, chips, planer shavings, black liquor, tops, trunk, bark, branches, leaves, roots, low-value trees, plywood, and particleboard) [18,40,56].

Forest residue use would provide environmental, economic, and health benefits in many countries because they are generated in huge quantities worldwide [18,34,39,41,56–58]. Wood, wood residues, and forest residues collectively contribute to 73% of biomass sources utilized in bioenergy, with wood representing 67% of the biomass, wood residues 5%, and forest residues 1% [45]; consequently, the forestry sector holds the predominant stake in biomass energy [34]. The forest bioeconomy from wastes allows the conversion of forestry residues, which would otherwise be disposed of [45,59], into higher value bioproducts. The transformation of forest residues into a circular economy is an appealing aim towards reaching long-term sustainability [45].

2.3. Conversion Technologies

The biomass valorization to produce different fuels for energy production and other commodities is crucial for mitigating the disposal of biowastes into the environment and replacing petroleum-derived chemicals within the framework of the circular economy [47]. There are different conversion technologies to convert biomass into useful secondary energy carriers and high-value-added bioproducts (chemicals, biopolymers, enzymes, etc.) with a low carbon footprint [41]. These technologies can be divided into different categories (Figure 1) [15,22,45,60–63]. In thermochemical conversions, either external or internal energy triggers the biomass transformation effectively into fuels and chemicals in short periods in comparison with biochemical technologies [64]. During carbonization, which is similar to a slow pyrolysis, biomass is heated usually in an oxygen-limited rather than an oxygen-free environment to produce a highly carbonaceous material. The controlled presence of oxygen facilitates the partial combustion of the biomass, hence providing the necessary heat for the pyrolysis reactions. Moreover, some carbonization procedures operate at elevated pressures, reaching up to 1 MPa [65]. Combustion is a complete oxidation process in which biomass undergoes decomposition in the presence of oxygen to obtain heat that can be used for industrial and domestic heating and power generation. During this process, the carbon and hydrogen constituents of the biomass are converted into carbon dioxide (CO_2) and water vapor (H_2O), while inorganic ash remains as a solid byproduct [64,66]. Gasification involves the oxidation of biomass at high temperatures (800–1600 °C) using a gasifying agent such as air, steam, limited oxygen, or carbon dioxide. This process generates syngas, primarily composed of hydrogen (H_2) and carbon monoxide (CO), along with small proportions of carbon dioxide and methane. Additionally, gasification produces a liquid fraction containing oil and tar, as well as a solid fraction known as biochar. Gasification can be classified as either direct or indirect, depending on the utilization of a limited amount or no oxidant at all, respectively [15,22,60]. Hydrothermal liquefaction, also known as hydrous pyrolysis, is a thermochemical depolymerization process conducted within a sealed reactor in the presence of a suitable solvent (commonly water). Wet biomass converts into biocrude oil and facilitates the extraction of water-soluble organics from the biomass at moderate temperatures (200–400 °C) and high pressures (5–25 MPa) [64,67]. Pyrolysis, executed in the absence of oxygen within the temperature range of 250–600 °C, yields three distinct products: solid coal (biochar), condensable heavy-molecular-weight compounds (bio-oil), and non-condensable light-molecular-weight gaseous products containing syngas based on operational parameters and biomass characteristics [15,60,64]. Biomass torrefaction is commonly conducted in a non-oxidative environment at mild temperatures (200–300 °C) and moderate residence times (30–60 min), resulting in torrefied biomass (which typically exhibits an enhanced energy value) as the primary product [22].

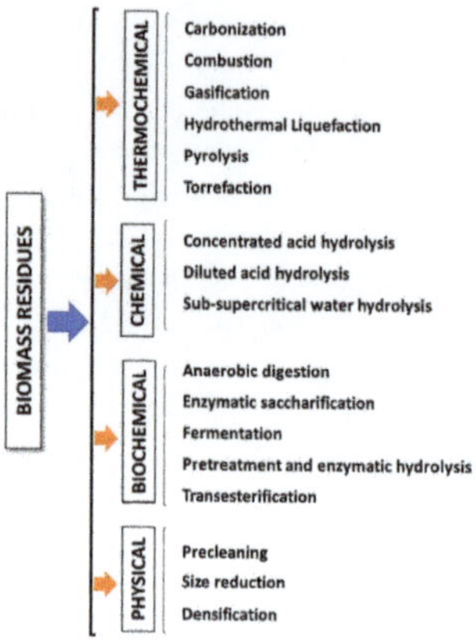

Figure 1. Biomass conversion technologies.

Biochemical conversions entail the use of microorganisms, enzymes, and bacteria to decompose biomass into gaseous (biogas) or liquid fuels (bioethanol). Fermentation, an anaerobic process, commences with sucrose undergoing hydrolysis by enzymes, leading to its conversion into fructose and glucose. Subsequently, fermentation, distillation, and dehydration are carried out to yield bioethanol. The remaining solid residue from the fermentation process can serve as cattle-feed [41,68]. Anaerobic digestion is a series of biological processes in which biomass undergoes degradation by microorganisms in the absence of oxygen to produce biogas, containing methane and carbon dioxide, which can be used to generate both electricity and heat [41]. Enzymatic saccharification involves converting liberated polymeric sugars into soluble monosugars. Bridging the gap between these stages requires comprehensive pretreatment processes (physical, chemical, or biological) to disrupt the lignin matrix, facilitating improved enzyme interaction with the bound polymeric sugars [69]. Enzymatic hydrolysis stands as the pivotal technology in a typical biochemical conversion process. Preceded by pretreatment and succeeded by microbial or inorganic catalyst conversion, enzymatic hydrolysis is the key process that releases monomeric sugars from the structural carbohydrates, cellulose, and hemicellulose in lignocellulosic biomass [70]. Transesterification is the process in which triacylglycerides are transformed into fatty acid methyl esters (biodiesel) with a catalyst in the presence of alcohol, typically methanol or ethanol [71]. In chemical conversion technologies, the biomass structure is altered using a suitable reactant to obtain valuable products [15,64]. Concentrated acid hydrolysis involves the hydrolysis of hemicelluloses and celluloses within lignocellulosic biomass using strong mineral acids like sulfuric, hydrochloric, nitric, or phosphoric acids in aqueous solutions at moderate temperatures. Diluted acid hydrolysis is conducted at elevated temperatures with a low concentration of acid [72]. Sub- and supercritical water hydrolysis emerge as clean and fast hydrolysis methods that uses water, either supercritical or subcritical, as the reaction medium to transform biomass into fermentable sugars. This technique does not produce solid residues [73]. Physical treatments focus on cleaning and reducing the size to increase the porosity of the biomass which serves as a preliminary step toward improving the efficiency of subsequent conversion processes. Densification is the typical mechanical processing method which involves compressing biomass using compaction forces with or without the use of binders to enhance the properties of solid fuels [45].

Thermochemical and biochemical conversion technologies are the two main approaches for biomass valorization into useful products such as bioenergy, biofuels, and bioproducts. Thermochemical processes can handle a wider variety of biomass, while biochemical methods offer a higher product selectivity and flexibility under mild conditions, making them suitable for producing diverse biofuels and biogas. Biochemical methods are frequently used in agricultural waste management, transportation fuel production, and bio-based chemical manufacturing [74]. Nevertheless, thermochemical processes are appropriated for large-scale energy production, converting biomass into heat, electricity, and biofuels, being commonly applied in industrial and power generation.

In most biochemical processes, the raw material must undergo various pretreatment stages to produce suitable intermediate products for microorganisms. As a result, different fractions of the feedstock are often discarded as waste or unutilized, leading to increased losses in the process and reducing the overall productivity per kilogram of raw material [75]. Thermochemical processes typically have a much faster response time compared to biochemical processes. Moreover, they are more efficient at breaking down organic compounds and generally involve lower purification and catalyst costs than biochemical methods, which supports their consideration as commercially viable. Although thermochemical processes offer faster conversion, they require complex operational parameters and cope with issues such as water content and energy costs. Conversely, biochemical processes are slower but more sustainable, producing fewer emissions and volatile compounds. However, they often deal with challenges like high costs in hydrogen production and algae growth. The integration of both methods could overcome individual limitations, although optimizing parameters (temperature and pressure) and reducing costs remain critical challenges for large-scale biomass valorization [76].

Pyrolysis is a thermal depolymerization of any organic material (carbon-based) in the absence of oxygen, which can be carried out on both pure and a mixture of materials (co-pyrolysis) [77]. In this thermochemical process, a multitude of reactions occur simultaneously and sequentially. It is widely recognized that biomass pyrolysis comprises three primary stages: (i) the evaporation of water or moisture from biomass, and (ii) the primary decomposition of lignocellulosic components through complex mechanisms, followed by (iii) secondary reactions involving cracking and recondensation/repolymerization to produce a stable solid [22]. The decomposition of biomass primarily occurs during the primary decomposition phase, leading to solid char formation (200–400 °C). Subsequently, secondary reactions continue within the solid matrix as the temperature increases. Hemicellulose decomposition, mainly represented by xylan, occurs between 250 and 350 °C, followed by cellulose decomposition (325–400 °C), yielding levoglucosan as the main pyrolysis product. Lignin, being the most stable component, undergoes decomposition at 300–550 °C [24]. Pyrolysis often yields products of distinct and frequently superior quality compared to the original residue. These products generally include solid residues like charcoal or bio-char, a condensable heavy-molecular-weight compound referred to as bio-oil, and a non-condensable, light-molecular-weight gaseous product containing gases such as hydrogen (H_2), carbon monoxide (CO), and carbon dioxide (CO_2) [60]. Pyrolysis conditions can be tailored according to the desired product (biochar, bio-oil, or non-condensable fraction). It has been found that high temperatures and short residence times tend to promote the formation of condensable products, whereas high temperatures and longer residence times favor the production of non-condensable gaseous products, largely due to the occurrence of secondary reactions. Solid products, on the other hand, are typically enhanced at lower temperatures. Thus, by adjusting parameters such as temperature and residence time, the composition of pyrolysis products can be modified towards desirable products [23]. For all these reasons, pyrolysis offers a unique opportunity to transform low-energy-density compounds into high-energy-density fuels, thereby enabling the efficient utilization of resources. Additionally, pyrolysis facilitates the recovery of valuable products from a wide range of waste, a task that is often challenging. This ability to extract value from waste materials underscores the versatility and potential of pyrolysis as a sustainable solution for resource recovery and energy production [78].

There are different technologies of biomass pyrolysis, and distinct methods, based on the process conditions for the desired product, can be implemented (Figure 2).

PYROLYSIS METHODS

	Temperature range (°C)	Residence time	Heating rate (°C/s)	Product yield (%)		
				Bio-oil	Biochar	Gas
Slow pyrolysis	300–900	5 min–few hours	0.08–10	20–40	25–50	10–35
Intermediate pyrolysis	350–650	2 s–16 min	1–70	35–50	25–40	20–30
Fast pyrolysis	400–850	0.5 s–2 min	5–200	60–75	10–25	10–30
Ultra-fast/flash pyrolysis	500–1300	< 0.5 s	1000–10,000	25–75	5–15	13–60

PYROLYSIS TECHNOLOGIES

Conventional pyrolysis
Microwave-assisted pyrolysis
Plasma pyrolysis
Hydropyrolysis
Autothermal pyrolysis
Solar pyrolysis
Vacuum pyrolysis
Catalytic pyrolysis
Co-pyrolysis

Figure 2. Pyrolysis methods and technologies [22,60,78–81].

Pyrolysis methods can be categorized as slow, intermediate, fast, and flash pyrolysis. Slow pyrolysis yields biochar as its principal product, facilitated by moderate temperatures, low heating rates, and prolonged residence time. The carbonization of feedstock occurs during slow pyrolysis at a temperature of between 300 and 900 °C and a heating rate of 0.08–10 °C/s. Fast pyrolysis operates at temperatures of 400–850 °C, aiming for the comprehensive transfer of thermal energy to the carbonaceous feedstock. The rapid removal of pyrolytic vapors characterizes this method due to its high heating rate (5–200 °C/s) within the initial few seconds. The predominant product of fast pyrolysis is oil, comprising up to 75% of the output, typically of superior quality and quantity compared to slow pyrolysis. Flash pyrolysis, executed at temperatures of 400–850 °C, features an exceptionally high heating rate exceeding 1000 °C/s, making it suitable for the production of liquid products while minimizing gas and biochar generation. Intermediate pyrolysis operates within a temperature range of 350–650 °C, with a heating rate of 1–70 °C/s. The residence time of pyrolytic vapor in this method spans from 2 s to 16 min, resulting in product yields similar to slow pyrolysis [22,60,78–81].

Distinct pyrolysis technologies exist, including conventional pyrolysis, microwave-assisted pyrolysis, plasma pyrolysis, hydropyrolysis, autothermal pyrolysis, solar pyrolysis, vacuum pyrolysis, catalytic pyrolysis, and co-pyrolysis. In conventional pyrolysis, heat is transferred from the surface to the core of the material. This heat transfer process occurs through conduction and convection, which are relatively slow and energy-inefficient methods. Due to the non-selective heating nature of conventional pyrolysis, it often requires longer periods and more energy to complete. Additionally, achieving the desired pyrolysis temperature necessitates a higher temperature gradient within the sample. Common feedstocks used in conventional pyrolysis plants include agricultural and forestry residues, scrap tires, and waste plastics. Through pyrolysis, these waste materials undergo thermal decomposition to produce non-condensable gases, bio-oil, and biochar [82]. The challenge with conventional pyrolysis lies in its endothermic nature, requiring a high heat flux from an external source. Autothermal pyrolysis addresses this by utilizing part of the heat from feedstock or pyrolysis products to meet the heat demand, allowing for process scale-up. By introducing limited oxygen into the reaction zone, chemical reactions, rather than the available heat, control the process. This method alters pyrolysis products, with vapor products often combusting to meet heat requirements. Autothermal pyrolysis, using air as the fluidizing gas, achieves a several-fold process intensification by providing energy through the partial oxidation of pyrolysis products, simplifying the reactor design and reducing capital costs [80]. A distinct pyrolysis approach is represented by microwave pyrolysis in which microwaves serve as the primary heat source, enabling uniform heating throughout the material. By inducing molecular rotation, microwaves efficiently heat polar compounds like water molecules, reducing the activation energy required for thermal decomposition. This method offers advantages over conventional pyrolysis, facilitating the quick and consistent heating of irregular waste fragments from within, without the need for agitation or controllability. Microwave heating is energy-effective and allows for rapid startup and shutdown, making it increasingly favored for waste decomposition. However, its applicability depends on the microwave absorption capacity of the materials, with some requiring additives known as susceptors (Al, Fe, $CaCO_3$, SiC, MgO, Fe_3O_4, KOH, CaO, Na_2CO_3, etc.) to enhance absorption [78,82]. Plasma pyrolysis, an emerging technique, offers a unique approach to pollutant degradation by generating ionized materials. This method has garnered significant attention for its potential in reducing pollution from waste decomposition and minimizing the emission of hazardous and toxic substances. Plasma technology is broadly categorized into two main groups, low-temperature (gas discharge plasma) and high-temperature (fusion plasma), with the latter being predominantly employed in waste treatment applications. Despite its numerous advantages, including the reduced emissions of hazardous compounds, rapid heating rates, and robust installation, plasma pyrolysis faces certain limitations such as high-power consumption, limited economic efficiency, and reduced endurance, which hinder its widespread adoption in the industry [78]. Hydropyrolysis, a distinct method from conventional pyrolysis, employs pressurized hydrogen (>10 MPa) to convert biomass into pyrolytic oil of higher quality. It inhibits char formation, resulting in bio-oil with a low oxygen content and infused hydrogen. The hydrocarbons produced exhibit enhanced stability, free from unwanted olefins and polynuclear aromatics. The presence of hydrogen facilitates effective bond cleavage within biomass components, enabling the production of fungible hydrocarbons. Unlike conventional pyrolysis, hydropyrolysis does not require a separation into organic and aqueous phases, allowing for the generation of a variety of specialty chemicals. Recently, the focus has shifted to fast hydropyrolysis, performed at high heating rates in a hydrogen atmosphere, yielding two liquid phases: an organic phase containing hydrocarbons, and an aqueous phase with char and permanent gases. The process, catalytic or non-catalytic, can be enhanced with a hydrotreating unit for deoxygenation, resulting in upgraded volatiles [23,81,83]. Solar pyrolysis utilizes concentrated solar energy (high-powered dish receivers or concentrators that enables the rapid attainment of the initial pyrolysis temperature) as the heat source, resulting in shorter heating times compared to fossil fuel heating methods. Moreover, the heating rate can be controlled effectively. Solar pyrolysis exhibits high reactivity due to the presence of functional sites, which reduces the residence time and shortens condensation reactions. Consequently, the biochar contains higher levels of oxygen and hydrogen and a lower carbon content. This method is conducted over a wide range of temperatures (150–2000 °C), heating rates (5–450 °C/s), and heat flux intensities (0.01–12 MW/m^2). The primary products of the solar pyrolysis of biomass include bio-oil (25–78 wt.%) and syngas (1.4–63 wt.%), with biochar (8–29 wt.%) being a minor product. Vacuum pyrolysis is conducted under reduced below atmospheric pressure to replicate an inert environment, eliminating the need for sparging inert gases like nitrogen or argon. Typically, vacuum pyrolysis operates at pressures of 0.5–50 kPa and moderate temperatures (400–600 °C). Prior to pyrolysis, a vacuum pump evacuates air from the reactor. During pyrolysis, the released volatiles rapidly diffuse towards the pump due to the pressure gradient between the reactor and the pump. Compared to conventional pyrolysis, vacuum pyrolysis consumes less energy as there is no need to heat inert gas. Vacuum pyrolysis converts biomass waste into bio-oil and biochar with a considerably high heating value (22.4–40 MJ/kg). The bio-oil obtained from the vacuum pyrolysis of biomass primarily consists of polycyclic macromolecular compounds [22]. Catalytic pyrolysis involves the use of catalysts to facilitate the process. These catalysts improve product quality, reduce process temperatures, and minimize energy requirements. By increasing the speed of cracking reactions, catalysts promote the production of lighter compounds and enhance gas generation. Catalytic pyrolysis may lead to a reduction in bio-oil production; nevertheless, its quality is typically higher [26]. The effectiveness of each catalyst depends on its specific catalytic properties (acidity, surface area, pore volume, and pore size). A diverse range of catalysts has been employed including metal oxides (CaO, NiO, Ni_2O_3, MgO, γ-Al_2O_3,

Fe_2O_3, etc.), zeolites (ZSM-5, HZSM-5, EDTA-HZSM-5, FeZSM-5, zeolite-β, natural zeolite, etc.), carbon materials (activated carbon), transition metals, and other catalysts like red mud, CO–Mo/Z, $Ca(OH)_2$, and $Al(OH)_3$ [78,84]. Co-pyrolysis is a variation of the pyrolysis process which involves organic compounds introduced to enhance the products quality [85]. Waste pyrolysis often produces components with low hydrogen-to-carbon ratios, making them unsuitable as fuel. Adding substances with higher hydrogen-to-carbon ratios, like biomass, which is rich in hydrogen, improves the hydrogen content of products [60]. This leads to a lower activation energy, better yield, and higher product quality, and reduces environmental pollutants' emissions during pyrolysis [78,86].

Biomass pyrolysis is a complex process in which three distinct fractions can be yielded: solid, liquid, and gas, as explained above. The solid fraction is a carbonaceous residue with a fine structure and a large specific surface area (biochar) [84] generated during primary and secondary pyrolysis reactions. However, the resulting biochar often possesses low quality due to the presence of ash and other impurities within its micropores [87]. Enhancing the structure of biochar to refine it makes possible that this can be used in several applications such as solid fuels to produce electricity and heat, raw materials for gasification, activated carbon or carbon nanofilaments fabrication, the removal of contaminants (heavy metals, dyes, pharmaceuticals, detergents, polyaromatic hydrocarbons, herbicides, and pesticides), catalysis, energy storage, gas adsorption, soil conditioner and enrichment, potential electrode material in biosensors, and food processing [8,22,23,30,61,62,88–90]. Biochar's composition includes ash with minerals like Al, Ca, Cl, K, Mg, P, and Si, while its formation involves the decomposition of tar into methane, hydrogen, water, and other hydrocarbons [1]. A low temperature, prolonged residence time, and slow heating rate enhance the biochar generation. These operational conditions facilitate the breakdown of weaker bonds while preserving stronger ones. Consequently, the rearrangement reaction is encouraged, leading to the creation of a solid biochar with structural stability. This process inhibits the formation of volatile compounds and enables the retention of a significant amount of energy within the biochar [88]. Char properties (physical, chemical, and mechanical characteristics) are influenced by both the type of feedstock used and the conditions under which pyrolysis occurs [24]. Bio-oil, derived from biomass pyrolysis, because of the inherent moisture in the raw material and the water generated from secondary reactions during bio-oil storage, results in two separate phases: an aqueous phase and an organic phase. This liquid fraction also contains remnants of cellulose, hemicellulose, and lignin, which either vaporized in the pyrolysis environment or were carried as small droplets entrained from the reactor [23]. Bio-oil is a dark brown fluid, in which the moisture content is typically 15–35 wt.% [24], alongside a really complex mixture of more than 230 organic chemicals [91] characterized by highly oxygenated organic compounds including C1–C4 light oxygenates, furans, pyrans, anhydrosugars, phenols, benzenes, PAHs, aldehydes, ketones, organic acids, esters, ethers, alcohols, alkenes, nitrogen compounds, miscellaneous oxygenates, aliphatic hydrocarbons, and solid particles [1,5,52,92]. This fluid finds extensive application as a transportation fuel, for power generation, and in various industries such as the production of bio-plastics, phenolic resins, polyurethane materials [93,94], food (flavoring), pharmaceutical, paint industries, wood preservation [84], and flavoring [23]. The complex composition of bio-oil, as well as its physical properties, including low pH, high oxygen content, and low viscosity, contribute to their instability, creating challenges for direct utilization and necessitating upgrading before commercial use [95]. Achieving a high oil yield is insufficient; understanding and optimizing the bio-oil composition are crucial [12]. Efforts in the catalytic upgrading of bio-oils aim to enhance their quality. Upgrading bio-oils to reduce the oxygen content and enhance the similarity to crude oil is crucial for broader industrial adoption, highlighting their potential as a renewable and versatile resource. Finally, the non-condensable gases' fraction in pyrolysis primarily originates from the cracking and decomposition of large molecules formed during the early stages of pyrolysis. It includes carbon dioxide, carbon monoxide, hydrogen, low-carbon hydrocarbons, nitrogen oxides, sulfur oxides, and other compounds [66]. The pyrolytic gases' composition of lignocellulosic biomass depends on feedstock components (cellulose, hemicellulose, and lignin, with approximate weight percentages of 35–50%, 25–30%, and 15–30%, respectively). Hemicellulose, due to its higher carboxyl content, yields more CO_2 during pyrolysis. Cellulose, on the other hand, yields higher amounts of CO, primarily due to the thermal cracking of carbonyl and carboxyl groups. Lignin, with its higher presence of aromatic rings and methoxyl functional groups, releases more H_2 and CH_4 during pyrolysis. Additionally, non-condensable gases can be recycled into the reactor to aid fluidization or utilized for process heat in large-scale operations (autothermal pyrolysis) [23,96].

3. Microwave-Assisted Pyrolysis (MAP)

Microwave-assisted pyrolysis utilizes microwaves to heat biomass wastes, offering rapid, targeted, and energy-efficient heating compared to conventional methods like furnace heating, which has drawn significant interest, supported by a series of reviews outlined in Zhang et al.'s study [97]. In conventional pyrolysis, heat is generated externally and transferred to the material through convection, conduction, and radiation, limiting efficiency due to surface temperature constraints and material properties. Heat moves from the surface to the core, creating an inward-flow temperature gradient [88]. In contrast, microwave pyrolysis combines microwave radiation with conventional pyrolysis to convert electromagnetic energy into kinetic energy. Unlike conventional pyrolysis, microwave pyrolysis generates heat from the interior outward, ensuring a uniform heat distribution within the sample particles. Additionally, microwave pyrolysis leads to the formation of micro-plasma, containing hot spots with significantly higher temperatures than the average, which stimulates catalytic reactions and increases hydrogen production. Moreover, microwave power induces non-thermal effects that reduce the activation energy of reactions, promoting chemical cracking processes and enhancing operational efficiency compared to conventional pyrolysis [98]. Because microwave pyrolysis' ability to reach high temperatures and heating rates, it is considered a fast, energy-efficient, and time-saving process compared to conventional methods [28], reducing production costs and increasing the product's yield [22,99] and quality [77].

Microwaves (MWs) are non-ionizing electromagnetic waves with frequencies between 300 MHz to 300 GHz and wavelengths ranging from 0.001 to 1 m. The potential of microwave (MW) heating is constrained by the inability to adjust the frequency over a broad range in standard microwave generators, such as magnetrons. The generation frequency of magnetrons is strictly determined by the strength of their magnetic field and their geometric dimensions [100]. Additionally, the Federal Communications Commission (FCC) has reserved the frequencies of 915 MHz and 2450 MHz for scientific, industrial, and medical aims, preventing interference with cellular phone and telecommunications frequencies [61]. Nevertheless, most researchers rely on commercial 2.45 GHz magnetrons [100]. Microwave heating occurs through the dissipation of electromagnetic waves within the irradiated material. The amount of power dissipated depends on the dielectric properties of the material and the local time-averaged strength of the electric field. Although microwaves generate volumetric heating, the energy distribution within the irradiated material is uneven due to the non-uniform electric field distribution. This unevenness depends on the geometry of the object and its dielectric properties [101]. MW-assisted heating involves three main mechanisms: dipolar reorientation, ionic conduction, and interfacial polarization. Dipolar molecules like water and some organic compounds in biomass attempt to realign with the rapidly alternating electrical field generated by microwaves. As the applied field oscillates, the dipolar molecules continuously rotate to follow these oscillations, causing the conversion of electromagnetic energy into heat within the material through dielectric loss and molecular friction [102]. In ionic conduction, ions contained in a material (such as salts or minerals in biomass) move back and forth due to the changing electric field, leading to an electric current which encounters internal resistance due to collisions between the charged species and neighboring molecules or atoms. This ionic conduction has a significantly higher impact on heat production compared to dipolar polarization [103].

Interfacial polarization, combining dipolar reorientation and ionic conduction, creates positive and negative charges at material interfaces, altering the field distribution and facilitating heating [60]. Different materials interact uniquely with MWs: insulators (glass, ceramics, and plastics) that transmit MWs, conductors or reflective materials (bulk metals) that reflect MWs, and dielectric or absorbers (carbonaceous materials, metal powders, SiC, and water) that absorb MWs to generate heat [28]. The critical electrical parameter governing the interaction between a material and electromagnetic field is its complex permittivity. This permittivity consists of two components: the dielectric constant, which reflects the material's ability to store electric field energy, and the relative loss factor, which indicates how much energy is dissipated as heat. Power converted from microwave energy to heat depends on the relative loss factor, the frequency, and the local electric field intensity. As microwaves travel through a material, their intensity decreases due to energy dissipation as heat [102]. This decay is characterized by the penetration depth (d_p) which is a crucial property used to categorize materials exposed to MWs. It refers to the depth at which the incident radiation inside the material decreases to 37% of its surface value. The penetration depth magnitude (d_p) indicates whether the material reflects ($d_p \approx 1$–7 μm), transmits ($d_p \approx 3$–300 m), or absorbs ($d_p \approx 0.1$–10 cm) microwaves [64]. Furthermore, the penetration depth depends on both the loss factor and the dielectric constant. Different materials interact uniquely with MWs: insulators (glass, ceramics, and plastics) that transmit MWs, conductors or reflective materials (bulk metals) that reflect MWs, and dielectric or absorbers (carbonaceous materials, metal powders, SiC, and water) that absorb MWs to generate heat [28]. The dielectric properties of a material, including dielectric loss (tan δ) and dielectric constant, are crucial in determining its ability to convert electromagnetic energy into heat (insulators (tan $\delta < 0.1$), reflectors (tan δ 0.1–0.5), and absorbers (tan $\delta > 0.5$)) [2]. The complexity of microwave heating arises from the fact that these dielectric properties change with temperature and moisture [30] and these variables change significantly during MAP. In contrast to conventional heating, MW heating operates in a volumetric mode; heating uniformly warms entire volumes of solids, suspensions, or liquids on an industrial scale, ensuring a consistent distribution of electromagnetic energy [62]. Consequently, MW radiation offers notable advantages, including the following: (a) rapid heat transfer and shorter reaction times, (b) selective and uniformly distributed volumetric heating, (c) straightforward operation and energy efficiency, (d) reduced degradation or formation of side products, and (e) increased safety and automation levels [14,66].

Because of the distinctive heating mechanism of microwaves, various limitations and advantages have been reported in comparison with conventional pyrolysis (Table 1).

Table 1. Microwave-assisted pyrolysis vs. conventional pyrolysis [22,28,30,97,104].

Microwave-Assisted Pyrolysis	Conventional Pyrolysis
Dependent on material properties	Less reliant on material properties
Energy conversion	Energy transfer
Generation of hot spots	Absence of hot spots
Selective heating	Non-selective heating
Rapid	Slower
Accurate and controlled heating	Less controllable
Uniform and volumetric heating within the core at the molecular level	Surface-level heating through conduction, convection, and radiation
Moisture in biomass feedstocks may increase heating rate	Moisture in biomass feedstocks may decrease heating rate
Improved efficiency in electricity conversion	Decreased efficiency in electricity conversion
Reduced thermal inertia and quicker response	Higher thermal inertia and slower response
Relatively difficult temperature measurement	Easier temperature measurement
Microwave absorbers and catalysts needed	Absorbers are not required and catalysts sometimes needed
Less feedstock pre-treatment needed	Feedstock pre-treatment is required and its cost is high

Table 1. *Cont.*

Microwave-Assisted Pyrolysis	Conventional Pyrolysis
Not well-established	Well-established (easy scale-up)
More complicated reactor design	Simple reactor design
High-quality products	Poor-quality products
Cost-saving	Less cost-saving
Formation of arcs and plasma resulting from thermal runaway	Absence of arcs and plasma
Huge capital investment	Lower capital investment
Increase the pre-exponential factor, and decrease the activation energy	

Table 1 shows that temperature measurement during MAP is relatively difficult. Contact methods like thermocouples are commonly used for temperature measurement; however, their use is limited by several factors. To avoid interference from the MW electric field, thermocouples must be positioned perpendicular to the field vector and shielded with a grounded covering. This shielding prevents current flow along the conductors and reduces the risk of discharges [105–107]. Nevertheless, this requirement makes thermocouples suitable primarily for single-mode systems where the electric field direction is predictable. The presence of conductive thermocouple elements distorts the electric field and, due to their high thermal conductivity, can affect the heating of the sample by dissipating heat. Thermocouples are best suited for measuring large sample temperatures, but accurate contact with the material being measured is required, which is challenging for low-density materials. Additionally, metal wires can interfere with the MW field, potentially causing sparks or discharges and affecting temperature measurements by reflecting MW radiation and altering the thermal field. These factors can result in measurement errors and unreliable data [100]. There are authors [108] who demonstrated that positioning thermocouples at the bottom of the cavity to avoid direct exposure to microwave (MW) irradiation significantly reduced interference. This setup provided more reliable and accurate temperature readings with minimal noise, as well as improved the biomass material's absorbance due to a uniform electric field distribution. Their study emphasized the need to shield metallic thermocouples and avoid direct exposure to the MW electric field. Optical methods for measuring temperature rely on fiber-optic lines and various physical principles, including amplitude and phase-based techniques [109]. Common methods include optical pyrometry, light absorption, and interferometry. However, fiber-optic sensors suffer from a time lag of 8–10 s, which complicates real-time monitoring and control. In addition, these methods require calibration to ensure accuracy, as external factors like the surrounding atmosphere and the low thermal conductivity of quartz tubes can affect temperature readings. Calibration methods involve comparing optical sensor readings with conventional thermocouples or using reference materials to improve accuracy. Fiber-optic sensors are immune to electromagnetic interference but typically measure temperatures only up to 300 °C. Therefore, calibration compares fiber-optic measurements to infrared pyrometer data. Additionally, gases present during MW heating can affect the apparent emissivity of the sample, influencing infrared-based temperature readings. Overall, the proper calibration and consideration of environmental factors are essential for accurate temperature measurement in MW systems [100]. Other methods used to obtain the temperature during MAP include non-contact pyrometry which is a common method for measuring temperature by detecting thermal radiation from an object, typically in the near-infrared and visible ranges. Its advantage is that it does not require direct contact with the sample, avoiding interference with the electric field and temperature. Pyrometry is based on the thermal radiation of a blackbody, and real objects deviate from this due to their emissivity, which often must be measured experimentally. Two main types of pyrometry exist: radiation pyrometry, which measures infrared radiation intensity using devices like infrared thermometers, and optical pyrometry, which relies on the color of radiation, typically used in brightness and color pyrometry methods. Brightness pyrometry measures radiation intensity at a specific wavelength, while color pyrometry uses the ratio of intensities at two wavelengths to estimate the temperature [100]. Pyrometry measurements only provide surface temperatures, limiting their effectiveness to detect hot and cold spots within a material's core. A single-spot measurement may be misleading if significant inhomogeneity is suspected, making pyrometer data less suitable for automated temperature-controlled microwave processes. Pyrometers are best for controlling homogeneous temperatures or at predefined critical points. Adding more pyrometers and combining their data through interpolation can offer a cost-effective alternative to thermal imaging. For systems with inhomogeneous heating like MAP, single-spot measurements provide limited information and require calibration based on factors like material and surface conditions. Despite these limitations, pyrometers allow fast, non-invasive temperature measurement, making them useful in automated microwave applications [110,111]. Spectral pyrometry, a more advanced technique, measures the full radiation spectrum over many wavelengths, improving accuracy even when the object's emissivity is unknown. This method is beneficial in MW systems where optical fibers transmit radiation data to small spectrometers. However, the low signal level in spectral pyrometers, caused by the distribution of radiation across thousands of elements, results in longer exposure times or larger observation areas compared to brightness pyrometers. This makes it complicated to establish what temperature is measured when there are significant temperature variations or non-uniform distribution within the observation area, which is what happens during MAP [112]. It is observed that new approaches, which combine the best qualities of known methods or innovative solutions, are required in order to improve temperature measurement during MAP.

One of the advantages of microwave pyrolysis as gathered from Table 1 is this type of pyrolysis can lead to increased product yields or improved product quality. For instance, bio-oils produced via microwave pyrolysis contain light hydrocarbons, lack polycyclic aromatic hydrocarbons (PAHs), and are rich in phenols, unlike the hydrophilic and corrosive oxygen-rich liquid produced in traditional pyrolysis. The gas products of microwave pyrolysis mainly consist of hydrogen (H_2) and carbon dioxide (CO_2), serving as a foundation for chemical product synthesis such as ammonia, alcohol, and acetic acid [98]. Furthermore, biochar obtained from microwave pyrolysis exhibits higher fixed carbon content compared to that from conventional pyrolysis. Additionally, microwave-produced biochar typically possesses a greater surface area, enhancing its suitability for adsorbing contaminants, retaining water, and other environmental applications [8,15]. There are studies that have shown that incorporating microwave heating into the pyrolysis process has a more negative environmental impact compared to conventional pyrolysis [113], whereas other authors have obtained a reduction in energy consumption when MAP was implemented in comparison with conventional processes [114]. Increased yields or improved product quality obtained during MAP can deal with the reduced attractiveness of MAP compared to conventional pyrolysis. Co-pyrolysis, catalyst use, and the integration of renewable energy sources like solar energy in MAP have further shown decreased potential emissions [115,116]. Despite the higher capital costs of MAP compared to conventional pyrolysis, techno-economic analyses have suggested that MAP processes have promising economic potential, particularly in bio-oil production. Studies have reported reduced bio-oil production costs. As microwave technologies mature, the economic viability of MAP is expected to improve, especially with the potential for higher biochar prices due to the enhanced quality. On the other hand, it should be highlighted that most techno-economic analyses for MAP rely on data from process simulations and lab-scale experiments, leading to uncertainty, particularly regarding the cost of MAP reactor systems, as there are few commercial systems for benchmarking. Product yields and conversion efficiency may change as the process is scaled up. Additionally, electricity plays a critical role in MAP economics, making cheap electricity essential for its viability [117].

Most biomass resources have a low microwave absorption capacity; consequently, enhancing heating rates is required during the microwave pyrolysis of biomass, adding microwave absorbers [118]. Microwave absorbers, also referred to as susceptors or receptors, are materials with a high loss tangent value (>0.5), enabling the efficient conversion of radiant energy into heat [64]. Different compounds such as Fe_3O_4, CuO, V_2O_5, and carbon can undergo significant heating under microwave radiation, reaching temperatures exceeding 700 °C in less than a minute [54]. One approach to utilizing microwaves for biomass pyrolysis involves mixing the biomass with an efficient microwave receptor such as water, SiC, K_3PO_4, NaOH, KOH, zeolites, clays, fly ash, metal powder (Fe, Co, Ni, Cu, Al, steel slag), powdered metal oxides (MgO, NiO, CaO, CuO, Fe_2O_3, and Al_2O_3), SiO_2, bauxite residue, metal salts ($MgCl_2$, $ZnCl_2$, $FeCl_3$, $CaCO_3$, Na_2CO_3, K_2CO_3, Na_2HPO_4, and NaH_2PO_3), or carbon-based materials (biochar, activated carbon, coke, and graphite) [52,53,84,104,119–122]. In addition to serving as an energy transfer medium, the presence of microwave absorbers in microwave-assisted pyrolysis can play, in some cases, a catalytic role, enhancing the selectivity towards desired products in the reaction [84,92,123]. The catalyst introduction can be carried out either in an in situ or ex situ configuration. In the in situ catalytic microwave pyrolysis, biomass and catalyst samples are mixed directly beforehand in the reactor before the experiment, enabling simultaneous biomass decomposition and vapor upgrading. In the ex situ setup, a catalytic reactor, independently controlled, is positioned downstream of the microwave pyrolysis reactor for pre-upgrading vapors [7,60]. Different catalysts have been investigated to be incorporated into microwave-assisted pyrolysis systems, giving rise to microwave-assisted catalytic pyrolysis [7]. These catalysts can be categorized into four groups: zeolites (clinoptilolite, Zeolite Socony Mobil-5 (ZSM-5) that can be synthesized with different Si/Al ratios, HZSM-5, EDTA-HZSM-5, SIO_2-HZSM-5, and FeZSM-5), metal oxides/salts/alkaline (γ-Al_2O_3, CaO, MgO, CuO, Fe_2O_3, NiO, Ni_2O_3, ZnO, ZrO_2, TiO_2, $MgCl_2$, $AlCl_3$, $CoCl_2$, $ZnCl_2$, Na_2HPO_4, K_2PO_4, $Fe_2(SO_4)_3$, NaOH, KOH, $K_2Cr_2O_7$, and H_3BO_3), carbonaceous materials (activated carbon, graphite, and char), and clays (their interlamellar cations (Na^+, K^+, and Ca^{2+}...) can be readily substituted by other cations or molecules) [64,84,118].

The microwave-assisted pyrolysis of biomass is a complex process influenced by different factors such as reactor design (type of reactor, type of microwave, and catalyst position), reaction material (feedstock type, susceptor, catalyst, feedstock composition, susceptor-to-feedstock ratio, catalyst-to-feedstock ratio, feedstock particle size, pre-treatment conditions, and type of co-processing feedstocks), and reaction conditions (microwave power, heating rate, temperature, residence time, carrier gas, flow rate of purging gas, microwave power switching frequency, mixing intensity, and pressure) [6,77,98,104,124,125]. It has been found that not all of these factors influencing the microwave-assisted pyrolysis possess the same importance [125]. The reactor design should be tailored to the desired products (fixed-bed configurations enhance gas yield and fluidized bed reactors facilitate liquid products) and catalyst location requires us to consider the cost and desired products. Furthermore, the temperature has a critical role and its best value depends on the type of feedstock; generally, a temperature increase produces a significant gain in gas yield, whereas an opposite trend occurs for the bio-liquid yield [60]. A raised microwave power and residence time increase gas yield, which is also favored by strong polar materials like NaOH. Additionally, the catalyst ratio adjustment improves the catalytic effects [98]. It has been demonstrated that the most effective strategy to optimize the MAP process is implementing strategies that consider the interaction between several independent factors. A well-optimized MAP reduces the energy, catalysts, and absorbent needs, and improves the quality and quantity of the desired product (biochar, bio-liquid, or non-condensable gases) [6].

MWs are not only used in pyrolysis, but also in pretreatments of biomass (microwave drying, and microwave-assisted Organosolv pretreatment) for enhancing pyrolytic yields [126–129], because it produces autohydrolysis, causing the separation of hemicellulose and lignin from cellulose [130,131]. Moreover, MWs have gained attention as a potential alternative to conventional activation methods due to its advantages, including the reduced activation time, uniform interior heating, high heating rate, selective heating, precise control over the process, absence of direct contact between the heating source and materials, and decreased equipment size and waste [89,132].

4. MAP of Forest Biomass

Forest biomass (FB) consists of lignocellulosic biomass containing three main components, hemicellulose (10.5–44.4 wt.%), cellulose (17.0–62.9 wt.%), and lignin (11.2–48.40 wt.%), along with minor amounts of ash and extractives [2,64,133–142]. Different types of FB have been used in microwave-assisted pyrolysis as Figure 3 shows. Considering the works that were evaluated in this review, approximately 20% have studied the microwave pyrolysis efficiency of biomass from pine. Secondly, the biomass most analyzed (14% of articles) was named wood pellets/wastes/sawdust without specifying the FB species. Douglas fir (12%) and spruce (9%) were the other two forest biomass species whose microwave-assisted pyrolysis has been most studied.

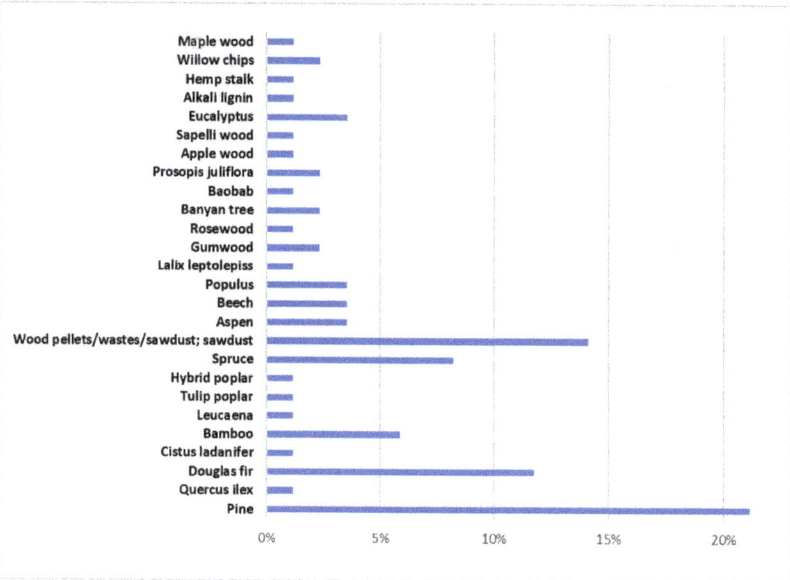

Figure 3. Forestry biomass that has been studied in bibliography.

Product yields from the microwave-assisted pyrolysis of FB are influenced by various operation conditions, including microwave power, biomass/absorber–catalyst ratio, temperature, biomass size, and residence time.

Using regression analysis, a linear relationship between the heating rate and microwave power level was identified by Huang et al. [143], concluding that a higher microwave power results in a faster heating rate. Heating rates significantly impact biomass pyrolysis, as demonstrated by Dong and Xiong [144], who analyzed the pyrolysis kinetics of bamboo using both conventional and microwave heating methods. Their experiments showed that heating rates significantly impact biomass pyrolysis, with microwave pyrolysis requiring a much lower activation energy compared to conventional methods, which suggests that microwave heating, due to its volumetric heating style, is more efficient and could be a more promising technology for biomass pyrolysis. Huang et al. [143] also established a logarithmic relationship between the maximum temperature and the microwave power used during pyrolysis. The implementation of distinct power levels during microwave-assisted pyrolysis not only impacts the maximum temperature, but also product yields. Martin et al. [145] concluded that microwave power influences product yields in MAP. These authors utilized six distinct power levels (1000, 500, 400, 300, 200, and 100 W) to pyrolyze a mix of biomass (*Pinus pinaster*) and 5% of absorber (activated carbon), finding that, at 200 and 100 W, the microwave-assisted pyrolysis did not take place. Furthermore, biochar production diminished as microwave power increased being lower than 20% at 1000 W. On the other hand, the gas yield rose when microwave power grows, reaching a yield superior to 40% at 1000 W. Finally, liquid production achieved its maximum value (58%) at 400 W, and, from that, the power suffered a decrease as microwave power rose. Parvez et al. [146] also implemented different microwave powers (2700, 2400, and 2100 W) to carry out MAP of pine wood at different temperatures (600, 500, and 400 °C), obtaining an increasing gas yield (from 47.3% to 71.4%) with rising pyrolytic powers (from 2100 W to 2700 W). Conversely, char (33.1% to 19.3%) and bio-oil (19.6% to 9.3%) yields went down with power growth (2100 W to 2700W). Nhuchhen et al. [147] conducted experiments at three different powers (3000, 2500, and 2000 W), as well. However, they studied the behavior of wood pellets from spruce sawdust, proving that the product yield distribution was affected by microwave power. The biochar yield decreased as the microwave power increased, with the reduction being more significant at a higher power. That behavior can be attributed to the high heating rates observed when higher microwave powers are implemented. The microwave power levels of 500, 400, and 300 W were applied during microwave pyrolysis experiments carried out by Huang et al. [139]. In that work, the three-phase product distribution was also affected by the power level. For instance, it was seen that, during the microwave pyrolysis at 300 W of bamboo leaves, around 33 wt.% of the products were solid-phase, 47 wt.% were liquid-phase, and 20 wt.% were non-condensable gases. Nonetheless, when the microwave power was 500 W, the solid, liquid, and gas yields were 22 wt.%, 42 wt.%, and 36 wt.% respectively. As those authors concluded [139], this change of product distribution indicates that higher microwave powers enhance the non-condensable gas yield, possibly due to the self-gasification of the solid phase produced during pyrolysis. Sellamuthu et al. [148] evaluated the effect of microwave power (550, 650, and 750 W) on the bio-char yield percentage when Adansonia kilima wood chips were pyrolyzed. They showed that the power augmentation of microwave-assisted pyrolysis resulted in a considerable decrease in the biochar yield. Du et al. [135], who pyrolyzed sawdust and rice straw with ionic liquids at five different power levels (160–800 W) under microwave irradiation, found that, when the microwave power was lower than 480 W, a small yield of bio-oil was produced. Nevertheless, when the microwave power was increased, the bio-oil production rose from less than 10% to more than 30%. The bio-oil yield, obtained in the microwave-assisted pyrolysis of wood from Banyan tree trunks, also grew significantly when heating power rose from 600 to 1200 W [149]. The bio-oil production increased from 26.3 to 36.4 wt.%, as well as the non-condensable gas yield, which was augmented from 31.2 to 34 wt.%, whereas the biochar one diminished from 42.5 to 29.6 wt.%. A comparable effect of increasing microwave power (from 100 to 800 W) on non-condensable gas and bio-oil yields was observed by Khelfa et al. [150]. The highest oil and gas yields during the microwave-assisted pyrolysis of pine wood sawdust were achieved at the maximum studied power (800 W), whereas the char production systematically diminished with increasing power. Those results were in agreement with the ones obtained by Wu et al. [151] who saw an increase in bio-oil and non-condensable gas yields from 43.23 to 47.10 wt.% and from 8.39 to 9.68 wt.%, respectively, with increasing microwave power (600 to1200 W) during wood biomass pyrolysis. Lin et al. [152] focused their study on the influence of plasma power (1000, 900, and 800 W) on the obtained gases from banyan leaves which were pyrolyzed into a microwave plasma system. The gaseous products which were produced consisted of H_2, N_2, CO_2, and CO, with smaller amounts of CH_4 and formaldehyde. As the microwave power increased, the volume fraction of H_2 increased and CO diminished. Specifically, the H_2 production rates were 18.48, 20.05, and 20.44 mg-H_2/g-biomass for power levels of 800, 900, and 1000 W, respectively, reflecting a roughly 10% increase in hydrogen production as the microwave power rises by 25%. The assessment of all these works shows that the three phases (biochar, bio-oil, and non-condensable gases) obtained during microwave-assisted pyrolysis are significantly influenced by microwave power which can be simply managed, allowing us to easily optimize the reaction requisites of microwave-assisted pyrolysis.

There are authors who have studied the impact of pyrolysis temperatures on product yields instead of microwave power levels, which influence heating rates and maximum temperatures, altering primary and secondary reactions. For instance, Zhang et al. [153] examined the effect of temperature (400, 500, 600, 700, and 800 °C) on the product distribution during the microwave-assisted pyrolysis of Aspen wood. They concluded that a pyrolysis temperature of 500 °C allowed them to maximize bio-oil production. They also deduced that high pyrolysis temperatures promoted pore formation in biochar. Furthermore, they saw that the compositions of bio-oil and non-condensable gases were influenced by the operation temperature. The phenol concentration diminished as the temperature rose, while the hydrocarbon and ketone concentration increased. In the case of the composition of non-condensable gases, it was noticed that, as the pyrolysis temperature rose, the hydrogen concentration increased as well, while the CO_2 concentration decreases. Additionally, the CO concentration reached a minimum at 600 °C and the methane concentration was the maximum at 700 °C. Similar results of the temperature effect on non-condensable gases were gathered by Wang et al. [154]. These authors studied the behavior of Douglas fir when it was pyrolyzed, considering different operation temperatures (721, 700, 650, 600, and 579 °C) and biochar catalyst/Douglas fir mass ratios (4.4, 4, 3, 2, and 1.6). They noted that higher temperatures promoted an increase in gas yield and a drop in bio-oil production. Additionally, these authors found that the hydrogen concentration increased significantly with temperature, although it was constrained by thermodynamic factors. Huo et al. [155] also studied the behavior of Douglas fir during microwave-assisted pyrolysis. However, these authors investigated MgO and activated carbon (from corncob using phosphoric acid for its thermochemical activation) catalysts. They considered the impact of the experimental temperature (600, 550, 500, 450, and 400 °C), activated carbon/Douglas fir ratio, and MgO/activated carbon ratio on product yields. In this study, it was observed that the bio-oil yield reached a maximum yield of 53.0% at 500 °C, while char and coke yields decreased at temperatures between 500 °C and 600 °C, whereas gas yield increased. These behaviors were seen when the MgO/activated carbon ratio was 0.8 and the activated carbon/biomass ratio was 1:1. Moreover, it was detected that the bio-oil composition varied with temperature. Phenols increased from 38.7% to 42.2% as the temperature increased. The selectivity of cellulose-derived furans peaked at 450 °C and then declined. The main target gas (CO) of gaseous products increased from 5.8% to 10.4% at the highest temperature. Wallace et al. [156] evaluated the consequences of temperature (348.4 ± 33.9, 398.8 ± 38.7, 427.5 ± 31.2, 528.7 ± 13.2, 604.2 ± 10.8 and 659.8 ± 60.0) on biochar characteristics when hemp-stalk and a mix of spruce and fir softwood chip were pyrolyzed, applying microwave radiation. The experiments demonstrated that, as the pyrolysis temperature increased, the biochar carbon content also grew, whereas the char yield was reduced due to the accelerated volatile release. Higher temperatures also lower the surface pore size in biochar. Moreover, it was found that the temperature and heating rate were key factors influencing the mechanical properties of biochar. Dutta et al. [157], who also assessed the impact of pyrolysis temperature (250, 290, and 330 °C) on biochar properties and production, pyrolyzed maple wood using microwave-assisted pyrolysis. They used as the absorber char from willow wood, studying the dopant ratio effect on biochar yield and its properties. Furthermore, the influence of the operation time on the pyrolysis process was tested. These authors found that biochar production was also reduced with increasing temperature. Moreover, it was demonstrated that only the temperature significantly affected the biochar yield in comparison with the doping ratio and time, whose impact was minimal. On the other hand, it was concluded that the quadratic term of the doping ratio and temperature are critical factors affecting both the exothermic energy and volatile matter content in biochar. Zhou et al. [158] conducted microwave-assisted pyrolysis on wood sawdust and characterized the obtained char. They evaluated the influence of the processing temperature (750, 650, 550, and 450 °C) on biochar production. Char yield decreased from 36 to 26 wt.% as temperature rose, while bio-oil yield dropped sharply from 36.13% to 14.85 wt.% and the non-condensable gas production rose from 9.10% to 28.07 wt.%. Moreover, it was concluded that higher temperatures led to an increased carbon content in biochars. Additionally, it was inferred that the process temperature during pyrolysis affected the adsorption capacities of chars, with the optimal performance observed at specific temperatures for methylene blue (MB), crystal violet (CV), and iodine. Shi and Wang [159] assessed the pyrolysis process of cellulose, xylan, a mixture of cellulose and lignin from pine wood, and pine wood, considering the effect of pyrolysis temperature (300, 500, and 700 °C) on the three phase products' distribution. At 700 °C, char yields were 16.7% for xylan, 16.1% for cellulose, 18.6% for the mixture, and 20.3% for pine wood, whereas, at 350 °C, these char yields were higher, particularly for the mixture and pine wood with minimal gas production from cellulose. Wang et al. [160] pyrolyzed pine sawdust at different temperatures (400, 500, 600, 700, and 800 °C) using conventional pyrolysis and microwave-assisted pyrolysis. The results showed that, during MAP, less biochar and more hydrogen and carbon monoxide in a non-condensable fraction were produced. As it has been observed in other works, higher temperatures during MAP led to a reduced biochar yield. These authors also found that MAP chars had smaller pores and smoother surfaces at higher temperatures, resulting in a lower reactivity compared to conventional pyrolysis chars. Additionally, the study observed that MAP could produce high-quality liquid products at lower temperatures by inhibiting secondary reactions, containing unique volatile compounds not found in conventional pyrolysis. Nzediegwu et al. [161] also compared the impact of conventional pyrolysis and microwave-assisted pyrolysis; however, they studied the differences in biochar properties from sawdust of white spruce that had been pyrolyzed at three distinct temperatures (500, 400, and 300 °C). At 500 °C, the biochar yields from both methods became similar because conventional pyrolysis required more time for heat transfer through conduction. MAP biochars generally had a higher pH than those from conventional pyrolysis. Nevertheless, the higher heating effects of MAP did not significantly alter the Gross Calorific Value (GCV) of the biochars. It was also detected that the thermal stability of biochar increased with pyrolysis temperature, leading to a higher carbon content. Biochar produced at 500 °C possessed a higher carbon stability and energy content, whereas those at 300 °C had higher energy yields. The biochar yield using MAP varied significantly with pyrolysis temperature, ranging from 45.3% at 300 °C to 24.7% at 500 °C.

Reaction time has been also selected as independent variable, which can impact product distribution in microwave-assisted pyrolysis, by several authors [95,148,162–170]. For instance, Bu et al. studied the impact of reaction time and temperature on the behavior of Douglas fir during MAP [162,166]. The microwave power was established at 700 W and a commercial activated carbon (GAC 830 PLUS) was used. These authors obtained the maximum gas and liquid yields when the reaction time was 12 min. Although it was found that the retention time impacts products, its influence was not as critical as the catalyst/biomass ratio and reaction temperature. Yang et al. pyrolyzed Douglas fir sawdust, taking into account four retention times (41, 34, 27, and 20 min), as well as different microwave powers and preparation conditions of activated carbon from corn stover with phosphoric acid [167]. These authors noted that the reaction time, as well as phosphoric-acid-to-corn-stover ratio, significantly impacted the final temperature. On the other hand, the irradiation time enhanced biomass devolatilization, which influenced the activated carbon yield negatively. Finally, these authors obtained the optimal conditions to maximize the phenolic compound production, which required that they control the reaction time, microwave power, and activated carbon preparation. Ren et al. evaluated five different reaction times during the MAP of Douglas fir pellets. They conclude that syngas and bio-oil yields increased with higher temperatures and longer retention times, achieving maximum yields of syngas and bio-oil at 471 °C and 15 min. Moreover, the control of the reaction time and temperature allowed them to achieve the maximum yield of specific phenolic chemicals [165]. Some years later, these authors pyrolyzed Douglas fir pellets again but employing as catalyst a commercial activated carbon (GAC 830 PLUS) impregnated with iron powder [95]. The pyrolysis in that work was carried out at 700 W, and the reaction time as well as temperature were chosen as independent variables. They studied different retention times (13.66, 12, 8, 4, and 2.34 min), obtaining the maximum bio-oil yield (37.1 wt.%) at 450 °C and 8 min as the reaction time. The syngas yield (from 31–44 wt.%) increased with longer retention times and higher temperatures, peaking at 550 °C and 12 min, whereas the biochar yield (16.33 wt.%) was the lowest at those conditions. Furthermore, coking on the catalyst decreased with higher temperatures and longer reaction times, likely due to coke decomposition. Sellamuthu et al. pyrolyzed Adansonia kilima wood chips using microwave heating to synthesize the high-quality activated carbon used for lead (II) cation removal from wastewater [148]. They evaluated the influence of three process variables: reaction time (10, 12.5, and 15 min), concentration of an activating agent of K_2CO_3, and microwave power. In that work [148], it was concluded that all the process variables that had been considered impacted the removal percentages of Pb (II) cations obtained by the activated carbon produced. Miura et al. [164] pyrolyzed wood blocks using microwave technology. They used two different ovens and evaluated yields considering diverse irradiation times (3, 6, 7, 9, 10, 11, 12, 12.5, 15, and 18 min). They found that the microwave irradiation time influences char and volatile (tar) productions as well as char properties (pore radius and specific surface area). They found that there was a specific irradiation time in which the tar yield was maximized; however, it was diminished when different times to that one were applied. Additionally, it was concluded that the pore radius decreased, and the specific surface area grew with rising radiation time until reaching a stabilization time, after which char properties were not altered.

Even though microwave-assisted pyrolysis can deal with larger feedstocks than conventional pyrolysis, not having strict particle size constraints, size particle control can effectively modify the final product distribution of microwave-assisted pyrolysis, as well as its energy efficiency. For instance, Miura et al. [164] studied cylindrical wood blocks of Lalix leptolepiss of different sizes (diameters and heights of 300, 100, 80, and 60 mm) to evaluate the impact of wood size on microwave-assisted pyrolysis. It was concluded that the char yield was correlated with an equation where the electric power consumption per weight was inversely proportional to the diameter square, which means that a smaller wood block possesses a higher electric power consumption per unit weight than a higher one during microwave-assisted pyrolysis [164]. Higher specific power consumptions when biomass with smaller sizes is pyrolyzed were also reported by Vorhauer-Huget et al. [171]. The influence of feedstock size on MAP was also examined by Fricler et al. [85]. These authors tested different particle sizes (2000, 1000, 800, 250, and 140 μm) of pine sawdust mixed with straw, rice husk, or wheat bran, which were pyrolyzed in a microwave reactor at 700 W. It was concluded that there was a relationship between the particle size and pyrolyzed gas composition that increased the concentration of CO_2 and combustible gases as the biomass size increased from 140 to 800 μm. Smaller particles (140 μm) had a lower porosity and higher bulk density, obstructing the volatiles' liberation and limiting thermal decomposition, which produced the highest char yield and the lowest bio-oil and gas yields. The highest gas yield was achieved by pyrolyzing 250 μm particles, while 800 μm particles produced the highest bio-oil yield and a better gas quality (higher CO and lower CO_2 concentrations). Increasing the particle size to 1000–2000 μm slowed the oil and gas liberation, increased the char yield, and ended up in an incomplete pyrolysis, obtaining more residual coke. Consequently, Fricler et al. [85] verified that the biomass size allows us to define the pyrolysis product distribution. Nonetheless, there were authors such as Klinger et al. [172] who pyrolyzed large pellets (with a thickness of 5–6 mm) of thirty different biomass materials (residue, herbaceous, woody, waste, and blended materials) and smaller pellets (with thicknesses of 2–3 mm), concluding that the sample size studied in their work had a minimal influence on the liquid yields.

Forestry biomass has poor microwave absorption due to its low dielectric properties. To improve heat generation and transfer (by conduction) during the microwave pyrolysis of biomass, materials with higher dielectric properties named absorbers are added. Most of the articles analyzed in this review that use absorbents choose activated carbon [95,140,145,150,162,166,167,172–178], SiC [135,140,174,176,177,179–183], or biochar [145–147,154,156,157,160,174,181,184], with the biochar being from different biomass depending on the authors. For instance, Martin et al. [145] utilized as absorbent the biochar that had been produced in their continuous pilot plant from Pinus pinaster at 960 W. Furthermore, these authors also evaluated the impact of a different absorbent (DARCO G-60) which is a commercial activated carbon. Their results showed that the biochar generates lower liquid and gas yields in comparison with those obtained using activated carbon. Consequently, more biochar had to be added to induce pyrolysis reactions as effectively as activated carbon. Dong et al. [181] also employed biochar as a microwave absorber. They produced two different biochar that were obtained from rice husk and rice husk combined with $Fe(NO_3)_3{\cdot}9H_2O$. Additionally, these authors selected SiC as the absorber for the control group. In that work, it was found that both biochar selected were able to achieve 400 °C (pyrolysis temperature used in the study) in a time that was approximately six times less than the one needed by SiC. Shang et al. [174] demonstrated that strong absorbers such as K_2CO_3, SiC, activated carbon, NaOH, and the produced coke from the microwave process influence the fractional yields, acting as catalysts. It was found that K_2CO_3 and NaOH generated the products primarily consisting of gases, whereas SiC resulted in a higher solid product yield and lower liquid and gas yields. In the work carried out by Dutta et al. [157], biochar from willow wood was used as the absorber. These authors concluded that the ratio of biochar (16–32%) did not significantly affect the biochar yield obtained during microwave-assisted pyrolysis. However, they had concluded previously that biochar use increased biochar yield compared to microwave-assisted pyrolysis without biochar as an absorbent. The impact of biochar on biomass MAP was also analyzed by Ellison et al. [184], who pyrolyzed biomass with a specified amount of biochar from tallow tree and cane bagasse obtained in induction pyrolysis. They observed that pyrolysis temperatures were not achieved without a microwave absorber. Biochar use significantly affected the absorbed power and, consequently, pyrolysis temperatures, which increased four times when the biochar content grew from 0% to 10%. Nevertheless, a rise in biochar from 10% to 20% did not show a significant difference in temperatures. In their work, Wang et al. [154] concluded that, at the same temperature, the biochar (from the microwave-assisted pyrolysis of nanocellulose powder) ratio had effects on product yields in comparison with the MAP without absorber, increasing bio-oils and reducing gas yields. Nonetheless, as in the work carried out by Ellison et al. [184], it was found that the biochar/biomass ratio variable does not have a significant effect on production yields. Some authors such as Undri et al. [185] have assessed the impact of the blending procedure between biomass and microwave absorbers on the final products' properties. They compared three different mixing methods between two absorbers (Fe and carbon) and wood pellets, resulting in different heat distributions and thus affecting the final product characteristics only when Fe was used.

Not only has the effect of absorbers on microwave-assisted pyrolysis been studied, but also the impact of catalysts on the distribution and composition of products (biochar, liquid phase, and non-condensable gases). Different catalysts have been tested in MAP. For instance, Lestinsky et al. [186] pyrolyzed spruce sawdust using microwave radiation and catalysts (char from sawdust, and sawdust char doped with metal ions of nickel, cobalt, and iron). The pyrolysis was carried out in a microwave reactor at 400 W for 20 min. These authors found that the use of metallic catalysts, particularly cobalt and nickel, reduced the liquid yield, while the non-condensable fraction rose from 31.21 wt.% to more than 50 wt.%. Additionally, the pyrolysis gas composition was influenced by the type of catalyst, doubling the hydrogen content in comparison with the char alone when cobalt and nickel catalysts were used. These catalysts also reduced the concentrations of CO_2, C_xH_y, and CH_4. On the other hand, catalysts significantly reduced the amount of oil-phase in the liquid product compared to pyrolysis without using catalysts while the liquid composition remained largely unchanged. Li et al. [187] tested the influence of iron additives as well; nonetheless, they investigated the catalytic effect of $Fe(OH)_3$, $Fe_2(SO_4)_3$, and Fe_2O_3. The biomass utilized for microwave pyrolysis was pine sawdust. The pyrolysis was carried out at 480 W for 60 min and the biochar was obtained from microwave-assisted pyrolysis without additives and with iron additives. The results evidenced that $Fe_2(SO_4)_3$ and Fe_2O_3 promoted pyrolysis at lower temperatures, indicating less catalytic activity and a higher required heat for pyrolysis. Additionally, it was found that the Fe_2O_3 encouraged non-condensable gas production by 6.37% due to its catalytic properties, while $Fe_2(SO_4)_3$ enhanced bio-oil yield by 80%. The $Fe(OH)_3$ increased biochar yield by 23.05%. An iron additive ($Fe(NO_3)_3$) was also implemented during the microwave pyrolysis of wood wastes by Guo et al. [188], who also studied the influence of $ZnCl_2$ at distinct pyrolysis temperatures (800, 700, 600, 500, and 400 °C). These authors concluded that the pyrolysis temperature significantly affects the product distribution as has been verified by other authors. Guo et al. [188] observed that the bio-oil yield peaked at 600 °C but decreased at higher temperatures, while the biochar yield declined with increasing temperature and the content of gas phase increases up to 800 °C. On the other hand, the $Fe(NO_3)_3$ and $ZnCl_2$ additives enhanced biochar production. $ZnCl_2$ also favored the condensation of aromatic hydrocarbons into macromolecules, while $Fe(NO_3)_3$ increased biochar yield and decreased bio-oil yield. $ZnCl_2$ was noted for enhancing furfural production at low temperatures. Furthermore, the presence of $Fe(NO_3)_3$ and $ZnCl_2$ affected the production of other compounds like ethylbenzene, p-xylene, and m-xylene, which are important for chemical applications. Finally, it was found that the use of $Fe(NO_3)_3$ and $ZnCl_2$ generated the existence of Fe_3O_4 and ZnO in the biochar, indicating potential applications in photocatalysis and adsorption. Different additives were used by Li et al. [189] who investigated the microwave-assisted pyrolysis of *Eucalyptus* wood combined with different ratios of MoO_3 (3:1, 2:1, and 1:1) and distinct nitrogen sources (ammonium chloride, soybean straw, and food waste digestate) to produce nitrogen-rich biochar and bio-oil. The addition of MoO_3 increased gas yields and decreased solid yields, with the highest liquid yield achieved at a 2:1 *Eucalyptus*-wood-to-MoO_3 ratio. Soybean straw at 20% increased liquid yield, while excessive food waste digestate led to more solid products. The effect of the catalysts on microwave-assisted pyrolysis has also been analyzed through the activation energy because it can define the variation of the reaction rate parameters during the pyrolysis process, as found by Liu et al. [190]. Recent research has increasingly focused on the interaction between microwaves and metals, confirming that microwave–metal interactions can reduce energy consumption and intensify chemical processes. In studies like the one carried out by Li et al. [191], the interaction between microwave radiation and iron-based needle metals, during microwave-assisted pyrolysis of forestry wastes, has been evaluated. The presence of iron-based needle metals reduced the initial temperature for pyrolysis by 141 °C and increased reaction rates. The addition of more than one iron-based needle metal had a diminishing return on the heating efficiency and pyrolysis rate. An increased microwave power also elevated bed temperatures and heating rates, further accelerating the pyrolysis process. Finally, it is necessary to note that it has been tested that, under particular conditions, like the presence of metal catalysts and chlorine and low pyrolysis temperatures during microwave-assisted pyrolysis, dibenzofurans, polychlorinated dibenzo-p-dioxins, and naphthalenes are generated [192]. That study verified that feedstock composition as well as MAP conditions altered the production of these chemicals; consequently, it is possible to more safely conduct microwave pyrolysis if variables impacting product yields are studied in detail.

Different feedstocks combined with a varied range of additives and operation conditions cause distinct effects on the product yields, as can be gathered from Figure 4.

It can be observed that the biochar production generated in the articles that have been considered in this review varies from 3.37% to 98%. The forest species whose yields are over 90% combining different operation conditions have been pine, beech, baobab, and willow, with the highest obtained production (98.1%) being the one using biomass from willow. On the other hand, the MAP of biomass from Douglas fir has generated the lowest biochar yield (3.37%) under specific conditions. Additionally, it has been seen that the MAP of biomass from pine has generated a biochar production of 7% [184], whereas different conditions in the MAP of biomass from pine have increased biochar productions to 98% [150]. This variability in production underlines the need for more detailed studies of the influence of operation conditions, which have been demonstrated to have a decisive impact on the biochar yields from the same forest species. Similar behaviors have been detected for the bioliquid fraction. The productions found range from 0% (beech and bamboo) to 70% (pine). The MAP of biomass from pine has generated bio-oil yields from 2% [150] to 70% [172]. Significant production variations in the bioliquid phase have also been obtained with biomass from beech, obtaining values as low as 0% [185] and as high as 58% [129]. By comparison with the biochar productions, it can be concluded that, up to now, the maximum bio-oil productions are lower under the analyzed conditions. Finally, non-condensable gas productions have also experienced huge variations from 0% (biomass from pine) to 82.1% (gumwood), depending on the operation conditions and feedstocks. Non-condensable gas yields during the MAP of biomass from pine have fluctuated from 0% [150] to 78.5% [140]; similar tendencies have been seen on the feedstock denominated as wood pellets/wastes/sawdust without specifying the species, where the gas fraction changes from 3% [183] to 70% [179].

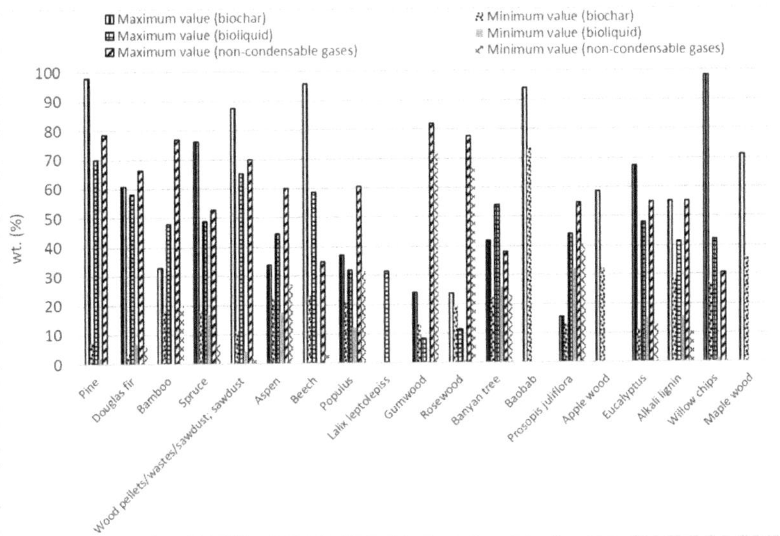

Figure 4. Impact of feedstock and operation conditions on biochar, bio-oil, and non-condensable gas yields.

MAP operation conditions not only impact yields, but also the biochar, bio-liquid, and non-condensable gas composition. Figure 5 displays the composition of biochars from different feedstocks. It is noticed that biochars possess carbon contents ranging from 29.98% (eucalyptus) to 93.96% (pine). It has been also detected that the biochars that have shown the highest variation of carbon content are those from eucalyptus in which the carbon concentration fluctuates from 29.98% to 80.24% [189].

Oxygen has turned out to be the component present in biochar with the second highest concentrations, with its contents ranging from 1.59% (biomass from pine) to 65.97% (biomass from eucalyptus). Nitrogen and hydrogen contents in biochars are low, being less than 4.05% and 7.17% respectively.

Although the bio-oil composition is complex, the different chemicals that can been detected are usually classified in different groups such as phenol and derivatives, kethone/aldehydes and derivatives, furan and derivatives, acids, esters, alcohols, naphthalene and derivatives, alkene, benzene and derivatives, nitrogenated compounds, and others. Figure 6 shows that phenols, kethone/aldehydes, and furans are the main types of chemicals in bio-oil from MAP. However, the content of these chemicals can vary enormously depending on operation conditions and feedstocks. For instance, the content of phenol and derivatives has fluctuated from 4% [159] to 78% [93] during the MAP of pine, while, in the MAP of Douglas fir, it varied from 2% [175] to 96% [154]. Variations in kethone/aldehydes (0.5–37%) [95,165] and furans (1–33%) [155,175], when operation conditions are modified, have also been found; however, these variations are lower.

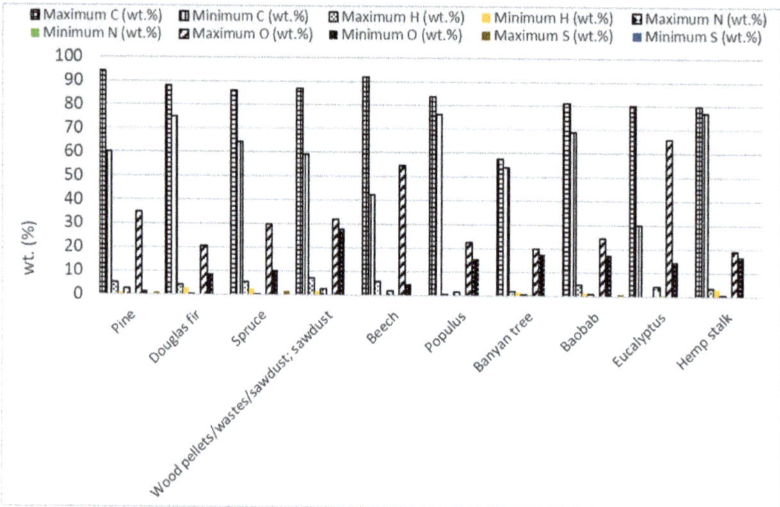

Figure 5. Impact of feedstock and operation conditions on biochar composition.

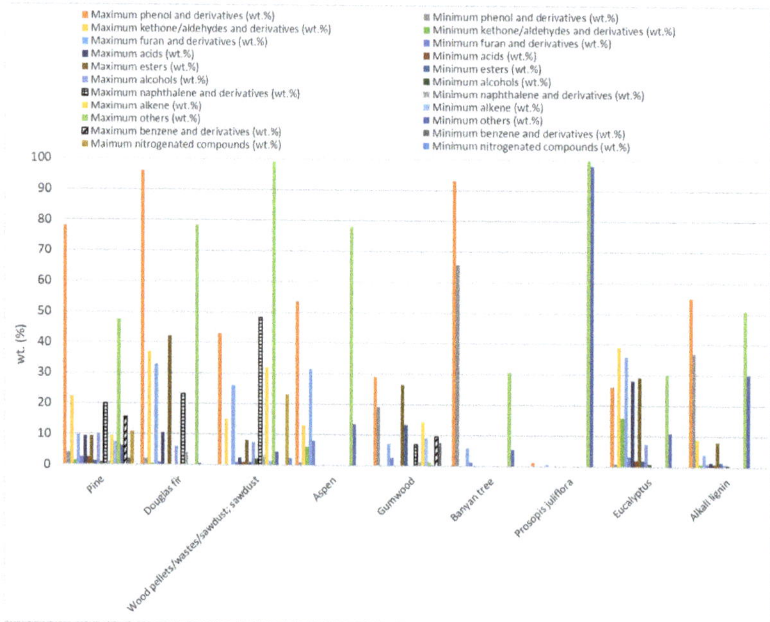

Figure 6. Impact of feedstock and operation conditions on bio-liquid composition.

The non-condensable gases' fraction of MAP mainly consists of hydrogen (H_2), carbon monoxide (CO), and carbon dioxide (CO_2). As it can be seen in Figure 7 the highest hydrogen content has been produced in the MAP of Douglas fir, in which a concentration of 85% was measured [154]. Nevertheless, when changing operation conditions, the hydrogen content during the MAP of Douglas fir was only 1% [155]. Significant variations in the CO concentration (0.07–51%) were also obtained during the MAP of pine [85,159] and feedstock denominated as wood pellets/wastes/sawdust, in which they were 7.85–41.79% [151,174]. Similar differences were measured in CO_2 content when biomass from pine (0.1–47.6%) [159,172] and feedstock denominated as wood pellets/wastes/sawdust (10.3–49%) [151,188] were pyrolyzed.

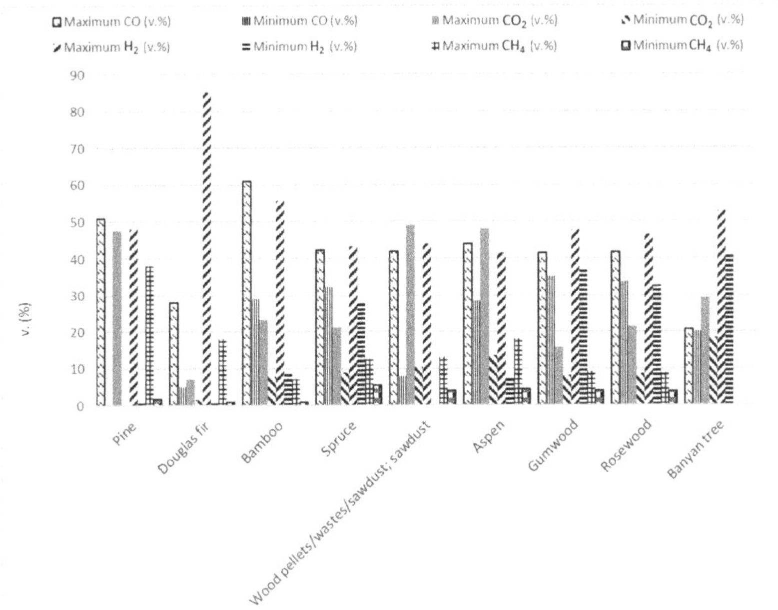

Figure 7. Impact of feedstock and operation conditions on non-condensable gas composition.

Studying the variations of biochar, bioliquid, and non-condensable gas yields, as well as in their composition, the highest differences have usually been found in those biomasses from forest species that have been the most studied. Consequently, more detailed studies about the combination of different additives, feedstocks, and specific operation conditions are required to reach desired yields and compositions using different feedstocks.

5. Future Directions

Despite the fact that MAP has emerged as an effective solution for biomass waste conversion, offering rapid, uniform heating and improved product quality by minimizing secondary reactions, it faces challenges such as temperature control and measurement, which are critical for optimizing pyrolysis performance. High temperatures can cause secondary reactions, breaking down condensable gases into non-condensable ones, while low temperatures may result in incomplete biomass decomposition, reducing the bio-oil yield. Accurate temperature measurement and control are difficult due to environmental interference, nonlinearity, and parameter uncertainty. Intelligent optimization algorithms have shown better performance than traditional controllers in managing these challenges. Additionally, as biomass materials lose moisture at high temperatures, they become less effective at absorbing microwave energy, so further research is needed on the dielectric properties of materials at elevated temperatures to enhance heating efficiency [22]. Furthermore, bio-oils have several drawbacks that obstruct their use as fuel, including a high moisture content, low heating value, thermal instability, and corrosiveness due to low pH and oxygenated compounds. The moisture and suspended char in bio-oil can lead to separation, filtration issues, and increased viscosity, which, in turn, cause operational challenges like equipment erosion, combustion inefficiencies, and pipeline blockages. The study and implementation of techniques such as hot vapor filtration and catalytic upgrading are needed to address these issues, improve combustion properties, and reduce the viscosity and other problematic challenges that have been found during bio-oil use [23]. The comprehensive process modeling of various conversion pathways is essential in order to scale up MAP for industrial use. This requires extensive research on parameters and kinetics, especially since existing studies on the dielectric properties of different forestry biomass feedstocks are limited. For instance, Vorhauer-Huget et al. analyzed the dielectric properties of several wood samples before and after pyrolysis. Measurements were conducted at room temperature in the 2–3 GHz range. Their results emphasized the importance of understanding secondary reactions and the reduced microwave penetration depth in samples with high loss factors [171]. McKeown et al. [193] measured the dielectric properties of pine and hardwood pellets at microwave frequencies across a temperature range of 10 °C to 50 °C and moisture contents varying from 4.9% to 16.0%. Wang et al. [194] presented a method for accurately determining the dielectric properties of different biomass (bamboo, poplar wood, rice straw, and corn straw) using paraffin wax and a dielectric mixing rule. Ellison et al. [195] characterized the dielectric properties of the pulverized biomass of pine sawdust, Chinese tallow tree wood, live oak, and energy cane bagasse, and biochar mixtures in the range from 0.5 GHz to 20 GHz at room temperature. These authors concluded that dielectric loss and constant suffer a quadratic rise with increasing biochar content. Although important, these results are not precise enough to establish the dielectric behavior of those feedstocks during microwave-assisted pyrolysis, in which higher temperatures are implemented. This was shown by Luo et al. [93], who highlighted how the dielectric properties of wood sawdust changed with temperature, affecting the microwave heating behavior and altering reaction pathways. Consequently, studies such as those carried out in agriculture biomass on dielectric properties [196,197] has to be carried out by studying forestry biomass. Expanding the research to include a variety of feedstocks will provide valuable data for accurate modeling. Optimizing key factors such as microwave power, reaction duration, temperature, pressure, and catalyst loading can enhance heating efficiency and reduce energy consumption. Understanding the dielectric properties of biomass and fine-tuning these parameters will promote more sustainable MAP by minimizing the release of undesirable substances and maximizing resource and energy utilization without compromising overall efficiency. Additionally, identifying the catalytic mechanisms is crucial for determining the suitability of different microwave absorbers/catalysts for various conversion processes on a larger scale [60]. Conducting more techno-economic assessments of MAP for FB is needed in order to assess the economic viability of the process. The primary products intended for sale are bio-oil and biochar, while syngas is utilized for power generation, contributing to the self-sustainability of the operation. Additionally, the search for new applications for the products from the MAP of forestry biomass as the purification of residual streams might promote the full-scale implementation of this technology, contributing to the bio economy.

6. Conclusions

The microwave-assisted pyrolysis (MAP) bioconversion route of forestry biomass is reviewed in this paper, as an attempt to discover the main species that have been studied and variables influencing this method. It was concluded that around 20% of the reviewed articles focused on evaluating the efficiency of MAP using biomass derived from pine, 14% concentrated on biomass categorized as wood pellets/wastes/sawdust without specifying the species, 12% studied biomass from Douglas fir, and 8% of the works evaluated biomass from spruce, with those being the most studied forestry biomass. Additionally, it was found that MAP is significantly influenced by various operation conditions such as microwave power, temperature, biomass size, and the use of absorbers/catalysts. Research has shown that increasing microwave power leads to higher heating rates, faster pyrolysis, and varied product yields. Higher microwave power typically results in increased gas yields and decreased biochar production, with optimal bio-oil yields observed at moderate power levels. Similarly, the pyrolysis temperature plays a crucial role, with higher temperatures generally promoting an increase in gas yield and reducing biochar yield. The composition of bio-oil and non-condensable gases is also temperature-dependent, with specific compounds being favored at different temperatures. Reaction time is another critical factor, with longer times enhancing gas and bio-oil yields while reducing biochar yield. Additionally, biomass particle size affects energy efficiency and product distribution, with smaller particles leading to higher char yields due to a limited volatile release. The use of absorbers, such as activated carbon, SiC, and biochar, improves microwave absorption and heat generation, further influencing the efficiency and outcomes of the pyrolysis process. Overall, the choice of operating conditions in MAP can be tailored to optimize the yields of desired products, making it a versatile and efficient method for biomass conversion. MAP feasibility, although promising due to its faster heating rate, shorter reaction time, lower energy consumption, higher product quality, and lower production cost compared to conventional pyrolysis, still presents some challenges such as precise temperature measurement and control; more detailed studies of the dielectric properties of forestry biomass at high temperatures are needed in order to improve efficiency during microwave heating.

Author Contributions: Conceptualization, I.F. and S.F.P.; writing—original draft preparation, I.F. and S.F.P.; writing—review and editing, J.F.-F. and T.L.; funding acquisition, S.F.P. All authors have read and agreed to the published version of the manuscript.

Funding: This research was funded by the Government of Cantabria through "Contrato Programa" which provides funding for the implementation of "Bridge Projects 2023" (Development of forestry bio economy through microwave-assisted stepwise pyrolysis, project PID2022-138142OB-I00).

Data Availability Statement: Data sharing is not applicable.

Conflicts of Interest: The authors declare no conflicts of interest.

References

1. Sridevi, V.; Surya, D.V.; Reddy, B.R.; Shah, M.; Gautam, R.; Kumar, T.H.; Puppala, H.; Pritam, K.S.; Basak, T. Challenges and opportunities in the production of sustainable hydrogen from lignocellulosic biomass using microwave-assisted pyrolysis: A review. *Int. J. Hydrogen Energy* **2023**, *52 Pt A*, 507–531. [CrossRef]
2. Hoang, A.T.; Nižetić, S.; Ong, H.C.; Mofijur, M.; Ahmed, S.F.; Ashok, B.; Bui, V.T.V.; Chau, M.Q. Insight into the recent advances of microwave pretreatment technologies for the conversion of lignocellulosic biomass into sustainable biofuel. *Chemosphere* **2021**, *281*, 130878. [CrossRef] [PubMed]
3. International Energy Agency. World Energy Outlook 2023 | Enhanced Reader. 2023. Available online: https://www.oecd-ilibrary.org/docserver/827374a6-en.pdf?expires=1727372493&id=id&accname=guest&checksum=D09E08F6E8794E5772E10F1FABA66EE0 (accessed on 25 September 2024).
4. Motasemi, F.; Ani, F. A review on microwave-assisted production of biodiesel. *Renew. Sustain. Energy Rev.* **2012**, *16*, 4719–4733. [CrossRef]
5. Mohamed, B.A.; Bilal, M.; Salama, E.S.; Periyasamy, S.; Fattah, I.M.R.; Ruan, R.; Awasthi, M.K.; Leng, L. Phenolic-rich bio-oil production by microwave catalytic pyrolysis of switchgrass: Experimental study, life cycle assessment, and economic analysis. *J. Clean. Prod.* **2022**, *366*, 132668. [CrossRef]
6. Mutsengerere, S.; Chihobo, C.H.; Musademba, D.; Nhapi, I. A review of operating parameters affecting bio-oil yield in microwave pyrolysis of lignocellulosic biomass. *Renew. Sustain. Energy Rev.* **2019**, *104*, 328–336. [CrossRef]
7. Zhu, H.; Saddler, J.; Bi, X. An economic and environmental assessment of biofuel produced via microwave-assisted catalytic pyrolysis of forest residues. *Energy Convers. Manag.* **2022**, *263*, 115723. [CrossRef]
8. Selvam, S.M.; Paramasivan, B. Microwave assisted carbonization and activation of biochar for energy-environment nexus: A review. *Chemosphere* **2022**, *286*, 131631. [CrossRef]
9. Zhang, J.; Tahmasebi, A.; Omoriyekomwan, J.E.; Yu, J. Direct synthesis of hollow carbon nanofibers on bio-char during microwave pyrolysis of pine nut shell. *J. Anal. Appl. Pyrolysis* **2018**, *130*, 142–148. [CrossRef]

10. Makepa, D.C.; Chihobo, C.H.; Manhongo, T.T.; Musademba, D. Life-cycle assessment of microwave-assisted pyrolysis of pine sawdust as an emerging technology for biodiesel production. *Results Eng.* **2023**, *20*, 101480. [CrossRef]
11. Arias, A.; Costa, C.E.; Feijoo, G.; Moreira, M.T.; Domingues, L. Process modeling, environmental and economic sustainability of the valorization of whey and eucalyptus residues for resveratrol biosynthesis. *Waste Manag.* **2023**, *172*, 226–234. [CrossRef]
12. Farag, S.; Fu, D.; Jessop, P.G.; Chaouki, J. Detailed compositional analysis and structural investigation of a bio-oil from microwave pyrolysis of kraft lignin. *J. Anal. Appl. Pyrolysis* **2014**, *109*, 249–257. [CrossRef]
13. Penín, L.; López, M.; Santos, V.; Alonso, J.L.; Parajó, J.C. Technologies for Eucalyptus wood processing in the scope of biorefineries: A comprehensive review. *Bioresour. Technol.* **2020**, *311*, 123528. [CrossRef]
14. Hassan, S.S.; Williams, G.A.; Jaiswal, A.K. Emerging technologies for the pretreatment of lignocellulosic biomass. *Bioresour. Technol.* **2018**, *262*, 310–318. [CrossRef] [PubMed]
15. Saravanan, A.; Yaashikaa, P.R.; Kumar, P.S.; Thamarai, P.; Deivayanai, V.C.; Rangasamy, G. A comprehensive review on techno-economic analysis of biomass valorization and conversional technologies of lignocellulosic residues. *Ind. Crop. Prod.* **2023**, *200*, 116822. [CrossRef]
16. Iglesias, S.P.; Miyazaki, M.R.; Mariano, A.P.; Franco, T.T. Techno-economic assessment of bio-oil produced from Eucalyptus forestry residues. *Ind. Crop. Prod.* **2021**, *171*, 113936. [CrossRef]
17. Wang, X.; Morrison, W.; Du, Z.; Wan, Y.; Lin, X.; Chen, P.; Ruan, R. Biomass temperature profile development and its implications under the microwave-assisted pyrolysis condition. *Appl. Energy* **2012**, *99*, 386–392. [CrossRef]
18. Konstantinavičienė, J. Assessment of Potential of Forest Wood Biomass in Terms of Sustainable Development. *Sustain.* **2023**, *15*, 13871. [CrossRef]
19. Gonçalves, M.; Freire, F.; Garcia, R. Material flow analysis of forest biomass in Portugal to support a circular bioeconomy. *Resour. Conserv. Recycl.* **2021**, *169*. [CrossRef]
20. Chen, X.; Che, Q.; Li, S.; Liu, Z.; Yang, H.; Chen, Y.; Wang, X.; Shao, J.; Chen, H. Recent developments in lignocellulosic biomass catalytic fast pyrolysis: Strategies for the optimization of bio-oil quality and yield. *Fuel Process. Technol.* **2019**, *196*, 106180. [CrossRef]
21. Salema, A.A.; Ani, F.N. Pyrolysis of oil palm empty fruit bunch biomass pellets using multimode microwave irradiation. *Bioresour. Technol.* **2012**, *125*, 102–107. [CrossRef]
22. Foong, S.Y.; Liew, R.K.; Yang, Y.; Cheng, Y.W.; Yek, P.N.Y.; Wan Mahari, W.A.; Lee, X.Y.; Han, C.S.; Vo, D.V.N.; Van Le, Q.; et al. Valorization of biomass waste to engineered activated biochar by microwave pyrolysis: Progress, challenges, and future directions. *Chem. Eng. J.* **2020**, *389*, 124401. [CrossRef]
23. Dhyani, V.; Bhaskar, T. A comprehensive review on the pyrolysis of lignocellulosic biomass. *Renew. Energy* **2018**, *129*, 695–716. [CrossRef]
24. Wang, S.; Dai, G.; Yang, H.; Luo, Z. Lignocellulosic biomass pyrolysis mechanism: A state-of-the-art review. *Prog. Energy Combust. Sci.* **2017**, *62*, 33–86. [CrossRef]
25. Papari, S.; Hawboldt, K. A review on the pyrolysis of woody biomass to bio-oil: Focus on kinetic models. *Renew. Sustain. Energy Rev.* **2015**, *52*, 1580–1595. [CrossRef]
26. Mohamed, B.A.; Ellis, N.; Kim, C.S.; Bi, X. Microwave-assisted catalytic biomass pyrolysis: Effects of catalyst mixtures. *Appl. Catal. B Environ.* **2019**, *253*, 226–234. [CrossRef]
27. Fan, J.; Shuttleworth, P.S.; Gronnow, M.; Breeden, S.W.; Clark, J.H.; Macquarrie, D.J.; Budarin, V.L. Influence of Density on Microwave Pyrolysis of Cellulose. *ACS Sustain. Chem. Eng.* **2018**, *6*, 2916–2920. [CrossRef]
28. Ge, S.; Yek, P.N.Y.; Cheng, Y.W.; Xia, C.; Wan Mahari, W.A.; Liew, R.K.; Peng, W.; Yuan, T.Q.; Tabatabaei, M.; Aghbashlo, M.; et al. Progress in microwave pyrolysis conversion of agricultural waste to value-added biofuels: A batch to continuous approach. *Renew. Sustain. Energy Rev.* **2021**, *135*, 110148. [CrossRef]
29. Tabakaev, R.; Kalinich, I.; Dimitryuk, I.; Asilbekov, A.; Astafev, A.; Ibraeva, K.; Shanenkov, I.; Mostovshchikov, A.; Chumerin, P. Experimental study of microwave processing of pine nut shells into a high-calorie gas: Main results and physicochemical features. *J. Anal. Appl. Pyrolysis* **2023**, *176*, 106264. [CrossRef]
30. Li, J.; Dai, J.; Liu, G.; Zhang, H.; Gao, Z.; Fu, J.; He, Y.; Huang, Y. Biochar from microwave pyrolysis of biomass: A review. *Biomass Bioenergy* **2016**, *94*, 228–244. [CrossRef]
31. Akbari, M.; Kumar, A. The development of data-intensive techno-economic models for the comparison of renewable natural gas production from six different biomass feedstocks for the decarbonization of energy demand sectors. *Fuel* **2024**, *358*, 130107. [CrossRef]
32. Patel, P.; Vaezi, M.; Billal, M.M.; Kumar, A. Development of data-intensive techno-economic models for the assessment of a biomass, waste heat, and MSW integrated waste-to-electricity facility. *Resour. Conserv. Recycl. Adv.* **2023**, *20*, 200188. [CrossRef]
33. Tisserant, A.; Hu, X.; Liu, Q.; Xie, Z.; Zhao, W.; Cherubini, F. Biochar and Its Potential to Deliver Negative Emissions and Better Soil Quality in Europe. *Earth's Futur.* **2023**, *11*, e2022EF003246. [CrossRef]
34. França, L.C.d.J.; e Silva, C.S.J.; Mucida, D.P.; da Costa, J.S.; Gomide, L.R. Towards renewable energy projects under sustainable watersheds principles for forest biomass supply. *Biomass Bioenergy* **2023**, *176*, 106916. [CrossRef]
35. Brunel, C.; Farnet Da Silva, A.M.; Lerch, T.Z.; Gros, R. Influence of tree residue retention in Mediterranean forest on soil microbial communities responses to frequent warming and drying events. *Eur. J. Soil Biol.* **2023**, *118*, 103541. [CrossRef]

36. Freer-Smith, P.; Bailey-Bale, J.H.; Donnison, C.L.; Taylor, G. The good, the bad, and the future: Systematic review identifies best use of biomass to meet air quality and climate policies in California. *GCB Bioenergy* **2023**, *15*, 1312–1328. [CrossRef]
37. Natural Resources Canada. *Is Forest Bioenergy Good for the Environment?* Natural Resources Canada: Ottawa, ON, Canada, 2010.
38. Serra, R.; Niknia, I.; Paré, D.; Titus, B.; Gagnon, B.; Laganière, J. From conventional to renewable natural gas: Can we expect GHG savings in the near term? *Biomass Bioenergy* **2019**, *131*, 105396. [CrossRef]
39. Jordan, M.; Meisel, K.; Dotzauer, M.; Schröder, J.; Cyffka, K.F.; Dögnitz, N.; Schmid, C.; Lenz, V.; Naumann, K.; Daniel-Gromke, J.; et al. The controversial role of energy crops in the future German energy system: The trade offs of a phase-out and allocation priorities of the remaining biomass residues. *Energy Rep.* **2023**, *10*, 3848–3858. [CrossRef]
40. Rijal, P.; Bras, P.; Garrido, S.; Matias, J.; Pimentel, C.; Carvalho, H. Residual Forestry Biomass Supply Chain: A Mapping Approach. *Int. J. Ind. Eng. Manag.* **2023**, *14*, 244–256. [CrossRef]
41. Negi, H.; Suyal, D.C.; Soni, R.; Giri, K.; Goel, R. Indian Scenario of Biomass Availability and Its Bioenergy-Conversion Potential. *Energies* **2023**, *16*, 5805. [CrossRef]
42. Gan, X.; Guo, B.; Ma, Z.; Fang, M.; Yan, Y.; Liu, W. The Effect of Forest Growth Rate on Climate Change Impacts of Logging Residue Utilization. *Atmosphere* **2023**, *14*, 1270. [CrossRef]
43. Watanabe, M.D.B.; Hu, X.; Ballal, V.; Cavalett, O.; Cherubini, F. Climate change mitigation potentials of on grid-connected Power-to-X fuels and advanced biofuels for the European maritime transport. *Energy Convers. Manag. X* **2023**, *20*, 100418. [CrossRef]
44. Hu, J.; Jåstad, E.O.; Bolkesjø, T.F.; Rørstad, P.K. Impact of large-scale Bio-CCS deployment on forest biomass competition and forest industry production. *Biomass Bioenergy* **2023**, *175*, 106896. [CrossRef]
45. Ibitoye, S.E.; Mahamood, R.M.; Jen, T.C.; Loha, C.; Akinlabi, E.T. An overview of biomass solid fuels: Biomass sources, processing methods, and morphological and microstructural properties. *J. Bioresour. Bioprod.* **2023**, *8*, 333–360. [CrossRef]
46. Rehan, M.; Amir Raza, M.; Ghani Abro, A.; M Aman, M.; Mohammad Ibrahim Ismail, I.; Sattar Nizami, A.; Imtiaz Rashid, M.; Summan, A.; Shahzad, K.; Ali, N. A sustainable use of biomass for electrical energy harvesting using distributed generation systems. *Energy* **2023**, *278*, 128036. [CrossRef]
47. Srivastava, R.K.; Shetti, N.P.; Reddy, K.R.; Nadagouda, M.N.; Badawi, M.; Bonilla-Petriciolet, A.; Aminabhavi, T.M. Valorization of biowastes for clean energy production, environmental depollution and soil fertility. *J. Environ. Manag.* **2023**, *332*, 117410. [CrossRef]
48. Zhou, C.H.; Xia, X.; Lin, C.X.; Tong, D.S.; Beltramini, J. Beltramini Catalytic conversion of lignocellulosic biomass to fine chemicals and fuels. *Chem. Soc. Rev.* **2011**, *40*, 5588–5617. [CrossRef]
49. Wang, F.; Ouyang, D.; Zhou, Z.; Page, S.J.; Liu, D.; Zhao, X. Lignocellulosic biomass as sustainable feedstock and materials for power generation and energy storage. *J. Energy Chem.* **2021**, *57*, 247–280. [CrossRef]
50. European Union. Infographics on Biomass Sources and Uses in the EU-27; European Commission: 2022. Available online: https://knowledge4policy.ec.europa.eu/publication/infographics-biomass-sources-uses-eu-27-2017-data_en (accessed on 25 September 2024).
51. Mujtaba, M.; Fernandes Fraceto, L.; Fazeli, M.; Mukherjee, S.; Savassa, S.M.; Araujo de Medeiros, G.; do Espírito Santo Pereira, A.; Mancini, S.D.; Lipponen, J.; Vilaplana, F. Lignocellulosic biomass from agricultural waste to the circular economy: A review with focus on biofuels, biocomposites and bioplastics. *J. Clean. Prod.* **2023**, *402*, 136815. [CrossRef]
52. Morgan, H.M.; Bu, Q.; Liang, J.; Liu, Y.; Mao, H.; Shi, A.; Lei, H.; Ruan, R. A review of catalytic microwave pyrolysis of lignocellulosic biomass for value-added fuel and chemicals. *Bioresour. Technol.* **2017**, *230*, 112–121. [CrossRef]
53. Al Shra'Ah, A.; Helleur, R. Microwave pyrolysis of cellulose at low temperature. *J. Anal. Appl. Pyrolysis* **2014**, *105*, 91–99. [CrossRef]
54. Wang, N.; Tahmasebi, A.; Yu, J.; Xu, J.; Huang, F.; Mamaeva, A. A Comparative study of microwave-induced pyrolysis of lignocellulosic and algal biomass. *Bioresour. Technol.* **2015**, *190*, 89–96. [CrossRef] [PubMed]
55. European Commission Forests. Available online: https://international-partnerships.ec.europa.eu/publications/forest-partnerships-factsheets_en (accessed on 20 August 2024).
56. Chiang, L.E.; Castro, F.A.; Molina, F.A. Socioeconomic and environmental benefits of substituting firewood with charcoal briquettes produced from biomass residues in the Forestry Belt in Chile. *Energy Sustain. Dev.* **2023**, *77*, 101341. [CrossRef]
57. Wang, R.; Cai, W.; Yu, L.; Li, W.; Zhu, L.; Cao, B.; Li, J.; Shen, J.; Zhang, S.; Nie, Y.; et al. A high spatial resolution dataset of China's biomass resource potential. *Sci. Data* **2023**, *10*, 384. [CrossRef] [PubMed]
58. Silva, J.P.; Teixeira, S.; Teixeira, J.C. Characterization of the physicochemical and thermal properties of different forest residues. *Biomass Bioenergy* **2023**, *175*, 106870. [CrossRef]
59. Węgiel, A.; Jakubowski, J.; Molińska-Glura, M.; Polowy, K.; Węgiel, J.; Gornowicz, R. Effect of logging residue removal and mechanical site preparation on productivity of the subsequent Scots pine (*Pinus sylvestris* L.) stands. *Ann. For. Sci.* **2023**, *80*, 5. [CrossRef]
60. Arpia, A.A.; Chen, W.H.; Lam, S.S.; Rousset, P.; de Luna, M.D.G. Sustainable biofuel and bioenergy production from biomass waste residues using microwave-assisted heating: A comprehensive review. *Chem. Eng. J.* **2021**, *403*, 126233. [CrossRef]
61. Motasemi, F.; Afzal, M.T. A review on the microwave-assisted pyrolysis technique. *Renew. Sustain. Energy Rev.* **2013**, *28*, 317–330. [CrossRef]

114. Seow, Y.X.; Tan, Y.H.; Kansedo, J.; Tan, I.S.; Chin, B.L.F.; Mubarak, N.M.; Bin Mohiddin, M.N.; Yek, P.N.Y.; Chan, Y.S.; Abdullah, M.O. Pyrolysis assessment of palm kernel shell waste valorization to sulfonated magnetic biochar from techno-economic and energy perspectives. *Discov. Appl. Sci.* **2024**, *6*, 398. [CrossRef]
115. Muniyappan, D.; Lima, G.R.; Pereira, A.O.; Gopi, R.; Ramanathan, A. Multivariate combined optimization strategy and comparative life-cycle assessment of biomass and plastic residues via microwave co-pyrolysis approach towards a sustainable synthesis of renewable hydrocarbon fuel. *J. Environ. Chem. Eng.* **2023**, *11*, 111436. [CrossRef]
116. Fodah, A.E.M.; Abdelwahab, T.A.M. Process optimization and technoeconomic environmental assessment of biofuel produced by solar powered microwave pyrolysis. *Sci. Rep.* **2022**, *12*, 12572. [CrossRef]
117. Foong, S.Y.; Chan, Y.H.; Yek, P.N.Y.; Lock, S.S.M.; Chin, B.L.F.; Yiin, C.L.; Lan, J.C.W.; Lam, S.S. Microwave-assisted pyrolysis in biomass and waste valorisation: Insights into the life-cycle assessment (LCA) and techno-economic analysis (TEA). *Chem. Eng. J.* **2024**, *491*, 151942. [CrossRef]
118. Ren, X.; Shanb Ghazani, M.; Zhu, H.; Ao, W.; Zhang, H.; Moreside, E.; Zhu, J.; Yang, P.; Zhong, N.; Bi, X. Challenges and opportunities in microwave-assisted catalytic pyrolysis of biomass: A review. *Appl. Energy* **2022**, *315*, 118970. [CrossRef]
119. Zhang, Y.; Chen, P.; Liu, S.; Fan, L.; Zhou, N.; Min, M.; Cheng, Y.; Peng, P.; Anderson, E.; Wang, Y.; et al. Microwave-Assisted Pyrolysis of Biomass for Bio-Oil Production. In *Pyrolysis*; InTech: Houston, TX, USA, 2017.
120. Fernandes, A.; Cruz-Lopes, L.; Esteves, B.; Evtuguin, D.V. Microwaves and Ultrasound as Emerging Techniques for Lignocellulosic Materials. *Materials* **2023**, *16*, 7351. [CrossRef]
121. Ravikumar, C.; Senthil Kumar, P.; Subhashni, S.K.; Tejaswini, P.V.; Varshini, V. Microwave assisted fast pyrolysis of corn cob, corn stover, saw dust and rice straw: Experimental investigation on bio-oil yield and high heating values. *Sustain. Mater. Technol.* **2017**, *11*, 19–27. [CrossRef]
122. Thue, P.S.; Adebayo, M.A.; Lima, E.C.; Sieliechi, J.M.; Machado, F.M.; Dotto, G.L.; Vaghetti, J.C.P.; Dias, S.L.P. Preparation, characterization and application of microwave-assisted activated carbons from wood chips for removal of phenol from aqueous solution. *J. Mol. Liq.* **2016**, *223*, 1067–1080. [CrossRef]
123. Giorcelli, M.; Das, O.; Sas, G.; Försth, M.; Bartoli, M. A review of bio-oil production through microwave-Assisted pyrolysis. *Processes* **2021**, *9*, 561. [CrossRef]
124. Yang, Y.; Shahbeik, H.; Shafizadeh, A.; Masoudnia, N.; Rafiee, S.; Zhang, Y.; Pan, J.; Tabatabaei, M.; Aghbashlo, M. Biomass microwave pyrolysis characterization by machine learning for sustainable rural biorefineries. *Renew. Energy* **2022**, *201*, 70–86. [CrossRef]
125. Pritam, K.; Puppala, H.; Palla, S.; Suriapparao, D.V.; Basak, T. A two-step hybrid multi-criteria approach to analyze the significance of parameters affecting microwave-assisted pyrolysis. *Process Saf. Environ. Prot.* **2023**, *171*, 975–985. [CrossRef]
126. Zheng, A.; Zhao, K.; Sun, J.; Jiang, L.; Zhao, Z.; Huang, Z.; Wei, G.; He, F.; Li, H. Effect of microwave-assisted organosolv fractionation on the chemical structure and decoupling pyrolysis behaviors of waste biomass. *J. Anal. Appl. Pyrolysis* **2018**, *131*, 120–127. [CrossRef]
127. Wang, X.; Chen, H.; Luo, K.; Shao, J.; Yang, H. The influence of microwave drying on biomass pyrolysis. *Energy and Fuels* **2008**, *22*, 67–74. [CrossRef]
128. Liang, J.; Xu, X.; Yu, Z.; Chen, L.; Liao, Y.; Ma, X. Effects of microwave pretreatment on catalytic fast pyrolysis of pine sawdust. *Bioresour. Technol.* **2019**, *293*, 122080. [CrossRef]
129. Karthäuser, J.; Biziks, V.; Frauendorf, H.; Mai, C.; Militz, H. Vacuum Low-Temperature Microwave-Assisted Pyrolysis of Technical Lignins. *Polymers* **2022**, *14*, 3383. [CrossRef]
130. Venegas-Vásconez, D.; Arteaga-Pérez, L.E.; Aguayo, M.G.; Romero-Carrillo, R.; Guerrero, V.H.; Tipanluisa-Sarchi, L.; Alejandro-Martín, S. Analytical Pyrolysis of Pinus radiata and Eucalyptus globulus: Effects of Microwave Pretreatment on Pyrolytic Vapours Composition. *Polymers* **2023**, *15*, 3790. [CrossRef]
131. Yang, X.; Cui, C.; Zheng, A.; Zhao, Z.; Wang, C.; Xia, S.; Huang, Z.; Wei, G.; Li, H. Ultrasonic and microwave assisted organosolv pretreatment of pine wood for producing pyrolytic sugars and phenols. *Ind. Crop. Prod.* **2020**, *157*, 112921. [CrossRef]
132. Duan, X.; Srinivasakannan, C.; Peng, J.; Zhang, L.; Zhang, Z. Comparison of activated carbon prepared from Jatropha hull by conventional heating and microwave heating. *Biomass Bioenergy* **2011**, *35*, 3920–3926. [CrossRef]
133. Mašek, O.; Budarin, V.; Gronnow, M.; Crombie, K.; Brownsort, P.; Fitzpatrick, E.; Hurst, P. Microwave and slow pyrolysis biochar—Comparison of physical and functional properties. *J. Anal. Appl. Pyrolysis* **2013**, *100*, 41–48. [CrossRef]
134. Merckel, R.D. Fast and Microwave-Induced Pyrolysis Bio-Oil from Eucalyptus Grandis: Possibilities for Upgrading. Master's Thesis, University of Pretoria (South Africa), Pretoria, South Africa, 2014.
135. Du, J.; Liu, P.; Zuo-Hua, L.; Da-Gui, S.; Chang-Yuan, T. Fast pyrolysis of biomass for bio-oil with ionic liquid and microwave irradiation. *J. Fuel Chem. Technol.* **2010**, *38*, 554–559. [CrossRef]
136. Wan, Y.; Chen, P.; Zhang, B.; Yang, C.; Liu, Y.; Lin, X.; Ruan, R. Microwave-assisted pyrolysis of biomass: Catalysts to improve product selectivity. *J. Anal. Appl. Pyrolysis* **2009**, *86*, 161–167. [CrossRef]
137. Bartoli, M.; Rosi, L.; Giovannelli, A.; Frediani, P.; Frediani, M. Bio-oil from residues of short rotation coppice of poplar using a microwave assisted pyrolysis. *J. Anal. Appl. Pyrolysis* **2016**, *119*, 224–232. [CrossRef]
138. Lo, S.L.; Huang, Y.F.; Chiueh, P.T.; Kuan, W.H. Microwave Pyrolysis of Lignocellulosic Biomass. *Energy Procedia* **2017**, *105*, 41–46. [CrossRef]

139. Huang, Y.F.; Chiueh, P.T.; Kuan, W.H.; Lo, S.L. Effects of lignocellulosic composition and microwave power level on the gaseous product of microwave pyrolysis. *Energy* **2015**, *89*, 974–981. [CrossRef]
140. Shi, K.; Yan, J.; Menéndez, J.A.; Luo, X.; Yang, G.; Chen, Y.; Lester, E.; Wu, T. Production of H2-Rich Syngas From Lignocellulosic Biomass Using Microwave-Assisted Pyrolysis Coupled With Activated Carbon Enabled Reforming. *Front. Chem.* **2020**, *8*, 3. [CrossRef]
141. Wauts, J. Catalytic Microwave Pyrolysis to Produce Upgraded Bio-Oil. Master's Thesis, University of Pretoria (South Africa), Pretoria, South Africa, 2016; pp. 1–111.
142. Gronnow, M.J.; Budarin, V.L.; Mašek, O.; Crombie, K.N.; Brownsort, P.A.; Shuttleworth, P.S.; Hurst, P.R.; Clark, J.H. Torrefaction/biochar production by microwave and conventional slow pyrolysis—Comparison of energy properties. *GCB Bioenergy* **2013**, *5*, 144–152. [CrossRef]
143. Huang, Y.F.; Chiueh, P.T.; Kuan, W.H.; Lo, S.L. Product distribution and heating performance of lignocellulosic biomass pyrolysis using microwave heating. *Energy Procedia* **2018**, *152*, 910–915. [CrossRef]
144. Dong, Q.; Xiong, Y. Kinetics study on conventional and microwave pyrolysis of moso bamboo. *Bioresour. Technol.* **2014**, *171*, 127–131. [CrossRef] [PubMed]
145. Martín, M.T.; Sanz, A.B.; Nozal, L.; Castro, F.; Alonso, R.; Aguirre, J.L.; González, S.D.; Matía, M.P.; Novella, J.L.; Peinado, M.; et al. Microwave-assisted pyrolysis of Mediterranean forest biomass waste: Bioproduct characterization. *J. Anal. Appl. Pyrolysis* **2017**, *127*, 278–285. [CrossRef]
146. Parvez, A.M.; Afzal, M.T.; Jiang, P.; Wu, T. Microwave-assisted biomass pyrolysis polygeneration process using a scaled-up reactor: Product characterization, thermodynamic assessment and bio-hydrogen production. *Biomass Bioenergy* **2020**, *139*, 105651. [CrossRef]
147. Nhuchhen, D.R.; Afzal, M.T.; Dreise, T.; Salema, A.A. Characteristics of biochar and bio-oil produced from wood pellets pyrolysis using a bench scale fixed bed, microwave reactor. *Biomass Bioenergy* **2018**, *119*, 293–303. [CrossRef]
148. Sellamuthu, S.; Chowdhury, Z.Z.; Khalid, K.; Shibly, S.M.; Rahman, M.M.; Rana, M.; Badruddin, I.A.; Khaleed, H.M.T.; Kamangar, S.; Johan, M.R.B.; et al. Mathematical Modelling and Optimization for Facile Synthesis of Structured Activated Carbon (ACs) from Adansonia kilima (Baobab) Wood Chips Integrating Microwave-Assisted Pyrolysis for the Elimination of Lead (II) Cations from Wastewater Effluents. *Molecules* **2023**, *28*, 6640. [CrossRef]
149. Zhao, Z.; Jiang, Z.; Lin, L.; Qiu, R.; Yan, K. Synthesis of alkoxyphenols-rich bio-oil by microwave-assisted catalytic pyrolysis of wood over MoS2 catalyst. *Renew. Energy* **2023**, *219*, 119491. [CrossRef]
150. Khelfa, A.; Rodrigues, F.A.; Koubaa, M.; Vorobiev, E. Microwave-assisted pyrolysis of pine wood sawdust mixed with activated carbon for bio-oil and bio-char production. *Processes* **2020**, *8*, 1437. [CrossRef]
151. Wu, C.; Budarin, V.L.; Gronnow, M.J.; De Bruyn, M.; Onwudili, J.A.; Clark, J.H.; Williams, P.T. Conventional and microwave-assisted pyrolysis of biomass under different heating rates. *J. Anal. Appl. Pyrolysis* **2014**, *107*, 276–283. [CrossRef]
152. Lin, Y.C.; Wu, T.Y.; Jhang, S.R.; Yang, P.M.; Hsiao, Y.H. Hydrogen production from banyan leaves using an atmospheric-pressure microwave plasma reactor. *Bioresour. Technol.* **2014**, *161*, 304–309. [CrossRef]
153. Zhang, Y.; Cheng, S.; Wang, B.; Shi, C.; Nie, Y. Microwave –assisted pyrolysis aspen wood for production of valuable products under different temperatures. *Arab. J. Chem.* **2023**, *16*, 105187. [CrossRef]
154. Wang, C.; Lei, H.; Zhao, Y.; Qian, M.; Kong, X.; Mateo, W.; Zou, R.; Ruan, R. Integrated harvest of phenolic monomers and hydrogen through catalytic pyrolysis of biomass over nanocellulose derived biochar catalyst. *Bioresour. Technol.* **2021**, *320*, 124352. [CrossRef] [PubMed]
155. Huo, E.; Duan, D.; Lei, H.; Liu, C.; Zhang, Y.; Wu, J.; Zhao, Y.; Huang, Z.; Qian, M.; Zhang, Q.; et al. Phenols production form Douglas fir catalytic pyrolysis with MgO and biomass-derived activated carbon catalysts. *Energy* **2020**, *199*, 117459. [CrossRef]
156. Wallace, C.A.; Afzal, M.T.; Saha, G.C. Effect of feedstock and microwave pyrolysis temperature on physio-chemical and nano-scale mechanical properties of biochar. *Bioresour. Bioprocess.* **2019**, *6*, 33. [CrossRef]
157. Dutta, B.; Gariepy, Y.; Raghavan, G.S.V. Effects of process parameters and selective heating on microwave pyrolysis of lignocellulosic biomass for biochar production. *Can. Biosyst. Eng. Genie Biosyst.* **2016**, *57*, 323–332. [CrossRef]
158. Zhou, C.; Deng, Z.; Zhang, Y.; Li, X.; Liu, Y.; Fu, J.; Chen, L.; Yuan, Y.; Jin, Y.; Dai, J.; et al. Pyrolysis of typical solid wastes in a continuously operated microwave-assisted auger pyrolyser: Char characterization, analysis and energy balance. *J. Clean. Prod.* **2022**, *373*, 133818. [CrossRef]
159. Shi, X.; Wang, J. A comparative investigation into the formation behaviors of char, liquids and gases during pyrolysis of pinewood and lignocellulosic components. *Bioresour. Technol.* **2014**, *170*, 262–269. [CrossRef]
160. Wang, X.H.; Chen, H.P.; Ding, X.J.; Yang, H.P.; Zhang, S.H.; Shen, Y.Q. Properties of gas and char from microwave pyrolysis of pine sawdust. *BioResources* **2009**, *4*, 946–959. [CrossRef]
161. Nzediegwu, C.; Arshad, M.; Ulah, A.; Naeth, M.A.; Chang, S.X. Fuel, thermal and surface properties of microwave-pyrolyzed biochars depend on feedstock type and pyrolysis temperature. *Bioresour. Technol.* **2021**, *320*, 124282. [CrossRef]
162. Bu, Q.; Lei, H.; Ren, S.; Wang, L.; Holladay, J.; Zhang, Q.; Tang, J.; Ruan, R. Phenol and phenolics from lignocellulosic biomass by catalytic microwave pyrolysis. *Bioresour. Technol.* **2011**, *102*, 7004–7007. [CrossRef]
163. Bu, Q. Catalytic Microwave Pyrolysis of Biomass for Renewable Phenols and Fuels. Ph.D. Thesis, Washington State University, Pullman, WA, USA, 2013.

164. Miura, M.; Kaga, H.; Sakurai, A.; Kakuchi, T.; Takahashi, K. Rapid pyrolysis of wood block by microwave heating. *J. Anal. Appl. Pyrolysis* **2004**, *71*, 187–199. [CrossRef]
165. Ren, S.; Lei, H.; Wang, L.; Bu, Q.; Chen, S.; Wu, J.; Julson, J.; Ruan, R. Biofuel production and kinetics analysis for microwave pyrolysis of Douglas fir sawdust pellet. *J. Anal. Appl. Pyrolysis* **2012**, *94*, 163–169. [CrossRef]
166. Bu, Q.; Lei, H.; Ren, S.; Wang, L.; Zhang, Q.; Tang, J.; Ruan, R. Production of phenols and biofuels by catalytic microwave pyrolysis of lignocellulosic biomass. *Bioresour. Technol.* **2012**, *108*, 274–279. [CrossRef] [PubMed]
167. Yang, Z.; Lei, H.; Zhang, Y.; Qian, K.; Villota, E.; Qian, M.; Yadavalli, G.; Sun, H. Production of renewable alkyl-phenols from catalytic pyrolysis of Douglas fir sawdust over biomass-derived activated carbons. *Appl. Energy* **2018**, *220*, 426–436. [CrossRef]
168. Moen, J.; Yang, C.; Zhang, B.; Lei, H.; Hennessy, K.; Wan, Y.; Liu, Y.; Le, Z.; Chen, P.; Ruan, R. Catalytic microwave assisted pyrolysis of aspen. *Int. J. Agric. Biol. Eng.* **2009**, *2*, 70–75. [CrossRef]
169. Yang, X.; Cui, C.; Zheng, A.; Zhao, Z.Z.; Wang, C.C.C.; Xia, S.; Huang, Z.Z.; Wei, G.; Li, H.H.; Karthäuser, J.; et al. Pyrolysis of oil palm empty fruit bunch biomass pellets using multimode microwave irradiation. *J. Anal. Appl. Pyrolysis* **2023**, *15*, 175–179.
170. Suriapparao, D.V.; Vinu, R. Biomass waste conversion into value-added products via microwave-assisted Co-Pyrolysis platform. *Renew. Energy* **2021**, *170*, 400–409. [CrossRef]
171. Vorhauer-Huget, N.; Seidenbecher, J.; Bhaskaran, S.; Schenkel, F.; Briest, L.; Gopalkrishna, S.; Barowski, J.; Dernbecher, A.; Hilfert, L.; Rolfes, I.; et al. Dielectric and physico-chemical behavior of single thermally thick wood blocks under microwave assisted pyrolysis. *Particuology* **2024**, *86*, 291–303. [CrossRef]
172. Klinger, J.L.; Westover, T.L.; Emerson, R.M.; Williams, C.L.; Hernandez, S.; Monson, G.D.; Ryan, J.C. Effect of biomass type, heating rate, and sample size on microwave-enhanced fast pyrolysis product yields and qualities. *Appl. Energy* **2018**, *228*, 535–545. [CrossRef]
173. Mamaeva, A.; Tahmasebi, A.; Tian, L.; Yu, J. Microwave-assisted catalytic pyrolysis of lignocellulosic biomass for production of phenolic-rich bio-oil. *Bioresour. Technol.* **2016**, *211*, 382–389. [CrossRef]
174. Shang, H.; Lu, R.R.; Shang, L.; Zhang, W.H. Effect of additives on the microwave-assisted pyrolysis of sawdust. *Fuel Process. Technol.* **2015**, *131*, 167–174. [CrossRef]
175. Bu, Q.; Lei, H.; Wang, L.; Wei, Y.; Zhu, L.; Liu, Y.; Liang, J.; Tang, J. Renewable phenols production by catalytic microwave pyrolysis of Douglas fir sawdust pellets with activated carbon catalysts. *Bioresour. Technol.* **2013**, *142*, 546–552. [CrossRef] [PubMed]
176. Chen, L.; Yu, Z.; Xu, H.; Wan, K.; Liao, Y.; Ma, X. Microwave-assisted co-pyrolysis of Chlorella vulgaris and wood sawdust using different additives. *Bioresour. Technol.* **2019**, *273*, 34–39. [CrossRef] [PubMed]
177. Shi, K.; Yan, J.; Luo, X.; Lester, E.; Wu, T. Microwave-assisted pyrolysis of bamboo coupled with reforming by activated carbon for the production of hydrogen-rich syngas. *Energy Procedia* **2017**, *142*, 1640–1646. [CrossRef]
178. Makepa, D.C.; Chihobo, C.H.; Ruziwa, W.R.; Musademba, D. Microwave-assisted pyrolysis of pine sawdust: Process modelling, performance optimization and economic evaluation for bioenergy recovery. *Heliyon* **2023**, *9*, e14688. [CrossRef]
179. Zhou, N.; Zhou, J.; Dai, L.; Guo, F.; Wang, Y.; Li, H.; Deng, W.; Lei, H.; Chen, P.; Liu, Y.; et al. Syngas production from biomass pyrolysis in a continuous microwave assisted pyrolysis system. *Bioresour. Technol.* **2020**, *314*, 123756. [CrossRef] [PubMed]
180. Parvez, A.M.; Wu, T.; Afzal, M.T.; Mareta, S.; He, T.; Zhai, M. Conventional and microwave-assisted pyrolysis of gumwood: A comparison study using thermodynamic evaluation and hydrogen production. *Fuel Process. Technol.* **2019**, *184*, 1–11. [CrossRef]
181. Dong, Y.; Tian, B.; Guo, F.; Du, S.; Zhan, Y.; Zhou, H.; Qian, L. Application of low-cost Fe-based catalysts in the microwave-assisted pyrolysis of macroalgae and lignocellulosic biomass for the upgradation of bio-oil. *Fuel* **2021**, *300*, 120944. [CrossRef]
182. Chen, M.; Wang, J.; Zhang, M.; Chen, M.; Zhu, X.; Min, F.; Tan, Z. Catalytic effects of eight inorganic additives on pyrolysis of pine wood sawdust by microwave heating. *J. Anal. Appl. Pyrolysis* **2008**, *82*, 145–150. [CrossRef]
183. Borges, F.C.; Du, Z.; Xie, Q.; Trierweiler, J.O.; Cheng, Y.; Wan, Y.; Liu, Y.; Zhu, R.; Lin, X.; Chen, P.; et al. Fast microwave assisted pyrolysis of biomass using microwave absorbent. *Bioresour. Technol.* **2014**, *156*, 267–274. [CrossRef]
184. Ellison, C.R.; Hoff, R.; Mărculescu, C.; Boldor, D. Investigation of microwave-assisted pyrolysis of biomass with char in a rectangular waveguide applicator with built-in phase-shifting. *Appl. Energy* **2020**, *259*, 114217. [CrossRef]
185. Undri, A.; Abou-Zaid, M.; Briens, C.; Berruti, F.; Rosi, L.; Bartoli, M.; Frediani, M.; Frediani, P. Bio-oil from pyrolysis of wood pellets using a microwave multimode oven and different microwave absorbers. *Fuel* **2015**, *153*, 464–482. [CrossRef]
186. Lestinsky, P.; Grycova, B.; Pryszcz, A.; Martaus, A.; Matejova, L. Hydrogen production from microwave catalytic pyrolysis of spruce sawdust. *J. Anal. Appl. Pyrolysis* **2017**, *124*, 175–179. [CrossRef]
187. Li, L.; Cao, K.; Cai, D.; Zhang, Z.; Zhao, Z.; Yu, M.; Zhang, L.; Zhang, Q.; Zou, G.; Wang, C. Influences of iron additives on microwave-assisted pyrolysis of woody biomass and microwave-induced discharge with spherical bio-char. *Energy* **2023**, *276*, 127549. [CrossRef]
188. Guo, H.; Qin, X.; Cheng, S.; Xing, B.; Jiang, D.; Meng, W.; Xia, H. Production of high-quality pyrolysis product by microwave–assisted catalytic pyrolysis of wood waste and application of biochar. *Arab. J. Chem.* **2023**, *16*, 104961. [CrossRef]
189. Li, M.; Yu, Z.; Bin, Y.; Huang, Z.; He, H.; Liao, Y.; Zheng, A.; Ma, X. Microwave-assisted pyrolysis of eucalyptus wood with MoO_3 and different nitrogen sources for coproducing nitrogen-rich bio-oil and char. *J. Anal. Appl. Pyrolysis* **2022**, *167*, 105666. [CrossRef]
190. Liu, C.; Liu, X.; He, Y.; An, X.; Fan, D.; Wu, Z. Microwave-assisted catalytic pyrolysis of apple wood to produce biochar: Co-pyrolysis behavior, pyrolysis kinetics analysis and evaluation of microbial carriers. *Bioresour. Technol.* **2021**, *320*, 124345. [CrossRef]

191. Li, L.; Tan, Y.; Sun, J.; Zhang, Y.; Zhang, L.; Deng, Y.; Cai, D.; Song, Z.; Zou, G.; Bai, Y. Characteristics and kinetic analysis of pyrolysis of forestry waste promoted by microwave-metal interaction. *Energy* **2021**, *232*, 121095. [CrossRef]
192. Gao, Q.; Budarin, V.L.; Cieplik, M.; Gronnow, M.; Jansson, S. PCDDs, PCDFs and PCNs in products of microwave-assisted pyrolysis of woody biomass—Distribution among solid, Liquid and gaseous phases and effects of material composition. *Chemosphere* **2016**, *145*, 193–199. [CrossRef]
193. McKeown, M.S.; Trabelsi, S.; Tollner, E.W. Effects of temperature and material on sensing moisture content of pelleted biomass through dielectric properties. *Biosyst. Eng.* **2016**, *149*, 1–10. [CrossRef]
194. Wang, C.; Ouyang, S.; Shen, Z.; Cai, B.; Zhao, C.; Peng, H.; Zhang, Y. Research on the determination method of biomass dielectric properties based on mixing rules. *Biomass Convers. Biorefinery* **2024**, 2–11. [CrossRef]
195. Ellison, C.; McKeown, M.S.; Trabelsi, S.; Boldor, D. Dielectric properties of biomass/biochar mixtures at microwave frequencies. *Energies* **2017**, *10*, 502. [CrossRef]
196. Motasemi, F.; Salema, A.A.; Afzal, M.T. Dielectric characterization of corn stover for microwave processing technology. *Fuel Process. Technol.* **2015**, *131*, 370–375. [CrossRef]
197. Fan, X.; Li, B.; Zi, W.; Kang, M.; Wu, H.; Bian, J.; Sun, M.Y. Microwave dielectric characterization and loss mechanism of biowaste during pyrolysis. *Energy Convers. Manag.* **2024**, *301*, 118075. [CrossRef]

energies

MDPI

Article

Analysis of the Effects of Biomass Stabilization Under Varying Thermal Conditions with Respect to the Quality Characteristics of Compost Transformation Products

Mariusz Adamski [1], Marcin Herkowiak [2,*], Przemysław Marek [2], Kamila E. Klimek [3] and Magdalena Kapłan [4]

1 Department of Biosystems Engineering, Poznań University of Life Sciences, 50 Wojska Polskiego Str., 60-627 Poznań, Poland; mariusz.adamski@up.poznan.pl
2 Institute of Technology and Life Sciences, National Research Institute, Falenty, Al. Hrabska 3, 05-090 Raszyn, Poland; p.marek@itp.edu.pl
3 Department of Applied Mathematics and Computer Science, University of Life Sciences in Lublin, 28 Gleboka Str., 20-612 Lublin, Poland; kamila.klimek@up.lublin.pl
4 Institute of Horticulture Production, University of Life Sciences in Lublin, 28 Gleboka Str., 20-612 Lublin, Poland; magdalena.kaplan@up.lublin.pl
* Correspondence: m.herkowiak@itp.edu.pl

Abstract: The aim of this work was to analyze the influence of thermal conditions and the presence of biomass in the chamber on the composting process. The work analyzed the process of the aerobic decomposition of grass, the inoculating fraction and the structure-forming fraction. The analysis covered the batch composting process using veterinary biomass in the treatment chamber. Observations of the process included the following: determining the rate of mineralization, process temperatures, pH, process gas concentrations, chemical composition, physical properties of the compost, and the maturity of the compost. In all analyzed samples, the composting process works correctly in terms of thermal parameters; the obtained fresh compost, after the thermophilic phase has ended, requires action be taken with reference to the values of the seed generation index and the respiration activity (AT4) parameter. After the thermophilic phase, after 60 days of composting, it was noted that for P1 (Probe 1) and P2 (Probe 2) mixtures, the seed germination level decreased below 10%. The AT4 parameter for the P1 and P2 compost samples was between 29.8 and 26.2 mg $O_2 \cdot g^{-1}$. The improvement of the germination level to values in the 30% to 40% range for the maturing compost samples was caused by the thermal conversion of biomass with the regulation of air and water conditions. The phytotoxicity of the compost was overcome, while an improvement in the value of the AT4 index was achieved.

Keywords: grass; composting; phytotoxicity; compost maturity

1. Introduction

The biological stabilization of biomass is an important part of biomass management for energy and fertilizer purposes, and helps reduce the burden of organic waste on the environment. Biological stabilization primarily contributes to reducing atmospheric emissions of greenhouse gases such as CO_2 and CH_4 [1].

Composting has a definite number of advantages over, for example, biogasification from the perspective of waste management and environmental benefits. These benefits include the increased stabilization of organic matter, the minimization of the amount of waste to be landfilled, and the destruction of pathogens [2,3].

Compared with anaerobic digestion and the preparation of solid biofuels, the aerobic composting process is characterized by low energy consumption and odor generation, and produces environmentally harmless organic fertilizers [4,5].

Initial moisture content and air velocity have a significant impact on the efficiency of the composting process [6]. The ideal composting conditions are a carbon to nitrogen ratio of between 20 and 40 in the composted material, a moisture content of between 50% and 65%, an adequate supply of oxygen, small particle sizes, and sufficient empty space through which air can flow. Optimizing the parameters that regulate aerobic biodegradation can reduce the maturation time of the compost [7].

One of the processes occurring during composting is humification. It involves the decomposition of multi-molecular organic substances (e.g., carbon dates, proteins and fats) by microorganisms. As a result of this decomposition, humic substances and acids are produced [8]. The high degree of humification of composting products promotes plant growth and stabilizes heavy metals. This promotes useful applications of compost, such as using them as organic fertilizers and soil reclamation products [9].

Various nitrogen transformations take place during composting. Most nitrogen is metabolized to ammonia (NH_3), and its insoluble compounds are converted into soluble salts. Nitrification also occurs, in the course of which nitrifying bacteria convert NH_4^+ into NO_3^-. However, the main nitrogen losses result from ammonia emissions, while a small portion can escape through salts (xNO_3) or acid drainage [10].

During the composting process, the so-called "concentration effect" can be observed, which causes an increase in selenium and heavy metal concentrations as the process continues. This effect is due to the loss of mass caused by the mineralization of organic matter [11]. Mineralization is the conversion of an organic compound into its non-organic components [12].

One can also notice an increase in the density of the composted mixture as the volume decreases. This trend is related to the duration of the process. The highest density and lowest volume are to be expected when composting is completed. When composting many types of biomass, it is normal for the pH value of the composted material to increase [13]. However, during the first few days of composting, the pH may decrease [10].

1.1. Composting Technological Process

Under optimal conditions, composting proceeds in three phases, in which temperature is an important variable, a critical parameter as well as an indicator of the composting stage. During the different phases of composting, different communities of microorganisms play a dominant role. There is a mesophilic phase, a thermophilic phase and a maturation phase.

In the first phase, there is a rapid decomposition of soluble and easily degradable compounds by mesophilic microorganisms. These compounds include organic matter and nitrogenous compounds. Following decomposition, there is an increase in temperature [14].

In the next phase, due to an increase in temperature, mesophilic microorganisms are displaced by thermophilic organisms. The decomposition of fats, proteins and complex carbohydrates (including major plant structural molecules such as cellulose and hemicellulose) occurs. This phase can last from a few days to several months. The occurrence of the thermophilic phase indicates that the composting process is proceeding properly, during which temperatures should exceed 60 °C for at least 48 h. This not only accelerates the rate of decomposition of organic matter, but also kills significant amounts of pathogens in the composted material [15].

The maturation phase, on the other hand, occurs while high-energy compounds are depleted as the temperature begins to drop. This process can last up to several months.

In the case of materials with a high moisture content, such as sewage sludge, it is possible to assist the process by additional heating. In this way, the faster evaporation of water and degradation of organic matter is possible.

Raising the composting temperature supports the reproduction of thermophilic bacteria, which releases heat during the decomposition of biomass. This creates a positive feedback effect [16].

1.2. Composting Methods

Composting can be done in three ways, such as the batch method and the continuous method.

In the batch method, the process begins by placing all the substrate in the composter.

In the batch method, the composting substrate is delivered to the composting bin gradually and in a systematic manner. This allows the process to maintain a higher temperature, which contributes to the efficiency of the process.

The third method involves the continuous addition of substrate to the composter. In this way, the material gradually matures, and it is possible to collect compost in a continuous manner [17].

1.3. Control Parameters of the Composting Process

Parameters such as electric conductivity, carbon/nitrogen ratio, moisture content, AT4 test and seed germination index are used to correctly evaluate compost.

The electrical conductivity of the compost helps determine the salt content. The increase in salinity increases with the degradation of organic matter during composting. The higher the value of electrical conductivity, the higher the salinity, and thus the more negative the impact on crop yields and plant growth [14]. According to HKORC standards, the electrical conductivity of mature compost should be less than or equal to 4 $mS \cdot cm^{-1}$ [18]. The salinity of compost can be reduced by using a mineral called zeolite [19].

The carbon/nitrogen ratio helps determine the maturity of the compost. Compost with a C/N ratio of less than or equal to 20 is considered mature [18,20].

Controlling the moisture content of the compost is crucial due to the fact that too low a water content (below 35%) cannot ensure the proper reproduction of microorganisms [21,22].

The AT4 parameter is one of the indicators of the biological activity of waste. The AT4 test is a short-term test that determines respiration activity, the parameter of which is the amount of oxygen consumed by microbial processes over 96 h per gram of dry matter. As the organic matter content depletes (the amount decreases steadily as a result of the mineralization process), the value of the AT4 parameter decreases [23]. A high value of the parameter indicates the biological activity of the material and disqualifies its further use. This parameter is measured indirectly by determining the pressure drop in a sealed vessel. This decrease is caused by the occurring aerobic processes, as a result of which carbon dioxide is produced. The generated carbon dioxide is chemically bound, which leads to a drop in pressure. According to Austrian standards OE-NORM S2027-4;2012 [24], sample preparation is a complex process consisting of many steps. Among other things, the sample must have the right size and moisture content. The oxiTop device from Xylem Water Solutions Polska, Warsaw, was used for AT4 analysis. During the test itself, the sample must be continuously aerated, and measurements of oxygen consumption must take place at least every hour.

The seed germination index is the most commonly used parameter for assessing the level of phytotoxicity of a composting product, and directly tests the effect on seed germination and seedling growth. It is considered an indicator of the maturity of the compost [14]. In order for compost to be used as a soil additive, it should be characterized by a seed germination capacity of at least 50%, in a mixture consisting of dry compost matter and distilled water in the ratio of solids (compost):water equal to 1:5. Compost that meets this condition can be considered mature [18,23].

Scientific research focuses on the phytotoxicity of waste substrates, which after the composting process become useful as substances that initiate plant growth by providing nutrients [25].

However, compost can also exhibit phytotoxicity after the aerobic stabilization process. The possibility of assessing the suitability of compost for fertilizer purposes is linked to the question of whether the composting process can be controlled to obtain a valuable fertilizer substrate [26].

In conclusion, it should be noted that compost formed from waste biological matter is not only a valuable food substrate for plants. The inhibitory effect of compost can be realized. The conversion of the inhibitory features of compost to the form of a valuable plant food fraction makes demands on composting technology [27].

The search for research results on the composting process of grasses has resulted in information on the composting process directed at the bioremediation of waste substances. There is a certain niche of information on composting, especially of grasses, within the framework of reducing the share of structural additives. The authors of the study placed particular emphasis on reducing the mass share of fractions such as straw. They were limited to inoculating the mixture with mature compost, self-generated on a grass basis in the preceding research period. As a result of the tests, the mixture experienced strong compaction, and conditions changed to anaerobic. A review of the industry literature advises the supplementation of structuring components in such a case. The research assumptions ruled out supplementing the structuring agents. This created a research situation that was difficult to compare with current examples of research results in the literature. At the same time, the authors solved the problem of changing the conditions of the composting process to an unfavorable one by proposing a two-stage process ending with the use of a propagator. The scientific publications used in the discussion, according to the article's authors, reflect well the similarity of the research to their own research question [28].

The scale of generation of organic waste consisting of grass and leaf cuttings is significant, whether in a single municipality, a country or a continent. It is estimated that a single urban municipality produces 300 Mg to 500 Mg of such waste per year. Biomass in a significant amount is now composted in industrial facilities. An example is Poland, where 16 provinces are divided into 314 counties and 2477 municipalities. In the total number of municipalities, we can identify 302 urban municipalities, including 66 cities with county rights, 711 urban–rural municipalities and 1464 rural municipalities [29,30].

Statistics show that in Poland, from 2007 to 2020, the amount of biodegradable waste increased to 1.9 million Mg. The dynamics of change in the last 4 years show a level of deviation from this level of only 4% year-on-year. In 2023, compost plants processed 1.6 million Mg of organic biomass. A significant share of this mass, during the summer and autumn, was grass and leaves [29,30].

The main difference in the authors' research idea from previous research in publications published in professional journals is the reduction in the proportion of the structural fraction to the necessary minimum. The adopted amount of structural fraction and its volume allow for efficient aerobic stabilization in the thermophilic phase. However, the research task included the maximum use of grass as the main substrate in the composting process. Therefore, in the mesophilic phase of the composting process, the occurrence of the disruption of air–water conditions in the structure of the sample was observed, and a solution to the problem was proposed.

The purpose of the study is to analyze the batch composting process using vertical biomass flow in the reaction chamber. The analysis is concerned with determining the effects of the temperature of the aerobic decomposition process on the dynamics and level of mineralization of biomass and mineral–organic and structure-forming admixture components. The scope of observations also included the determination of the chemical composition and physical properties of the compost, as well as the determination of the degree of maturity of the compost using the seed germination index and the AT4 parameter.

2. Materials and Methods

2.1. Materials

The main substrate used in the study was grass. Straw was used as a filler (structuring agent) to retain excess moisture content and reduce air voids. The testing material consists of a mixture of grass varieties with the following mass shares:

1. *Perennial ryegrass* (*Lolium perenne* L.)—25% m/m;

2. *Oldenburg ryegrass* (*Lolium* × *boucheanum* Kunth)—15% m/m;
3. *Multifloral ryegrass, Italian ryegrass,* tares (*Lolium multiflorum*)—40% m/m;
4. *Westerwold ryegrass* (*Dutch ryegrass*) (*Lolium multiflorum* v. *westerwoldicum*)—20% m/m.

Perennial ryegrass (*Lolium perenne* L.), along with other varieties, is a species of plant in the panicle family. It is found in Europe, northern Africa and western Asia. In central Europe it is a common species, both growing in the wild and also being a cultivated plant [31]. The test material at coordinates N 52.50, E 16.58 was obtained by cutting a swath of a selected grass mixture.

Mature compost was also added to the composted mixture to improve the rate of degradation of organic matter. The compost added to the fresh material was expected to inoculate the mixture with bacterial cultures characteristic of the aerobic decomposition of biomass. This type of treatment is common in the industrial composting process [32,33]. The different substrates are shown in Figure 1.

(a) (b) (c)

Figure 1. Samples used for testing: (**a**) grass, (**b**) straw, (**c**) mature compost. Author: Mariusz Adamski.

Such parameters as dry matter, organic dry matter material and elemental content (such as carbon and nitrogen) were determined in the basic research.

For the determination of dry matter, the drying weight method was used in accordance with PN EN 12880:2004 [34]. The dry residue is here obtained by drying with conduction to minimize losses through uncontrolled removal from the crucible as a result of forced drying air flow.

For the determination of dry organic matter, we followed the method of thermal mineralization in a muffle furnace in accordance with PN-75/C-04616/01 [35] and PN-EN 12879:2004 [36]. Thermal mineralization involves the destruction of the organic structure by ashing the sample at high temperatures (usually 550 °C). Under these conditions, the organic substance is completely decomposed, carbonized and oxidized by atmospheric oxygen to CO_2 and H_2O.

The tests were carried out using the elemental analysis method on Elementar's VarioMax instrument from Elementar Analysensysteme GmbH Germany Langenselbold, Hesse. The fully automatic Vario Max analyzer is used for simultaneous analysis in CNS (C-carbon, N-nitrogen, S-sulphur) and CN (C-carbon/N-nitrogen) modes. The apparatus allows for the automatic insertion of samples into the combustion tube from a 60-position macro-sample feeder with crucibles (steel or ceramic) and the automatic removal of crucibles with ashes after the analysis is completed. The basic principle of quantitative CNS/CN analysis is high-temperature combustion with the catalytic afterburning of samples of organic compounds and many inorganic compounds in both solid and liquid states.

Based on our own research, the dry matter content, dry organic matter, carbon content and nitrogen content were determined in the above substrates. It was noted that the carbon content of the mature compost is close to the value that Chang et al. [7] obtained in their study. The results of the study are shown in Table 1.

Table 1. Parameters of the substrates used.

Parameter	Unit	Grass	Straw	Mature Compost
Dry matter	% m/m	30.8 ± 0.1	92.1 ± 0.1	46.7 ± 0.1
Organic dry matter	% m/m	80.7 ± 0.1	96.8 ± 0.1	37.6 ± 0.1
Carbon content	% m/m	57.82 ± 1.44	56.02 ± 1.40	40.01 ± 1.00
Nitrogen content	% m/m	3.1 ± 0.08	0.69 ± 0.02	0.64 ± 0.02

Knowing the properties of the various substrates, their ratio in the mixture was selected so that it would be optimal for composting. For this purpose, a spreadsheet available on the Cornell Waste Management Institute website was used. It was determined that the weight of the mixture would consist of 74% grass, 4% straw and 22% compost. It was calculated that such a mixture would be characterized by a C to N ratio of 22.3 to 1. This means that the carbon to nitrogen ratio of the compost mixtures was adequate for the process to proceed properly [13]. The mixture was characterized by a pH value of 6.2. The assumed weight of the mixture and its elements, as well as the actual weights of the mixtures in the individual experiments, are shown in Table 2.

Table 2. Weights of individual substrates and total weight of mixtures.

Substrate	Unit	Conception	Sample 1	Sample 2
Grass	kg	100	100.2	100.0
Straw	kg	5	4.8	4.7
Mature compost	kg	30	29.8	29.6
Total	kg	135	134.7	134.4

2.2. *Substrate Feeding and Mixing*

It was assumed that the substrate would be fed in a batch method. According to this method, the ingredients are added at the beginning and then gradually and systematically in controlled sequences. The details of the method of substrate administration were suggested by the research that Wang et al. [17] conducted. It was decided that at the beginning, as much substrate would be fed as the volume of the composter chamber allowed; this was 55% of the assumed amount of substrate. The further addition of substrate began after the composted material fell to the bottom of the composter (below the aeration system). Three doses of 15% each (of the total amount of substrate) were added, every 8 days.

Before feeding the new material, the substrate in the bioreactor was mixed. After substrate addition was completed, mixing was performed every 8 days, with the first such mixing preventing the material from falling below the aeration system.

The test material was stirred, in order to evenly aerate the composted material, homogenize the compost mixture and equalize bacteria access the nutrients. Mixing always followed the measurements.

2.3. *Test Stand*

The study was carried out in a closed composter (thermally insulated), at the Institute of Biosystems Engineering of the University of Life Sciences in Poznań.

The decomposition of organic matter was carried out in a chamber with a total volume of 220 dm^3. The chamber contains a system for regulating the amount of injected air, which supports the rapid conversion of biodegradable waste into mineral fractions by microorganisms. A specified amount of waste was fed from the top of the bioreactor, and then covered with a lid to allow the condensation of steam and the drainage of excess moisture. The accumulation of water in the closed composting system results in the preservation of moisture levels, which are necessary to sustain the vital functions of aerobic bacteria. If there was too much water, a 16 mm-diameter hole was drilled in the bottom of the chamber to drain the excess liquid into an external tank. The maturing compost was taken from the bottom of the composter.

A propagator that heats the material was also used to test the material. The heating was performed using infrared. The propagator is shown in Figure 2. Compost with a disturbed air–water ratio is directed into the propagation chamber due to the strong hydration of the maturing compost. Mixed and dehumidified monosubstrate, with temperature maintained at the lower range of the thermophilic phase of the composting process, converts to form a non-phytotoxic fraction.

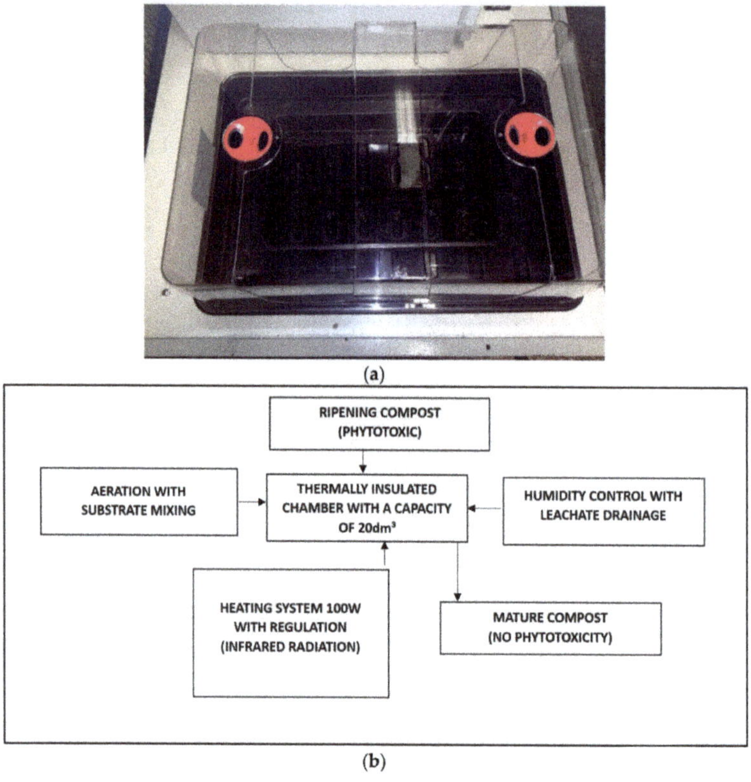

Figure 2. Propagator (**a**) and propagator operation diagram (**b**). Author: Mariusz Adamski.

During the decomposition of organic matter by aerobic microorganisms, odor nuisance substances, otherwise known as odors, are produced. Those emitted include organic and inorganic sulfides, ammonia, hydrogen sulfide, amines, fatty acids, and aromatic and aliphatic hydrocarbons [33]. During the P2 experiment, a system was used to capture selected gases generated during composting. A hose was connected to the outlet port of the lid, with the help of which a pump sucked the gases from the composter. The amount of air supplied to the measuring heads was regulated by the operation of the flow meter. Before the measurement of the gases, the air stream was dried and cleaned of fine impurities. The actual measurement of volatile substances was carried out with the measurement heads of the following gases: ammonia and carbon dioxide.

In the experiment carried out, the following samples were specified to start the experiments of the composting process using reaction chambers: P1, where the temperature of the air entering the compost chamber depended on atmospheric conditions, and P2, where the temperature of the air entering the chamber was stabilized at 20 °C.

On the 50th day of the process, additional laboratory samples were separated from the bulk samples P1 (Probe 1) and P2 (Probe 2) and placed in a special container (propagator), where the temperature of the compost was maintained at 30 °C and the moisture content and density were regulated. Samples conditioned in propagators were designated as P1O (Probe 1, O series tested in propagator) and P2O and P2 (Probe 2, O series tested in propagator). Samples P1 and P2, maintained in thermally insulated compost chambers, were not regulated for moisture and temperature. The thermal processes and moisture levels of the P1 and P2 samples resulted from changes in the phases of the composting process.

Sample P1 was exposed to the weather during the 4-month composting process. Sample P2 was subjected to the composting process under laboratory conditions, and the temperature of the air flowing into the reaction chamber stabilized at 20 °C. After the thermophilic phase, P1O and P2O samples of 2000 g each were separated from P1 and P2 samples, respectively. These samples were taken as representative of pooled samples P1 and P2 in the mesophilic phase.

The structure of the experiment is shown in Figure 3.

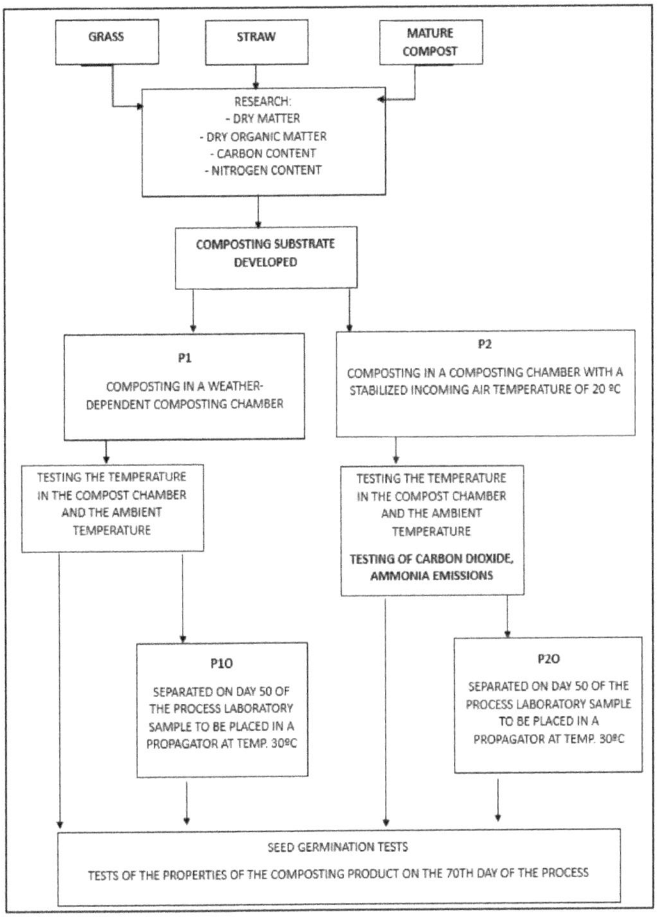

Figure 3. Research scheme.

3. Results and Discussion

A significant decrease (by more than half) in volume was observed during the P1 and P2 experiments. Similar decreases also followed the administration of subsequent doses of substrate. The decrease in volume is due to the digestion process, in which organic matter is broken down into elements that can be absorbed by microorganisms, thus turning the size of the organic matter into small particles, resulting in a decrease in the volume of the compost [37]. As nutrients are metabolized, the decomposition of organic matter results in the release of moisture, which contributes to a reduction in the weight and volume of the composted fraction. Nur et al. [38] found that the volume reduction is caused by the decomposition of lignin, hemicellulose, fat and cellulose. The decomposition process produces new materials, such as carbon dioxide, water and heat energy, which are released into the environment. The resulting material reduction of about 40–60% depends on the basic raw material used in the process.

3.1. Process and Ambient Temperature

Temperature determines the rate of microbial reactions occurring in the composted mixture, and it also indicates the changes occurring in the substrate mixtures. This parameter is considered a major indicator of the occurrence of process disturbances [39–42]. Figures 4 and 5 show the changes in the average temperature value of composted substrates over the first 60 days of the experiment, in both bioreactors. The figures also mark the limits of the occurrence of the thermophilic phase [10]. In all the graphs presented in this paper, the lines are only visualizations of the trends of changes occurring during the process.

Analyzing the data in Figures 4 and 5, it can be seen that it only took two days for the temperature inside the bioreactor of experiment P2 to rise to a value of about 65 °C. Meanwhile, in the P1 experiment, the temperature rose very slowly for the first 6 days, and only on the next day did it begin to increase intensively, until the 8th day of the process, when the temperature reached about 75 °C. The high temperature inside the bioreactors lasted for several days, after which it began to decrease.

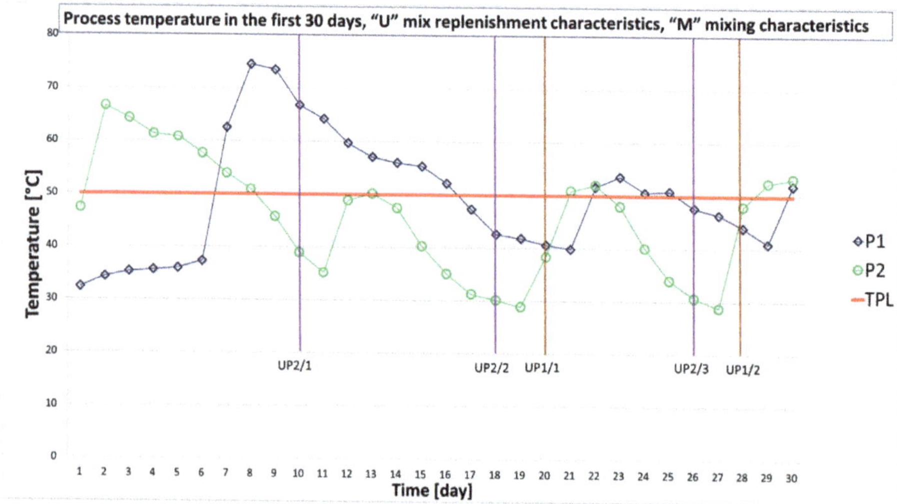

Figure 4. Temperature of the process in the first 30 days: P1—experiment no. 1, P2—experiment no. 2, TPL—thermophilic phase limit (source: own elaboration).

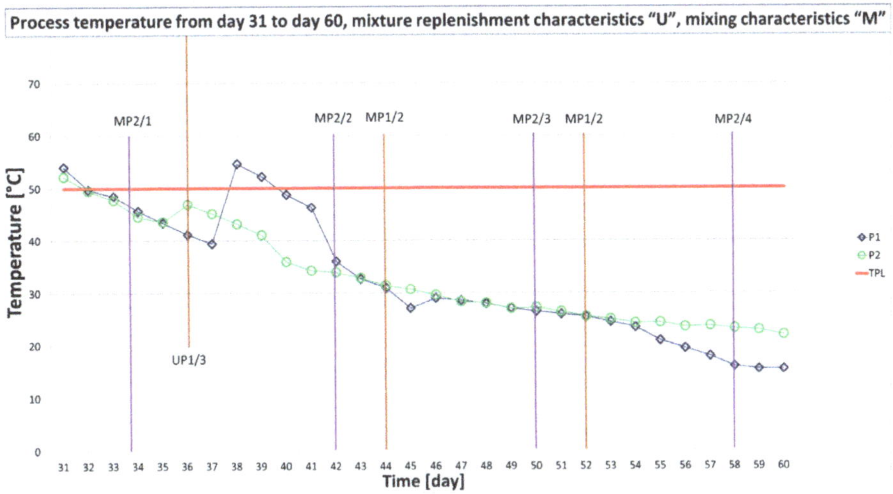

Figure 5. Process temperature from day 31 to 60: P1—experiment no. 1, P2—experiment no. 2, TPL—thermophilic phase limit (source: own elaboration).

The apparent delay (6 days) in achieving thermophilic phase temperature levels above 50 °C may be due to the impact of changes at different times of day (night, day). With the passage of time, the compost pile is naturally compacted; from day 1 to day 6, the air exchange in the P1 experiment space was dynamic, and the periodic decrease in temperature can be attributed to strong gravitational circulation drawing in cooled air during the night period. Sample P2, unlike sample P1, was stratified under conditions of stable temperature independent of the time of day, at about 20 °C. In this range of observation of temperature changes, the difference between the controlled conditions and the natural conditions of the composter operation became apparent.

Temperature values in both bioreactors differed during the first thermophilic phase, but in subsequent thermophilic phases following the start of substrate addition, the temperature values were similar. During both experiments, the addition of substrate caused an initial drop in temperature and then a rapid increase in temperature; a period of 24 h between the onset of new conditions and the response of the bioreactor may indicate a natural delay in the response to the increase. In the final stage of the study, a correlation was observed between the ambient temperature and the temperature of the composted material. During this period, the temperature of the composted material was about 2 °C higher than the ambient temperature. The lowest temperature values for both bioreactors arose at the end of the experiment.

After first stirring, the temperature increased slightly from the time the substrate addition was completed (day 44 for experiment P1 and day 34 for experiment P2). During the next stirring and all subsequent stirrings, no increase in temperature was observed, indicating that the composting process was completed and no further intensive decomposition of organic substrates would occur [41].

Temperature measurements in different layers showed the highest temperature in both experiments in the highest layer of compost. During substrate addition, the temperature even rose above 60 °C for 48 h. This indicates that the composting process was proceeding properly. This order of temperature allows for the rapid decomposition of organic matter, and also allows the inactivation of a significant number of pathogens in the composted material [15,41].

Other researchers who have conducted experiments related to the composting process of different types of biomass have reported similar temperature values in their studies [13,37,42,43]. However, a study by [43] proved that a temperature between 60 °C and 65 °C provides suitable conditions for the greatest activity of aerobic microorganisms involved in the decomposition of organic matter.

To indicate the effect of the environment on the operation of experiments P1 and P2, the thermal characteristics of the composting process of samples P1 and P2 were approximated, in conjunction with the characteristics of the ambient temperature at the time of measurement (Figures 6 and 7).

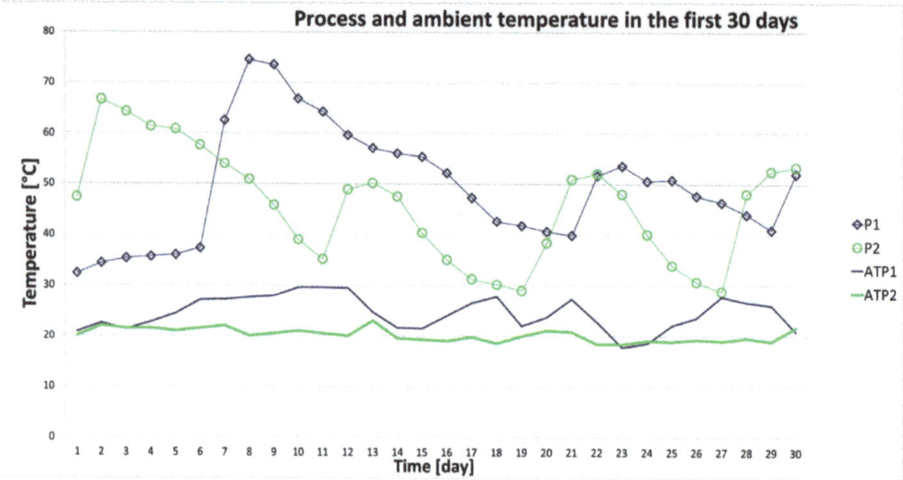

Figure 6. Process temperature from day 1 to 30: P1—experiment no. 1, P2—experiment no. 2, TOP1—ambient temperature P1, ATP2—ambient temperature P2 (source: own elaboration).

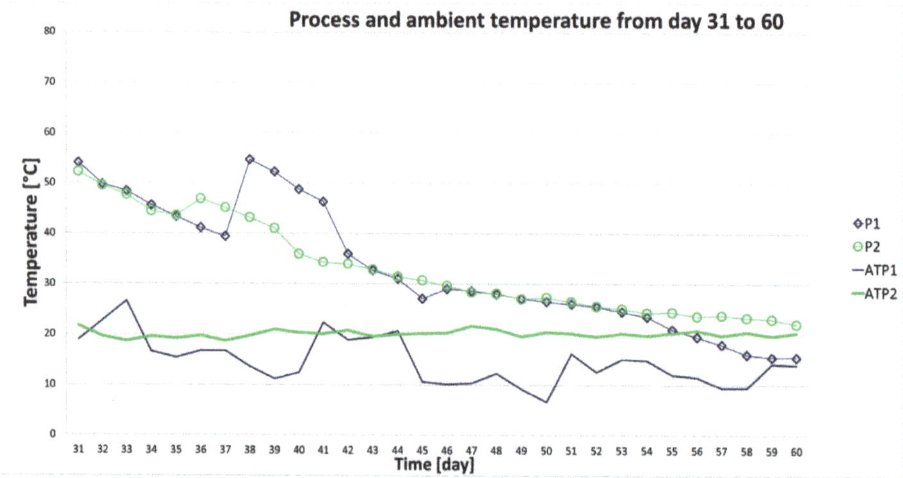

Figure 7. Process temperature from day 31 to 60: P1—experiment no. 1, P2—experiment no. 2, TOP1—ambient temperature P1, ATP2—ambient temperature P2 (source: own elaboration).

3.2. Carbon Dioxide and Ammonia Emissions

One of the gases produced during the composting process is carbon dioxide [44]. Carbon dioxide emissions from the composter are shown in Figures 8 and 9. Increases in carbon dioxide emissions were observed after each addition of substrate, and decreases were observed thereafter. Carbon dioxide emissions are an indicator of the activity of microorganisms and their provocation of organic matter degradation [44–46]. The tests were performed for sample P2 due to the possibility of connecting the nozzle to the gas measuring line.

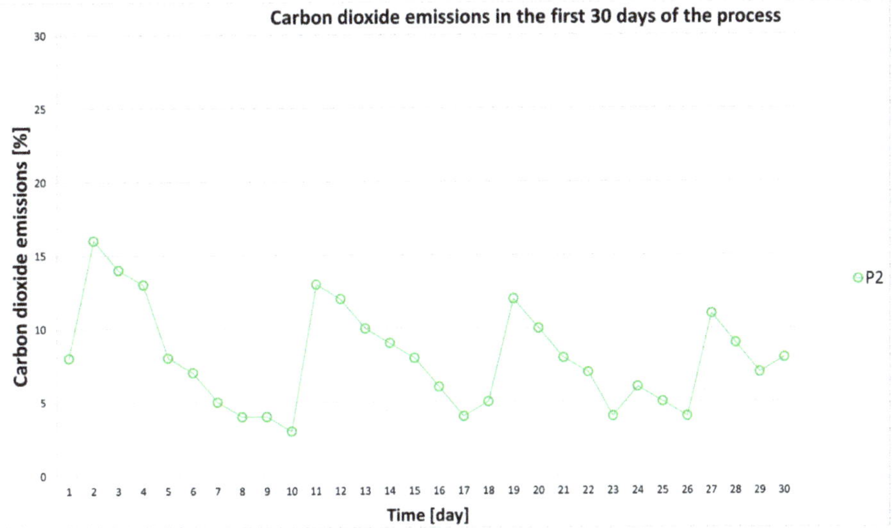

Figure 8. Carbon dioxide emissions in the first 30 days of the process (source: own elaboration).

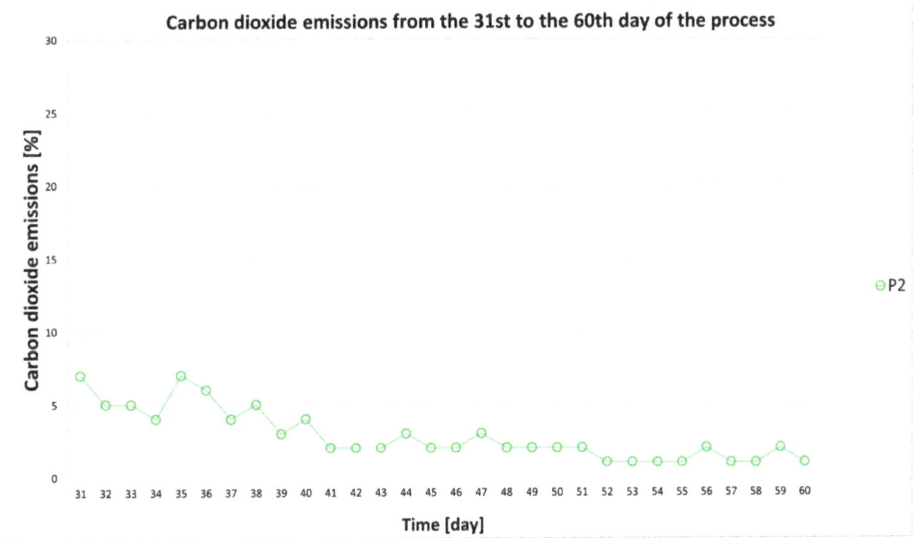

Figure 9. Carbon dioxide emissions from the 31st to the 60th day of the process (source: own elaboration).

The decomposition of proteins (mainly amino acids) during the composting process produces ammonia [41]. Ammonia emissions are shown in Figures 10 and 11.

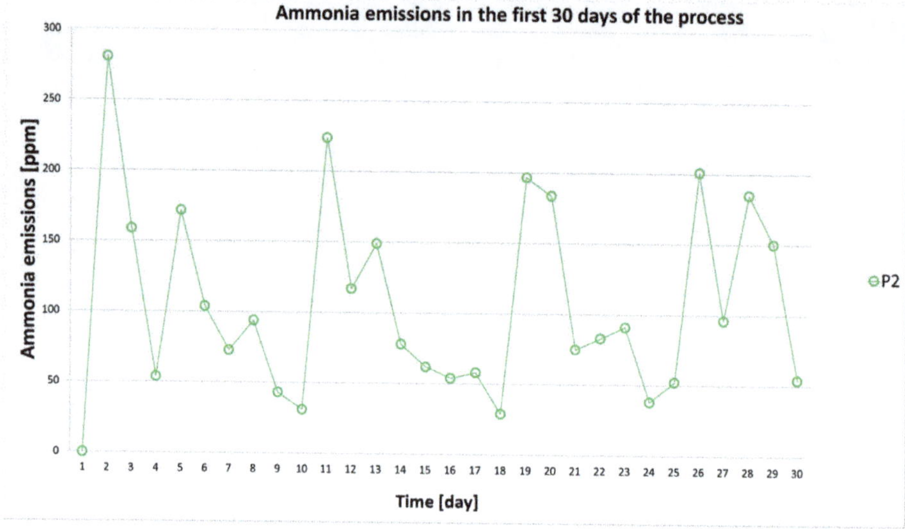

Figure 10. Ammonia emissions in the first 30 days of the process (source: own elaboration).

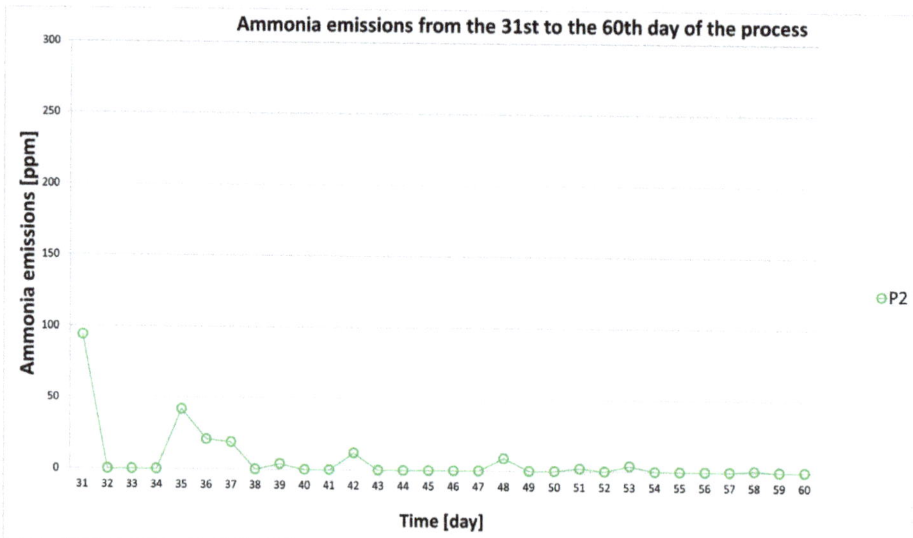

Figure 11. Ammonia emissions from the 31st to the 60th day of the process (source: own elaboration).

Similar to carbon dioxide emissions, the largest increases in ammonia emissions occurred after substrate addition, followed by decreases. Ammonia emissions are inevitable during composting, when organic matter is mineralized and pH and temperature are high. The release of NH_3 from the compost mix creates odor problems, and the loss of nitrogen from the compost mass results in a reduction in the nutrient content of the final product [14]. Komilis and Ham [47] noted that about 65% of the initial nitrogen from food waste can volatilize as NH_3.

3.3. Seed Germination Test

The seed germination rate is an important and frequently studied parameter used to assess the phytotoxicity level of the final composting product and directly examine the effect of compost on seed germination [14,48,49]. Seed germination rate studies were conducted in accordance with the 2021 HKORC recommendations [18]. According to these recommendations, seed germination should be at least 50% in order for the compost to be considered mature. The seed germination test was performed on dishes using a test sample and a control (Figure 12).

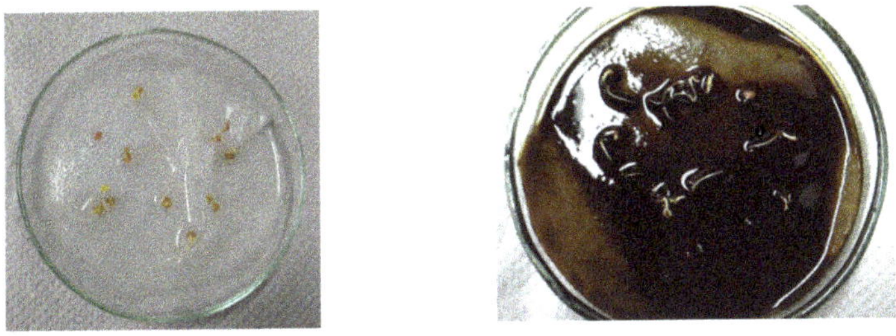

(**a**) (**b**)

Figure 12. Seed germination rate testing in accordance with HKORC recommendations from 2021 (**a**) control, (**b**) sample P2 (Author: Mariusz Adamski).

Initially, the germination rate of seeds in samples P1 and P2 increased and then began to decrease. In the heated samples, the rate was initially (55 days) lower than in the unheated samples, but later it only increased. From 60 days, a similar increase was also observed in sample P2, whereas in the P1 experiment, the germination rate dropped to zero and remained at this level until the end of the experiment (Figure 13). Similar changes in the germination rate as in samples P2, P1O and P2O were observed in the studies performed by Chan et al. [14].

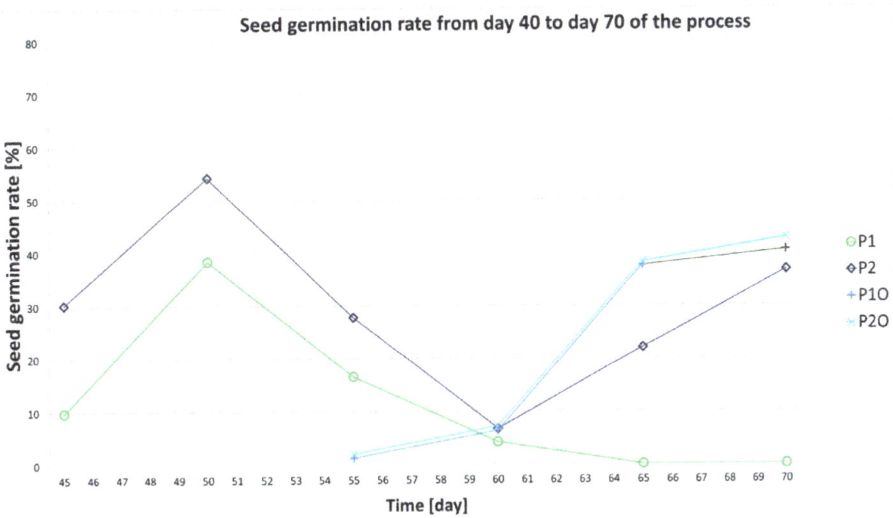

Figure 13. Seed germination rate from day 40 to day 70 of the process (Author: Mariusz Adamski).

3.4. Properties of the Obtained Compost

The obtained composting products were tested on the 70th day of the process. The results are presented in Table 3.

Table 3. Properties of the composting product on the 70th day of the process (source: own elaboration).

Parameter	Units	Initial Mixture	P1	P2	P1O	P2O
Dry matter	% m/m	23.8	29.1 ± 0.1	38.8 ± 0.1	40.9 ± 0.1	41.3 ± 0.1
Dry organic matter	% m/m	68.9	51.5 ± 0.1	42.8 ± 0.1	40.9 ± 0.1	40.3 ± 0.1
Carbon content	% m/m	27.7	11.51 ± 0.26	11.02 ± 0.25	10.64 ± 0.24	10.09 ± 0.23
Nitrogen content	% m/m	1.3	0.81 ± 0.02	0.79 ± 0.02	0.82 ± 0.02	0.81 ± 0.02
C to N ratio	-	21.3	14.41 ± 0.12	13.72 ± 0.12	13.21 ± 0.12	12.89 ± 0.12
Phosphorus content mg	mg P_2O_5 $(100\ g)^{-1}$	-	13.3 ± 0.1	13.5 ± 0.1	13.6 ± 0.1	13.4 ± 0.1
Potassium content	% m/m	-	0.31 ± 0.1	0.32 ± 0.1	0.31 ± 0.1	0.33 ± 0.1
AT4	mg O_2 g^{-1}	-	29.8 ± 0.1	26.2 ± 0.1	25.4 ± 0.1	24.1 ± 0.1

Analysis of the data from Table 3 allows us to observe changes in relation to the input mixture, typical for the composting process, in all the tested samples. Changes in many of the parameters were the smallest in sample P1, which may suggest that the compost produced in this experiment is less mature compared to the products of the other experiments (P2, P1O, P2O).

The increase in dry matter content indicates a decrease in water content. Since water is essential for the reproduction of microorganisms, water content can also be used to characterize the composting process [29]. The water content decreases because the decomposition of a large amount of organic matter by thermophilic bacteria releases a large amount of heat, which accelerates water evaporation [50]. Usually, a water content level below 37% can be considered as a criterion for the end of composting, because too low a water content can cause the inactivation of microorganisms [51]. In none of the tested samples is the dry matter content higher than 50%, from which it is concluded that the water content in the samples is not lower than 50%. The smallest increase in dry matter content occurred in sample P1.

The decrease in the content of dry organic matter indicates the western mineralization of the material. Similar decreases were also observed by Piotrowska-Cyplik et al. [40], as well as Bojarski et al. [13]. The smallest decrease in the content of dry organic matter occurred in the P1 experiment. The content of organic matter in all samples was higher than in the case of compost from the municipal composting plant.

The carbon to nitrogen (C to N) ratio is one of the parameters used to evaluate the composting process [52]. The simultaneous consumption of carbon and nitrogen organic substrates indicates the effectiveness of the composting process. The decrease in the total organic carbon content during the composting process results from the mineralization of organic carbon by microorganisms, which use it as an energy source [53]. Nitrogen is not consumed at the same rate as coal, and its content may even increase due to the concentration effect [14]. This consumption (carbon and nitrogen) leads to a decrease in the C to N ratio. A decrease in the C to N ratio below 20 indicates the correct course of the composting process [20]. Such a decrease was noted in all products of the conducted experiments; the decrease was the smallest in sample P1.

The nitrogen, phosphorus and potassium contents of compost affect the soil's nutrient content. The contents of these substances in the tested samples are similar to in compost from a municipal composting plant; hence, it is assumed that they have a similar positive effect on the soil's nutrient content, which benefits plants.

According to German standards, the AT4 parameter of mature compost should be less than 16 [54]. The AT4 parameter value of the tested samples was higher than 16 but lower than 40, which, according to the discussed standard, classifies them as fresh compost. The AT4 parameter had the highest value in the P1 experiment.

The research assumptions propose a formula for bench testing without repetition for the P1 and P2 samples on a fractional technical scale. As a result of the search for research results on the effects of ambient temperature and mixture composition on the parameters of the composting process, the possibility of the multiplication of samples based on the use of reaction chambers that maintain the thermal conditions of the thermophilic phase through implemented heating systems was observed. In such a case, it is possible to move from the level of analysis of two samples to research using a series of experiments (three to five repetitions). The authors intend to use the aforementioned research solution in the next growing season to enable a more complete statistical evaluation of the research results.

4. Conclusions

Studies of the composting process using cut grass were carried out in summer and winter. Knowing the initial characteristics of the materials used in terms of chemical composition and physical properties, the mixtures were balanced to determine the effects of modifying the composting process in terms of temperature and humidity. The comparison of the composting process for experiments P1 and P2 made it possible to identify a number of observations and conclusions based on the observation of the parameters of both experiments:

1. In the analyzed experiments (P1 and P2), the composting process proceeded properly, as evidenced by temperature changes characteristic of the phases of the composting process. It was possible to distinguish the initial phase, the thermophilic phase and the maturation phase. The correctness of the process is also confirmed by the following parameters observed in the experiments—reduction in volume, reduction in moisture content, increase in mineral (inorganic) content, and other parameters;
2. The results obtained at the final stage of the compost maturation process in the experiments (low seed germination, high value of the AT4 parameter and high pH value) indicate that the compost obtained in the study requires further maturation before it can be used, for example, as organic fertilizer;
3. The seed germination rate increased in the final phase of the experiment in all samples except P1. It is presumed that in the P1 experiment, the observed factors, such as changes in temperature during the diurnal cycle and low air temperature during the compost maturation period, may have had an inhibitory effect and stopped the maturation process, limiting the functioning of microorganisms (mineralization). The cessation of unfavorable properties in the experiment and the P1 sample was observed due to modifications in the process temperature and moisture content of the mixture in the propagator (P1O experiment);
4. Observations and comparisons of the initial phase in the P1 and P2 experiments show that maintaining a constant temperature of the air flowing into the reaction chamber (20 °C) in the P2 experiment allowed for accelerating the onset of the thermophilic phase by 5 to 6 days. A negative effect of low air temperature on seed germination rate was also observed (experiment P1);
5. The positive effect of separating the maturing fraction of the compost and initiating modification of the moisture content, compaction and temperature of the mixture was observed for the P2 sample transferred to the propagator (P2O). The parameters of both indicators, seed germination and AT4, were improved, preceding the positive effects on maturing compost in the P2 chamber;
6. The P2 experiment confirmed the thesis of the validity of the regulatory measures taken. Also, the P1 experiment, despite the isolation of the P1O sample during the period of low air temperatures and the apparent inhibition of compost maturation (unfavorable germination index results) in the compost chamber, responded positively to the application of temperature and humidity control in the propagator (P1O);

7. The thesis of appropriately undertaken control measures was also confirmed by the results of an experiment conducted with compost maturing under outdoor temperature conditions. It is particularly noteworthy that the P1 sample showed complete phytotoxicity at the final stage of maturation, while the application of the propagator and modification of thermal and moisture conditions led to the seed germination index of the P1O sample practically equaling that of the P2O sample.

This indicates the effectiveness of the proposed procedures for accelerating the maturation of compost using the propagator. The basic solution to the problems of the composting process, such as high moisture content, anaerobic conditions and high compaction of the pile, is the introduction of a larger amount of a structure-forming additive (such as straw). In the conducted research, it was assumed that mainly a monosubstance (grass cutting), with admixture of an inoculation fraction (mature compost) and a structuring fraction (straw), would be used, pre-balancing the mixture from the point of view of the proper initiation of the composting process. According to the authors of the article, the proposed technological treatments significantly reduced the need to increase the admixture of additives that regulate, for example, air–water relations in the pile, which contributes to maintaining the proportion of grass fraction in mature compost, and does not escalate the process quantitatively.

Author Contributions: Conceptualization, M.A. and M.H.; methodology, M.A. and M.H.; software, M.A; validation, M.A., M.H. and P.M.; formal analysis, M.A. and M.H.; investigation, M.A.; resources, M.A., M.H., P.M., K.E.K. and M.K.; data curation, M.A., M.H., P.M.; writing—original draft preparation, M.A., M.H., P.M., K.E.K. and M.K.; writing—review and editing, M.A., M.H., K.E.K. and M.K.; visualization, M.A., M.H. and P.M.; supervision, M.A., M.H. and P.M.; project administration, M.A.; funding acquisition, M.A. All authors have read and agreed to the published version of the manuscript.

Funding: Statutory funds were provided by the University of Life Sciences in Lublin (TKZ/S/45/2024 and OIP/S/51/2024).

Data Availability Statement: The data presented in this study are available on request from the corresponding author.

Conflicts of Interest: The authors declare no conflicts of interest.

References

1. Lou, X.F.; Nair, J. The impact of landfilling and composting on greenhouse gas emissions—A review. *Bioresour. Technol.* **2009**, *100*, 3792–3798. [CrossRef] [PubMed]
2. Bhattacharyya, P.; Chakrabarti, K.; Chakraborty, A.; Nayak, D.C.; Tripathy, S.; Powell, M.A. Municipal waste compost as an alternative to cattle manure for supplying potassium to lowland rice. *Chemosphere* **2007**, *66*, 1789–1793. [CrossRef] [PubMed]
3. Rutkowska, B.; Schröder, P.; Mench, M.; Rineau, F.; Szulc, W.; Szulc, W.; Pobereżny, J.; Tiideberg, K.; Niedziński, T.; Loit, E. Role of phytolith occluded carbon of cereales plants for climate change mitigation. *J. Water Land Dev.* **2024**, 50–58. [CrossRef]
4. Maroušek, J.; Gavurová, B. Recovering phosphorous from biogas fermentation residues indicates promising economic results. *Chemosphere* **2022**, *291 Pt 1*, 133008. [CrossRef]
5. Kupryś-Caruk, M.; Lisowski, A.; Chomontowski, C. The effect of silage additive on the kinetics of biogas production from lignocellulosic perennial crops. *J. Water Land Dev.* **2023**, 58–66. [CrossRef]
6. Luangwilai, T.; Sidhu, H.S.; Nelson, M.I. One-dimensional spatial model for self-heating in compost piles: Investigating effects of moisture and air flow. *Food Bioprod. Process.* **2018**, *108*, 18–26. [CrossRef]
7. Chang, J.I.; Tsai, J.J.; Wu, K.H. Thermophilic composting of food waste. *Bioresour. Technol.* **2006**, *97*, 116–122. [CrossRef]
8. Kulikowska, D.; Sindrewicz, S. Effect of barley straw and coniferous bark on humification process during sewage sludge composting. *Waste Manag.* **2018**, *79*, 207–213. [CrossRef]
9. Hui, K.; Tang, J.; Cui, Y.; Xi, B.; Tan, W. Accumulation of phthalates under high versus low nitrogen addition in a soil-plant system with sludge organic fertilizers instead of chemical fertilizers. *Environ. Pollut.* **2021**, *291*, 118193. [CrossRef]
10. Margaritis, M.; Dimos, V.; Malamis, D.; Loizidou, M. An experimental investigation of the composting process in an innovative home composting System: The influence of additives. *Clean. Mater.* **2023**, *8*, 100185. [CrossRef]
11. Wang, Z.; Zhao, M.; Xie, J.; Wang, Z.; Tsui, T.-H.; Ren, X.; Zhang, Z.; Wang, Q. Insight into the fraction variations of selenium and their effects on humification during composting. *Bioresour. Technol.* **2022**, *364*, 128050. [CrossRef] [PubMed]

12. Poffenbarger, H.; Coyne, M.S.; Frye, W.W. Nitrogen in Soils/Cycle. In *Reference Module in Earth Systems and Environmental Sciences*; Elsevier: Amsterdam, The Netherlands, 2018; ISBN 9780124095489.
13. Bojarski, W.; Czekała, W.; Nowak, M.; Dach, J. Production of compost from logging residues. *Bioresour. Technol.* **2023**, *376*, 128878. [CrossRef] [PubMed]
14. Chan, M.T.; Selvam, A.; Wong, J.W.C. Reducing nitrogen loss and salinity during 'struvite' food waste composting by zeolite amendment. *Bioresour. Technol.* **2016**, *200*, 838–844. [CrossRef] [PubMed]
15. Wolna-Maruwka, A.; Dach, J. Effect of type and proportion of different structure-creating additions on the inactivation rate of pathogenic bacteria in sewage sludge composting in a cybernetic bioreactor. *Arch. Environ. Prot.* **2009**, *35*, 87–100.
16. Wang, Y.; Ma, F.; Zhu, T.; Liu, Z.; Ma, Y.; Li, T.; Hao, L. Electric heating promotes sludge composting process: Optimization of heating method through machine learning algorithms. *Bioresour. Technol.* **2023**, *382*, 129177. [CrossRef]
17. Wang, X.; Pan, S.; Zhang, Z.; Lin, X.; Zhang, Y.; Chen, S. Effects of the feeding ratio of food waste on fed-batch aerobic composting and its microbial community. *Bioresour. Technol.* **2017**, *224*, 397–404. [CrossRef]
18. HKORC. 2021. Compost and Soil Conditioner Quality Standards. Hong Kong Organic Resource Centre. Available online: https://hkorc-cert.org/hkorc-content/uploads/2021/12/ORC-%E2%80%93-Compost-and-Soil-Conditioner-Quality-Standards-2021-1.pdf (accessed on 14 September 2024).
19. Turan, N.G. The effects of natural zeolite on salinity level of poultry litter compost. *Bioresour. Technol.* **2008**, *99*, 2097–2101. [CrossRef]
20. Wang, X.; Selvam, A.; Wong, J.W.C. Influence of lime on struvite formation and nitrogen conservation during food waste composting. *Bioresour. Technol.* **2016**, *217*, 227–232. [CrossRef]
21. Aung, K.; Jiang, Y.; He, S.Y. The role of water in plant–microbe interactions. *Plant J.* **2018**, *93*, 771–780. [CrossRef]
22. Karadag, D.; Özkaya, B.; Ölmez, E.; Nissilä, M.E.; Çakmakçı, M.; Yıldız, Ş.; Puhakka, J.A. Profiling of bacterial community in a full-scale aerobic composting plant. *Int. Biodeterior. Biodegrad.* **2013**, *77*, 85–90. [CrossRef]
23. Sidełko, R.; Siebielska, I.; Janowska, B.; Skubała, A. Assessment of biological stability of organic waste processed under aerobic conditions. *J. Clean. Prod.* **2017**, *164*, 1563–1570. [CrossRef]
24. OE-NORM S2027-4; 2012 ÖNORM S 2027-4-Beurteilung von Abfällen aus der mechanisch-biologischen Behandlung-Teil 4: Stabilitätsparameter-Atmungsaktivität (AT4). 2012. Available online: https://www.austrian-standards.at/de/shop/onorm-s-2027-4-2012-06-01~p2422598 (accessed on 10 December 2024).
25. Batista-Barwinski, M.J.; Venturieri, G.A.; Miller, P.R.M.; Testolin, R.C.; Niero, G.; Somensi, C.A.; Almerindo, G.; Ariente-Neto, R.; Radetski, C.M.; Cotelle, S. Swine slaughterhouse biowaste: An environmental sustainability assessment of composting, amended soil quality, and phytotoxicity. *Environ. Technol.* **2024**, *45*, 1404–1411. [CrossRef] [PubMed]
26. Liu, Y.; Liu, J.; Cheng, H.; Luo, Y.; Oh, K.; Meng, X.; Zhang, H.; Liu, N.; Chang, M. Seedling Establishment Test for the Comprehensive Evaluation of Compost Phytotoxicity. *Sustainability* **2022**, *14*, 1920. [CrossRef]
27. Slávíková, M.; Báreková, A.; Tátosová, L.; Ducsay, L. Phytotoxicity Testing of Composts from Biodegradable Municipal Waste. *J. Ecol. Eng.* **2022**, *23*, 83–88. [CrossRef]
28. Lelicińska-Serafin, K.; Rolewicz-Kalińska, A.; Manczarski, P. Challenges in the Valorization of Green Waste in the Central European Region: Case Study of Warsaw. *Energies* **2024**, *17*, 5056. [CrossRef]
29. Rolewicz-Kalińska, A.; Lelicińska-Serafin, K.; Manczarski, P. The Circular Economy and Organic Fraction of Municipal Solid Waste Recycling Strategies. *Energies* **2020**, *13*, 4366. [CrossRef]
30. Statistics Poland, Agriculture and Environment Department. Environment 2024 Statistical Analyses. Supervised by Marta Wojciechowska ISSN 0867-3217, GUS 2024. Available online: https://stat.gov.pl/ (accessed on 14 September 2024).
31. Cao, Y.H.; Zhao, X.W.; Nie, G.; Wang, Z.Y.; Song, X.; Zhang, M.X.; Hu, J.P.; Zhao, Q.; Jiang, Y.; Zhang, J.L. The salt-tolerance of perennial ryegrass is linked with root exudate profiles and microflora recruitment. *Sci. Total Environ.* **2024**, *916*, 170205. [CrossRef]
32. Guo, A.; Gu, J.; Wang, X.; Zhang, R.; Yin, Y.; Sun, W.; Tuo, X.; Zhang, L. Effects of superabsorbent polymers on the abundances of antibiotic resistance genes, mobile genetic elements, and the bacterial community during swine manure composting. *Bioresour. Technol.* **2017**, *244*, 658–663. [CrossRef]
33. Chen, J.; Chen, T.-B.; Gao, D.; Lei, M.; Zheng, G.-D.; Hong-Tao Liu, H.-T.; Guo, S.-L.; Cai, L. Reducing H_2S production by O_2 feedback control during large-scale sewage sludge composting. *Waste Manag.* **2011**, *31*, 65–70. [CrossRef]
34. *PN-EN 12880:2004*; Charakterystyka Osadów Ściekowych: Oznaczanie Suchej Pozostałości i Zawartości Wody. Polski Komitet Normalizacyjny: Warszawa, Poland, 2004.
35. *PN-75/C-04616/01*; Woda i Ścieki-Badania Specjalne Osadów. Oznaczanie Zawartości Wody, Suchej Masy, Substancji Organicznych i Substancji Mineralnych w Osadach Ściekowych. Polski Komitet Normalizacyjny: Warszawa, Poland, 1975.
36. *PN-EN 12879:2004*; Charakterystyka Osadów Ściekowych-Oznaczanie Strat Przy Prażeniu Suchej Masy Osadu. Polski Komitet Normalizacyjny: Warszawa, Poland, 2004.
37. Agustina, W.; Triyasa, S. Composting of grass clippings using different commercial microbial activators. *IOP Conf. Ser. Earth Environ. Sci.* **2020**, *462*, 012002. [CrossRef]
38. Nur, H.S.; Anja, M.; Hamim, H. Utilization of Potential Cellulolytic and Xilanolytic Bacteria for Decomposition of Rice Straw. *J. Tanah. Trop.* **2009**, *14*, 71–80.
39. Akyol, Ç.; Ince, O.; Ince, B. Crop-based composting of lignocellulosic digestates: Focus on bacterial and fungal diversity. *Bioresour. Technol.* **2019**, *288*, 121549. [CrossRef] [PubMed]

40. Meng, X.; Yan, J.; Zuo, B.; Wang, Y.; Yuan, X.; Cui, Z. Full-scale of composting process of biogas residues from corn stover anaerobic digestion: Physical-chemical, biology parameters and maturity indexes during whole process. *Bioresour. Technol.* **2020**, *302*, 122742. [CrossRef] [PubMed]
41. Czekała, W.; Janczak, D.; Pochwatka, P.; Nowak, M.; Dach, J. Gases Emissions during Composting Process of Agri-Food Industry Waste. *Appl. Sci.* **2022**, *12*, 9245. [CrossRef]
42. Zhao, Y.; Li, W.; Chen, L.; Meng, L.; Zhang, S. Impacts of adding thermotolerant nitrifying bacteria on nitrogenous gas emissions and bacterial community structure during sewage sludge composting. *Bioresour. Technol.* **2023**, *368*, 128359. [CrossRef] [PubMed]
43. Piotrowska-Cyplik, A.; Chrzanowski, Ł.; Cyplik, P.; Dach, J.; Olejnik, A.; Staninska, J.; Czarny, J.; Lewicki, A.; Marecik, R.; Powierska-Czarny, J. Composting of oiled bleaching earth: Fatty acids degradation, phytotoxicity and mutagenicity changes. *Int. Biodeterior. Biodegrad.* **2013**, *78*, 49–57. [CrossRef]
44. Shan, G.; Li, W.; Liu, J.; Zhu, L.; Hu, X.; Yang, W.; Tan, W.; Xi, B. Nitrogen loss, nitrogen functional genes, and humification as affected by hydrochar addition during chicken manure composting. *Bioresour. Technol.* **2023**, *369*, 128512. [CrossRef]
45. Sołowiej, P.; Pochwatka, P.; Wawrzyniak, A.; Łapiński, K.; Lewicki, A.; Dach, J. The effect of heat removal during thermophilic phase on energetic aspects of biowaste composting process. *Energies* **2021**, *14*, 1183. [CrossRef]
46. Hwang, H.Y.; Kim, S.H.; Shim, J.; Park, S.J. Composting process and gas emissions during food waste composting under the effect of different additives. *Sustainability* **2020**, *12*, 7811. [CrossRef]
47. Komilis, D.P.; Ham, R.K. Carbon dioxide and ammonia emissions during composting of mixed paper. yard waste and food waste. *Waste Manag.* **2006**, *26*, 62–70. [CrossRef]
48. Luo, Y.; Liang, J.; Zeng, G.; Chen, M.; Mo, D.; Li, G.; Zhang, D. Seed germination test for toxicity evaluation of compost: Its roles. problems and prospects. *Waste Manag.* **2008**, *71*, 109–114. [CrossRef] [PubMed]
49. Kebibeche, H.; Khelil, O.; Kacem, M.; Harche, M.K. Addition of wood sawdust during the co-composting of sewage sludge and wheat straw influences seeds germination. *Ecotoxicol. Environ. Saf.* **2019**, *168*, 423–430. [CrossRef] [PubMed]
50. Ma, F.; Wang, Y.; Quan, H.; Wang, Z.; Zhao, C.; Li, X.; Liang, B.; Zhang, H.; Hao, L.; Tong Zhu, T. Exploring the humification process of municipal sludge in hyperthermophilic composting through metagenomic and untargeted metabolomic. *Bioresour. Technol.* **2023**, *387*, 129575. [CrossRef] [PubMed]
51. Liang, C.; Das, K.C.; McClendon, R.W. The influence of temperature and moisture contents regimes on the aerobic microbial activity of a biosolids composting blend. *Bioresour. Technol.* **2003**, *86*, 131–137. [CrossRef] [PubMed]
52. Zhou, Y.; Selvam, A.; Wong, J.W.C. Chinese medicinal herbal residues as a bulking agent for food waste composting. *Bioresour. Technol.* **2018**, *249*, 182–188. [CrossRef]
53. Wong, J.W.C.; Fang, M. Effects of lime addition on sewage sludge composting process. *Water Res.* **2000**, *34*, 3691–3698. [CrossRef]
54. Chaher, N.E.H.; Chakchouk, M.; Nassour, A.; Nelles, M.; Hamdi, M. Potential of windrow food and green waste composting in Tunisia. *Environ. Sci. Pollut. Res.* **2021**, *28*, 46540–46552. [CrossRef]

energies

MDPI

Article

Oxidative Pyrolysis for Variable Heating Output with Wood Pellets

Franco Corinto Cavalloni, Joris Strassburg, Daniel Lustenberger and Timothy Griffin *

Institute of Biomass and Resource Efficiency, University of Applied Sciences Northwestern Switzerland, 5210 Windisch, Switzerland; franco.cavalloni@fhnw.ch (F.C.C.); joris.strassburg@fhnw.ch (J.S.)
* Correspondence: timothy.griffin@fhnw.ch

Abstract: A carbon-negative heating system can be realized by pyrolyzing wood pellets, burning the product gas, and storing the produced biochar. Oxidative pyrolysis simplifies the reactor design by replacing an external heat supply with internal oxidation driven by a sub-stoichiometric "primary" air supply. Previous studies have only examined the influence of primary air supply on biochar yield and heating power in a continuous pyrolysis reactor within a limited fuel–air spectrum. In this work, an oxidative pyrolysis reactor, with a nominal heating power of 15 kW, was investigated with the aim to vary the useful heat output and biochar yield over a wide range and still produce biochar of the highest quality in accordance with the EBC (European Biochar Certificate) guidelines. This study demonstrated that within an air flux range of 0.03–0.14 kg/(m^{2}s), there is a linear relationship between air flux and both wood flux and useful heat, resulting in a power output range of 4–30 kW. The useful heat output could be varied by a factor of three in less than 15 min, verifying concept feasibility as a central heating system to meet the variable heating demands of both single and multi-household applications. The biochar yield was observed to range from 12% to 24% of the incoming wood mass flow, meeting the EBC Feed Plus quality standards at all conditions. Depending on the operating point, up to 40% of the biomass's heating value is stored in the biochar.

Keywords: oxidative pyrolysis; biochar; demand-driven; pyrolysis heater; smoldering

1. Introduction

Given the increasing CO_2 concentration in the atmosphere, "negative emissions" technologies removing CO_2 from the atmosphere will become a focus of climate policy [1]. One of the available negative emission technologies is biochar from sustainable wood resources, which stores the CO_2 that plants extract from the atmosphere in a permanent form [2]. Biochar is a safe and feasible way to store CO_2 before it even enters the atmosphere [3]. Biochar is also a stable form of carbon and can, e.g., be used in agriculture by being applied to the soil, where it can contribute to humus formation in the long term as a soil conditioner and increase the water retention capacity of the soil [3,4]. During biochar production, heat can be generated from the combustion of the product gas, making biochar heating systems net energy suppliers. The startup Pyronet GmbH is developing these systems, which can be used for private, commercial, and agricultural applications [5,6].

Biochar can be produced from a wide range of biomass waste streams via pyrolysis. Pyrolysis is an endothermic process in which biomass is oxidized by the previously bound oxygen in the absence of air at a temperature between 450 °C and 700 °C. This process produces biochar, pyrolysis oil, and pyrolysis gas, with the product distribution dependent

https://doi.org/10.3390/en18071702

on the process conditions. Up to 50% of the carbon content of the original biomass can be stored in the biochar [7].

A pyrolysis reactor requires a heat supply to maintain high temperatures for endothermic reactions. One option is to use an external heat supply for allothermic pyrolysis. This external heat transfer becomes increasingly difficult as the reactor size increases. The increasing heating demand exceeds the possible supply through heat input due to the volumetric heating requirement versus surface heat supply [8].

Another option is autothermal pyrolysis, also known as oxidative pyrolysis, which is classified as an advanced pyrolysis process and has only been studied to a limited extent [9]. Using a reactor with a fixed bed and a co-current flow of biomass fuel and air, the pyrolysis process can be operated without an external heat supply [8,10–16]. This leads to fewer components and, overall, a simpler reactor design. The air supply can vary with an air excess ratio (lambda) of 0.1–0.3 for a controlled and stable process. The following sections present findings from previous studies on oxidative pyrolysis.

In an experimental study, the EEI (Escuola Ingenieria Industrial) research team at the University of Vigo, Spain, investigated aspects of the pyrolysis front as a function of the air supply in a fixed-bed co-current reactor using the TLUD (Top Lit Up Draft) principle. With a low air supply, in this case below an air excess ratio of 0.5, the reactions are limited only by the oxygen supplied, as shown in Figure 1. In this regime, the conversion rate of the supplied biomass to biochar and product gas is mass-transfer-limited and is described by a linear relationship with respect to the oxygen mass flow flux. At a certain air supply, the reactions are no longer mass-transport-limited, and not all oxygen is consumed, leading to residual oxygen reaching the produced biochar, which is further oxidized to product gas. This process must occur slowly as the overall conversion rate does not significantly increase in this regime. At stoichiometric combustion (lambda 1), the regime changes, and the conversion rate decreases, as excess air cools the process [10].

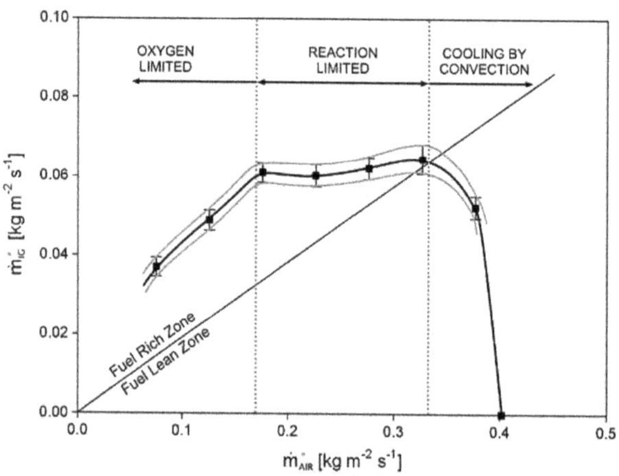

Figure 1. Relationship between air flow rate and ignition mass flux for brassica pellets, including the representation of different combustion phases, measurement uncertainties, and the stoichiometric line. (Reprinted from [10] with permission from Elsevier).

Further investigations showed that the pyrolysis front thickness is of the same order of magnitude as the particle size of the initial biomass. With a stationary and sufficiently high air supply for a stationary pyrolysis front, the pyrolysis front thickness is almost independent of the air supply. For wood pellets, the thickness is approximately 2 cm. The investigation showed that with a very low air supply, the smoldering layer increases, the pyrolysis front does not reach a stationary state, and, therefore, the thickness of the front increases [11].

The research was continued by the CIRAD research team from Montpellier, France, on a "downdraft continuous fixed bed reactor". The reactor is operated with oxidative pyrolysis and has a wood turnover of 3.3 kg/h. The oxidation zone (pyrolysis front) and the pyrolysis products were analyzed. In this study, it was shown again on a specific reactor that the wood flux increased almost linearly with the increase in air flux in a low range of 0.014–0.029 kg/(m^2s). Temperature measurements distributed over a height in the pyrolysis reactor showed that the pyrolysis front forms a surface and is orientated at a 90° angle to the air supply [14]. The lambda value for the three examined air fluxes with wood pellets remained constant, while it slightly increased with wood chips as the air supply increased. The biochar yield was not dependent on lambda. The biochar yield decreased as the pyrolysis temperature increased due to the higher air supply [14].

The research group RE-CORD (Renewable Energy Consortium for Research and Development) from Scarperia e San Piero in Italy carried out similar tests on a ten-fold larger co-current fixed-bed pyrolysis reactor with wood chips. It was confirmed that oxidative pyrolysis is scalable up to a wood turnover of 36.5 kg/h. Furthermore, the entire system was balanced, and it was determined that of the calorific value, 36% (69 kW) remains stored in the biochar and 52% (103 kW) is the thermal energy output [15].

The research group at the Institute of Thermal Engineering at the HES-SO (University of Applied Sciences and Arts Western Switzerland) investigated oxidative pyrolysis with wood chips and wood pellets at low air flows with a range of 0.008–0.017 kg/(m^2s). It was shown that the pyrolysis front speed is inversely proportional to the bulk density of the biomass. For wood chips, the speed was 30 cm/h at a bulk density of 244 kg/m^3 [16].

A recent study by Gao et al. has demonstrated the effect of an atmosphere containing 11% oxygen and 89% nitrogen compared to a 100% nitrogen atmosphere on the properties of biochar. The process was slow pyrolysis with a heating rate of 20 °C/min and a residence time at a final temperature of 30 min. Oxygen reduced the carbon content while oxidizing hydrogen species on the biochar surface. This resulted in a higher C/H ratio and a lower C/O ratio. An oxygen-rich environment could promote carbonization reactions, thereby enhancing crystallinity. Another influence of oxygen was observed in increasing the hydrophilicity of biochar, indicating that oxidation improved its ability to retain water [17].

In previous studies, the primary air supply in a continuous pyrolysis reactor has only been investigated in a very limited range and without focusing on meeting a variable heating demand. Our group has already conducted in-depth research in the field of biochar in recent years, investigating the influence of primary air on the pyrolysis process. In the current study, an oxidative pyrolysis reactor, with a nominal heating power of 15 kW, is investigated with the aim to vary the useful heat output and biochar yield over a wide range and still produce biochar of the highest quality in accordance with the EBC (European Biochar Certificate) guidelines. The parameters for operation with maximum flexibility are determined for a continuous, fixed-bed downdraft pyrolysis reactor intended for use as a heating system and are compared with measurements reported in the literature.

2. Oxidative Pyrolysis

To gain a better understanding of the process of generating heat and biochar without an external heat source, we explain the mechanisms of controlled oxidative pyrolysis.

While oxidative pyrolysis is not a conventional pyrolysis process, it has similarities with slow pyrolysis. The temperature is in a range of 500–800 °C; the residence time in all known setups is at least 10 min, and the heating rate is slow (max 10 °C/s) but not constant because of the acceleration at higher temperatures.

The presence of a moving pyrolysis front, in which pyrolysis reactions propagate through a fixed bed, is a characteristic of oxidative pyrolysis. Through a self-sustained heat supply via oxidation, the propagation of this front depends only on the primary air mass flux, as long as the process is oxygen limited. For the controlled movement of the pyrolysis front and high biochar quality, the primary air is directed from the fresh biomass to the pyrolysis front, known as a co-current reactor. (Co-current means that the mass flows of biomass and air come from the same direction.) Another variant is a counter-current reactor, where the primary air must first pass through the hot biochar, leading to lower biochar yields and about ten times slower pyrolysis front speed [18,19]. This study examined a co-current fixed-bed reactor.

2.1. Terms

To help in understanding the following discussion, we define important terms here. The pyrolysis temperature is the average value of the highest temperature measurement in the pyrolysis reactor over a stationary period of useful output. It is assumed that the biochar was exposed to this temperature over a longer period.

The air flux is the reactor cross-sectional area-specific primary air mass flow that is supplied to the pyrolysis zone.

The wood flux is the reactor cross-sectional area-specific pellet mass flow that is introduced into the pyrolysis reactor.

The speed of the pyrolysis front (v pyr) describes the relative speed in relation to the fresh pellets.

Lambda pyrolysis describes the molar ratio of the supplied primary air to the supplied pellets in relation to the stoichiometric amount of air required for its complete combustion.

The biochar yield is determined by the discharged biochar mass in relation to the input pellet mass over a stationary period.

The useful heat describes the sensible heat power output that is transferred to the heating water after the heat exchanger of the combustion chamber. The flow and return temperatures and the volume flow of the heating water are used to determine this.

2.2. Pyrolysis Front

In the considered control volume, as shown in Figure 2, all mass flows are visible (incoming and outgoing). The primary air and biomass mass flows are incoming, while the pyrolysis gas and biochar mass flows are outgoing. The pyrolysis front moves toward the primary air, with its speed determined almost solely by the primary air mass flux, as all oxygen reacts through partial oxidation.

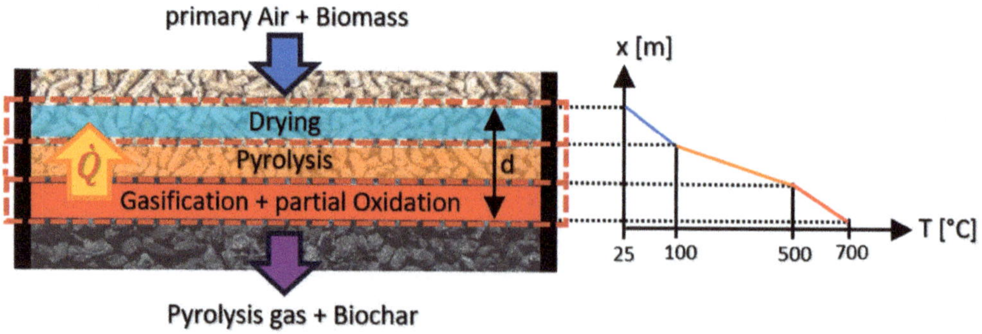

Figure 2. Schematic division of the pyrolysis front into three layers with a temperature profile (d = thickness of pyrolysis front [m]; $\dot{Q}$ = heat flux between layers [W]).

At the highest temperature, oxidation reactions with oxygen from the primary air take place simultaneously with the gasification reactions. This results in increased exothermic oxidation reactions of, for example, CO to CO_2 or H_2 to H_2O, which raises the temperature of the preceding unreacted biomass by heat transfer, represented by the heat flux $\dot{Q}$. In this preceding layer, the temperature is sufficient for endothermic gasification reactions and increased production of CO and H_2. The greater the temperature, the more long-chain molecules transition into the gas phase and can react with each other and with oxygen from the primary air. In the first layer, biomass drying occurs. The subdivision of the layers is not distinct, and they can overlap [18].

For a better understanding, an attempt was made to qualitatively illustrate the progression of the pyrolysis front using a two-layer pellet model, as shown in Figure 3. The process was divided into seven steps. At the start, the pellet is outside of the area of influence of the pyrolysis front.

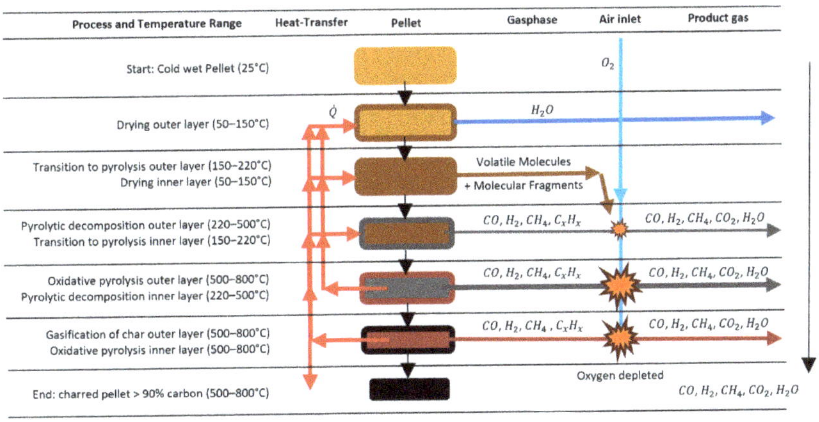

Figure 3. Progression of a single pellet in the pyrolysis visualized with a two-layer pellet model.

As soon as heat is transferred from the pyrolysis front to the pellet, the outer layer begins to dry, and H_2O evaporates. At 150 °C, irreversible changes in the structure of the biomass begin by breaking up the hydrogen and carbon bonds [19]. Between 150 and 220 °C, volatile molecules and molecular fragments are released and transition into the gas phase [20]. This process closely resembles torrefaction. The temperatures at this stage are still too low for oxidative reactions with air oxygen.

Pyrolytic decomposition starts from a temperature of approximately 220 °C, at which more molecular fragments are released, which are further decomposed, and CO, H_2, CH_4, and C_xH_x are produced in endothermic reactions [21]. These do not yet react with the air oxygen due to the temperature being below 300 °C. Between 300 and 500 °C, the gas formation rate increases significantly, and more CO, H_2, and CH_4 are produced; in this temperature range, oxidation reactions occasionally take place [8,20,21].

Above 500 °C, the reaction rate increases, and all of the air oxygen is consumed at the pyrolysis front. There is no further oxygen available for gasification of the formed biochar. Depending on the oxygen supply, a temperature between 500 and 800 °C is established due to exothermic oxidation reactions. Part of the released heat is transferred to the preceding pellets and drives the pyrolysis front. The main part of the released heat is contained in the pyrolysis gas. The oxygen supply determines how much heat is released, and this heat determines the reaction rate of the endothermic reactions. The conditions described here correspond to those of "slow pyrolysis", with relatively low heat rates and enough time for the intermediates released in the pyrolysis reactions to further react to form CO, H_2, and CH_4.

3. Laboratory Test Facility and Methodology

The core element of the experimental setup, as shown in Figure 4, is the pyrolysis reactor with the "fixed bed downdraft" principle and autothermal/oxidative pyrolysis. Product gas flows from the pyrolysis reactor into the combustion chamber, and hot biochar is discharged at the bottom. The product gas is further oxidized in the combustion chamber. The useful energy of the process is contained in the heating value and sensible heat of the exhaust gas as well as the sensible heat of the biochar.

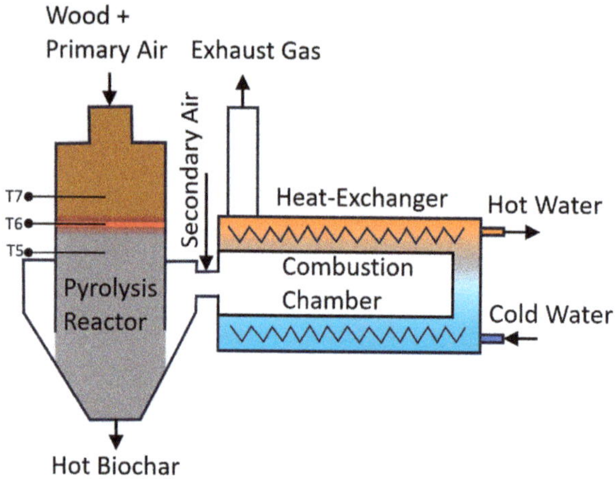

Figure 4. Schematic visualization of the pyrolysis heater test facility.

The pyrolysis reactor is equipped with several temperature sensors (T5, T6, and T7) for process monitoring and control. Additionally, the primary and secondary air volume flows, the mass loss in the biomass feed container (fuel mass flow), and biochar quantity are measured. The pyrolysis reactor is insulated with a 100 mm thick ceramic fiber. Additional parameters of the pyrolysis reactor are displayed in Table 1. Wood pellets certified by the ENplus® (Austria) A1 Standard were used as fuel in this study and fed by a screw conveyor from a container. The water content of the pellets is 8.5%, and the calorific value dry is 18.5 MJ/kg. The hot biochar is discharged from the reactor at the bottom. The sensible heat from the hot biochar as it cools down can be utilized as useful heat. The combustion chamber is surrounded by a shell and a tube heat exchanger through which the exhaust gas flows. Into the combustion chamber, a secondary air flux, with a range of 7–62 m^3/h, is introduced to oxidize the product gas completely. The generated useful heat is measured by the temperature increase in the water circuit at the system's inlet and outlet and the water volume flow.

Table 1. Overview of the parameters of the pyrolysis reactor of the laboratory test facility.

Designation	Unit	Value
Reactor cross-sectional area	m^2	0.026
Reactor height	m	0.37
Reactor volume	m^3	0.0079
Thermal insulation thickness	m	0.1

Table 1. *Cont.*

Designation	Unit	Value
Distance between temperature measurements (T5, T6, and T7)	m	0.06
Biomass input range	kg/h db	2.3–9.6
Primary air flux range	m^3/h	2–11
Temperature range	°C	500–800

3.1. Course of the Measurement

In nominal operations, the pyrolysis front moves upward at a speed dependent on the reaction rate and biochar yield. These two variables are, in turn, dependent on the oxygen supply within the mass-transport-limited regime. Due to the volume shrinkage during the conversion of the pellets to biochar (blue), the pellet level in the reactor continuously decreases. The empty volume in the reactor is monitored by a level sensor and regularly refilled with fresh pellets (green), ensuring continuous reactor filling. The steps described are shown in Figure 5.

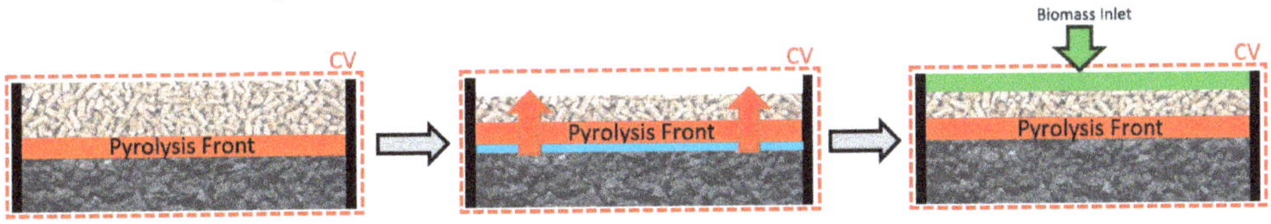

Figure 5. Movement of the pyrolysis front (red arrow) through the production of biochar.

The process described would cause the pyrolysis front to move upward through the reactor.

As shown in Figure 6, to maintain the pyrolysis front at a fixed position, biochar is discharged (purple) when the temperature at the measurement point T6 exceeds 400 °C until the temperature drops again. The temperature drops because the pyrolysis front sinks in the reactor due to the reduced biochar volume. Pellets are refilled (green) in the empty volume at the top of the reactor during biochar discharge. Due to the volume shrinkage during the pyrolysis of the pellets to biochar, the volume of the produced biochar is smaller than the volume of the pellets consumed.

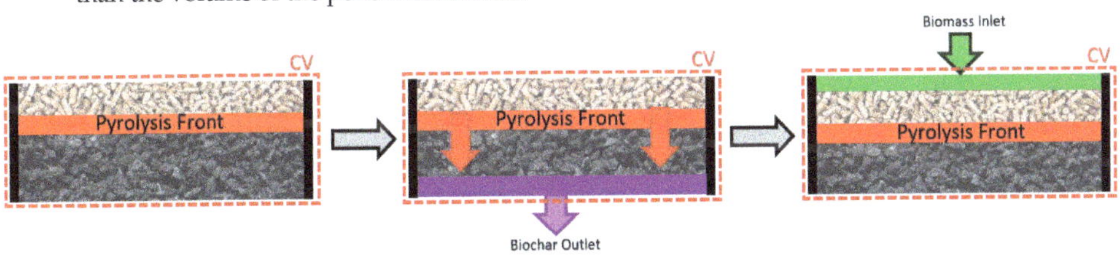

Figure 6. Movement of the pyrolysis front (red arrow) due to the discharge of biochar.

In a stationary state, the discharged biochar corresponds to the yield relative to the introduced pellets, assuming that the bulk density of the biochar remains constant. However, the bulk density of the biochar in the pyrolysis reactor before discharge is a variable that cannot be measured.

The course of a measurement using three temperatures in the pyrolysis reactor is shown in Figure 7.

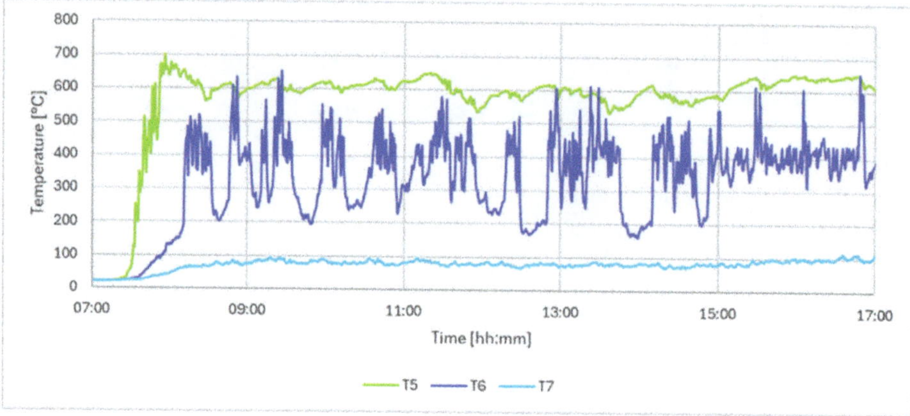

Figure 7. Temperatures in the pyrolysis reactor during a test measurement.

The pyrolysis front is held at the point of T6, and the biochar discharge is controlled. T5 is located below the pyrolysis front in the produced biochar, and T7 is located above the pyrolysis front in the still cold pellets.

3.2. Calculations for Characterization

In Figure 8, the system used to calculate the mass balance is shown and Table 2 describes the symbols used in the mass and energy balances. It is evident that three mass flows enter, and two mass flows exit the system boundaries. For the balance, it is assumed that the process is stationary.

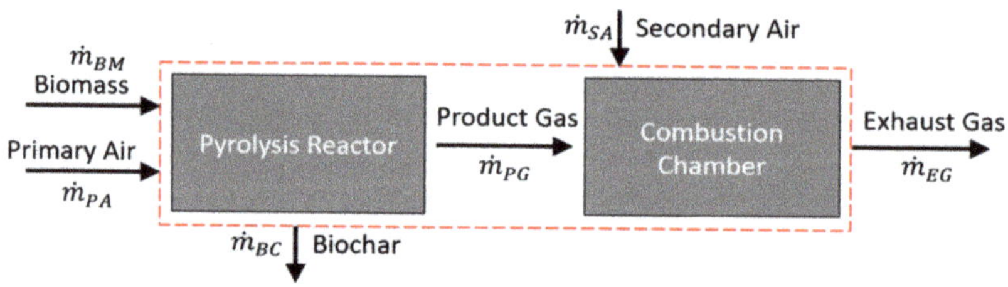

Figure 8. System for the mass balance. (Red dashed box represents the control volume).

Table 2. Symbols used in calculations with the corresponding unit.

Designation	Unit	Symbol
Mass flow of biochar	kg/s	$\dot{m}_{BC}$
Mass flow of dry biomass	kg/s	$\dot{m}_{BM}$
Mass flow of exhaust gas	kg/s	$\dot{m}_{EG}$
Mass flow of primary air	kg/s	$\dot{m}_{PA}$
Mass flow of secondary air	kg/s	$\dot{m}_{SA}$
Difference in mass over time	kg	Δm
Timeframe	s	Δt
Volume flow of primary air	m^3/s	$\dot{V}_{PA}$
Density of air (primary and secondary)	kg/m^3	ρ_A

The formula for the mass balance is as follows:

$$\frac{dm_{KV}}{dt} = \dot{m}_{BM} + \dot{m}_{PA} + \dot{m}_{SA} - \dot{m}_{BC} - \dot{m}_{EG} = 0, \quad (1)$$

For comparability, the mass balance is based on dry biomass (100%).

To determine the mass flow of the biomass fed in, the weight loss of the daily pellet container over time was measured and converted to dry mass flows. Pellets are continuously transported from this container into the pyrolysis reactor using a screw conveyor.

$$\dot{m}_{BM} = \Delta m \div \Delta t = \frac{(m_{t2} - m_{t1})}{(t_2 - t_1)}, \quad (2)$$

The volume flow of primary and secondary air was measured directly with a hot-wire anemometer.

$$\dot{m}_{PA} = \dot{V}_{PA} \times \rho_A, \quad (3)$$

The mass flow of the produced biochar was determined over the measurement duration by weighing the produced biochar mass.

$$\dot{m}_{BC} = \frac{\Delta m_{BC}}{(t_2 - t_1)}, \quad (4)$$

The combustion gas mass flow was determined using the mass balance. Due to tars in the exhaust gas at some operation points, the volume flow could not be measured directly.

$$\dot{m}_{EG} = \dot{m}_{BM} + \dot{m}_{PA} + \dot{m}_{SA} - \dot{m}_{BC}, \quad (5)$$

Figure 9 shows the system used to perform the energy balance.

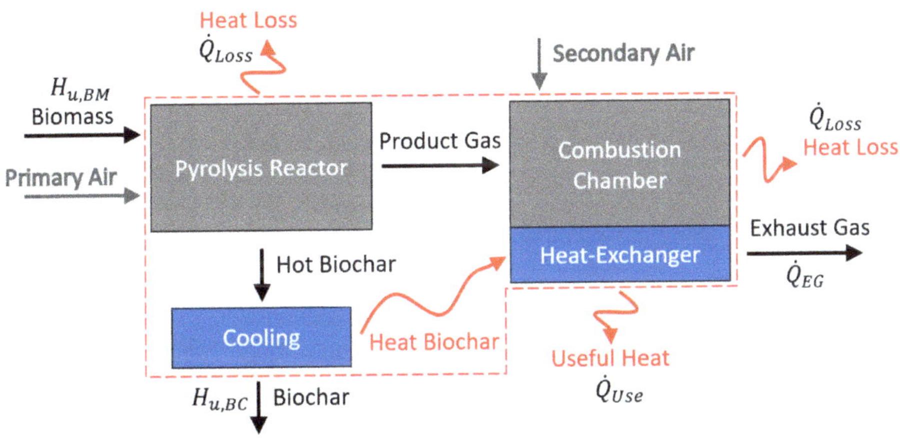

Figure 9. System for the energy balance. (Red dashed box represents the control volume).

The calorific values of the biomass ($H_{u,BM}$) and the produced biochar ($H_{u,BC}$), along with the useful heat ($\dot{Q}_{Use}$), were measured to evaluate the overall energy efficiency and performance of the pyrolysis process. The reference temperature of the environment (T_{Env}) is 25 °C so that the calorific values can be used. The difference from an ideal balance under stationary conditions is the loss of the exhaust gas ($\dot{Q}_{EG}$) and the loss of heat via the housing of the system ($\dot{Q}_{Loss}$). Complete combustion is assumed in the combustion chamber.

The energy balance is as follows:

$$\frac{dE_{KV}}{dt} = \dot{m}_{BM}\, H_{u,BM} - \dot{m}_{BC}\, H_{u,BC} - \dot{Q}_{EG} - \dot{Q}_{Loss} - \dot{Q}_{Use} = 0, \quad (6)$$

The energy for pyrolysis (which is provided by the internal oxidation reactions) can be calculated using the following formula, considering the heating of the biomass and the reaction enthalpy [8]:

$$\dot{Q}_{Pyr} = \dot{m}_{BM}\,\left(c_{p,BM}\,\left(T_{Pyr} - T_{Env}\right) + \Delta h_{pyr}\right), \quad (7)$$

The reaction enthalpy of pyrolysis (Δh_{pyr}) is determined using a differential thermal analysis (DTA). This has been documented in various publications. A linear relationship dependent on the biochar yield (Y_{BC}) was found as follows [8]:

$$\Delta h_{pyr}\left[\frac{kJ}{kg}\right] = 553 - 3142\; Y_{BC}, \quad (8)$$

The heat losses through the exhaust gas depend on the exhaust gas temperature (T_{EG}) with respect to the heat capacity of the exhaust gas ($c_{p,EG}$).

$$\dot{Q}_{EG} = \dot{m}_{EG}\; c_{p,EG}\;\left(T_{EG} - T_{Env}\right), \quad (9)$$

For comparability, the energy balance is based on dry biomass (100%).

The velocity of the pyrolysis front (v_{PF}) was calculated using the relationship between the measured wood flux (ϕ_w) and wood bulk density (ρ_w) [14].

$$v_{PF} = \frac{\phi_w}{\rho_w} \quad (10)$$

4. Results

To meet the variable heating demands of both single and multi-household applications, a central heating system must work at a broad range of operating points. Three different operating points from the pyrolysis heating system are characterized in Table 3 to provide an overview of the range of key figures at a nominal load (100%), partial load (30%), and maximal load (200%). The air flux was varied in the range of 0.027–0.144 kg/m^2/s. At an air flux of 0.144 kg/m^2/s, the limit of the used suction fan was reached.

Table 3. Overview of the parameters of characterization (* accuracy is ± 2%).

Designation	Unit	Partial Load	Nominal Load	Maximal Load
Air flux	kg/(m^{2}s)	0.027	0.075	0.144
Pyrolysis temperature	°C	520	724	809
Wood flux	kg/(m^{2}s) db	0.025	0.052	0.102
Biomass input	kg/h db	2.3	4.9	9.6
Velocity pyrolysis front	cm/h	16	33	65
Lambda pyrolysis	-	0.20	0.26	0.25
Biochar yield *	kg/kg db	23.7%	14.2%	12.4%
Useful heat	kW kW/m^2	4.1 158	15.5 596	29.5 1132

The air flux drives the oxidation reactions at the pyrolysis front, and the exothermic reactions determine the pyrolysis temperature at this front. Consequently, the pyrolysis temperature dictates the rate at which fresh biomass converts to product gas and biochar, thereby defining the wood flux. Figure 10 shows the influence of air supply in the pyrolysis reactor on the temperature and the influence of the pyrolysis temperature on the wood flux.

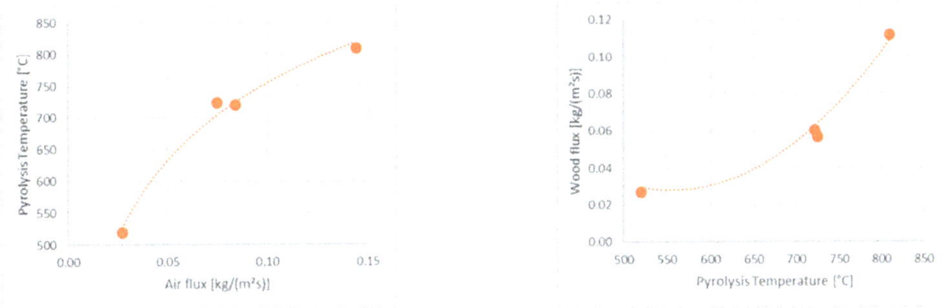

(**a**) (**b**)

Figure 10. Orange dots represent measurements conducted in this study. (**a**) Influence of air flux on pyrolysis temperature. (**b**) Influence of pyrolysis temperature on wood flux.

Although only three air flux conditions were measured, it appears that the influence of air flux on the pyrolysis temperature decreases as the flux is increased. The increased air supply has a partially cooling effect due to the mass of cold air. While a higher temperature in the pyrolysis reactor increases the heat transfer to the environment, the influence of the air flux is reduced.

Higher pyrolysis temperature positively affects the amount of converted wood mass. It is assumed that the speed of the pyrolysis front toward fresh pellets increases with higher temperatures. As long as the reactions are oxygen limited, wood conversion is directly dependent on the air supply, as shown in Figure 11.

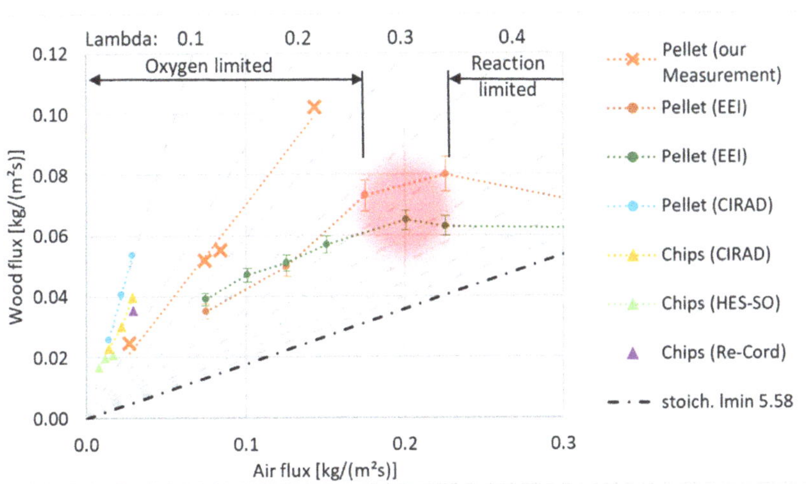

Figure 11. Influence of the air flux on the wood flux (grey lines show a constant lambda). Comparison with the literature data (EEI [11], CIRAD [14], HES-SO [16], RE-CORD [15]).

The CIRAD measurements showed that with a low air flux and an optimized reactor (well-insulated, preheated air supply) for oxidative pyrolysis, the wood flux is linearly dependent on the air flux. In the RE-CORD measurement, a similar characteristic of the chips was measured in a ten-fold larger reactor than in CIRAD but with a similar design.

The EEI measurements of the pellets showed, however, that a maximum for the wood flux can be reached at an air flux of 0.3 kg/(m^2s). This maximum depends on the reactor type and design. EEI employed a basic reactor design without insulation, leading to increased heat loss that had to be compensated for by oxidation reactions. Consequently, these measurements required a higher air flux to achieve a wood flux comparable to that of other reactors. The area of the transition from the oxygen-limited to the reaction-limited regime is marked in red in Figure 11. It is not yet clear where this transition is for other reactor types.

The measurements of this study at a maximum load showed a wood flux above 0.1 kg/(m^2s), and the curve did not flatten. This suggests that the maximum for stable oxidative pyrolysis is higher. Once the maximum is reached, reactions are no longer limited by the oxygen supplied but by the reaction kinetics.

It is assumed that at the maximum, the regime shifts from predominantly pyrolysis to biochar gasification, significantly reducing the biochar yield. This determines the modulation limit for the pyrolysis reactor. The exact point of this limit is not yet determined. Attempts to determine the maximum power in this study still produced biochar, but the measured yield is inaccurate, so no absolute statements can be made.

The influence of pyrolysis temperature on biochar yield is shown in Figure 12, where we compare our results and those of CIRAD [14] to values in the literature for non-oxidative pyrolysis with an external heat supply [7]. (Only data for temperatures above 400 °C are shown, where biochar is formed with a high carbon content.) The literature values (measurements and calculations) show a decrease in biochar yield with the increasing pyrolysis temperature, as more molecules are released and transition to the gas phase.

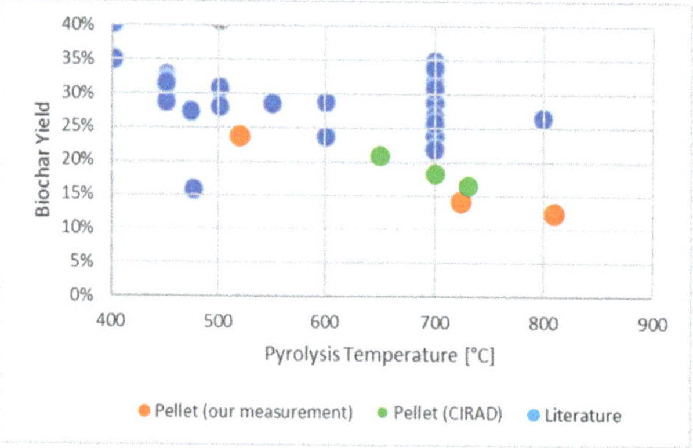

Figure 12. Influence of pyrolysis temperature on biochar yield [7,14].

The biochar yield measurements by IBRE and CIRAD are slightly lower than the literature values. Adding oxygen in the pyrolysis reactor for oxidative pyrolysis produces more pyrolysis gas, oxidizes more biochar, and produces less biochar than conventional pyrolysis with an external heat supply [8,12].

The biochar yields of CIRAD are slightly higher than those of this study with comparable temperatures. This is caused by the different setups with the following impact factors: reactor geometry, air preheating, residence time, and thermal insulation.

was stored in the biochar. At a partial load with 165 kW/m^2, the proportion of heat losses increases, with 35% converted to useful heat and 43% stored in the biochar. Regarding the limit of oxidative pyrolysis and the transition to gasification at higher air flux rates, it was shown that at an air flux of 0.144 kg/m^2/s and a specific useful power of 1132 kW/m^2, oxidative pyrolysis is stable and achieves a biochar yield of 12% ± 2%. Uncertainty remains regarding biochar yield at higher temperatures and the limit of oxidative pyrolysis and the oxygen-limited regime, as shown by EEI [10].

The influence of the input biomass characteristics on biochar quality was analyzed for wood pellets. It is expected that the quality of biochar from wood chips will be similar depending on the pyrolysis temperature; this must be verified by further measurements. For alternatives to wood, it must be determined whether oxidative pyrolysis is stable and what the influence on the biochar is. Initial investigations have already been carried out by HES-SO with LADR (lignocelulosic anaerobic digestion residues). LADR has an ash content of 20%, while wood chips have an ash content of 0.5%. These investigations showed that the yield and composition at 600 °C are similar to those of wood chips. There were differences in the carbon content, which was only around 50% for LADR, and in the tar content in the pyrolysis gas, which was 30% higher for LADR [16].

The advantage of oxidative pyrolysis compared to allothermal pyrolysis is that the process depends almost exclusively on one variable, the primary air, which can be increased or reduced quickly. The pyrolysis front limits the pyrolysis process to a specific cross-sectional area, making it significantly easier to scale with primary air than by heat transfer through the reactor outer wall, as is common in allothermal pyrolysis [8]. By dividing the heating system into two stages, a pyrolysis reactor and a combustion chamber, the solid phase of the biomass remains in the pyrolysis reactor, and only the gas phase enters the combustion chamber, resulting in low particle emissions. One potential issue is the relatively high front temperature in oxidative pyrolysis of up to 700 °C, which may impact performance when using alternative biomasses with high ash contents and low ash melting points, such as grain waste products.

A heating system driven by oxidative pyrolysis offers flexibility in applications. The broad range of variable heating output allows for a choice between more heat and lower biochar yield or lower heat and higher biochar yield. However, the disadvantage of oxidative pyrolysis compared to allothermal pyrolysis is the reduced biochar yield due to partial oxidation for heat generation, as shown in Figure 12.

The precise boundaries of the stable oxidative pyrolysis range, along with the variability in wood flux and biochar yield, remain as areas requiring further investigation. Further studies are essential to fully understand these parameters and optimize the process for different biomass types and operating conditions. A potential application could involve utilizing a single reactor capable of operating in oxidative pyrolysis mode during periods of moderate heat demand and switching to full gasification for short durations when high heat demand is required. Additionally, the influence of air flux and pyrolysis temperature on the lambda value in the pyrolysis reactor is not yet fully understood and should be a part of future research.

Author Contributions: T.G. and F.C.C. designed and developed the study. D.L. and J.S. performed the experiments. J.S., D.L. and F.C.C. evaluated the measurement. F.C.C. wrote the original draft preparation. T.G. supervised the study and finalized the manuscript with F.C.C. Portions of the work were carried out during the master's studies of F.C.C. All authors have read and agreed to the published version of the manuscript.

Funding: This research was funded by the Swiss Innovation Agency (Innosuisse) as part of the project "Klimapositive Heizung mit Pflanzenkohleproduktion" with reference number 104.371 IP-EE.

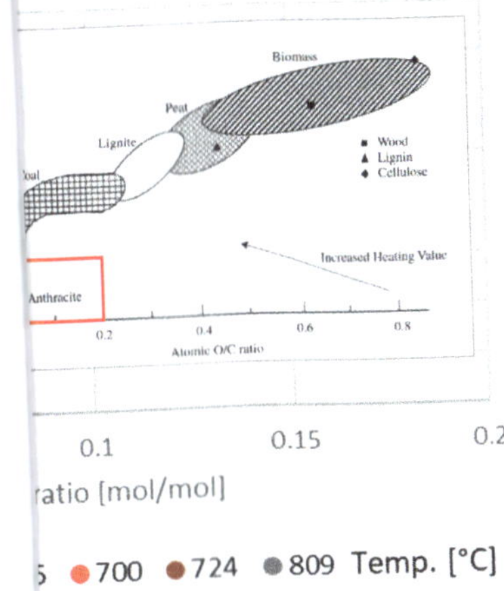

produced biochar at different pyrolysis temperatures
diagram from source [22], which is reprinted from [22]

ly captured carbon.

Partial Load	Nominal Load
520	724
165	621
33.4	34.2
23.7%	14.2%
44%	27%

pyrolysis have provided data on the performance
llets. In oxidative pyrolysis, thermal energy is
exothermic oxidation reactions with primary air.
air mass flux on the process of oxidative pyrolysis
ange of 0.027 to 0.144 kg/m^2/s, the air excess ratio
constant with a range of 0.2–0.26. Maintaining a
r flux results in an increased movement velocity
t, confined to a small region within the pyrolysis
creased as long as the reactions are limited by the
nearly dependent wood flux in the range of 0.025
of the investigated pyrolysis heater.
reactor produces biochar, which meets the quality
har Certificate) Feed Plus limits for all conditions.
ting, the pyrolysis temperature (T5 in Figure 7)
iochar quality. It was shown that with a constant
perature range of 520 °C to 724 °C, the biochar
a nominal load of 621 kW/m^2, 62% of the pellet
at from combustion of the product gas, and 26%

4.1. Mass and Energy Balance

The mass balance is reported in Figure 13.

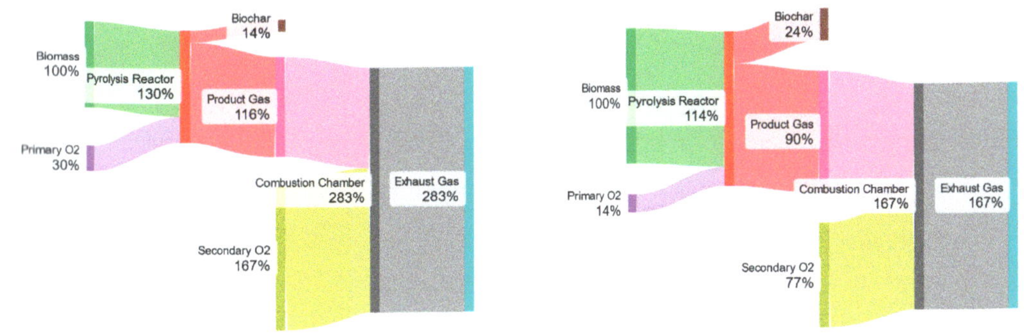

Figure 13. Mass balance at (**a**) nominal load of 15 kW (100% = 4.8 kg/h); (**b**) partial load of 4 kW (100% = 2.3 kg/h).

Approximately twice as much oxygen is required at a nominal load as compared to a partial load. This leads to more than three times the useful heat output. The higher primary oxygen mass flux leads to a higher pyrolysis temperature, which results in a reduced biochar yield. The energy balance associated with the mass balance is shown in Figure 14.

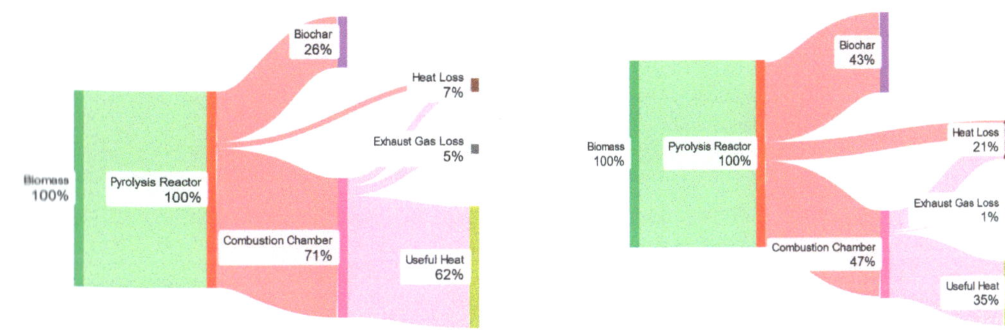

Figure 14. Energy balance at (**a**) nominal load of 15 kW (100% = 25.0 kW); (**b**) partial load of 4 kW (100% = 11.9 kW).

The energy requirement for oxidative pyrolysis to maintain the pyrolysis temperature was calculated using Equation (7). This energy is not lost but stored in the hot biochar and pyrolysis gas. The energy for pyrolysis was found to be between 6 and 12% of the heating value contained in the dry biomass. As a comparison, 10% was estimated in the literature, and RE-CORD indicated 10% for a 50 kW reactor [15,18].

By reducing the power by reducing the air flux, more energy from the biomass is stored in the biochar, and less energy is transferred as useful heat in the heat exchanger. At a nominal load, 11.5 MJ/kg of useful heat, or 62% of the wood heating value, is produced per kilogram of pellets. At a partial load, 6.4 MJ/kg of useful heat, or 35% of the wood heating value, is produced per kilogram of pellets. This means that at a partial load, about 56% less useful heat is produced than at a nominal load. The relationship between the

However, daily and seasonal variations limit the capacity of renewables to fully replace fossil electricity. Therefore, solutions to meet the daily and seasonal storage of electricity at global and local scales are urgently required [16] to speed the transition towards renewables [17]. Therefore, the role of GH_2 as an energy carrier capable of being stored in normal tanks becomes crucial compared to lithium–ion batteries [18]. The impacts of the latter are increasingly raising concerns and currently can only be partially mitigated by solutions integrating Circular Economy (CE) principles [19] and the consideration of social and environmental justice dimensions in policies or industrial strategies [20].

Aside from hydro, wind, and solar PV, GH_2 can be produced from other renewable feedstocks including biomasses not in competition with food production, such as household food, crop residues, livestock manure, and aquatic biomass [6]. With that, GH_2 production fully meets the principles of a CE [21].

The implementation of a future GH_2 economy faces several challenges due to the current higher production costs of GH_2 [10], the need to improve the efficiency and maturity of electrolyzers, and the investments needed for storage and transportation [6,22]. There are also uncertainties concerning the environmental impacts of decommissioning electrolyzers and some renewable energy technologies [23], as well as the contribution of GH_2 to the mitigation of other impact categories beyond GWP, such as abiotic resource depletion [24]. For these reasons, it is important to strengthen the research towards improving its economic competitiveness and environmental soundness [22,25]. According to Bloomberg forecasts, GH_2 should be able to compete with grey and blue hydrogen in Brazil, China, India, and some EU countries before the year 2050, depending on the scale up of more efficient electrolyzers. In the year 2023, the forecasts have been updated, limiting the 2050 competitiveness to only grey H_2 in China and India [24].

Recent literature [23–28], listed and analysed in Appendix A, Table A1, reviewed LCA studies concerning the production of GH_2 through electrolysis, shedding light on some life cycle impacts depending on the renewable sources used, the types of water electrolysis technologies, and the storage systems, before finally pointing out that wind and solar are the most favorable sources for GH_2 production, with potentially lower contributions to GWP when using wind power compared to solar PV [23,25].

These results have been confirmed by further literature reviews (Appendix A, Table A1, [29,30]) analysing the environmental impacts of GH_2 production as well as the economic impacts and policies supporting GH_2 development in some countries, including the EU [29]. However, the prospects could be more favorable in the future for solar-powered GH_2 due to the progressive decarbonization of the grid, leading to a reduction of the carbon intensity of the solar cells [23].

Other authors have explored the literature concerning the technological characteristics and environmental impacts of GH_2 production, comparing them with other hydrogen production patterns identified with different colours [9–11,31–39], integrating the analysis of economic impacts [11,34,35] and, in some cases, mapping the geographical areas most suitable for renewable energies and GH_2 development [11,33]. In areas rich with these sources, e.g., South Africa, GH_2 production could be used both for domestic use and exports [40–42]. Many countries have adopted hydrogen strategies and roadmaps [31,43,44], and some studies have also offered an overview of the latter and the current capacity and status of hydrogen projects within the framework of net zero emissions targets of the year 2050 [44], along with the analysis of the environmental impacts of GH_2 [43] and other hydrogen colors [31,44].

Some authors also focused on the comprehensive assessment of GH_2 production processes [21,45,46], storage and distribution modalities and their economic costs [47], and the analysis of specific electrolyzers such as AEL [48] that expand on basic principles and designs, materials for the components of the electrolyzers, the integration of renewable energy sources, electrolyzer costs, and environmental impacts [48].

A few literature reviews investigated the social aspects of hydrogen [49] or the social impacts of GH_2 production, storage, transportation, and distribution [50–52], integrating the analysis of the social impacts with that of the environmental impacts [52,53]. The low number of literature reviews confirms that social aspects/impacts are not yet included in most environmental and especially technical/economic assessments [50].

The economic costs are only one of several dimensions on how to decide GH_2 projects and strategies. Social aspects (particularly those related to the local population involved in the production of GH_2) should be much more integrated for a more comprehensive analysis useful for decision making [54]. There is a need for more research on the process of governance in the implementation of GH_2 technologies and their social acceptance [50].

This study aimed to contribute to the closing of this gap in the literature, evaluating the environmental and social impacts of GH_2 in the whole life cycle, as well as to improve our understanding of the effectiveness of GH_2 in reducing the contribution of energy systems to current environmental and social challenges.

The analysis is performed in light of the CE transition and the adoption of its principles aimed at changing the way products, materials, and food are designed, produced, and consumed [55] via a new model of circular local development [55]. The goal is to address the main research question that is as follows: **Is Green Hydrogen an environmentally and socially sound solution for decarbonizing energy systems in the transition to a circular economy?**

The novelty of the manuscript is as follows:

- Creation of a state-of-the-art framework composed of previous literature reviews useful to understanding the main current trends and gaps in the literature of GH_2;
- Strengthening the existing literature reviews on GH_2, enriching the understanding of the environmental and social impacts of the life cycle of GH_2 with the analysis of the most recently published articles;
- Assessment of the environmental and social impacts of GH_2 within the framework of the CE model and its principles to conceptualize the areas of relationship between CE and GH_2. This would clarify how GH_2 aligns with the CE framework and show how its implementation is consistent with the CE principles such as resource efficiency, closed-loop production, and local development.

This study develops over the next four sections. Section 2 analyses the relation between CE and GH_2, which is useful to understand its role in CE transition and the relevance of its principles in enabling a more environmentally and socially sound decarbonisation pathway. Section 3 provides details on the type of data used and the method adopted to extract them from the main international database. Section 4 presents the results of the narrative literature review, while Section 5 discusses the main results and presents the conclusions of this study.

2. Theoretical Framework: Circular Economy and Green Hydrogen

There is a strict link with the production of GH_2 and the CE model [47]. The CE promotes the vision of a new socio-economic system as well as sustainable production and consumption patterns [56] according to the three pillars of sustainable development [57]. Such a model suggests the elimination of waste and pollution by design and the continuous circulation of products and materials at their highest value by means of the most appropriate CE principles (such as "reduce, repair, reuse, recycling" and so on) [58] to tackle the need for extracting new raw materials [59].

The CE also promotes energy efficiency, lower energy use, and energy transition towards renewables [47,60]. These sources of energy continuously regenerate beyond human timescales and are part of the processes of the geobiosphere driven by the global power of solar, geothermal, and tidal energy [61]. Within this framework, the circulation of energy and the production of hydrogen from renewable sources may contribute to local and global decarbonization goals [62], avoiding the extraction and consumption of fossil fuels and related impacts, thereby supporting the independence from them [63].

On the other hand, the CE is a great opportunity to better ascertain the value of all local physical resources, be they natural or mined in the local urban systems [64]. Some authors provided evidence that the spatial distribution of renewable energies should be taken into account in decarbonization strategies, because some areas are richer in renewables compared to other areas or their geographical locations are more favorable towards solar or wind power [65]. This may likely have effects on GH_2 production and its environmental efficiency [66]. GH_2 can be considered a renewable fuel (a carrier of other renewable energies) because it can be produced indefinitely [22]. As a fuel, it is also more energy efficient per unit of mass compared to diesel and gas: the production of electricity with fuel cells powered by hydrogen achieves an efficiency of up to 65%, which is much higher than the efficiency achieved by, for example, a coal power plant (up to 35%) [22] or electricity generated from natural gas (45%) and oil (38%) [67].

The application of CE principles to renewable technologies (including green hydrogen), from the design stage, is absolutely required to reduce their impacts [68]. As a matter of fact, renewable technologies rely on finite materials for the production and store of energy [69], and for this reason, their lifecycle cannot be based on a linear approach [70]. This implies the designing of renewable technologies towards longer lifetimes, easier maintenance and repair, and the application of standardization and modularity in their design for disassembly at end-of-life [71]. Other CE solutions are full repowering, design for reuse and recycling, and, more broadly, the rethinking of the economics of renewable technologies [69] and supporting their development via policies and regulations such as those adopted in the European Union [72]. The adoption of certification schemes for encouraging more responsible supply chains in line with CE principles [10,73–75] under the EU Renewable Energy Directive and for renewable hydrogen [72], as well as the adoption of decision-making frameworks and standardized guidelines for reuse (e.g., for solar panels) [76], are also very important patterns for reducing the impacts on the whole life cycle of renewable energies and GH_2 [72].

3. Materials and Methods

This study mainly collected secondary data and information from the most recent international literature, although other sources and particular reports published by key stakeholders in the field of CE (such as the Ellen Mac Arthur Foundation) and the energy sector have been considered (European Hydrogen Observatory and International Energy Agency).

The method of the narrative literature review adopted by other scholars in the field of hydrogen research [10] has been considered appropriate for pursuing the goal of this study. The review has been conducted while taking into account the suggestions/guidelines for this type of narrative literature review [77,78]. In particular, Green et al. [77] emphasizes that the results should be presented in a "*condensed format that typically summarizes the contents of each article*". Moreover, narrative literature reviews have the advantage of offering a broad overview of a topic, going through its history and evolution. Finally, these reviews "*may be an excellent venue for presenting philosophical perspectives in a balanced manner*" [77].

However, this type of literature review may be criticized when compared to the rigorousness of the method followed in systematic literature reviews [77]. In this regard, the method followed in this study has tried to reduce biases by considering guidelines from systematic literature reviews [78] in order to provide more reliable results and conclusions.

The keywords and query used for the searches, the database searched, and the criteria for selecting the articles are shown below. The method followed is more transparent and may be reproducible [78], adding scientific quality to the narrative literature review performed in this study.

Clarivate Web of Science has been selected as the main database to search the relevant literature in accordance with other authors who performed bibliometric analyses concerning hydrogen research [6].

The overall process started with the identification of the following keywords and query for searching the literature:

- Green Hydrogen AND Literature review;
- Green Hydrogen AND Circular economy;
- Green Hydrogen AND Environmental impacts;
- Green Hydrogen AND Life cycle assessment;
- Green Hydrogen AND Social life cycle assessment;
- Green Hydrogen AND Social impacts.

The criteria followed for the inclusion and exclusion of the literature are the temporal timeframe (2022 to 2025) and the coverage of the selected literature of the topics useful to address to the main research questions and understand if GH_2 could be an environmentally and socially sound solution for the main applications (as a fuel, heat producer, and a feedstock for other chemical products) in future energy systems.

Table 1 summarises the keywords, the query, the results for each query, the criteria for the inclusion/exclusion of the searched articles, and the number of selected articles. The initial number of articles resulting from all the six queries was 514, and the total number of selected articles after the screening of the abstracts and content of the articles, as well as the removal of duplicates (28), resulted in 73. Some articles (20) listed in the note at the bottom of Table A2 of Appendix A have been extracted from the references of the selected articles and integrated into the articles retrieved from the search on Web of Science, because they have been considered useful for this study. The total number of articles is therefore 93. The latter are all listed on the basis of the year of publication in Table A2 of Appendix A and presented according to the author/s, main goal, method, and results.

Table 1. Description of the most important aspects used in searching and selecting the sample of articles of this study.

Keywords	Query	Results from All Databases	Inclusion	Exclusion	Number of Selected Articles
Green Hydrogen; Literature review	"Green Hydrogen" AND "Literature review"	63	Years 2022–2025 (28 March 2025)	Years before 2022	9
			Literature reviews of LCA studies on GH_2, reviews considering the analysis of the environmental impacts of GH_2 life cycle, or environmental and social impacts and economic costs of GH_2 life cycle	Reviews only focused on financial or cost analysis of GH_2 production/life cycle	
			Only literature reviews as document type	Document types beyond literature reviews	
Green Hydrogen; Circular economy	"Green Hydrogen" AND "Circular economy"	94	Years 2022–2025 (28 March 2025)	Years before 2022	10
			Articles/Reviews focused on LCA/analysis of environmental and social impacts and economic costs of GH_2 production within CE framework	Articles/Reviews only focused on financial or cost analysis of GH_2 production within CE framework	

Table 1. *Cont.*

Keywords	Query	Results from All Databases	Inclusion	Exclusion	Number of Selected Articles
Green Hydrogen; Environmental Impacts	"Green Hydrogen "AND "Environmental impacts"	109	Years 2022–2025 (28 March 2025) Articles/Reviews focused on LCA or analysis of environmental impacts of GH_2 with other methods and eventually including the analysis of social impacts and cost of GH_2 production	Years before 2022 Articles/Reviews focused only on financial or cost analysis of GH_2 production	24
Green Hydrogen; Life cycle Assessment	"Green Hydrogen" AND "Life cycle Assessment"	239	Years 2022–2025 (28 March 2025) Articles/Reviews focused on LCA of GH_2 life cycle or comparative LCA of GH_2 and other hydrogen colours	Years before 2022 Articles/Reviews not focused on LCA of GH_2 life cycle or comparative LCA of GH_2 with other hydrogen colours	51
Green Hydrogen; Social Life cycle Assessment	"Green Hydrogen" AND "Social Life cycle Assessment"	2	Years 2022–2025 (28 March 2025) Articles focused on social life cycle assessment of GH_2 production/life cycle	No criteria for the exclusion	2
Green Hydrogen; Social Impacts	"Green Hydrogen" AND "Social Impacts"	7	Years 2022–2025 (28 March 2025) Articles focused on social life cycle assessment or analysis of the social impacts of GH_2 life cycle with other methods	Years before 2022 Articles not focused on the analysis of social impacts of GH_2 life cycle	5
Duplicates					28
Total					73

Figure 1 shows that more than half of the selected articles consist of literature reviews and life cycle assessment studies. This latter method is, in some cases, combined to Life Cycle Costing or Emergy Accounting. The rest of the sample are studies that apply different methods, such as techno-economic and environmental analysis, environmental and cost analysis, and social life cycle assessment.

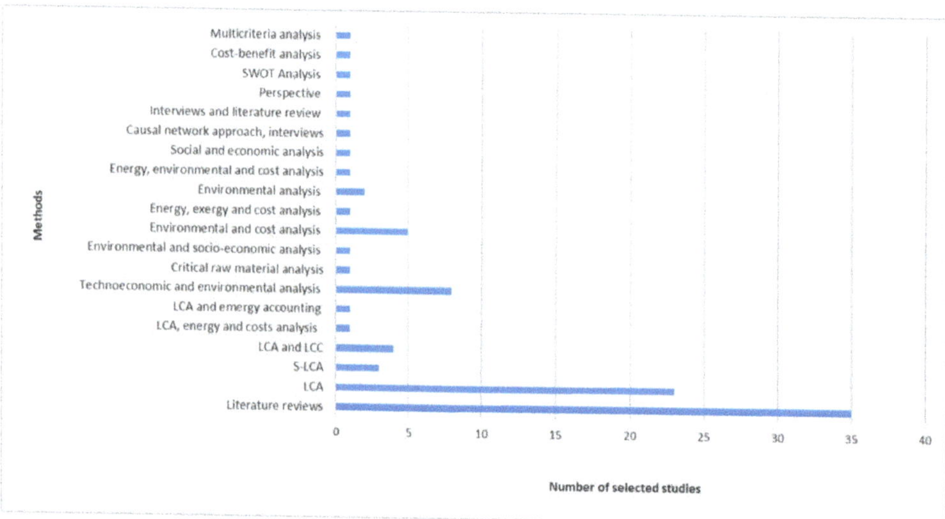

Figure 1. Distribution of all the selected articles according to their method of analysis.

4. Results

4.1. Hydrogen and Green Hydrogen Production

Hydrogen is a very abundant element in the universe, composing 75% by weight according to [79], while also being widespread on Earth, where is never available in its molecular form (H_2) due to its high reactivity. Instead, it is a fundamental constituent of complex molecules of biological origin, organic salts, and water [80].

In order to separate hydrogen from the other elements, it must be "extracted" by providing energy (be it renewable, fossil, or nuclear) for the separation process leading to different economic costs and environmental impacts, depending on the source of extraction [9]. For this reason, hydrogen is considered an energy vector and not an energy source as with wind or solar [81].

Currently, most of the hydrogen worldwide is obtained by using fossil fuels as feedstocks, the processes of steam reforming of natural gas, and coal gasification [82]. Figure 2, listing the colours attributed to hydrogen according to the kind of energy needed for its production, shows that, in the case of both brown/black hydrogen and grey hydrogen, their production is accompanied with the release of CO_2 emissions. A fraction of the latter can be captured and stored underground in the production process of blue hydrogen [83].

Colour	Fuel	Process	Products
Brown/Black	Coal	Steam reforming or gasification	$H_2 + CO_2$ (released)
White	N/A	Naturally occurring	H_2
Grey	Natural Gas	Steam reforming	$H_2 + CO_2$ (released)
Blue	Natural Gas	Steam reforming	$H_2 + CO_2$ (% captured and stored)
Turquoise	Natural Gas	Pyrolysis	$H_2 + C$ (solid)
Red	Nuclear Power	Catalytic splitting	$H_2 + O_2$
Purple/Pink	Nuclear Power	Electrolysis	$H_2 + O_2$
Yellow	Solar Power	Electrolysis	$H_2 + O_2$
Green	Renewable Electricity	Electrolysis	$H_2 + O_2$

Figure 2. List of the colours of hydrogen depending on the fuel needed for their production. Source: [84].

Turquoise hydrogen is also produced from natural gas as blue and grey hydrogen but involves the process of methane pyrolysis, which produces turquoise hydrogen and solid carbon. The latter can be sold and reused in various production processes. Going further, red and purple/pink hydrogen are produced with electrolysis or thermochemical processes by using nuclear electricity [83].

Yellow hydrogen is obtained by means of the water electrolysis process driven by solar electricity, while GH_2 is also from water electrolysis driven by renewables other than solar. The carbon footprint of GH_2 approaches closer to zero compared to the other types of hydrogen derived from fossil fuels, which release CO_2 in the production stage [23,31,32,44,83].

For the purposes of determining the environmental impacts of hydrogen, it is appropriate not to limit the analysis of direct emissions in the usage stage, as shown in Figure 2. Therefore, when GH_2 is burned as a fuel in a traditional engine or in fuel cells, the upstream hydrogen production process should also be considered [31]. The impacts of these different steps will be further analysed in Section Policy and Practical Recommendations.

Renewable energies such as wind, solar PV, geothermal, hydro, tidal, and biomass can be used to produce the electricity required for the electrolysis process [32], as well as to decompose water into its constituent elements (hydrogen and oxygen): "*Electrolytic water splitting is driven by passing the electrical current through the water, where conversion of the electrical energy to chemical energy takes place at the electrode-solution interface through charge transfer reactions in a unit called an electrolyzer. Water reacts at the anode to form oxygen and protons, whereas a hydrogen evolution reaction takes place at the cathode*" [85].

Agricultural residues, by means of biological fermentation, electrolysis, and thermochemical processes, can also be used to produce GH_2. Residues have a relevant content of lignocellulose, leading to a higher H/C ratio compared to fossil fuels. For example, H/C of wood biomass ranges from 1.7 to 2, while that of bituminous coal comprised between 0.1 and 0.3 [21]. The higher the H/C ratio, the higher the energy efficiency of the fuel and the lower the CO_2 emissions from combustion [86]. Furthermore, GH_2 can also be derived from food and transportation industries.

Some authors have also highlighted the benefits of producing GH_2 by using part of the treated effluents from a wastewater plant for the production of GH_2 by water electrolysis and electricity produced from solar PV panels. In a case study located in Oman, the economic analysis, in particular, shows that reusing the wastewater instead of discharging it into the sea generates further revenues due to the selling of the produced GH_2. The revenues would cover the initial investment costs within a short amount of time [87].

Given the variety of renewable sources from which GH_2 can be produced, some authors suggested to further characterise the greenness of GH_2 into new colours such as "dark green" for GH_2 coming from the electrolysis process and "light green" obtained from other production approaches [46].

The European Commission has defined renewable hydrogen as a fuel of non-biological origin and identified two criteria to assess its renewability (additionality and temporal and geographic correlation). The first criteria: "*Additionality ensures that increased hydrogen production goes hand in hand with new renewable electricity generation capacity. Hydrogen producers must therefore conclude power purchase agreements with new and unsupported renewable electricity generation capacity*". while the second ensure that hydrogen is produced "*when and where renewable electricity is available to avoid the demand for renewable electricity for hydrogen production incentivize more fossil electricity generation*" [88].

Some authors [89] underline that, in the transition to GH_2, the role of blue hydrogen is important since it could use the existing infrastructure of natural gas accelerating the general adoption of hydrogen. These authors have compared, by means of LCA, different pathways for the life cycle of blue hydrogen (obtained by the process of steam reforming and two new processes such as Partial Oxidation of Methane and Auto-Thermal Reforming) and found that the production of hydrogen via auto-thermal reforming and delivery of ammonia to consumers has the lowest contributions to GWP, i.e., 2.12 kg CO_2 eq./kg of hydrogen. The study also jointly evaluated the life cycle contributions to GWP of blue hydrogen pathways with that of GH_2 (produced with water electrolysis and electricity sourced by concentrated solar PV collectors) and found that the integration of GH_2 strongly reduces the impacts in all the considered pathways that provide H_2 as a final product.

Concerning the levelized production costs of the different hydrogens, in the year 2023, the costs of grey hydrogen by SMR was the most financially competitive in the EU (being 3.76 €/kg of hydrogen), while the average costs of blue hydrogen by SMR with carbon capture were 4.41 €/kg of H_2. Figure 3 shows the levelised production costs of GH_2 in the EU. These ranged from 4.13 €/kg of H_2 to 9.30 €/kg of H_2, with an average of 6.61 €/kg of H_2 and a median of 6.20 €/kg of H_2 [90].

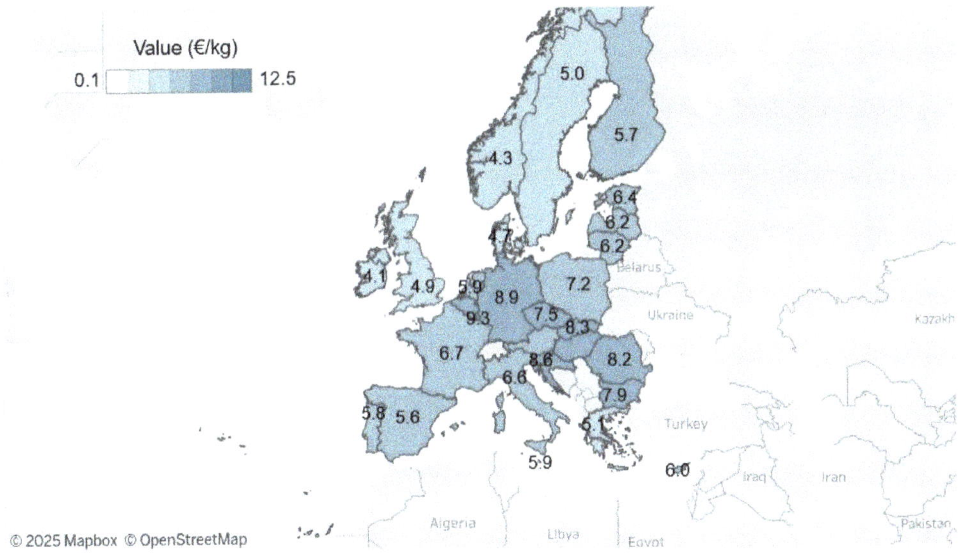

Figure 3. Distribution of the levelized costs of production of GH_2 in the EU. Source: [90].

It is also important to note that the levelized production cost is a financial indicator [91] and that the "lower" values of the indicators for grey and blue hydrogen do not account for the negative external costs related to the CO_2 released into the natural environment and the society involved in the production process. Conversely, the positive externalities, due to the avoidance of CO_2 emissions, are not accounted for in the "highest" value of levelized cost of production of GH_2. Therefore, the levelized production costs should be integrated by further analysis for a more comprehensive accounting of the social costs of the different hydrogens [92].

For the production of GH_2, four types of electrolysers were mainly analysed by the literature: proton-exchange membrane (PEM), alkaline water electrolysis (AWE), solid oxide electrolysis cell (SOEC), and anion exchange membrane (AEM) [9,45,48,83]. The latter are a less consolidated technology compared to the former three [9,83]. Each type of electrolyser has its advantages and limitations in terms of Faradaic efficiency, hydrogen purity, costs, and materials used for their manufacturing. For example, the PEM electrolyser has a high efficiency in terms of hydrogen produced per unit of electricity used in the reaction and achieves the highest purity (99.99%) compared to the other three, but its manufacturing requires precious and expensive materials such as platinum and iridium [9]. The high costs of the catalyst prevent a larger-scale application of the PEM electrolyzer [83]. AWE is made of less costly materials but has a lower efficiency and the hydrogen purity is lower (99.5%) [83].

The IEA [93] underlined those financial decisions for investments (FIDs) in carbon capture and storage involved in large-scale projects for the production of hydrogen from fossil fuels has grown, particularly in North America and EU in the year 2023. The IEA [93] also reported that FID for increasing electrolyser capacity have achieved 6.5 GW worldwide in the year 2024. China is emerging as an important player in electrolyser capacity, accounting about 40% of the global FIDs, and is further strengthening its manufacturing role, that is, 60% of the global electrolysers' manufacturing capacity. However, final investments' decisions for electrolysis projects have reached more than 2 GW in the EU as well, while in India, the decisions have reached a capacity of 1.3 GW.

4.2. Storage and Distribution of Hydrogen and Green Hydrogen

After production, the life cycle of hydrogen encompasses several processes before it can be used for its final applications. The latter affect the ways hydrogen can be stored and distributed [94]. Purification processes could be also required in case hydrogen is used, for example, in fuel cells. Hydrogen can be stored in different ways (liquid, compressed gas, or bounded in physical or chemical terms in a suitable solid-state material) and the choice of one method over the other depends on various factors such as the geographical location, weight constraints, the demand [94], costs, environmental impacts, and safety issues [47]. Therefore, each storage method has its advantages and disadvantages [47].

Currently, hydrogen is mainly stored aboveground in small and medium metal tanks by the industry in the form of compressed gas or in liquid form stored at very low temperatures (so-called cryogenic storage) [95]. Moreover, hydrogen is mainly produced and consumed on-site without the need to be delivered by transportations means or infrastructures [93]. The transport requirements could increase in the future due to the particularly higher demand of GH_2 and the need to link locations of GH_2 production where there is a high availability of renewable resources with locations where there is a demand for consuming GH_2.

Gas compression has the advantage of reducing the volume of hydrogen but entails high energy costs and the use of cylinders that can safely support high pressures, which are typically made of high-cost materials [94]. There is an increasing interest in storing large quantities (up to several tons) of hydrogen in gaseous form in unused geological repositories [11,47,94]. Underground storage of hydrogen in gaseous form in salt caverns, aquifers, depleted hydrocarbon deposits, as well as in specially designed rock caverns, is possible for hydrogen at pressures up to about 150 bar [94]. Salt caverns, in particular, have proven to be a viable storage solution [11,47] and are used to store hydrogen in the United States and United Kingdom [93].

The distribution of hydrogen as compressed gas or in liquid and solid forms on the road by truck is considered optimal for small stations [94] and is the most common distribution means today [82]. However, this option seems to have a higher environmental impact compared to the pipeline distribution [47].

The transport of hydrogen as compressed gas or in liquid and solid forms by ship might be a cost-efficient solution for hydrogen (in the form of ammonia, methanol, and LOHC or liquid hydrogen) over long distances [93], but this transport option has not yet achieved a commercial scale and has the disadvantage of relevant energy losses [61]. Ideally, the distribution in pipeline networks under a gaseous state seems essential in the future and in view of the potential large-scale use of hydrogen [47]. The EU is supporting the development of hydrogen infrastructure by means of the European Hydrogen Backbone [95]. The latter is an initiative joining together 33 energy operators in the field of energy infrastructures. The mission is speeding the decarbonisation of the EU while recognising the relevant role of hydrogen infrastructures (existing and new ones) in driving the development of a pan-European, low-carbon hydrogen market [95].

In this regard, the existing natural gas distribution networks could serve for the transport of hydrogen–natural gas blends or pure hydrogen [47]. In this case, the distribution of hydrogen up to distances of 2500–3000 km at quantities of hydrogen above 200 kt per year seems the less costly way of distribution for repurposed pipelines; currently, about 5000 km of hydrogen pipelines are already functioning around the world [93]. However, in the case of dedicated new pipelines, they require very high capital investments [47].

Finally, some authors have analysed and compared the performances of hydrogen batteries [96] and fuel cells and lithium–ion batteries [18] for storing energy for short and longer periods in PV solar systems [96]. Hassan et al. [96] compared lithium–ion batteries and hydrogen batteries while considering their technical and financial performances. Their results seem favour the lithium–ion batteries that would also have better financial performances due to the lower payback period compared to hydrogen batteries. However, green hydrogen batteries have a longer service life for a longer energy storage than lithium–ion batteries [18].

4.3. Final Uses of Hydrogen

Figure 4 shows that hydrogen can be used for the following three main purposes:

1. Fuel for the most adopted transportation means such as cars, buses, airplanes, trains, and tractors [97]. As a fuel, it can also be stored for electricity production in order to help satisfy the peak demand [98].
2. Production of heat needed in industrial processes and residential and commercial buildings [97]. In particular, the use of GH_2 in "hard-to-abate" systems would be relevant, i.e., in industrial sectors where electrification is technically difficult and not very competitive. This is the case with the production of steel, ceramics, or cement, as well as in the chemical industry and foundries. These sectors need a large amount of energy, and it is difficult to reduce their GHG emissions [99].
3. Raw material for the production of chemicals such as ammonia (needed for the production of fertilizers) and methanol used in the chemical industry to produce other compounds, fuels, and additives [22,82,97], as well as in refineries to remove impurities and upgrade heavy oil fractions into lighter products [22,82].

In the year 2023, the data from [100] provided evidence that hydrogen consumed in the EU amounted to 7.9 Mt. The highest share of total hydrogen demand was used by refineries (57%) and by the ammonia industry (25%). Globally, these two sectors accounted for 82% of all the hydrogen demand. The rest of the hydrogen is used in the chemical industry (11%) and other industrial uses (3%) and emerging hydrogen applications (less than 0.1%) as clean hydrogen in mobility, steel production, and e-fuels [93].

4.4. Case Studies of Application of Green Hydrogen

Table 2 lists several cases of application of GH_2 as a fuel and as a heat producer selected from the analysed literature. Each case study is analysed according to its goal/main focus, key findings, success factors, and limitations. These cases evaluated the environmental impacts of the use of GH_2 by means of different methods such as LCA [17,60,101] or its technical and financial performances [102]. Some studies combined LCA and Emergy Accounting [103] or technical/energy/exergy analyses [104], while others assessed the technical and financial viability of GH_2 projects, adding to the evaluation of the environmental benefits [98,105]. Other studies also investigated the optimal geographical location of GH_2 projects [106] or the preferences for different types of fuel, including GH_2 [107], and the financial analysis of GH_2, integrating them with citizens perception assessment [108]. In general, the cases show the environmental benefits of using GH_2 as a fuel in transport [101,103] and industrial hard-to-abate sectors [105], contributing to the decarbonisation goals but sometimes worsening other impact categories such as TETP and FETP [101] or FD [103]. With regard to financial feasibility, the investigated cases show that it depends on local factors, such as the integration of a GH_2 plant with other uses (e.g., peaker plants) [98] or the availability of more tailored policy tools, or other factors such as the prices of natural gas [105]. In this regard, it is important to complement financial analyses with social cost–benefit analyses, as well as to more broadly evaluate the socio-economic feasibility and welfare of GH_2 projects.

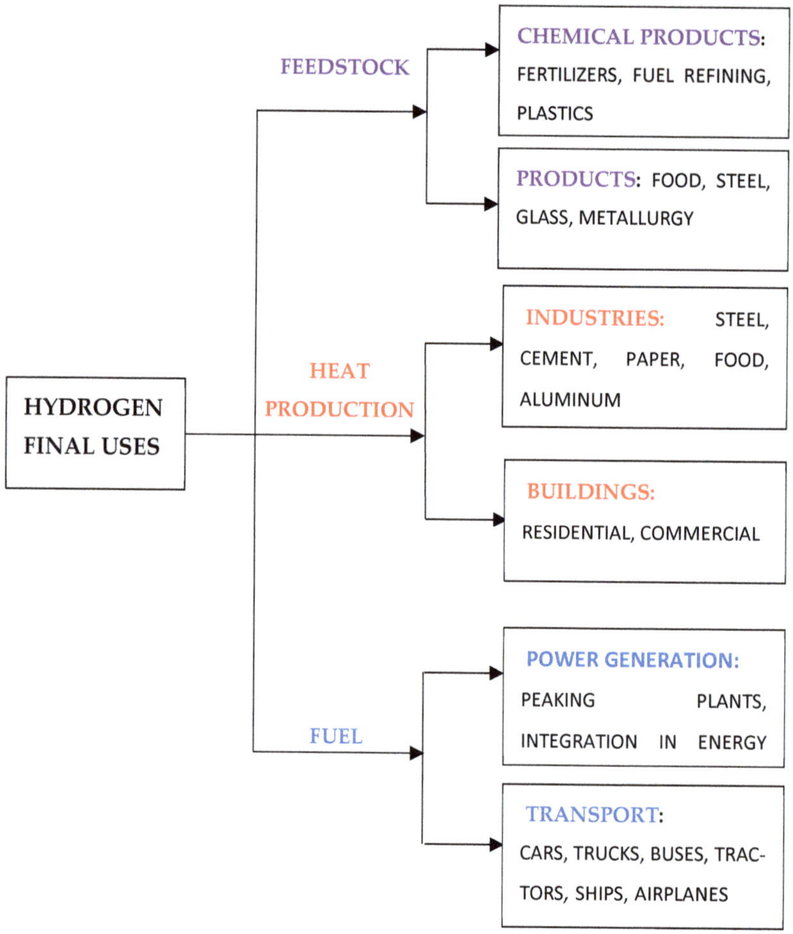

Figure 4. Potential final uses of hydrogen.

Table 2. Features of GH_2 case studies from literature.

	Authors and Geographical Scale	Case Study Goal	Key Findings	Success Factors	Limitations
Industrial processes involving GH_2	Pellegrini et al. [105], Cagliari (Italy)	Analysis of the techno-economic sustainability of integrating a solar photovoltaic to an electrolyser within an aluminium fluoride production plant. In particular, the goal is assessing the costs of converting natural gas to GH_2 for one of the burners of a chemical plant.	The proposed solution showed itself to be technically feasible and beneficial for the environment. In financial terms, the investment has a high payback period and a low profitability. The analysis also showed that energy efficiency certificates are not adequate to support the investments for GH_2, while the support up to 4.5 Euro/kg provided within the EU Hydrogen Bank Action would be more suitable. https://ec.europa.eu/commission/presscorner/detail/en/ip_23_5982 (accessed on 19 May 2025).	The research performs an important first step in the evaluation of the environmental and techno-economic viability of introducing GH_2 in hard-to-abate industries where there is the potential of achieving relevant reductions of CO_2 emissions.	The main limitation of the investments in power-to-hydrogen integration are the lack of financial feasibility. In that regard, further analysis are needed to explore more comprehensively the socio-economic benefits of such investments.
	Torrubia et al. [60], Germany	Evaluation of copper recovery in three types of waste electrical and electronic equipments, WEEEs (scraps, mix, and waste-printed circuit boards), through three scenarios for shredding, reduction, and oxidation, fire refining and electrorefining. The first one is driven by power from fossil fuels, and the second and third ones by GH_2 from renewable (wind) electricity.	The results show that the carbon footprint of copper recovered from WEEE is much lower than the carbon footprint of primary copper. The contribution to GWP from the recovery of copper in WEEE is further reduced by using GH_2 produced from renewable electricity from the grid (wind power) compared to the scenario where the electricity from the grid is produced from non-renewable energies.	The study shows the environmental benefits to GWP and other impact categories coming from the recycling of copper and the use of GH_2, underscoring the central role of energy transition and renewable energies in the CE model and transition.	The development of GH2 is facing several challenges and, therefore, the results of the study are conditioned by the improvement of the technical and economic viability of GH2 production, storage, as well as the limitations of renewable energies.
USE of GH_2 as a fuel for transport	Kugemann and Poladis [107], Gotland (Sweden)	The study developed a new integrated Multi Criteria Decision Analysis framework to classify several alternative fuel buses on the basis of technical, environmental, social, and economic dimensions.	The analysis assessed seven types of fuels for buses (bioethanol, biodiesel, biogas, plug-in hybrid electric biodiesel, plug-in hybrid electric biogas, battery electric, and hybrid electric hydrogen fuel cell). The results provide evidence that the most preferred fuel is biogas (particularly those produced by waste resources) and as buses that of Plug-in hybrid electric biogas. Battery electric and hybrid electric hydrogen fuel cell resulted in interesting alternatives from the perspective they are produced locally. However, even when assuming a further development of wind capacity and the improvement of the capacity of GH_2 towards the achievement of energy self-sufficiency, as well as the use of subsidies to support the development of GH_2, the most preferred option is for biogas as a fuel.	The study has been tailored on the basis of the local needs of the island and identified the most important aspects improving the awareness of the alternatives for buses and fuels. The analysis also enriched the evaluation process by enhancing the understanding of the different opinions of the stakeholders participating to the analysis.	Lack of standardisation of the method, particularly with regard to the social criteria and the focus of the latter dimension on negative impacts rather than positive. The analysis examines only some alternative buses and fuels most viable for the island. The ranking is based on individual preferences.

Table 2. *Cont.*

	Authors and Geographical Scale	Case Study Goal	Key Findings	Success Factors	Limitations
USE of GH_2 as a fuel for transport	Agostinho et al. [103], Brazil	This LCA and emergy accounting study analyses the impacts of replacing diesel with H_2 produced by means of grid electricity (83% from renewables such as hydro and biomass) in urban buses. Emergy accounting and life cycle assessment (LCA) are applied to obtain complementary environmental indicators.	The comparison of the hydrogen scenario (functional unit: km person) with other urban bus scenarios shows that the hydrogen scenario has worst performance compared to buses fuelled by diesel, electricity, LPG, and biodiesel in the fossil energy depletion category, but performs better compared to GWP, AP, and PMFP. The results by emergy accounting show that the operational stage requires 95.6% of total emergy use, with the highest shares being electricity (51.2%) and labour (44.4%). Hydrogen energy source has a higher renewability (41.49%) compared to diesel (0.05%) and biodiesel (18.83%), with lower emergy loading ratio and higher emergy sustainability index.	The study focuses on LCA and Emergy, considering them as individually and complementary to avoid the so-called "entropy trap" and trade-offs between the results of the LCA and that of emergy accounting.	The limitations reported by the authors regard the boundaries of the system that do not include the end-of-life for the materials or components such as batteries. Moreover, there is still a lack of LCA studies and emergy accounting research on the topic, preventing a broader discussion of the results of the study.
	Martelli et al. [101], Italy	Analysis and comparison of the life cycle impacts of a traditional diesel-powered tractor, a fuel cell hybrid tractor using grey hydrogen and a fuel cell hybrid tractor powered by GH_2.	The fuel cell hybrid tractor shows much lower impacts in all the investigated impact categories with the exception of FFP, where the impacts, compared to the traditional tractors, are lower by 4.55%. In the life cycle of a traditional tractor, the most critical step is the use stage, while in the fuel cell tractor is the manufacturing stage. The comparison with the fuel cell hybrid tractor using GH_2 with the traditional tractor shows a strong reduction of the contributions to GWP and other impact categories except TETP and FETP. The comparison between the fuel cell hybrid tractor powered with GH_2 and that using grey hydrogen shows a worsening performance in almost all the impact categories except GWP and FFP.	The literature still lacks a complete LCA from cradle to grave, and this study covers this void by improving the knowledge on the impacts of replacing traditional diesel-powered tractors with alternatives based on GH_2 and fuel cells.	Future research should further explore the impacts following the adoption of further mitigating strategies such as the design of fuel cells, the use of more sustainable materials and energy mix for GH_2 production.
Techno-economic viability of integrated processes of production and use of GH_2 from renewable sources	Bonesso et al. [108], Southern Italy	Analysis of the levelized cost of energy of GH_2 produced from an electrolysis plant with electricity sourced by wind. Perception of GH_2 suitability in citizens, assessed by means of a questionnaire survey.	The levelized cost of hydrogen resulted in 3.60 euro/kg of GH_2 in the base scenario, while in the alternative scenarios, it comprised between 3.20 and 4 euro/kg of GH_2. The sensitivity analysis revealed that the most significant factor influencing the LCOH is the capacity factor. Most of the respondents (72%) of the sample appeared to not know the difference between green and blue hydrogen. In the sample, those with the highest knowledge on hydrogen production were men of 35 years of age. However, provided the required knowledge about blue and green hydrogen, the respondents showed an increasing willingness to pay for the installation of the plants.	The study provides important information to policy makers and other stakeholders about the role of GH_2 in contributing to more sustainable and flexible energy systems.	One of the limits is that the methodological approach does not include the analysis of the environmental sustainability of GH_2 produced in the case study. Large-scale projects could require the verification of the results with a live experiment.

Table 2. *Cont.*

	Authors and Geographical Scale	Case Study Goal	Key Findings	Success Factors	Limitations
	Oyewole et al. [98], Johannesburg, Pretoria and Cape Town (South Africa)	The study evaluated the techno-economic viability of an energy system based on renewables (wind and solar PV) used as a peaker plant and an on-site green hydrogen refuelling station. The goal is to reduce the electricity costs of the system and amplify its reliability.	The assessment of the energy system was performed for three selected cities and showed itself to be a viable option for producing electricity and hydrogen fuel. The levelized cost of electricity at Johannesburg, Pretoria and Cape Town for a 2 MW grid resulted in 74.2 $/MWh, 76.3 $/MWh, and 50 $/MWh, respectively, and competes with the LCOE of natural gas plants generally used as peaking plants. The results also showed the CO_2 equivalent emissions and the related carbon taxes ($) avoided for the energy systems in the three cities. The levelized cost of hydrogen at Johannesburg, Pretoria and Cape Town energy systems resulted in 5.85 $/kg, 5.97 $/kg, and 4.45 $/kg, respectively.	The analysis showed the environmental and economic benefits of optimising the urban energy systems with the use of GH_2. The proposed solution based on the complementarity of both technologies helps to mitigate the high costs of production of GH_2 in a standalone station.	There are some methodological limits due to the assumptions made in the modelling of the study. For example, the progressive ageing of the components before the achievement of the service life is not considered in the model.
Techno-economic viability of integrated processes of production and use of GH_2 from renewable sources	Currie et al. [102], South Australia	Assessment of the financial viability of integrating GH_2 in an electrical grid of South Australia, enabling the presence of a higher share of renewable energies in the electrical system.	The analysis confirmed that GH_2 would improve the reliability and security of the electricity system in the presence of wind and solar variability or in the impossibility of storing the surplus energy from these sources in an efficient way. GH_2 would be produced in gaseous form by using surplus renewable electricity, a water aquifer, and a PEM electrolyser. The electricity would be produced by means of a gas turbine and be sent back to the grid. However, considering only financial indicators, the proposed project would not be viable, requiring economic subsidies.	The research shows how to improve the seasonal storage of the electrical system and through this, its reliability in satisfying the demand of electricity.	The study underlines limitations in the modelling of the research, such as the assumed storage type, data availability, and the limited focus on financial costs disregarding environmental and social impacts.
	Lykas et al. [104], Kythnos (Greece)	Analysis of energy, exergy, and financial performances of the production of GH_2 from excess solar electricity in a Greek island.	The analysis proposed a configuration involving a solar parabolic collector plant generating electricity and heat. The electricity in excess that cannot be received by the local grid is delivered to a PEM water electrolyser for the production of GH_2. The latter is assumed to be compressed, stored in tanks, and eventually used as a fuel for vehicles or ferries. The results showed that the proposed system can satisfy the annual diverse energy needs of the island, producing 210 MWh and 2356.5 kg of GH_2 with a good financial return, resulting in an investment payback period of less than 7 years and a positive net present value.	The innovativeness of the study is the application of two technologies for storing energy such as the solar thermal energy storage tank and the compressed hydrogen.	Future research can evaluate more broadly the environmental and social benefits of the proposed installation and investments.

Table 2. *Cont.*

	Authors and Geographical Scale	Case Study Goal	Key Findings	Success Factors	Limitations
Application of renewable energies for GH_2 production to provide heat and electricity in buildings	Gabbar and Ramadan. [97], Canada	Analysis and comparison of seven scenarios, among which a business-as-usual scenario meeting all the energy needs (electric, thermal, and gas) by means of the utility grid. The other energy scenarios evaluate the partial replacement of the grid supply by the local production of solar PV panels or wind turbine or fuel cells or their integration, as well as the use of a GH_2 system.	The scenario integrating wind turbine and solar PV panels covers until 63% of the electricity lighting needs of the building, avoiding the supply from the utility grid and the achievement of economic benefits. The integration of the green hydrogen–fuel propane blend system for the thermal needs further widens the economic benefits and generate environmental benefits in terms of CO_2 savings per year.	The study contributes in covering a relevant gap in the literature, showing the benefits of the integration of renewables and GH_2 production in buildings to replace fossil fuels	One of the possible limits could be the small scale of the project and the need for evaluating the potential scalability of the analysis and its effects on the results of the study.
	Nhien et al. [106] Swedish cities (Visby, Helsingborg, Lund and others)	This study assessed the potential of wind–fuel cell hydrogen energy system to generate a sustainable and reliable energy in meeting households' demand.	The system was centred on a wind farm integrated with a PEM electrolyser and reverse osmosis desalination units for the production of electricity, hydrogen, and freshwater. The system includes PEM fuel cells equipped with a hydrogen tank to meet the demand and overcome the intermittency of wind power. The analysis confirms the great potential of Sweden in electricity production from wind due to its abundance of coastlines, particularly in the cities of Visby and Helsingborg. The results show that, in the city of Visby and in the other analysed cities, the proposed energy system would provide a sustainable energy supply to 4500 households.	The analysis is particularly meaningful in social terms, showing the possibility of reducing the dependence on fossil fuels and rely on a renewable local source of energy.	Future research could consider the analysis of further areas with a high potential of wind power production and conduct an LCA in order to understand the environmental impacts of the energy system and its components.
	Meastre et al. [17], Spain	Monitoring the performances of a hybrid, solar PV-based GH_2 demonstrative pilot plant, which is useful to ensure the total annual coverage of electricity needs of a social housing unit.	The results show the environmental, economic, and social benefits achieved by the hybrid solar PV–hydrogen pilot plant after two years of monitoring. The economic benefits of reduced electricity costs are particularly relevant for the people living in social housing or at potential risk of energy poverty. The monitoring and control system implemented for the pilot plant allow the analysis of variables and energy performances. The control system has shown itself to work effectively since the electrolyser or the fuel cells enter into operation when there is a surplus of energy or when the batteries and the solar PV plant do not meet the energy demand of the social housing.	The study shows a modular and scalable solution that integrates the production of GH_2. It is possible to satisfy the electricity and thermal needs of isolated residential systems, adding value to the building and realize multiple benefits.	The analysis could be extended by assessing further impact categories beyond CO_2 and primary energy savings.

4.5. Environmental Impacts of the Life Cycle of Green Hydrogen

The Table A2 in Appendix A lists all the LCA studies included in our literature review. For each study, the goal and some results are provided. The next Sections 4.5.1–4.5.3 analyse in detail the goals of the LCA studies, the system boundaries, and the main results in terms of environmental impacts. Further key aspects such as the functional units, the impact assessment methods, and impact categories are dealt with in Section 4.5.4.

4.5.1. Goal and Scope of Reviewed GH_2 LCA Studies

"Goal and Scope" is the first step in an LCA study according to the ISO standards 14,040 and 14,044, and its relevance is crucial, since it defines the approach to be adopted. Some of the main aspects that must defined in this stage are the final objective and the context of the study, namely a number of important characteristics, such as the functional unit, the system boundaries, the type of data (foreground or background), the exclusion of some life cycle stages or inputs, and the selected impact indicators and characterization factors.

The goal of the selected LCA studies regards, in many cases, the assessment and comparison of the impacts of the production of various hydrogen pathways (or "colours") such as green, blue, and grey hydrogen [66,89,109–113]. Selected studies also aimed to identify the most environmentally sound electrolysis technology (AEL, PEM, SOEC, AEM) for GH_2 production [114–118] or to assess the impacts of integrating the production of GH_2 in industrial processes [60,119,120].

The goal of some LCAs was evaluating the impacts of hydrogen production by means of specific processes, such as photocatalysis [121], or substrates, such as biogas, or by means of chemical looping, dry reforming methane technology [122] as well as comparing the impacts of delivering hydrogen through various types of hydrogen storage (e.g., liquid and compressed hydrogen) and delivery options (by pipeline or by ship) [123]. Furthermore, the goals of other LCA studies were the evaluation of GH_2 produced by water electrolysis and renewable electricity such as a floating PV system [124], offshore or onshore wind and solar PV plants in Italy and the UK [125], an offshore wind plant located in the Eagen Sea [126], and the whole life cycle from cradle to grave using PEM electrolyser and wind electricity as well as AEL electrolyser and solar PV power [127].

Finally, another non-negligible goal of some reviewed LCA studies has been the comparison of the impacts of conventional and alternative fuels used in internal combustion engines in marine applications [128,129], or in urban buses such as the use of diesel and H_2 produced by using electricity from the grid in Brazil [103] or in Dubai [130]. The comparison of the impacts has also involved autonomous-driving, hydrogen-powered boats and long-haul tracking [131], traditional and fuel cells hybrid tractors powered by GH_2 [101], or systems for producing electricity such as a diesel-based energy system and a proposed RES-based energy system that also include the production of hydrogen [132].

A few LCA studies [110,133] have a specific goal associated with the circular economy. The first one was aimed at assessing the environmental impacts of the electrolyser on the manufacturing stage by considering the use of recycled and reused materials, while the second one assessed the environmental impacts of producing hydrogen by dark fermentation and electrolysis. In the case of the latter process, the use of treated wastewater was evaluated and two scenarios of electrolysis with/without a solar PV plant have been also considered.

4.5.2. System Boundaries of LCA Studies

Figure 5 shows the whole life cycle of GH_2, encompassing all the "cradle to grave" processes within the boundaries of the investigated hydrogen system. Basically, the whole life cycle consists of upstream and downstream phases before and after the operation stage of production of GH_2 by means of the water electrolysis process. The upstream processes comprise the extraction and processing of the raw materials and energy needed for the manufacturing of the renewable energy technologies (PV, hydro, and wind power plants), and the electrolyser required for the water electrolysis process. The downstream processes involve the end-of-life stages of renewable technologies and electrolyser, consisting of their decommissioning/dismantling and the reuse/refurbishing/repurposing of their components [133] along with the recycling of remaining materials [101,110,114,127,133]. Moreover, the LCA approach "cradle to grave" is still not fully circular because a fraction of solid waste is disposed of in landfill [114] or treated by the incinerator [133].

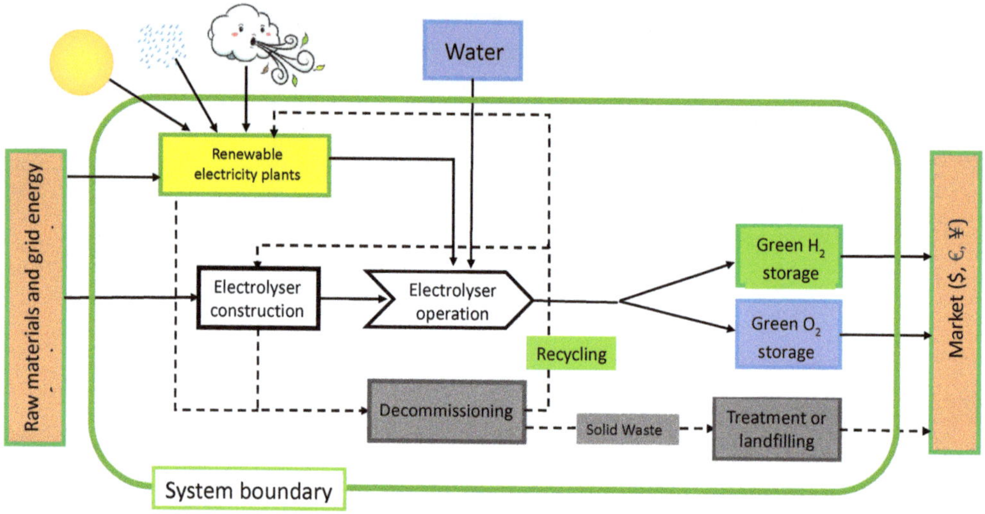

Figure 5. The life cycle of GH_2 according to the system boundary approach "cradle to grave".

Table 3 evidences that 9 out of 29 LCA studies of this literature review adopted a "cradle to grave" approach, involving all the stages of the life cycle of renewable technologies and the electrolysers, while 13 followed a "cradle to gate" approach excluding the use stage and end-of-life, 5 used a "cradle to use" procedure, and 1 used "Well to Wheel". Only one study considered the construction and end-of-life stages.

Table 3. Summary of some key aspects of the selected LCA literature.

Author/s	Publication Year	System Boundaries	F.U.	Impact Assessment	Impact Categories	Use of Other Methods/Indicators	Electrolyser	RE Source
[124]	2025	Cradle to gate	1 kg of GH_2	Intergovernmental Panel on Climate Change, 2021	GWP, particulate matter formation, NOx and SOx emissions, and water consumption	LCCA	PEM	Floating PV
[133]	2025	Construction and end-of-life recycling	5 MW of AWE system	ReCiPe (2016)	GWP, Mineral Resource Scarcity, Human Toxicity Potential for cancer (HTPc)	NO	AWE	NO
[121]	2025	Cradle to gate	0.01 MJ hydrogen energy	Environmental Footprint (Mid-point Indicator)	19 impact categories, including climate change	NO	NO	Solar energy (componds parabolic concentrator)

Table 3. *Cont.*

Author/s	Publication Year	System Boundaries	F.U.	Impact Assessment	Impact Categories	Use of Other Methods/Indicators	Electrolyser	RE Source
[128]	2025	Cradle to grave	1 kWh of engine energy output	ReCiPe 2016 Midpoint, CML-IA, IPCC	18 impact categories of ReCiPe 2016 Midpoint	NO		Onshore wind
[131]	2025	Cradle to grave	1 km of AHB travel	Environmental footprint 3.1	GWP100	Cost Assessment		
[125]	2025	Cradle to gate	1 kg of GH_2	Environmental Footprint 3.0	Acidification, fossil resource use, mineral and metal resource use	NO	AEL	Solar PV and offshore and onshore wind
[116]	2024	Cradle to grave	1 kg of GH_2		GHG emissions	Financial analysis	AEL, PEM	Solar PV
[114]	2024	Cradle to grave	1 kg of GH_2	ILCD 2011 Midpoint method	GWP and 15 impact categories	NO	ALE-P and aLE-C	Solar PV and wind energy
[89]	2024	Cradle to grave	1 kg of H_2	CML 2001–January 2016	GWP	NO		Concentrated photovoltaic collectors
[123]	2024	Cradle to utilization	Delivery of 1 Mt of H_2 in 1 year	EF impact assessment method (European Commission, 2021)	16 impact categories, including climate change	NO	PEM	Energy mix of the grid
[110]	2024	Cradle to gate		ReCiPe (H)	GWP100, FDP, WDP	NO	PEM	Solar PV
[119]	2024	Cradle to grave	300.000 t of RDF producible from residual MSW	ReCiPe 2016	All impact categories of ReCiPe 2016 Midpoint		PEM	Solar PV
[120]	2024	Cradle to grave	1 kg of H_2	CML 2001, ReCiPe 2016 Endpoint, and EPS 2015 d	GWP, FPMF, Asthma Cases, Water consumption, OFP, EP	NO	SOE	Solar PV
[126]	2024	Cradle to gate	1 kg of H_2	ReCiPe2016	GWP-100 years, GWP-20 years, FPMF-20 years, FPMF-100 years	Life cycle Costing Assessment	PEM	Offshore Wind
[127]	2024	Cradle to grave	1000 tons of H_2	ReCiPe 2016	GWP, WDP, FDP, MDP, ODP, HTP, TAP	NO	AEL, PEM	Solar PV and wind
[101]	2024	Cradle to grave	1 kg/vehicle/year	ReCiPe 2016 V. 1.03	10 impact categories of ReCiPe 2016	NO	PEM	Solar PV
[122]	2024	Cradle to gate	1 kg of H_2	Environmental Footprint 3.1 method	GWP, AP, FEP, ODP, POFP, ADP			
[109]	2024	Cradle to gate	1 kg of H_2	ReCiPe 2016 Midpoint (H)	All the environmental categories of Recipe, including GWP	Energy analysis (EROI), financial analysis, LCOH	AEC Alkaline electrolysis cell	Floating photovoltaic
[111]	2024	Cradle to gate	1 kg of H_2		GWP100 and GWP20	NO		Wind and solar energy
[112]	2024	Cradle to utilisation	1 kWh fuel use		Climate change, water use, human toxicity, eco-toxicity, resource depletion and land use	Environmental costs	PEM	European electricity mix with renewables for PEM
[129]	2024	Cradle to gate	5.3 EJ/year	ReCiPe 2016 Hierarchal	GWP100	NO	PEM	Renewable energies
[60]	2024	Cradle to utilisation	1 kg of copper	CML 2001–January 2016	ADP, abiotic depletion-fossil fuels, Acidification, Eutrophication, Freshwater Acquatic Ecotoxicity, GWP, HT, MT, OLD, PO, TT	NO		Wind
[117]	2024	Cradle to gate	1 MJ of hydrogen production at 1 bar pressure in Europe in 2023	IPCC, ImpactWorld+, EPS2000	CC, Human Health, Ecosystem Quality, Abiotic resource depletion	NO	AEL, AEM, PEM, and SOE	Solar PV, Wind, Hydro power

Table 3. *Cont.*

Author/s	Publication Year	System Boundaries	F.U.	Impact Assessment	Impact Categories	Use of Other Methods/Indicators	Electrolyser	RE Source
[103]	2023	Well to Wheel	Kilometer–person		Embodied fossil energy, GWP, AP, Particulate matter	Emergy accounting and emergy indicators		80% of the electricity mix of Brazil is sourced by renewables
[113]	2023	Cradle to gate	1 kg of H_2		GHG emissions		PEM	Solar PV, offsfore wind
[118]	2023	Cradle to gate	1 kg of H2, 99.8% purity, 1 bar pressure	IPCC GWP 100a	GHG emissions	Financial analysis (net present value and payback period	AEM	Solar PV
[66]	2023	Cradle to utilisation	500 Mt/yr of hydrogen	Planetary Boundaries framework LCIA, ReCiPe 2016 (Midpoint) Hierarchist, Environmental Footprint 3.0 (Midpoint), CML-IA baseline 4.8 from 2016 (Midpoint)	CC and other impact categories			Solar PV and wind
[115]	2023	Cradle to gate	1 kg of H_2		GWP, AP, ODP, EP	NO	AEL, PEM, SOEC	Onshore and offshore wind
[132]	2022	Cradle to utilisation	1 kWh of electricity	Environmental Footprint (EF) 3.0 method	CC, ODP, POFP, PM; AP, FEP, TEP, MEP, Ecotoxicity Freshwater, water use, Resource use, fossils, Resource use, minerals and metals.	Economic analysis already performed	Alkaline electrolyzer	Solar PV

Note: The empty cells refer to missing information in the articles.

The end-of-life stage of the electrolyser still represents a challenge for LCA practitioners and the recycling industry because of the lack of reliable data and the low level of maturity of the recycling processes [38]. Advancements in the end-of-life stages, particularly for the electrolyser, are necessary for a transition to the CE model and for an improvement in the impacts during the manufacturing stage of such technology.

4.5.3. Analysis/Interpretation of the Environmental Impacts

The analysis of the results of the selected LCAs shows that the production of GH_2 by water electrolysis using renewable electricity presents more favourable results if using wind electricity compared to solar PV.

In this regard, Tabrizi et al. [125] assessed the production of GH_2 in Italy and the UK, calculating a carbon footprint for wind-based hydrogen lower than solar-based GH_2. Moreover, the carbon footprint differs depending on the location. The values for wind off-shore-based hydrogen for the UK are lower (0.62–1.06 kg CO_2 eq./kg H_2) compared to Italy (0.8–1.36 kg CO_2 eq./kg H_2). For solar-based hydrogen, the carbon footprint for Italy is between 1.76 and 2.25 kg CO_2 eq./kg H_2 by using electricity from single-SI modules, while for the UK, it is between 2.65 and 2.77 kg CO_2 eq./kg H_2. The authors pointed out that the source of electricity of the electrolyser generates the highest contributions to the carbon footprint of hydrogen. In the case of solar PV-based GH_2, the PV modules are responsible for the large share of the carbon footprint of the PV plant due to the fact that the cells are mainly manufactured in China by means of the still coal-based electricity used in this country. Therefore, a reduction of the production of solar PV-based GH_2 could only happen following a decarbonisation of the Chinese electricity mix [125].

Other authors also confirmed that a large share of the contribution to GWP for the production of GH_2 comes from the production of electricity. Pawłowski et al. [118] evaluated the impacts of GH_2 produced by an AEM electrolyser and a 5 MW solar PV plant located in Poland, calculating a carbon footprint between 2.73 and 3.85 kg CO_2 eq/kg H_2 (PV tracker) as well as 3.01 and 4.34 kg CO_2 eq/kg H_2 (PV standing). Similarly, Ajeeb et al. [114] also found in their Portugal case study that the electricity mix (solar and wind) of the AEL electrolyser is responsible for the largest impacts in the production of GH_2, ranging from 92% for ODP, 96% for mineral fossil resource depletion, and 98% for the other investigated impact categories (including GWP).

Furthermore, Affandi et al. [116] showed that the GHG emissions of GH_2 produced using PEM electrolyser and solar PV electricity were between 2.26 and 4.46 kg CO_2 eq/kg GH_2, while those related to AEL electrolyser were between 2.61 and 5.15 kg CO_2 eq/kg GH_2. Zhang et al. [115], comparing the impacts of three water electrolysis technologies (AEL, PEM, and SOEC), calculated that onshore wind and the use of PEM for GH_2 production generate the lowest contributions to the selected impact categories (GWP, AP, EP, ODP), with values of 0.0936 kg CO_2 eq./kg of H_2 (GWP), 2.97×10^{-4} kg SO_2 eq./kg of H_2 (AP), 2.86×10^{-5} kg Phosphate eq./kg H_2 (EP), and 4.22×10^{-15} kg R11 eq./kg H_2 (ODP).

Other authors have compared the environmental impacts of hydrogen produced by means of different feedstocks and processes. Patel et al. [111] showed that the production of GH_2 from wind power generate the lowest emissions (0.6 kg CO_2 eq./kg H_2) compared to GH_2 produced from solar PV electricity (2.5 kg CO_2 eq./kg H_2), as well as the hydrogen from fossil fuels delivered in two pathways by pipeline from Russia and LNG routes from USA. In particular, grey hydrogen produced by SMR released 13.9 kg CO_2 eq./kg H_2 and 12.3 kg CO_2 eq./kg H_2, blue hydrogen produced by SMR-CCS emitted 7.6 kg CO_2 eq./kg H_2 and 9.3 kg CO_2 eq./kg H_2, and turquoise hydrogen emitted 6.1 kg CO_2 eq./kg H_2 and 8.3 kg CO_2 eq./kg H_2. Moreover, if the production of GH_2 occurs by using electricity from the grid connected to both solar and wind at the same percentage (50% wind and 50% solar), most of the contribution to GWP comes from solar electricity (1.2 kg CO_2 eq. per kg H_2) while wind has a lower impact (0.3 kg CO_2 eq. per kg H_2).

These results are in line with the study by De Kleijne et al. [113], who assessed that GH_2 from electricity by offshore wind has the lowest GHG footprint, ranging from 0.4–0.8 kg CO_2eq./kg of GH_2, while that from solar PV leads to a higher GHG footprint at 1.7–4.4 kg CO_2eq./kg of GH_2. The use of electricity from the present EU grid worsens the GHG footprint to 6.3–16.6 kg CO_2eq./kg of GH_2, assuming the transition towards a cleaner EU electricity mix reduces the GHGs footprint to 2.1–5.6 kg CO_2eq./kgH_2.

Shen et al. [112] assessed the impacts of four decarbonisation scenarios by the year 2050 for the European industry hard-to-abate sectors as well as transport. The four scenarios are as follows: blue hydrogen replacing fossil fuels, GH_2 replacing fossil fuels, decarbonisation without GH_2, and business as usual without changes in the energy mix. Their results show that GH_2 scenarios have much lower impacts than the other three scenarios in terms of CC, fossil resource depletion, photochemical ozone formation, and acidification. On the other hand, the performances of GH_2-based scenarios compared to the others are worse for land use, water scarcity, mineral resource depletion, particulate matter, eutrophication freshwater, human toxicity, ecotoxicity, and ionising radiation. The Shen et al. [112] study concluded with evidence that the production of electricity from renewables (wind and solar) will have to increase by 50% to support the decarbonisation of industry and transport by 2050, thus requiring further infrastructures. This would cause the increase of the contribution to other impact categories such as human health, mineral resource use, and ecosystem damage [112].

Finally, Mio et al. [109] found that in the comparison among grey, blue, grid, and green hydrogens, the latter has lower environmental contributions to GWP and to the other air pollutants. In other impact categories, the impacts of GH_2 production are higher (such as in the case of LOP and WCP) or lower than the grid electricity-based hydrogen that have resulted in the worst performances across almost all impact categories due to the impacts coming from the electricity mix of the grid. The impacts of GH_2 production to land use seem to not be a concern, particularly when mitigated by the installation of an offshore plant or rooftop solutions.

4.5.4. Most Relevant Factors Affecting the Environmental Impacts

As evident in the previous section, the environmental impacts of the life cycle of GH_2 depend on different factors such as the source of renewable electricity (e.g., wind or solar), the local context, and the type of electrolysis technology. Further influencing factors are the assumptions related to the efficiency and lifetime of the electrolyser [113].

In the life cycle of GH_2, the operation stage of the electrolyser, due to the source of electricity used for the electrolysis process, generates the highest contribution to GWP of GH_2 [114,116,134]. Some authors have underlined that, given the high contribution of the operativity stage of the electrolyser, it is difficult to identify the specific contribution of the manufacturing stage of the electrolyser as well as the impacts of the end-of-life within a recycling scenario. They underlined that if a production method of H_2 with lower impacts is used, both the impacts of manufacturing and end-of-life stages can be more visible and comparable to the operating stage of the electrolyser [134]. However, the results of the LCA when adopting a "cradle to grave" approach show that even in the case of production of GH_2, the impacts of manufacturing and end-of-life stages seems less relevant compared to the operation stage of the electrolyser [127]. Khan et al. [127] showed that the inclusion of the end-of-life recycling and the remanufacturing of the electrolyser with recycled materials lead to a reduction of the GWP of the whole life cycle of GH_2 by 5–12%, FDP by 5–8%, and by 3–20% for MDP [127].

Of course, regarding the end-of-life recycling scenario, the reduction of the impacts during the manufacturing stage of the electrolyser is significant for the AEL electrolyser compared to the PEM electrolyser due to the higher quantity of assumed recyclable materials. The contributions to GWP decrease by 70%, while those of FDP decrease by 65% and MDP decrease by 59% [127].

Hoppe and Minke [133] investigated the LCA impacts on the manufacturing and end-of-life stages of a 5 MW AWE, confirming the relevant reductions of GWP (−50%) obtained by manufacturing an AWE electrolyser (5 MW) with recycling materials compared to the use of virgin materials. The authors also underlined that about 77% of the materials comprising the electrolyser can be reused or recycled.

Wei et al. [117] focused on the analysis of the manufacturing process of four electrolysers depending on the materials and input used in the process. An AEL electrolyser with electricity sourced by hydro as well as an AEM electrolyser sourced by hydropower emerged as those with the highest potential contributions to CC at 0.083 kg CO_2 eq./MJ H_2 and 0.075 kg CO_2 eq./MJ H_2. The results reveal that in the manufacturing of 1 kW of AEL electrolyser and particularly in the anode production, nickel contributes to almost half of the total contributions to CC. On the other hand, the scenario with PEM (0.009 kg CO_2 eq./MJ H_2) and AEM (0.008 kg CO_2 eq./MJ H_2) electrolysers powered by wind electricity resulted in them having the lowest contribution to CC. With regard to the analysis of the other impact categories of the electrolyser system, such as human health, ecosystem quality, and abiotic stock resource impacts, the most impacting scenarios were those with AEL electrolyser and electricity sourced by solar PV and AEL with hydropower [117].

4.5.5. Analysis of the Results of Techno-Economic and Environmental Studies

The analysis of the environmental impacts has been also performed by further selected studies available in Table A2 of Appendix A. Most of the studies conducted a techno-economic analysis in combination with an environmental analysis in order to evaluate the technical and financial viability and environmental benefits of decarbonisation projects by integrating renewable energies and the production of hydrogen. Currie et al. [102], in an Australian case study, found that the integration of GH_2 in the electricity system improves the reliability and security of the system, overcoming the variability of renewable sources such as wind and solar PV and thus providing the opportunity of storing the energy surplus from these sources in an efficient way. However, in order to ensure its financial viability, the project should be supported by subsidies [102].

Oyewole et al. [98] analysed an energy system based on renewables (wind and solar PV) used as a peaker plant and an on-site GH_2 refuelling station. They showed that it is a viable option for producing electricity and hydrogen fuel at competitive costs compared with those of natural gas plants generally used as peaking plants in the urban context of three cities in South Africa [98]. In Morocco, Boulmrharj et al. [135] calculated the GHG emissions of a hybrid system for the production of hydrogen and electricity required for public transport and street lighting in three cities. The GHG_2 emissions were lower (0.017 kg CO_2eq./kWh) than the GHG emissions from the grid electricity [135].

With regard to the applications in buildings concerning the implementation of a hybrid PV–hydrogen demonstrative pilot plant in Spain, Maestre et al. [17] showed that it is possible to achieve independence from grid electricity, thus obtaining CO_2 emissions savings and socio-economic benefits in terms of avoided electricity costs [17].

Similarly, Hassan et al. [96], in a case study assessing the use of a commercial hydrogen battery in rooftop solar systems, indicated that hydrogen batteries reduce the dependence on grid electricity and its costs [96]. Moreover, the hydrogen battery, compared to the Li–Ion battery, increases the life of the battery and thereby achieving a longer duration of energy storage [51,96]. In terms of costs, Arsalis et al. [18] pointed out that those related to LIBs have decreased significantly within the last decade, while in the case of RHFC, the high costs of the electrolysers and fuel cells stacks limit their larger development [18].

Arcos and Santos [83] as well as Singhla et al. [136], assessing the CO_2 emissions and production costs of various hydrogen production methods, have found that grey, black/brown, and blue hydrogens are the least environmentally friendly [83,136] but have lower production costs (ranging from 0.67 to 2.05 USD/kgH_2, respectively) [83]. Green hydrogen is the most environmentally friendly but has one of the highest production costs, ranging from 2.28 to 7.39 USD/kgH_2 [83]. In the European context, the case study by Bonesso et al. [108] in Italy calculated a levelized cost of wind-based hydrogen at 3.60 euro/kg of GH2 in the base scenario, while in the alternative scenarios, it comprised between 3.20 and 4 euro/kg of GH_2 [108].

Awad et al. [130] suggested that the best solutions for the decarbonisation of the Dubai bus fleet would be to replace the diesel buses with those using mixed hydrogen as a fuel in order to benefit from the existing grey hydrogen being available at lower costs. Boulmrharj et al. [135], meanwhile, pointed out that the analysed hybrid systems producing hydrogen and electricity for transportation and lighting services in cities in Morocco would be profitable only in cities of medium size or are not profitable at all regardless of the size. They could become more profitable in the future due to the reduction of income taxes and the expected increase in selling prices of hydrogen and oxygen. The latter study has found some of the most relevant factors affecting the profitability of the hybrid systems [135].

Al-Ghussain et al. [137] have confirmed that the geographical location of renewables is a relevant factor influencing the carbon intensity (CI) of GH_2 and its costs [137]. The

Electrolysis Cells are not a mature electrolyser technology but have the advantage of allowing the use of impure water [144].

Another important aspect is the integration of the principle of renewability to the production of GH_2 and its whole life cycle. As we have seen in the previous sections, the source of electricity for feeding the electrolyser affects the renewability of GH_2. Thus, it is important to ensure the renewability of the source by reducing the impacts of the renewable technologies plants because their manufacturing could involve electricity produced from fossils [125] as well as having several social hotspots [140–142].

The principle of renewability goes hand in hand with that of localness since, for example, the carbon intensity of GH_2 depends on the local production of renewables [137] and these are not distributed equally [33] because some geographical areas are richer in wind [106,125] or solar [125] or hydro [11]. However, importing hydrogen from other countries has a higher carbon footprint [123,144], and the production of GH_2 at the local scale improves the security of the local electricity system or that of single buildings [53]. Clearly, the creation of a surplus at the local scale in cities [98] or islands [104] is easier than at the larger scale of a nation, and some studies show that the production of GH_2 is viable in technical and financial terms [98,104]. In the short term, de Kleijne et al. [113] suggested that the surplus of electricity should first be used to decarbonise the heat and transport sectors, while hydrogen production should be a second option [113]. Therefore, it is relevant to monitor and evaluate the impacts of GH_2 and adopt policies supporting research projects to understand the alignment of GH_2 development and technologies to the CE principles and the decarbonisation goals [43].

Table 4. Principles of CE and the strategies applying them to GH_2.

Principles of CE	Application to GH_2	Challenges
Circular Design	Innovative design approaches to reduce the use of natural resources in the electrolysers [133,142] and reduce/eliminate emissions and other forms of waste in the life cycle of GH_2 [43].	Costs and durability of catalysts in AWE electrolyzers [48]
	Considering for electrolysers the substitution of critical materials as well as designing their components for repair, reuse, refurbishing and their materials for recyclability [133].	
	Designing electrolysers' components for repair, reuse, and refurbishing [133].	
Reduction	Reduction in material intensity to improve the resource efficiency of the electrolysers [143]	Costs and durability of catalysts in AWE electrolyzers [48]
Reuse	Reuse of wastewater in the electrolysis process at the small local scale to avoid the use of freshwater [22,25] and competition with other water uses [46,145].	Production pathways of GH_2 and treatment of wastewater such as biological treatment. Anaerobic membrane bioreactors are characterised by a low hydrogen production yield and a low efficiency in COD removal [144].
	Tertiary effluents require low investments compared to seawater desalination systems and are a secure water supply compared to stormwater [46,145].	
	Reuse of oxygen from the electrolysis process for wastewater treatment and industrial processes. This strategy could reduce the cost of hydrogen and the environmental footprint of wastewater [22] because oxygen is re-used for the treatment of wastewater [46].	
	Recovery of water as a byproduct of the use of hydrogen for energy purposes (e.g., use in fuel cells, internal combustion engines). This process releases pure water independently from the type of water used as an input for the electrolysis process (freshwater, seawater, wastewater) [46].	
	Countries with scarce availability of water could find more appeal in importing hydrogen than ex-porting due to the fact they import both energy and water [46].	
	Solid Oxide Electrolysis Cell are an interesting solution for producing GH_2 as they can use impure water sources as feedstock [144]	
Recycling	About 77% of the materials contained in a AWE can be recycled or reused [133].	Most LCA studies do not include the recycling stage of precious materials of the electrolyzers due to lack of reliable data [38].
	Recycling the materials composing of electrolysers at the end-of-life [133].	

Table 4. *Cont.*

Principles of CE	Application to GH_2	Challenges
Renewability	Production of GH_2 from renewables (wind, solar, hydro) [25,146]; Buses using GH_2 as a fuel, compared with buses fuelled by diesel, electricity, LPG, and biodiesel, have a higher renewability and other better emergy indicators such as emergy sustainability index and environmental loading ratio [103].	• Improving the share of electricity from renewables to create a surplus of electricity for GH_2 production [10,33]; • The electrolyzers and renewables technologies are currently manufactured by using the electricity mix, with contributions from fossil energy [126].
Localness/local scale	GH2 produced from renewable electricity contributes to the diversification of the electricity mix from fossils, thereby improving energy security [53]. Implementation of small-scale hydro plants or using local biomass from waste treatment plants for GH_2 production [45].	Balance between distributed and centralised hydrogen systems [46].

5. Concluding Remarks

The goal of this review study is to understand if GH_2 could be an environmentally and socially sustainable option for decarbonising energy systems in the transition to CE. The research comprised the analysis of the environmental and social impacts of GH_2 in its life cycle, with special focus on the impacts associated with the use of GH_2 for its main three final applications (fuel, heat producer, and feedstock), by means of several case studies from the selected literature. The latter contained literature reviews as well as LCA, LCA and LCC, LCA and Emergy Accounting, Social LCA, environmental, technical, and financial and cost analysis studies. This way, the main research question has been evaluated on the basis of different approaches and dimensions of sustainable development.

The results from the assessed LCAs and environmental analyses show that the production of GH_2, compared to other hydrogen typologies (so-called "colours") from fossils or grid electricity, has lower contributions to GWP, FPMF, and TAP [109]. Contributions of GH_2 to other impact categories are mainly caused by the sources of electricity (wind, solar, or hydro) used for the electrolysis process and the impacts associated with the electricity mix and materials necessary for the manufacturing of the solar PV panels [125] and wind turbines [139]. Moreover, several LCAs point out that wind-based GH_2 has lower contributions to GWP than solar-based GH_2 [111,113,115,125,127]. With that, this study confirms the results from previous literature reviews [25].

Compared to previous literature reviews, this research has also integrated the results from the small number of existing S-LCA literature. The results show the complexity of the value chain of GH_2 and the uneven geographical distribution of the social impacts (particularly the child labour impact category) in the countries contributing to the extraction and provision of materials, as well as the components required for the manufacturing of electrolysers and renewable power plants, such as the wind turbines and the solar PV panels in the EU [142–144]. These results also show the dependence of the EU on foreign countries (particularly southern Africa and East and southern Asia) for the availability of key materials such as silicon, rare earths, titanium, platinum, and iridium [142] and the difficulties of applying traceability to the GH_2 life cycle and the concept of circular local development [55,64]. The creation of a domestic production industry for the components of the electrolysers and renewable power plants would mitigate the impacts, avoiding that the most impacting activities are outsourced outside the EU [140].

This study has also explored how GH_2 is coherent with the principles of CE such as circular design, reduction, reuse, recycling, renewability, and localness/local scale. The application of the CE principles is important in mitigating some of the impacts related to the use of natural resources and critical raw materials, the consumption of water in the electrolysis process, and the production of waste [48]. Some studies also confirm the

SOEC	Solid Oxide Electrolysis Cells
SOE	Solid Oxide Electrolyzer
AEM	Anion Exchange Membrane
IEA	International Energy Agency
FID	Financial Decisions for Investments
LOHC	Liquid Organic Hydrogen Carrier
CC	Climate Change
GWP	Global Warming Potential
FPMF	Fine Particulate Matter Formation
PM	Particulate Matter
FFP	Fossil Resource Scarcity
FETP	Freshwater Ecotoxicity Potential
FEP	Freshwater Eutrophication Potential
HTPc	Human Carcinogenic Toxicity Potential
HTPnc	Human Non-Carcinogenic Toxicity Potential
HTP	Human Toxicity Potential
IRP	Ionising Radiation Potential
LUP	Land Use Potential
TETP	Terrestrial Ecotoxicity Potential
TE	Terrestrial Ecotoxicity
FETP	Freshwater Ecotoxicity Potential
MT	Marine Aquatic Ecotoxicity
FD	Fossil Depletion
FDP	Fossil Depletion Potential
MDP	Metal Depletion Potential
ADP	Abiotic Depletion Potential
TAP	Terrestrial Acidification Potential
AP	Acidification Potential
FEP	Freshwater Eutrophication Potential
MEP	Marine Eutrophication Potential
TEP	Terrestrial Eutrophication Potential
ODP	Ozone Depletion Potential
OFP	Ozone Formation Potential
OLD	Ozone Layer Depletion
POFP	Photochemical Ozone Formation Potential
PO	Photochemical Oxidation
WDP	Water Depletion Potential
ETP	Ecotoxicity Potential
EP	Eutrophication Potential
WEEE	Waste from Electrical and Electronic Equipment
LCOE	Levelized Cost of Electricity
LCOH	Levelized Cost of Hydrogen

Appendix A.

Table A1. Previous literature reviews resulting from the search of the literature according to the procedure explained in Section 3 (Materials and Methods).

Title of the Study	Goals/Main Focus	Primary Findings	Timeframe of the Selected Literature
Lagioia et al. [10]	This narrative literature review focused on the analysis of production technologies of blue and green hydrogen, as well as their management and applications in view of the hydrogen goals of the EU.	Green hydrogen is the only type that could play a role in future decarbonisation scenarios. For this to happen, blue hydrogen could pave the way for GH_2, but there are uncertainties related to the development of CO_2 carbon capture. GH_2 produced by electrolysis is a mature technology. However, GH_2 is also constrained by the availability of a surplus in the demand of electricity from renewables. The supply chain of GH_2 should also be strengthened along with the use of certification schemes for CO_2 emissions. In order to meet the future targets of the EU for hydrogen, the authors suggested the acceleration of investments in hydrogen innovation and its use in hard-to-abate sectors rather than e.g., in light-duty transport where hydrogen use is considered not efficient.	Not specified
Guareiro et al. [11]	Overview of several aspects associated with GH_2, including the sources for its production, regulation, typologies of storage, transportation, and final uses. Critical analysis of the economic and environmental impacts, as well as the main challenges and opportunities it could have for chemistry.	In terms of the environmental impacts of GH2 production, the study shed light on the water required in the process (9 kg per kg of GH_2). The water required for production of grey and blue hydrogen is half of that needed for GH2 production. The latter process is also much more energy-intensive and its production is currently not financially competitive. However, the authors, by reviewing the literature, also reported the results of studies performed in Brazil where the production of GH_2, by using electricity from hydro and wind, achieved production costs comparable to that of grey hydrogen. The study also underlined the need for increasing the capacity of electrolysers for the development of a future hydrogen economy.	Not specified
Arsalis et al. [18]	Critical comparison of several characteristics (serice life, costs, recyclability, environmental impacts, safety issues, use and integration within energy systems) of two solar-powered energy systems: Lithium–Ion Batteries (PV-LIB) and Regenerative Hydrogen Fuel Cell (RHFC) energy systems.	The comparative analysis shows that, for example, hydrogen (within RHFC subsystems) produced through water electrolysis can be stored in high quantities in hydrogen storage units. The RHFC is then more suitable for long-term storage than short-term storage, while LIBs are more suitable for short-term storage. The refueling of RHFC subsystems is fast, requiring only a few minutes, while the recharging of LIBs is slow and needs several hours. In terms of costs, those related to LIBs are much decreased within the last decade while in the case of RHFC, the high costs of the electrolysers and fuel cell stacks limits their larger development.	Not specified
Sudalaimuthu and Sathyamurthy [21]	Overview of agro-waste for the production of GH_2 and focusing on the thermochemical method of production.	The review highlighted several key findings such as, among others, the high energy content of hydrogen from agro-waste by gasification process. The H/C ratio is high, supporting the use of agro-waste instead of fossil fuels for hydrogen production. Thermochemical process resulted in a viable option among the analysed processes for converting agro-waste to hydrogen. Cellulose and lignin, as constituents of agro-waste, contributed to deciding the reaction temperature of gasification and the reaction temperature directly related to energy efficiency, process design, and technological and economic feasibility.	Not specified
Mokrzycki and Gawlik, [22]	Review of various aspects of the green hydrogen economy, including its advantages and disadvantages, risks, environmental impacts, and future water availability for hydrogen production.	Very strict relation between renewables and GH_2 development. Costs, innovation in both technologies, GH_2 storage and investments in transport, as well as efficiency improvements are key factors driving a faster GH_2 development. A future acceleration to the green hydrogen economy will ultimately depend on widespread support and worldwide development as well as the willingness of the countries to perceive the goal of a future green hydrogen economy.	Not specified
Sola et al. [23]	Overview of the state of the art of LCA research on GH_2 in the last five years, taking into account GHG emissions and other environmental categories (such as water use, critical materials use, and land use) and issues related to the large-scale production of GH_2.	Increasing trend of LCA research on GH_2. EU studies accounts for half of the literature sample. GHs have to be produced from renewable electricity to be claimed as a low emissions energy carrier. Water consumption and land use (due to the use of renewable energies) are important hotspots that could limit GH_2 development in particular areas with scarcity of water, such as Saudi Arabia and Oman, that have high abundance of sun but are scarce in freshwater Rare metals and critical raw materials for electrolysers production (particularly PEM), such as iridium and platinum, and solar cells production serve as further hotspots Uncertainties related to the end-of-life disposal of electrolysers.	2019–2024

Table A1. *Cont.*

Title of the Study	Goals/Main Focus	Primary Findings	Timeframe of the Selected Literature
Hren et al. [36]	Comparative analysis of the environmental impacts (GHG, AP, and EP) and energy footprint of 11 technologies in the production of hydrogen, as well as its storage and transport stages.	The analysis of the GHG footprint of the technologies for hydrogen production reveals that the lowest contribution comes from glycerol and alcohol waste steam reforming and water electrolysis, but using electricity from the EU27 mix revealed the highest GHGs emissions are associated to the water electrolysis process. The latter has also the highest values for the energy footprint indicator, while for acidification and eutrophication, the values are similar to that of natural gas steam reforming and higher than glycerol and alcohol waste steam reforming technologies, which reported the lowest values. Regarding storage, the gaseous hydrogen resulted in a lower GHG footprint than liquified hydrogen, while the transport of gaseous hydrogen by the pipeline had the smallest GHG footprint.	Not specified
Mehmod et al. [37]	Analysis of the existing literature assessing the contribution to GWP, AP, EP, and resource depletion categories related to GH_2, as well as the evaluation of the technological maturity and efficiency of hydrogen production methods.	The contribution to GWP, AP, freshwater eutrophication, and resource depletion is the highest producing hydrogen from coal gasification, while the lowest is by producing GH_2 from wind electricity. Electrolysis technologies: AEM, PEM, SOE, and advanced biomass gasification require more comprehensive impact assessments to understand their scalability and potential environmental impacts.	Excluded the literature before the year 2014
Maniscalco et al. [38]	Comprehensive analyses of LCA studies on hydrogen production technologies and impacts on GWP and other impact categories.	The results from the analysis of the selected LCA literature showed that the contributions to GWP from GH_2 as well as that of blue, turquoise, and pink production are lower compared to hydrogen produced by steam reforming. Waste gasification has the worst performance compared to the different hydrogen colours. The contribution to AP has a similar pattern of GWP with a lower difference between yellow and green H_2 due to the use of materials in both solar PV and wind systems. The hotspot analysis showed that in the electrolysis stage, the source of electricity and the specific and precious materials used for the production of the electrolysers affect the contributions to AP in AEM and PEM (nickel and gold) and to ADP due to the use of iridium and titanium. The inclusion of the end-of-life stage and the recycling of materials used for the electrolysers would mitigate the impacts to ADP but at present, the authors underlined that most of the LCA studies do not consider the recycling of precious materials due to lack of reliable data. For example, the process of recovery of platinum still faces industrial maturity issues, as well as high energy and economic costs.	Last 5 years
Buffi et al. [39]	Review of production pathways for producing hydrogen from biomass or biomass-derived sources (biogas, liquid bio-intermediates, sugars) also focusing on their maturity level, as well as the energy and environmental performances of the production pathways emerged from LCA studies.	The type of feedstock biomass influences the energy required for the process of hydrogen production. *Energy requirements should be compared with each other under the same conditions of feedstock, production rates, and environmental conditions.* From the analysis of the LCA literature emerged some of the studies following the ISO standard, while others proposed different methological approaches. Moreover, LCA studies differed on the system boundaries ranging from cradle to gate to cradle to grave. There are also different allocation assumptions. Regarding impact assessment, most of the studies adopted midpoint assessment methods and impact categories such as GWP, AP, and ODP.	Not specified
Angelico et al. [40]	Review of GH_2 using the most cited papers and recent literature reviews published in energy- and hydrogen-focused journals.	The study showed the advantages and limitations of each GH_2 production technologies (PEM, AWE, SOEC, AEM), as well as their efficiency, purity, and costs. The most important methods for GH_2 production, storage and grid integration, environmental impacts, as well as challenges and research gaps are also reviewed along with the potential solutions to reducing the operating costs of GH_2 production, storage, and distribution and to overcome any related challenges.	Not specified
Koshikwinja et al., [41]	Evaluation of the potential hydrogen life cycle in Africa in the stages of production and final uses, policies, infrastructures, and hydrogen logistics.	Africa has a great potential of developing a future GH_2 economy due to large availability of renewables such as solar, wind, and hydro. The costs of GH_2 are expected to decrease strongly to 1.5 Euro/kg by 2050. The mapping of hydrogen policies shows that some countries have clear hydrogen roadmaps, such as Morocco and South Africa.	January 2000–15 May 2024
Dyantyi-Gwanya et al. [42]	Evaluation of the opportunities for a socio-economic development of GH_2 involving its production, storage, and use in South Africa.	The analysis reveals that GH_2 in South Africa has good prospects for GH_2, both for the domestic use and export to other countries. The availability of renewables has an important role in the development of a green economy and could drive the demand of GH_2, contributing to the reduction of the costs from 3.54 to 1.40 €/kgH_2 and thus facilitating its larger production, use, and export.	1990–2021

Table A1. *Cont.*

Title of the Study	Goals/Main Focus	Primary Findings	Timeframe of the Selected Literature
Singh [43]	Analysis on how India is approaching three key areas such as: feedstock system, circular economy integration, and environmental impacts of hydrogen that are important for achieving the climate goals of the country.	The analysis underscored the huge availability of feedstocks for GH_2 production of India playing a key role in its hydrogen mission. The government is committed to the green hydrogen mission, prioritising the circular economy transition across all the sectors. India has set initiatives to implement the circular economy model, while the transition toward the use of GH_2 is in its infacy and requires political support. The CE framework suggests the elimination of GHG emissions from hydrogen production. However, it is important to support with stringent regulations and technological innovation the reduction of the GHG emissions. Green hydrogen will play a key role within the circular economy transition in reducing the GHG emissions necessary for the success of the hydrogen mission and the pursuing of the climate goals.	Not specified
Tasleem et al. [44]	Review of H_2 colours, hydrogen strategies, and roadmaps worldwide, with ongoing large-scale advances in green, yellow, blue, white, and gold hydrogen production, with results from life cycle assessments, future perspectives, as well as challenges and solutions for hydrogen colours.	The study reported the results by [66] underlying the importance of taking into account the impacts of metal depletion and ecotoxicity due to the production of GH_2. Results from other studies pointed out that the impacts of hydrogen production are mainly due (90%) to the source of electricity for the production of hydrogen. In this regard, the LCA by [118] of a solar-to-H_2 plant, using AEM electrolysers in Poland, showed a low contribution to GWP of between 2.74 kg CO_2 eq./kg of hydrogen and 4.34 kg CO_2 eq./kg of hydrogen. Further results from LCA studies are provided.	Not specified
Anand et al. [45]	Provided an overview of GH_2 production pathways, their efficiencies and environmental sustainability, characteristics of electrolysers, the materials used for their production and challenges, and final uses of GH_2.	GH_2 as an increasingly important element in global decarbonisation scenarios. AEL, PEM, and SOE electrolysers technologies still face costs and efficiency issues despite their consolidation. Collaborations with countries rich in renewable energies (wind and solar) and countries with scarcity of renewables serve as a possible pathway for GH_2 development. Moreover, small-scale hydro plants or waste treatment plants serve as alternative for the production of GH_2 at the local scale.	Not specified
Squadrito et al. [46]	Overview of the stateof-the-art hydrogen production technologies, and exploration of the most relevant geopolitical and economic aspects and implications of the hydrogen economy, including the water aspects.	The study pointed out the need for achieving a balance between distributed hydrogen systems and centralised systems, considering the advantages of producing GH_2 locally, where it is water neutral (and CO_2 neutral), and compared to the production of GH_2 for the external market where transport generates carbon emissions. Furthermore, the energy and costs of producing GH_2 from non-traditional water sources such as using wastewater would be lower compared to the production of GH_2 with fresh water "*Distributed hydrogen production is a powerful tool for maximising the social utility of the hydrogen economy, reducing household energy bills, increasing energy system efficiency and resilience, and reducing the environmental impact of city services. All these aspects must be considered in future energy policies*".	Not specified
Revinova et al. [47]	Review of the literature investigating the environmental and economic impacts of hydrogen storage and transportation technologies.	The analysis revealed that salt cavern and transportation by pipeline are the most economically competitive for transport hydrogen over long distances. Hydrogen stored and transported in gaseous form and by pipeline has lower impacts compared to other ways of storage and transport.	Not specified
Dash et al. [48]	Thorough analysis of GH_2 with a particular focus on Alkaline Electrolyser, including its basic principles, materials of the components, the design of the electrolyser, integration of renewable energy sources, electrolyser costs, and environmental impacts.	The review reported the results of an Australian study evaluating a levelised cost of GH_2 of 9.6 USD/kg GH_2 produced while considering a solar electricity price between 38 and 65 USD/kg GH_2. The production of GH_2 by means of AEL requires a considerable amount of energy, water, and land for the installation of the electrolysis plants, as well as generates impacts in terms of land use and biodiversity loss. Moreover, the process generates waste such as spent electrodes and other materials.	Not specified
Almaraz et al. [49]	Identification of the social aspects concerning the development of the hydrogen economy, research gaps, and future research suggestions by using both bibliometric and systematic literature reviews.	The methodological approach identified, described and ranked 12 social aspects associated with hydrogen economy research. The most recurrent social aspects in the literature are accessibility, information, H_2 markets, acceptability, and policies and regulations. Despite their importance, responsibility and technological safety were less occurrent in the analysed literature. Therefore, gaps in the literature or challenges to fulfil are the scarce use of social life cycle assessment in hydrogen research.	2000–August 2023
Vallejos-Romero et al. [50]	Analysis of the research on GH_2 from the perspective of social impacts and challenges.	Social impacts of the life cycle of GH_2 from production to storage and transport as well as social and environmental impacts of GH_2 on local communities and indigenous groups are under research. During the investigation period, Europe and Asia mainly contributed to the research on social impacts of GH_2.	1997–2022

Table A2. *Cont.*

Authors	Year of Publication	Goals	Method	Some Results
Singh et al. [27]	2025	Analyse the production of GH_2 hydrogen from biogas/landfilling gas by means of conventional reforming processes, such as dry methane reforming, steam methane reforming, partial oxidation reforming, and autothermal reforming. The review also considered the contribution to GWP (CO_2 eq·kg^{-1} H_2) emissions, as well as the economic aspects of the processes and their efficiencies	Literature review	The results of the reviewed studies showed that carbon footprint of biomethane reforming was lower than fossil methane reforming. The carbon footprint of biomethane from steam reforming was 4.8 kg CO_2 eq./kg of hydrogen versus 11.2 kg CO_2 eq./kg of hydrogen produced from steam reforming of natural gas. In other studies, harmonised carbon footprint of GH_2 from biogas reforming resulted in 7.34 kg CO_2 eq./kg of hydrogen and the carbon footprint of GH_2 produced from water electrolysis and electricity sourced by wind, solar thermal, solar PV, and hydro was 5 kg CO_2 eq./kg of GH_2, while that of electricity from the grid was 32 kg CO_2 eq./kg of GH_2. Finally, a study assessed the production of hydrogen from agricultural residues such as cheese whey, sugar beet molasses, and wine vinasses with wastewater treatment plant sludge, and found that GHG emissions are the lowest for hydrogen produced from sugar beet molasses (3.56 kg CO_2 eq./kg of hydrogen) compared to the other two feedstocks. Further LCA studies are suggested to validate the results provided in the existing research.
Sola et al. [23]	2025	State-of-the-art LCA research on GH_2 in the last five years taking into account GHG emissions and other environmental categories (such as water use, critical materials use, and land use) and large-scale GH2 development	Literature review	Increasing trend of LCA research on GH_2. Water consumption and land use (due to the use of renewable energies) in important hotspots that could limit GH_2 development, particularly in areas scarce in water such as Saudi Arabia and Oman that have high abundance of sun but are scarce in freshwater. Rare metals and critical raw materials for electrolysers production (particularly PEM) such as iridium and platinum, as well as solar cells production, serve as further hotspots. Uncertainties related to the end-of-life disposal of electrolysers.
Tabrizi et al. [125]	2025	Analysis of the carbon footprint of GH_2 production through an AEL system with electricity sourced by PV or wind in Italy and the UK.	Life cycle assessment	The results show that the updating of the baseline scenario (to consider the advances in the solar PV and wind technologies) leads to a reduction in the carbon footprint for Italy and the United Kingdom. The contributions to GWP for GH_2 produced in Italy using solar PV (single-SI modules) electricity in the lower bound is 1.76 kg CO_2 eq./kg GH_2. The carbon footprint of onshore and offshore wind shows values in the lower bound that are well below 1 kg CO_2 eq./kg GH_2, both for Italy and the UK in all the scenarios and plants. The study also confirmed that the source of electricity (solar or wind) is the most relevant factor in determining the carbon footprint of GH_2.
Tasleem et al. [44]	2025	Review of H_2 colours, hydrogen strategies and roadmaps worldwide, currently ongoing large-scale advances in green, yellow, blue, white, and gold hydrogen production, results from life cycle assessments, future perspectives, as well as challenges and solutions for hydrogen colours.	Literature review	The study reported the results by [66], underlying the importance of taking into account the impacts of metal depletion and ecotoxicity due to the production of GH_2. Results from other studies pointed out that the impacts of hydrogen production are mainly due (90%) to the source of electricity for the production of hydrogen. in this regard, the LCA by [118] of a solar-to-H_2 plant, using AEM electrolysers in Poland, shows a low contribution to GWP between 2.74 kg CO_2 eq./kg of Hydrogen and 4.34 kg CO_2 eq./kg of hydrogen. Further results from LCA studies are provided.
Affandi et al. [116]	2024	Assessing the environmental and economic viability of three case studies of GH_2 production in Malaysia and Thailand by means of PEM and AEL electrolyser powered by a solar PV system from cradle to grave (extraction of raw materials, transportation, construction, use phase, disposal and recycling phase) Results are measured in terms of greenhouse gas emissions and levelised cost of hydrogen and other financial indicators.	Life cycle assessment	The GHG emissions of GH_2 produced using PEM electrolysers resulted comprised, depending on the case studies, between 2.26 and 4.46 kg CO_2 eq/kg GH_2 while that related to AEL electrolysers are between 2.61 and 5.15 kg CO_2 eq/kg GH_2. The LCOH ranges changed across the three case studies from \$5.64/kg H_2 to \$5.12/kg H_2 and \$510.82/kg H_2 (case with AEL electrolyser) to \$7.31/kg H_2, \$6.38/kg H_2 and \$14.23/kg H_2 (case with PEM electrolyser) and is higher than fossil-derived hydrogen (\$1–\$2/kg H_2). For the GHG emission, the most important factors are the energy mix and the specific context where the electrolysers plant operates, while the capital costs of the electrolysers and the capacity of the PV systems are relevant in affecting the LCOH of GH_2.

Table A2. *Cont.*

Authors	Year of Publication	Goals	Method	Some Results
Ajeeb et al. [25]	2024a	Reviewed the recent literature on GH_2 production in LCA studies, taking into account many factors such as the electrolysers technologies, renewable sources for GH_2 production, the life cycle inventories, the local scale of GH_2 production, storage, and environmental impacts.	Literature review	Wind and solar energy resulted in the most appropriate sources for the production of GH_2, with wind reporting a lower contribution to GWP compared to solar energy. PEM electrolysers have a potential contribution to GW lower than the other technologies such as ALEL and SOEL. Improvement of the energy efficiency and environmental impacts of the water electrolyser system can be realised by means of different strategies, including waste treatment strategies such as the use of oxygen in industrial processes or the reuse of materials of the electrolyser.
Ajeeb et al. [114]	2024b	Identified the most adequate and environmentally sound electrolysis hydrogen option between two ALE technologies such as ALE-Pressurised and ALE-Capillary sourced by renewables in Portugal. The production of GH_2 is based on 50% electricity from solar PV and 50% from wind in Portugal. The goal was to identify the most appropriate and environmentally sound electrolysis hydrogen production system between the two ALE technologies (ALE-P and ALE-C).	Life cycle assessment	The ALE-Capillary system resulted in the generation of lower impacts than the ALE-Pressurised within the 16 analysed environmental impacts categories. The source of electricity (the energy mix of wind and solar PV) in the life cycle of GH_2 generated most of the impacts ranging from 92% fpr ODP, 96% fpr mineral fossil resource depletion, and 98% for the other investigated impact categories (including GWP). The contribution to GWP resulted in 1.98 kg CO_2 eq./kg GH_2 for ALE-C and 2.39 kg CO_2 eq./kg GH_2 for ALE-P.
Al-Ghussain [137]	2024	Analysed the techno-economic feasibility and the life cycle GHG emissions of GH_2 production (wind and solar-based electricity) and the influence of geographical variations of wind and solar, ambient conditions, and PEM electrolyser size on the carbon intensity and costs of both renewables and GH_2.	Techno-economic and environmental analysis	The study underlined the importance of optimising the size of the PEM electrolyser and the use of hourly-based models for the purpose as well as for the analysis of the production costs and carbon intensity of GH_2. The geographical location of renewables influences the carbon intensity (CI) of GH_2 and costs. The CI ranges from 0.3 to 1.9 kg CO_2 eq./kg GH_2 CO_2 eq./kg GH_2 (wind-based electricity), with peaks of 4.34 CO_2 eq./kg GH_2 for central regions. GH_2 produced using solar electricity has a high variability ranging from 1.58 to 2.95 kg CO_2 eq./kg GH_2. GH_2 production costs comprised between 1.5 and 15 USD/kg GH_2 for wind-based systems and 3.0 and 5.2 USD/kg H_2 for the solar PV-based systems. Geographical differences and ambient conditions have an important role in affecting the costs of GH_2 as well as the CI of electricity from renewables and of GH_2.
Alghool et al. [89]	2024	Assessing the environmental contributions to GWP of 18 different blue hydrogen processes throughout their life cycle. The study also considered the comparison of the contribution to GWP with GH_2 produced by using solar collectors.	Life cycle assessment	The LCA results showed that the NH_3 process has the least environmental impact, releasing 2.12 kg CO_2 eq./kg H_2 compared to the other 18 blue hydrogen processes. NH_3 process produces hydrogen from ATR technology in combination with carbon capture and storage of the CO_2 with hydrogen that could be delivered as ammonia to consumers.
Almaraz et al. [49]	2024	Identification of the social aspects concerning the development of the hydrogen economy, research gaps, and future research suggestions by using both bibliometric and systematic literature reviews.	Literature review	The methodological approach identified, described, and ranked 12 social aspects associated to the hydrogen economy research. The most recurrent social aspects in the literature results are accessibility, information, H_2 markets, acceptability, and policies and regulations. Despite their importance, responsibility and technological safety are less occurrent in the analysed literature. Therefore, gaps in the literature or challenges to fulfil are the scarce use of social life cycle assessment in hydrogen research.
Arrigoni et al. [123]	2024	Evaluation and comparison of the potential environmental impacts of different hydrogen storage and delivery options for an industrial cluster located in northern Europe and expected to be operative after 2030.	Life cycle assessment	The hydrogen production process and the type of delivery affect the environmental performances of hydrogen delivery options. Lowest impacts are due to hydrogen produced on-site by relying on abundant renewable energies. The import of GH_2 can be less impactful than grey hydrogen or hydrogen produced by electrolysis with electricity sourced partially by the grid.

Table A2. *Cont.*

Authors	Year of Publication	Goals	Method	Some Results
Khan et al. [127]	2024	Evaluated the total life cycle impacts of GH_2 from cradle to grave including the production of the electrolyser, its transport and installation, and the end-of-life by means of a case study in Australia	Life cycle assessment	GH_2 produced by means of PEM electrolyser and wind electricity has lower impacts than GH_2 from AEL electrolyser and solar PV power. The most significant stage result was during the operation stage when considering GWP due to the use of electricity and the impacts of the construction of wind and solar PV plants. The impacts in the other categories such as water depletion, fossil fuel, and metal depletion saw relatively lower results in the case study. However, from the perspective of scalability, the impacts can be mitigated by manufacturing electrolysers with renewable energies and recycled materials and making them recyclable. Moreover, further mitigation strategies for reducing the impacts regard the life cycle of the electricity used for the electrolysis process and a more sustainable use of water.
Kugemann and Poladis [107]	2024	The study developed a new integrated Multi Criteria Decision Analysis framework to rank several alternative fuels for buses and types of buses on the basis of technical, environmental, social, and economic dimensions. The framework is applied to the public transport system of the island of Gotland (Sweden).	Multicriteria assessment	The results evidenced that the most preferred fuel for buses is biogas (particularly that produced by waste resources) andplug-in hybrid electric biogas. Battery electric and hybrid electric hydrogen fuel cell resulted in interesting alternatives from the perspective that they are produced locally. However, the most preferred option is for biogas as a fuel despite assuming a further development of wind capacity and the improvement of the capacity of GH_2 towards the achievement of energy self-sufficiency as well as the use of subsidies to support the development of GH_2.
Maestre et al. [17]	2024	Monitoring the performances of a hybrid solar PV-based GH_2 demonstrative pilot plant, useful to ensure the total annual coverage of electricity needs of a social housing unit.	Techno-economic and environmental analysis	The results showed the environmental, economic, and social benefits achieved by the hybrid solar PV–hydrogen pilot plant after two years of monitoring. The economic benefits such as avoided electricity costs are particularly relevant for the people living in social housing, as they are at potential risk of energy poverty. The monitoring and control system implemented for the pilot plant allow the analysis of variables and energy performances. The control system has shown to work effectively since the electrolyser or the fuel cells entered into operation when there is a surplus of energy or when the batteries and the solar PV plant do not meet the energy demand of the social housing.
Maniscalco et al. [38]	2024	Comprehensive overview of hydrogen production technologies and impacts generated to the different categories including GWP.	Literature review	The results from the analysis of the selected LCA literature showed that the impacts to GWP from GH_2 as well as that of blue, turquoise, and pink production are lower compared to hydrogen produced by steam reforming. Waste gasification has the worst performances compared to the different hydrogen colours. The contribution to AP has a similar pattern of GWP with a lower difference between yellow and GH_2 due to the use of materials in both solar PV and wind systems. The hotspot analysis shows that in the electrolysis stage, the source of electricity and the specific and precious materials used for the production of the electrolysers affect the contributions to AP in AEM and PEM (nickel and gold) and to ADP due to the use of iridium and titanium. The inclusion of the end-of-life stage and the recycling of materials used for the electrolysers would mitigate the impacts to ADP, but at present, the authors underlined that most of the LCA studies do not consider the recycling of the precious materials due to lack of reliable data. For example, the process of recovery of platinum still faces industrial maturity issues, as well as high energy and economic costs.

Table A2. *Cont.*

Authors	Year of Publication	Goals	Method	Some Results
Martelli et al. [101]	2024	Comparison of the life cycle impacts of a traditional diesel-powered tractor, a fuel cell hybrid tractor using grey hydrogen, and a fuel cell hybrid tractor powered by GH_2.	Life cycle assessment	The fuel cell hybrid tractor showed much lower impacts in all the investigated impact categories with the exception of FFP, where the impacts compared to the traditional tractors are lower by 4.55%. In the life cycle of a traditional tractor, the most critical step is the use stage, while in the fuel-cell tractor, it is the manufacturing stage. The comparison with the fuel cell hybrid tractor using GH_2 to the traditional tractor shows a strong reduction ointhe impacts to GWP and other impact categories except TETP and FETP. The comparison between the fuel cell hybrid tractor powered with GH_2 and that using grey hydrogen shows a worsening of the performances in almost all the impact categories except for GWP and FFP.
Martinez-Ramon et al. [122]	2024	Evaluating the environmental impacts and identification of the hotspots in the production of hydrogen from biogas by means of chemical looping dry reforming methane technology and compared with the green hydrogen performances (3.38 or 3 kg of CO_2 eq./kg of H_2).	Life cycle assessment	The results showed that the contribution to GWP is 10.76 kg CO_2 eq./kg of hydrogen while the potential contribution to the other impact categories are: 3.19E-02 kg mol H+ eq (AP), 1.35E-04 kg PO_4^{-3} eq. (FEP), 7.81E-06 kg Sb eq. (ADP), 9.92E-08 kg CFC-11 eq. (ODP), 4.52E-02 kg NMVOC eq. (POFP). The contribution to GWP, in particular, is well above 3 kg and cannot be qualified as GH_2 according to the Commission Delegated Regulation (EU) 2021/2139 of 4 June 2021. The contribution to GWP is dominated by the processes related to biogas production while the contributions to ADP are mainly caused by the production of copper and other metals used for the two catalysts (LTS and HTS).
Mertens et al. [143]	2024	The commentary focused on the analysis of mitigation strategies for reducing the supply chain risks vulnerability with regard to critical raw materials needed for the production of solar photovoltaics, wind turbines, Li–Ion batteries, and water electrolysers.	Critical raw material analysis	The mitigation strategies comprise material efficiency, material replacement, recycling and eco-design, relocation of mining activities, and renewable production activities. The analysis showed that the highest supply chain risk is for solar PV because of the important role of China in the manufacturing of solar PV panels.
Mio et al. [109]	2024	Assessment and comparison of the impacts of different methods for hydrogen production with the goal of identifying the most environmentally sustainable performance: GH_2 from water electrolysis with renewable electricity, grid hydrogen produced from water electrolysis and grid electricity, grey hydrogen from steam reforming of natural gas, blue hydrogen by steam reforming of natural gas and carbon capture and storage. The analysis considered the Energy Return on Energy Invested, the levelised cost of hydrogen and LCA assessment. The study evaluated the use of hydrogen for 2 trips per day of a medium size ferryboat to navigate full electric for 7 h in the Adriatic Sea.	Life cycle assessment, energy and costs analysis	The analysis showed that GH_2 produced by water electrolysis and renewable electricity (from floating solar OV system) achieved the best performances both with regard to the energy indicators and GWP (4.32 kg CO_2 eq./kg of GH_2 versus 1.26E+01 kg CO_2 eq./kg of grey hydrogen and 2.35E+01 kg CO_2 eq./kg of grid hydrogen), even if further improvement should be considered to reduce the impacts beyond the GWP such as that related to FETP, METP, TETP, HTP, ODP, SOP, LOP, and WCP. For these impact categories, the impacts of GH_2 production are higher than blue and grey hydrogen, while that of WCP is greater than the grid hydrogen.
Nhien et al. [106]	2024	Potentiality of wind–fuel cell hybrid systems in providing an adequate energy supply. Focus on the efficient hydrogen production/storage/use to minimise energy costs and maximise the renewable use in the local electricity grid.	Energy, environmental, and cost analysis	The optimal plant configuration for renewable and GH_2 production and the best geographical locations are identified. Calculated the potential annual electricity generation and benefits in terms of avoided CO_2 emissions and the created new green spaces.
Oyewole et al. [98]	2024	The study evaluated the techno-economic viability of an energy system based on renewables (wind and solar PV), used as a peaker plant and an on-site GH_2 refuelling station. This way, the complementarity of both technologies in the energy systems helps to mitigate the high costs of production of GH_2 that would have been incurred in a standalone station.	Techno-economic and environmental analysis	The assessment of the energy system was performed for three selected cities and showed itself to be a viable option for producing electricity and hydrogen fuel. The levelised cost of hydrogen at Johannesburg, Pretoria and Cape Town for a 2 MW grid resulted in 74.2 $/MWh, 76.3 $/MWh and 50 $/MWh, respectively, and are competitive with the LCOE of natural gas plants generally used as peaking plants. The results also showed the CO_2 equivalent emissions and the related carbon taxes ($) avoided for the energy systems in the three cities. The LCOE of hydrogen at Johannesburg, Pretoria and Cape Town energy systems resulted in 5.85 $/kg, 5.97 $/kg, and 4.45 $/kg, respectively.

Table A2. *Cont.*

Authors	Year of Publication	Goals	Method	Some Results
Shaya and Glöser-Chahoud [28]	2024	Assessment of the LCA studies focused on hydrogen production technologies such as AWE, PEMWE, SOEC, AEMWE, their environmental impacts, and recent technological progresses.	Literature review	From the review of LCA studies emerged a variability of impacts on GWP across the four technologies for hydrogen production and, in particular, for PEMWE where the impacts range from 0.5 kg CO_2 eq./kg of GH_2 to 30 kg CO_2 eq./kg of G H_2. The impacts of hydrogen production are mainly affected by the source of electricity used for the electrolysis process, the efficiency of the electrolyser, and the materials used for its manufacturing. The use of renewable electricity can reduce the contribution to GWP as well as the advancement in the materials of catalysts and membranes. The review also offers an interesting analysis of the advantages and disadvantages of each water electrolyser technology with the inclusion of critical raw materials, technological maturity, operational lifetime among others. In this regard, AEMWE is the only one within the four analysed water electrolyser technologies that is critical raw materials free.
Simoes and Santos [13]	2024	The study adopted the framework of a SWOT (Strengths, Weaknesses, Opportunities, and Threat) analysis to investigate the current state of affairs of GH_2 market.	SWOT Analysis	The main strength of GH_2 is its low environmental contribution to GWP. Moreover, a further strength are the global policies that have incentivised over the years the transition towards more clean and renewable energy sources such as the UN Development Goals, the Paris Agreement, and the European Green Deal. Policies and regulations, programmes, strategies and roadmaps favouring the increased use of GH_2 or hydrogen have been adopted in various countries such as Germany, Spain, Denmark, Japan, and Australia. However, internal factors related to the hydrogen technology and external factors are obstacles to GH_2 development such as the lack of international standards, carbon taxes as the primary tool for supporting decarbonisation, the future development of renewable energies, and the associated environmental and social impacts, and the problem of scarcity of freshwater.
Tomos et al. [129]	2024	Comparison of the life cycle GHG emissions of several fuel options having the highest potential of reducing the contributions to GWP such as green and blue hydrogen, green and blue ammonia, e-methanol, bio-methanol, fatty acid methyl ester (FAME) biodiesel, and bio-methane. All these alternatives are compared to heavy fuel oil that is mainly used in international shipping.	Life cycle assessment	The results show that GH_2, FAME biodiesel, and bio-methanol generated the lowest impacts to GWP compared to the other alternatives. These fuels have the best decarbonisation potentials. However, the use of these alternative fuels is still very low and there is a lack of initiatives aimed at accelerating their use, undermining their potential of decarbonizing on time the international shipping fleet.
Torrubia et al. [60]	2024	Evaluation of the impacts of copper recovery in three cases of easte electrical and electronic equipments, WEEE (scraps, mix, and waste printed circuit boards), through three scenarios: conventional, green hydrogen, and hydrogen from grid electricity in order to assess for the last case the impact of electricity mix for the production of hydrogen.	Life cycle assessment	The results show that the carbon footprint of copper recovered from WEEE is much lower than the carbon footprint of primary copper. The contribution to GWP from the recovery of copper in WEEE is further reduced by using GH_2 produced from renewable electricity from the grid (wind power) compared to the scenario where the electricity from the grid is produced from non-renewable energies.
Wang et al. [29]	2024	Analysis of the evolution and development of policies supporting GH_2 policies in the United States, the European Union, Japan, and China, and technological progress of electrolysis process from renewables as well as of GH_2 produced from solar PV and wind. The study also integrates the environmental, social, and economic benefits of the different hydrogen pathways.	Literature review	The USA has been the first to introduce policies supporting hydrogen development, but all the four investigated areas/countries can be considered early adopters of hydrogen policies worldwide. Moreover, they all have strategies and goals for incentivising GH_2 adoption. The results from the reviewed studies confirm that GH_2 from wind electricity generate the lowest contribution to GWP and AP and low energy costs (785 g CO2 eq./kg H_2 and 10.5 MJ/kg of H_2) while hydrogen produced by using wind and solar electrolysis had the highest economic costs. Out of the electrolyses technologies, AWE has advantages in terms of costs, maturity and wide application while PEM is adaptable to the variability of renewables but has high costs limiting its wider development. SOEC and AEM are still in the research and demonstration phase and have not yet been commercialised.

Table A2. *Cont.*

Authors	Year of Publication	Goals	Method	Some Results
Wei et al. [117]	2024	Comparison of the environmental emissions of four water electrolysers technologies such as AEL, AEM, PEM, and SOE. The assessment also expands on the materials and energy used for the manufacturing of each electrolyser technology for identifying their environmental contribution.	Life cycle assessment	The results reveal the impacts to the investigated impact categories generated in the manufacturing process of each water electrolyser technology as well as the factors that mainly contribute to the impacts of the production of hydrogen and the operativity of the electrolyser system. The results reveal that in the manufacturing process of the electrolysers materials such as nickel, steel, and platinum mainly contribute to most of the investigated impact categories. For example, in AEM, the bipolar plate is mainly made of steel and the latter contributes to almost 80% of the contributions to CC. With regard to the impacts at a system level in the operation stage, the results evidenced that the lowest impacts come from the production of GH_2 by AEM electrolyser using electricity from wind. On the contrary, AEL electrolyser with electricity from hydropower provided the highest contributions to CC. Further important categories considered in the LCA are: Human Health, Ecosystem Quality, and Abiotic Stock Resources.
Agostinho et al. [103]	2023	This LCA and Emergy Accounting study analyses the environmental impacts of replacing diesel with H_2 produced by means of grid electricity (83% from renewables such as hydro and biomass) in urban buses.Emergy Accounting and life cycle assessment (LCA) are applied to obtain complementary environmental indicators.	Life cycle assessment and emergy accounting	The comparison of the hydrogen scenario (functional unit: km person) with other urban buses scenarios shows that the hydrogen scenario has worst performances compared to buses fuelled by diesel, electricity, LPG, and biodiesel in the fossil energy depletion category, but performs better compared to GWP, AP, and PMFP. The results by emergy accounting show that the operational stage requires 95.6% of the total emergy use with the highest shares as electricity (51.2%) and labor (44.4%). Hydrogen energy source has a higher renewability (41.49%) compared to diesel (0.05%) and biodiesel (18.83%), lower emergy loading ratio and higher emergy sustainability index.
Akhtar et al. [140]	2023	This study performs a cradle to gate social life cycle assessment pf hydrogen produced from water electrolysis by means of solar PV and wind electricity in 7 countries. The main goal is identifying the social hotspots in the GH_2 supply chain. Moreover, a further goal is the comparison of the social hotspots with hydrogen produced from natural gas, and assess the impacts of GH_2 on SDGs for policy purposes and future recommendations for a large-scale development of a GH_2 economy	Social life cycle assessment	The results of the study show that the production of GH_2 in South Africa presents the highest risks to most of the social indicators and in particular child labor, fair salary, unemployment, association and bargaining rights and gender wage gap. The risk in the other countries can be mitigated by avoiding the imports of the renewable and electrolysis technologies from other countries and manufacturing them domestically. The implications for policy regard the adoption of better working conditions and of international regulations in order to avoid that more polluting activities are exported towards developing and emerging countries.
Aravindan and Kumar [30]	2023	Overview of GH_2 production from renewable sources (solar, geothermal, biomass, wind and technologies (e.g., water splitting: thermolysis, electrolysis and photolysis), production costs, environmental impacts and challenges in GH_2 development	Literature review	The results include the analysis of studies evaluating the contribution to GWP of hydrogen from solar, wind, biomass and geothermal sources and available technologies. In overall, wind-based production of hydrogen resulted to provide the lowest contribution to GWP ranging according to the selected studies from the value of 600 g CO_2 eq /kg of GH_2 to 970 g CO_2 eq./kg of GH_2. Production costs of GH_2 from biomass were the lowest (1–3 $/kg hydrogen) while that of solar the highest (7–10 $/kg GH_2). On the contrary the efficiency of solar and wind in GH_2 production (70–80%) is the highest compared to biomass and geothermal (50–60%).
Arcos and Santos [83]	2023	Analysis and technical and economic comparison of the different hydrogen production pathways in order to understand their current feasibility	Techno-economic and environmental analysis	Grey, black/brown and blue hydrogen are the least environmentally friendly but those with the lower production costs (ranging from 0.67 to 2.05 USD/kgH_2 respectively). Green Hydrogen is the most environmentally friendly but has one of the highest production costs ranging from 2.28–7.39 USD/kgH_2. Yellow hydrogen resulted with the highest production costs (6.06–8.81 USD/kg H_2) but these latter as the CO_2 emissions depend on the location as well as on the electricity mix of the specific location.

Table A2. *Cont.*

Authors	Year of Publication	Goals	Method	Some Results
Pawlowski et al. [118]	2023	Calculation of the carbon footprint and financial indicators of GH_2 produced by water electrolysis and AEM technology using electricity from solar PV panels. The PV plant with 5 MW of peak power is located in Poland.	Life cycle assessment and life cycle costing	The carbon footprint resulted comprised between 2.73 and 3.85 kg CO_2 eq./kg H_2. The financial analysis evidenced that the investment in the project is sustainable only in the presence of external subsidies. In this case, the net present value is positive and the payback period is 8 years.
Squadrito et al. [46]	2023	Overview of the state of the art of hydrogen production technologies, and exploration of the most relevant geopolitical and economic aspects and implications of the hydrogen economy development, including the water aspects.	Literature review	The study points out the need for achieving a balance between distributed hydrogen systems and centralised systems and considering the advantages of producing GH_2 locally, where it is water neutral (and CO_2 neutral), compared to the production of GH_2 for the external market where transport generates carbon emissions. Further, the energy and costs of producing GH_2 from non-traditional water sources such as e.g., using wastewater would be lower compared to the production of GH_2 with fresh water "*Distributed hydrogen production is a powerful tool for maximising the social utility of the hydrogen economy, reducing household energy bills, increasing energy system efficiency and resilience, and reducing the environmental impact of city services. All these aspects must be considered in future energy policies*".
Vallejos-Romero et al. [50]	2023	Analysis of the research on GH_2 from the perspective of social impacts and challenges	Literature review	Social impacts of the life cycle of GH_2 from production to storage and transport as well as social and environmental impacts of GH2 on local communities and indigenous groups are under researched. In the investigated period Europe and Asia mainly contributed to the research on social impacts of GH_2.
Weidner et al. [66]	2023	Assessment of the environmental impacts of different options to produce 500 Mt/yr of hydrogen involving green (from solar PV or wind power), blue and grey hydrogen pathways.	Life cycle assessment	The impacts to GW show the surpassing of the planetary boundaries for blue and grey hydrogen while not for green hydrogen produced by using wind electricity.
Zhang et al. [115]	2023	Identifying the most suitable and environmentally sound method for hydrogen produced from water electrolysis among the seven processes combining three types of electrolysers (AEL, PEM and SOEC) and onshore and offshore wind electricity	Life cycle assessment	The production of GH_2 by using electricity from onshore wind and PEM electrolyzer technology generate the lowest contribution to GWP by 0.0936 kg CO_2-eq. The impacts are also lower for such combination with regard to AP, ODP and EP impact categories.
Arsalis et al. [18]	2022	Critical comparison of several characteristics (service life, costs, recyclability, environmental impacts, safety issues, use and integration within energy systems) of two solar powered energy systems: Lithium–Ion Batteries (PV-LIB) and Regenerative Hydrogen Fuel Cell (RHFC) energy systems	Literature review	The comparative analysis shows that, for example, hydrogen (within RHFC subsystems) produced through water electrolysis can be stored in high quantities in hydrogen storage units. The RHFC is then more suitable for long-term storage than short-term storage, while LIBs are more suitable for short-term storage. The refuelling of RHFC subsystems is fast, requiring only a few minutes, while the recharging of a LIBs is slow and needs several hours. In terms of costs, those related to LIBs are much decreased within the last decade while in the case of RHFC the high costs of the electrolyzers and fuel cell stacks limits their larger development.
Barghash et al. [87]	2022	Evaluation of technical and economic feasibility of producing GH2 in Oman and exploration of the benefits of using treated effluent from wastewater treatment.	Cost–benefit analysis	The results of the study show the technical and economic feasibility of using treated effluents from wastewater for the production of GH_2. Moreover, the selling of GH_2 provides a source of economic revenue by which cover the initial investments costs.
Buffi et al. [39]	2022	Review of production pathways for producing hydrogen from biomass or biomass-derived sources (biogas, liquid bio-intermediates, sugars) while also focusing on their maturity level, as well as the energy and environmental performances of the production pathways emerged from LCA studies.	Literature review	The type of feedstock biomass influences the energy required for the process of hydrogen production. *Energy requirements should be compared with each other under the same conditions of feedstock, production rates and environmental conditions.* From the analysis of the LCA literature emerges that some of the studies followed the ISO standard while others proposed different methodological approaches. Moreover, LCA studies differ on the system boundaries ranging from cradle to gate to cradle to grave. There are also different allocation assumptions. Regarding impact assessment, most of the studies adopt midpoint assessment methods and impact categories such as GWP, AP and ODP.

Table A2. *Cont.*

Authors	Year of Publication	Goals	Method	Some Results
Gandiglio et al. [132]	2022	Evaluation and comparison of the environmental impacts of electricity in the island of Ginostra produced by two energy systems: a diesel-based energy system (Reference scenario) and a proposed RES-based energy system (Renewable scenario). The study also aims to identify the processes or components that mainly contribute to the environmental impacts and evaluate the possible future improvements.	Life cycle assessment	The renewable scenario is composed of a solar PV plant and a hydrogen battery energy storage. The results (referred to a time horizon of 25 years) underlined that for the production of 1 kWh, the impacts are higher for the current diesel-based system for most of the categories investigated. Therefore, the renewable scenario improves the current scenario in such categories. The exceptions are for EP, freshwater; Ecotoxicity potential, freshwater; Water use and Resource use, minerals and metals where the impacts of the renewable scenario are slightly higher than the reference scenario.
Guareiro et al. [11]	2022	Overview of several aspects associated with GH_2 including the sources for its production, regulation, typologies of storage, transportation, and final uses. Critical analysis of the economic and environmental impacts and the main challenges and opportunities it could have for chemistry.	Literature review	In terms of environmental impacts of GH_2 production, the study shed light on the water required in the process (9 kg per kg of GH_2). The water required for production of grey and blue hydrogen is half of that needed for GH_2 production. The latter process is also much more energy-intensive and, currently, its production is not financially competitive. However, the authors, by reviewing the literature, also reported the results of studies performed in Brazil where the production of GH_2 by using electricity from hydro and wind achieved production costs comparable to that of grey hydrogen. The study also underlined the need for increasing the capacity of electrolysers for the development of a future hydrogen economy.
Ishaq et al. [32]	2022	Analysis of the state-of-the-art hydrogen production processes using renewable energy sources and other sources, uses of hydrogen as a fuel, typologies of storage, transportation, distribution, and key challenges and opportunities for a wider development of Hydrogen.	Literature review	Hydrogen is getting increased interest beyond its conventional role in industrial sector for the production of ammonia and methanol since, for example, as a fuel, it offers the advantage of avoiding carbon dioxide emissions in the use stage. In particular, the production of GH_2 by using electricity from renewable energies is considered the best option to overcome the intermittence of renewables. However, despite these advantages, there are many challenges for its large-scale adoption such as those related to the development of an environmentally clean production system for hydrogen at a large scale, as well as the development of the needed infrastructures for its distribution, storage systems for stationery and vehicle applications, and reduction of the production costs.
Lykas et al. [104]	2022	Analysis of energy, exergy, and financial performances of the production of GH_2 from excess solar electricity in a Greek island.	Energy, exergy, and cost analysis	The analysis proposed a configuration involving a solar parabolic collector plant generating electricity and heat. The electricity in excess that cannot be received by the local grid is delivered to a PEM water electrolyser for the production of GH_2. The latter is assumed to be compressed, stored in tanks, and eventually used as a fuel for vehicles or ferries. The results showed that the proposed system can satisfy the annual diverse energy needs of the island, producing 210 MWh and 2356.5 kg of GH_2 with a good financial return, resulting in an investment payback period of less than 7 years and a positive net present value.

Note: The following have been extracted from the references of the selected articles and integrated to the articles retrieved from the search on Web of Science: [18,36,40,51,53,96–99,102–108,123,132,133,143].

References

1. Daewood, F.; Martin, A.; Shafiullah, G.M. Hydrogen production for energy: An overview. *Int. J. Hydrogen Energy* **2020**, *45*, 3847–3869. [CrossRef]
2. Mehmeti, A.; Angelis-Dimakis, A.; Arampatzis, G.; McPhail, S.J.; Ulgiati, S. Life cycle assessment and water footprint of hydrogen production methods: From conventional to emerging technologies. *Environments* **2018**, *5*, 24. [CrossRef]
3. Mehmeti, A.; McPhail, S.J.; Ulgiati, S. Life cycle inventory data and metrics for high-temperature fuel cells: A streamlined decision-support tool and case study application. *Energy* **2018**, *159*, 1195–1205. [CrossRef]
4. Ghisellini, P. Economia All'idrogeno. Ph.D. Thesis, University of Ferrara, Ferrara, Italy, 2006. (In Italian). [CrossRef]
5. Hoffman, P. *Era Dell'idrogeno*; Franco Muzzio Editore: Rome, Italy, 2005. (In Italian)

59. Bianchi, M.; Cordella, M. Does circular economy mitigate the extraction of natural resources? Empirical evidence based on analysis of 28 European economies over the past decade. *Ecol. Econ.* **2023**, *203*, 107607. [CrossRef]
60. Torrubia, J.; Parvez, A.M.; Sajjad, M.; Paz, F.A.G.; van den Boogaart, K.G. Recovery of copper from electronic waste: An energy transition approach to decarbonise the industry. *J. Clean. Prod.* **2024**, *485*, 144349. [CrossRef]
61. Brown, M.T.; Ulgiati, S. Emergy assessment of global renewable sources. *Ecol. Model.* **2016**, *339*, 148–156. [CrossRef]
62. Bompan, E.; Tavazzi, D. Walter Stahel: L'economia Circolare Deve Diventare Per Tutti. Available online: https://www.renewablematter.eu/walter-stahel-leconomia-circolare-deve-diventare-per-tutti (accessed on 2 March 2025).
63. Mutezo, G.; Mulopo, J. A review of Africa's transition from fossil fuels to renewable energy using circular economy principles. *Renew. Sustain. Energy Rev.* **2021**, *137*, 110609. [CrossRef]
64. Ghisellini, P.; Ulgiati, S. Evolution and current development of the circular economy concept with insights from worldwide cities and regions. In *Regions, Cities and the Circular Economy*; Bourdin, S., Torre, A., van Leeuwen, E., Eds.; Edward Elgar: Cheltenham, UK, 2024.
65. Zhu, X.; Wang, S.; Wang, L. Life cycle analysis of greenhouse gas emissions of China's power generation on spatial and temporal scale. *Energy Sci. Eng.* **2022**, *10*, 1083–1095. [CrossRef]
66. Weidner, T.; Tulus, V.; Guillen-Gosalbez, G. Environmental sustainability assessment of large-scale hydrogen production using prospective life cycle analysis. *Int. J. Hydrogen Energy* **2023**, *4*, 8310–8327. [CrossRef]
67. Graus, W.H.J.; Voogt, M.; Worrell, E. International comparison of energy efficiency of fossil power generation. *Energy Policy* **2007**, *35*, 3936–3951. [CrossRef]
68. Mulvaney, D.; Richards, R.M.; Bazilian, M.D.; Hensley, E.; Clough, G.; Shridhar, S. Progress towards a circular economy in materials to decarbonize electricity and mobility. *Renew. Sustain. Energy Rev.* **2021**, *137*, 110604. [CrossRef]
69. Ellen Mac Arthur Foundation. We Need to Talk About Renewables. 2022. Available online: https://www.ellenmacarthurfoundation.org/we-need-to-talk-about-renewables/part-1 (accessed on 4 March 2025).
70. Cerchione, R.; Morelli, M.; Passaro, R.; Quinto, I. Balancing sustainability and circular justice: The challenge of the energy transition. *J. Clean. Prod.* **2025**, *494*, 144942. [CrossRef]
71. Ghisellini, P.; Ncube, A.; Casazza, M.; Passaro, R. Toward circular and socially just urban mining in global societies and cities: Present state and future perspectives. *Front. Sustain. Cities* **2022**, *4*, 930061. [CrossRef]
72. Burchart, D.; Przytuła, I. Review of Environmental Life Cycle Assessment for Fuel Cell Electric Vehicles in Road Transport. *Energies* **2025**, *18*, 1229. [CrossRef]
73. Enel Green Power. Sustainable Solar Panels. 2021. Available online: https://www.enelgreenpower.com/stories/articles/2021/07/sustainable-solar-panels (accessed on 4 March 2025).
74. European Union. Voluntary schemes. 2024. Available online: https://energy.ec.europa.eu/topics/renewable-energy/bioenergy/voluntary-schemes_en (accessed on 4 March 2025).
75. Le, T.T.; Tran, P.Q.; Dhar, B.K. Circular economy and social life cycle assessment: The role of corporate renewable energy strategies, environmental justice, and environmental impacts. *J. Clean. Prod.* **2024**, *485*, 144387. [CrossRef]
76. Pivatto, M.; Laís, V.; Pereira, F.; Rüther, R. Circular solar economy: PV modules decision-making framework for reuse. *J. Clean. Prod.* **2025**, *493*, 144941. [CrossRef]
77. Green, B.N.; Johnson, C.D.; Adams, A. Writing Narrative Literature Reviews for Peer-Reviewed Journals: Secrets of the Trade. *J. Chiropratic Med.* **2006**, *5*, 101–117. [CrossRef]
78. Ferrari, R. Writing narrative style literature reviews. *Eur. Med. Writ. Assoc.* **2015**, *24*, 230–235. [CrossRef]
79. American Chemical Society. Hydrogen. 2020. Available online: https://www.acs.org/molecule-of-the-week/archive/h/hydrogen.html#:~:text=Hydrogen%20is%20the%20most%20abundant,of%20the%20abundance%20of%20water (accessed on 16 March 2025).
80. Greenwood, N.N.; Earnshaw, A. *Chemistry of the Elements*; Elsevier: London, UK, 1997. [CrossRef]
81. EneL Green Power. Hydrogen. 2025. Available online: https://www.enelgreenpower.com/learning-hub/renewable-energies/hydrogen (accessed on 6 March 2025).
82. European Hydrogen Observatory. Hydrogen Value Chain. 2025. Available online: https://observatory.clean-hydrogen.europa.eu/learn-about-hydrogen/hydrogen-basics/hydrogen-value-chains (accessed on 13 March 2025).
83. Arcos, J.M.M.; Santos, D.M.F. The Hydrogen Color Spectrum: Techno-Economic Analysis of the Available Technologies for Hydrogen Production. *Gases* **2023**, *3*, 2. [CrossRef]
84. Sustainable Northern Ireland. The Many Colours of Hydrogen. 2024. Available online: https://www.sustainableni.org/blog/many-colours-hydrogen (accessed on 16 March 2025).
85. Vozniuk, O.; Tanchoux, N.; Millet, J.-M.; Albonetti, S.; Di Renzo, F.; Cavani, F. Chapter 14—Spinel Mixed Oxides for Chemical-Loop Reforming: From Solid State to Potential Application. In *Studies in Surface Science and Catalysis*; Elsevier: Amsterdam, The Netherlands, 2019; Volume 178, pp. 281–302. [CrossRef]

86. Gopalakrishnan, B.; Khanna, N.; Das, D. Chapter 4—Dark-Fermentative Biohydrogen Production. In *Biohydrogen*; Pandey, A., Venkata Mohan, S., Chang, J.-S., Hallenback, P.C., Larroche, C., Eds.; Elsevier: London, UK, 2013.
87. Barghash, H.; Al Farsi, A.; Okedu, K.E.; Al-Wahaibi, B.M. Cost benefit analysis for green hydrogen production from treated effluent: The case study of Oman. *Front. Bioeng. Biotechnol* **2022**, *10*, 1046556. [CrossRef]
88. European Commission. Renewable Hydrogen. 2024. Available online: https://energy.ec.europa.eu/topics/energy-systems-integration/hydrogen/renewable-hydrogen_en (accessed on 13 March 2025).
89. Alghool, D.; Haouari, M.; Trucco, P. It is not the same blue: A comparative LCA study of blue hydrogen supply network pathways. *Int. J. Hydrogen Energy* **2024**, *81*, 214–224. [CrossRef]
90. European Hydrogen Observatory. Cost of Hydrogen Production. 2025. Available online: https://observatory.clean-hydrogen.europa.eu/index.php/hydrogen-landscape/production-trade-and-cost/cost-hydrogen-production (accessed on 13 March 2025).
91. Kabeyi, M.J.B.; Olanrewaju, O.A. The levelized cost of energy and modifications for use in electricity generation planning. *Energy Rep.* **2023**, *9*, 495–534. [CrossRef]
92. Cantuarias-Villessuzanne, C.; Weinberger, B.; Roses, L.; Vignes, A.; Brignon, J.-M. Social cost-benefit analysis of hydrogen mobility in Europe. *Int. J. Hydrogen Energy* **2016**, *41*, 19304–19311. [CrossRef]
93. IEA. 2024. Global Hydrogen Review. 2024. Available online: https://www.iea.org/reports/global-hydrogen-review-2024/executive-summary (accessed on 13 March 2025).
94. De Matteis, E.M.; Baricco, M. La via Dell'idrogeno: Dalla Produzione All'uso. *La Chimica e l'Industria* **2021**, *V*, 27–33. [CrossRef]
95. IEA. European Hydrogen Backbone Initiative. 2025. Available online: https://ehb.eu/ (accessed on 13 March 2025).
96. Hassan, M.U.; Bremner, S.; Menictas, C.; Kay, M. Assessment of hydrogen and Lithium-ion batteries in rooftop solar PV systems. *J. Energy Storage* **2024**, *86*, 111182. [CrossRef]
97. Gabbar, H.A.; Ramadan, A. Integrated Renewable Energy Systemsfor Buildings: An Assessment of the Environmental and Socio-Economic Sustainability. *Sustainability* **2025**, *17*, 656. [CrossRef]
98. Oyewole, O.L.; Nwulu, N.I.; Okampo, E.J. Techno-economic investigation of hybrid peaker plant and hydrogen refuelling station. *Int. J. Hydrogen Energy* **2024**, *49*, 509–529. [CrossRef]
99. Gandiglio, M.; Marocco, P. Mapping Hydrogen Initiatives in Italy: An Overview of Funding and Projects. *Energies* **2024**, *17*, 2614. [CrossRef]
100. European Hydrogen Observatory. Hydrogen Demand. 2025. Available online: https://observatory.clean-hydrogen.europa.eu/hydrogen-landscape/end-use/hydrogen-demand (accessed on 13 March 2025).
101. Martelli, S.; Martini, V.; Mocera, F.; Soma', A. Life Cycle Assessment Comparison of Orchard Tractors Powered by Diesel and Hydrogen Fuel Cell. *Energies* **2024**, *17*, 4599. [CrossRef]
102. Currie, G.; Behrens, E.; Bolitho, S.; Coen, M.; Wilson, T. Green Hydrogen Driven by Wind and Solar—An Australian Case Study. *Wind* **2024**, *4*, 6. [CrossRef]
103. Agostinho, F.; Serafim Silva, E.; Cezar da Silva, C.; Villas Boas Almeida, C.M.; Giannetti, B.F. Environmental performance for hydrogen locally produced and used as an energy source in urban buses. *J. Clean. Prod.* **2023**, *396*, 136435. [CrossRef]
104. Lykas, P.; Bellos, E.; Caralis, G.; Tzivanidis, C. Dynamic Investigation and Optimization of a Solar-Based Unit for Power and Green Hydrogen Production: A Case Study of the Greek Island, Kythnos. *Appl. Sci.* **2022**, *12*, 11134. [CrossRef]
105. Pellegrini, M.; Guzzini, A.; Saccani, C.; Lodi, B.; Rosati, J.; Tocco, D. Green hydrogen application in the hard-to-abate industry: A feasibility study. In Proceedings of the 14th European Conference on Industrial Furnaces and Boilers (INFUB-14), Algarve, Portugal, 2–5 April 2024.
106. Nhien, L.C.; Behzadi, A.; Assareh, E.; Lee, M.; Sadrizadeh, S. A new approach to wind farm stabilization and peak electricity support using fuel cells: Case study in Swedish cities. *Int. J. Hydrogen Energy* **2024**, *80*, 22–38. [CrossRef]
107. Kugemann, M.; Poladis, H. Multi-criteria decision analysis of alternative fuel buses: Methodology and case-study. *Int. J. Green Energy* **2024**, *21*, 2493–2513. [CrossRef]
108. Bonesso, F.; D'Adamo, I.; Gastaldi, M.; Giannini, M. Green hydrogen as a sustainable operations strategy: A socio-economic perspective. *Sustain. Oper. Comput.* **2025**, *6*, 1–14. [CrossRef]
109. Mio, A.; Barbera, E.; Pavan Massi, A.; Bertucco, A.; Fermeglia, M. Sustainability analysis of hydrogen production processes. *Int. J. Hydrogen Energy* **2024**, *54*, 540–553. [CrossRef]
110. Barghash, H.; Farsi, M.A.; Alrashdi, Z.; Okedu, K.E.; Al-Alawi, A. Achieving decarbonization considering green hydrogen production: Case study of Oman. *Results Eng.* **2024**, *23*, 102657. [CrossRef]
111. Patel, G.H.; Havukainen, J.; Horttanainen, M.; Soukka, R.; Tuomaala, T. Climate change performance of hydrogen production based on life cycle assessment. *Green Chem.* **2024**, *26*, 992. Available online: https://pubs.rsc.org/en/content/articlelanding/2024/gc/d3gc02410e (accessed on 19 May 2025). [CrossRef]
112. Shen, H.; Crespo del Granado, P.; Santos Jorge, R.; Löffler, K. Environmental and climate impacts of a large-scale deployment of green hydrogen in Europe. *Energy Clim. Change* **2024**, *5*, 100133. [CrossRef]